貿易英文撰寫實務

張錦源　著

三民書局

Drafting Technique of Business English for Foreign Trade

國家圖書館出版品預行編目資料

貿易英文撰寫實務 / 張錦源著. －－初版三刷. －－
臺北市：三民，2008
　　面；　　公分
ISBN 957-14-3889-8 （平裝）

1.商業書信 2.英國語言－應用文

493.6　　　　　　　　　　　　　　92022408

© 貿易英文撰寫實務

著作人	張錦源
發行人	劉振強
著作財產權人	三民書局股份有限公司 臺北市復興北路386號
發行所	三民書局股份有限公司 地址／臺北市復興北路386號 電話／(02)25006600 郵撥／0009998-5
印刷所	三民書局股份有限公司
門市部	復北店／臺北市復興北路386號 重南店／臺北市重慶南路一段61號

初版一刷　2004年2月
初版三刷　2008年6月
編　　號　S 550880
定　　價　新臺幣800元

行政院新聞局登記證局版臺業字第○二○○號

有著作權．不准侵害

ISBN　957-14-3889-8　（平裝）

http：//www.sanmin.com.tw　三民網路書店

序

邁入二十一世紀之後，國際經貿競爭日益激烈，各國為繁榮經濟，提高人民生活水準，莫不致力於國際市場的開拓。我國在政府與民間的共同努力下，對外貿易方面已有飛躍的進展，頗受世人的注目。從事國際貿易也成了青年人嚮往的職業。

做國際貿易，與外國人打交道，須靠共同的語文來溝通。而在各種語文中，無疑英文已成為國際性語文。因此，英文很自然地成為國際商務往來的共同語文。貿易英文 (Business English for Foreign Trade) 就是英文中的應用文，有志從事國際貿易工作的人，為求有效處理國際貿易事務，自需通曉貿易英文。

想成為一個能獨當一面的傑出國際貿易從業員，僅僅通曉英文是不夠的。因為國際貿易作業程序錯綜複雜，從事國際貿易工作者，若不具備豐富的國際貿易實務知識，儘管英文程度很好，寫出來的英文，難免隔靴搔癢，不得要領。所以有志從事國際貿易的人，不但需要學習貿易英文的寫作技巧，同時還要勤加研習國際貿易實務。

我服務於貿易界有年，日常接觸貿易英文的機會很多。在工作中，如遇到別人寫的貿易英文，有可資模仿、學習的措詞、語法，隨時加以摘錄，以供自己參考之用。

一方面由於這些筆記越積越多，翻閱不便，而且又易於散失；他方面覺得這些筆記也許可供有志從事國際貿易的青年人參考，我便不自揣譾陋，將這些筆記加以整理，付梓問世。

在撰寫方法上，我根據實務上的經驗，以「學英文做貿易，談貿易學英文」的方式，一邊談貿易英文的撰寫要領，一邊講解做貿易的方法。在內容安排方面，則按進出口貿易過程的先後，循序漸進，採用現行我國進出口貿易實務，做親切的說明。並在每章之後，搜集了豐富的例句，以供隨時參考應用。

由於本書是按進出口貿易過程先後編排，取材新穎，內容充實，講解親切，而且是配合我國對外貿易實況撰寫而成，讀者若能仔細研讀，勤加練習，相信不但可學習到貿易英文的撰寫技巧，同時還可領悟到做貿易的竅門，進而成為優秀的國際貿易從業員。

本書定稿後，即承三民書局劉董事長振強先生欣然同意出版，在此敬申衷心的謝忱。

著者　張錦源　謹識
2004 年 1 月於臺北市

貿易英文撰寫實務 目次

序

■　第三十三章　求才與求職 (Position Vacant & Position Wanted)　663

■　第三十四章　貿易社交信 (Social Letters in Foreign Trade)　717

■　第三十五章　標點符號的用法　751

第一章
緒論 (Introduction)

第一節　貿易英文的定義

「貿易英文」(Business English for Foreign Trade) 就是應用於國際貿易的「商用英文」(Business English; Commercial English)。所以，本質上「貿易英文」是「商用英文」的一種。正如英文應用在醫學或化學方面一樣，在「貿易英文」裡也使用各種貿易術語和慣用語，例如 FOB, CIF, L/C, D/P, D/A 或 B/L 等，但是在文法和修辭方面，「貿易英文」卻和普通英文並無兩樣。

想學好貿易英文，必須：

1. 要有良好的英文基礎：如果英文根基不好，就無法寫出好的貿易英文。所以，想學好貿易英文，首先應具備一般英文文法的常識與嫻熟英文的基本語法。

2. 要通曉貿易術語：貿易英文需使用各種貿易術語才能達成簡潔的目的。如不通曉貿易實務上的各種術語，必無法寫好貿易英文。

3. 要嫻熟貿易慣例與實務：從事貿易的當事人居於不同風俗、習慣的環境裡，想求貿易的順利進行，就必須熟悉貿易的慣例與實務。熟悉了貿易慣例與實務，才能寫出理想的貿易英文。

4. 要多閱讀別人所撰寫的貿易英文：擷取可資模仿的各種表現 (Expressions) 用例，以供運用於不同情形的需要。

第二節　貿易英文的範圍

如上所述，貿易英文是指運用於國際貿易的商用英文而言。然而，國際貿易所涉及範圍相當廣泛。除國際貿易業務本身之外，尚與銀行、外匯、倉儲、運輸、保險、行銷等等有密切的關係。所以貿易英文的範圍可包括：

1. 貿易書信 (Foreign trade correspondence)。

2.貿易契約 (Foreign trade contract)。

3.貿易單證 (Foreign trade documents)。

4.保險與航運 (Insurance and shipping)。

5.銀行與外匯 (Banking and foreign exchange business)。

6.電傳 (Teletransmission)。

7.貿易社交信 (Social letters in foreign trade)。

8.有關貿易的求才和求職 (Employment relating to foreign trade)。

9.其他 (Others)：包括商情報告、廣告及貿易單證等。

貿易書信裡，又包括下列各種：

1.尋找交易對象 (Looking for customers)。

2.招攬交易 (Trade proposal)。

3.徵信 (Credit inquiry)。

4.詢價 (Trade inquiry)。

5.答覆詢價 (Response to inquiry)。

6.報價 (Offer)。

7.接受 (Acceptance)。

8.推銷與追查 (Sales promotion and follow-up)。

9.訂貨 (Order)。

10.答覆訂貨 (Response to order)。

11.收款與付款 (Collection and payment)。

12.索賠與調處 (Claim and adjustment)。

13.挽回客戶 (Regaining lost customers)。

14.建立代理關係 (Establishment of agencyship)。

15.招標與投標 (Invitations and bids)。

16.通告 (Announcement) 等等。

商用英文信各部分圖解

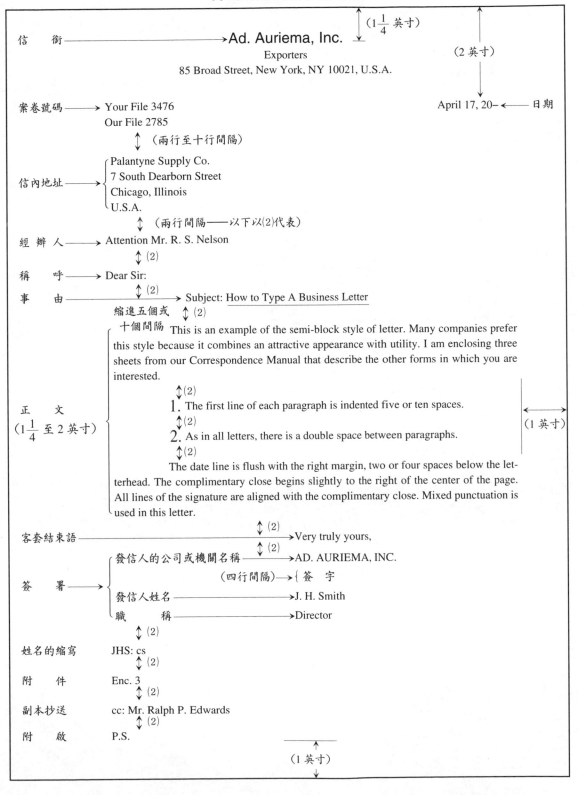

信　　銜 ————→ **Ad. Auriema, Inc.** $(1\frac{1}{4}$ 英寸)

Exporters

85 Broad Street, New York, NY 10021, U.S.A.

(2 英寸)

案卷號碼 ————→ Your File 3476　　　　　　　　　　　　　April 17, 20– ←—— 日期

　　　　　　　　Our File 2785

　　　　　　　　（兩行至十行間隔）

　　　　　　　　┌ Palantyne Supply Co.

信內地址 ————→│ 7 South Dearborn Street

　　　　　　　　│ Chicago, Illinois

　　　　　　　　└ U.S.A.

　　　　　　　　（兩行間隔——以下以(2)代表）

經 辦 人 ————→ Attention Mr. R. S. Nelson

　　　　　　　　(2)

稱　　呼 ————→ Dear Sir:

事　　由 ————————————→ Subject: How to Type A Business Letter

縮進五個或 (2)

十個間隔　This is an example of the semi-block style of letter. Many companies prefer this style because it combines an attractive appearance with utility. I am enclosing three sheets from our Correspondence Manual that describe the other forms in which you are interested.

正　　文

$(1\frac{1}{4}$ 至 2 英寸)

　　(2)

1. The first line of each paragraph is indented five or ten spaces.

　　(2)

2. As in all letters, there is a double space between paragraphs.

　　(2)

　　　The date line is flush with the right margin, two or four spaces below the letterhead. The complimentary close begins slightly to the right of the center of the page. All lines of the signature are aligned with the complimentary close. Mixed punctuation is used in this letter.

(1 英寸)

客套結束語 ————————————→ Very truly yours,

　　　　　　　　(2)

　　　　　　發信人的公司或機關名稱 ————→ AD. AURIEMA, INC.

簽　　署　　　（四行間隔）——→ { 簽　字

　　　　　　發信人姓名 ————————→ J. H. Smith

　　　　　　職　　稱 ————————→ Director

　　　　　　(2)

姓名的縮寫　　JHS: cs

　　　　　　(2)

附　　件　　Enc. 3

　　　　　　(2)

副本抄送　　cc: Mr. Ralph P. Edwards

　　　　　　(2)

附　　啟　　P.S.

(1 英寸)

第二章
商用英文信的結構 (The Structure of an English Business Letter)

一封商用英文信是由信文 (The letter) 與信封 (The envelope) 組合而成。

 ## 第一節　商用英文信的構成部分

一、商用英文信的構成部分

商用英文信的構成部分，可分為基本部分與附加部分。

1. 基本部分：這是一封信不可或缺的主要部分 (Main parts)，基本部分包括：

(1) Letterhead, Heading　　　信銜

(2) Date　　　發信日期

(3) Inside address　　　信內地址

(4) Salutation, Greeting　　　稱呼

(5) Body of the letter　　　正文，本文

(6) Complimentary close　　　客套結束語

(7) Signature　　　發信人簽署

2. 附加部分：一封信除上述基本部分之外，可能還有特別事項需附加的。例如下面幾項都是商用英文信中常出現的附加事項：

(1) Reference number　　　案號，參考文號

(2) Attention　　　特定受理人，經辦人

(3) Subject of the letter　　　事由，主旨

(4) Identification marks　　　鑑別符號

(5) Enclosure marks　　　附件符號

(6) Carbon copy notation 　　副本抄送單位記號

(7) Postscript, P.S. 　　　　附啟

(8) Continuation sheet 　　　續頁

現在就以上各項分別說明。

二、信　銜

信銜是信箋上頂端事先印好了的有關發信人 (Sender) 的：

(1) Firm's name 　　　　　商號名稱

(2) Address 　　　　　　　地址

(3) Fax 　　　　　　　　　傳真號碼

(4) E-mail 　　　　　　　　電子郵件信箱

(5) Telephone number 　　　電話號碼

(6) Line of business 　　　　行業種類

(7) Year of establishment 　創業年份

(8) Trade mark 　　　　　　商標

(9) Bankers 　　　　　　　往來銀行

(10) Capital 　　　　　　　資本額

(11) P.O. Box 　　　　　　　信箱號碼

信銜的設計宜簡潔、美觀、大方，不宜多占地位。一般商用英文信多用印妥信銜的信箋。如用沒有印妥信銜的信箋時，信銜應以打字機打上發信人的商號名稱及地址，其位置視採用那一種形體或格式 (Styles or forms) 而定。採 Indented form、Semi-block form 及 Modified block form 時，排在右邊，其末端靠右緣。如採用 Full block form 則自左緣起筆。

(Illustrations of Letterhead)

BURDA ENTERPRISES INC.
Exporters-Importers-Manufacturers

E-MAIL: burda@ms1.hinet.net
E-MAIL: burda1@ms22.hinet.net
5TH FL., 26 SEC. 3, JEN-AI ROAD
TAIPEI, TAIWAN, R.O.C.

REFERENCE BANK:
FUBON COMMERCIAL BANK
HUA NAN COMMERCIAL BANK

PHONE : (O2) 27O5-9286 (1O LINES)
FAX　 : (O2) 27O1-5235
　　　　　 27O1-5236
MODEM : (O2) 27O5-9289

三、發信日期

1.位置：商用英文信應寫明發信日期，其位置在信銜最後一行的下面兩行至四行之間。至於其起寫地方，要看信的格式而定。

在 Full block form 時，日期自左緣開始。採用其他格式時，日期可以放在中央，也可從中央稍靠右的地方開始，但其最後一個字母須與正文 (Body of the letter) 的右邊排整齊。

2.寫法：

⑴ October 12, 20–（美式）

⑵ 12th October, 20–（英式）

⑶ 12 October 20–（英美均用，尤其美軍）

注意事項：

①採第⑴式時，「日」的後面不可用表示序數的縮寫字 st, nd 或 rd 等符號 (never use the abbreviation st; nd and rd after the day of month of the date)，因此，October 12th, 2004 的寫法應避免。但在正文中如「日」未與月份併用，則不在此限。例：

In response to your inquiry of October 5. (not October *5th*)

Your order will be shipped on the *30th*. (right)

"30th" 因無月份，所以 "30th" 等於「同月第 30 日」之意。

②月份不宜用縮寫字，如 Jan., Feb., Mar., Aug.。

Wrong	Right
Apr. 2, 20–	April 2, 20–
9th Mar., 20–	9th March, 20–

③採第(1)、(2)式時，月與年之間，都要加上逗點 (Comma)，但第(3)式，則不可加逗點。

④以下列方法表示日期，均易混淆，宜避免。

英式	美式
12/10/2004	10/12/2004
12/10/04	10/12/04
12/10/'04	10/12/'04
12–10–2004	10–12–2004
12–10–04	10–12–04

（以上均表示 2004 年 10 月 12 日）

四、信內地址

通常包括：

1. Name of addressee　　收信人姓名
2. Title of addressee　　收信人尊稱（頭銜）、職銜
3. Name of the firm　　商號名稱
4. Full address　　地址

但有時可能缺第 1.，2.項。

1.位置：Inside address 應寫在日期或案號下面二行的地方。

TAIWAN TRADING CO., LTD.
EXPORTERS & IMPORTERS

OUR REF. 123　　　　　　　　　　　　　　　　　　October 12, 20–
MR. Charles H. Franklin
Vice President
Atlantic Trading Corp.
61 Broadway
New York, NY 10021
U.S.A.

但也有將 Inside address 放在信的左下角，低於簽署一、二行者。如：

Yours truly,

CHF: tp
Mr. Charles H. Franklin
Vice President
Atlantic Trading Corp.
61 Broadway
New York, NY 10021
U.S.A.

2.注意事項：

⑴Inside address 的內容與格式應與信封上所寫的郵遞地址 (Mailing address) 完全一致。

⑵收信人的尊稱：人名（或商號名稱）寫在第一行，收信人為個人或以人名為商號名稱時，應在其名稱之前，加以適當尊稱。例如 Mr., Miss, Mrs.，或 Messrs., Mmes. 等。如收信人為商號，而非以人名為商號名稱時，不需加任何尊稱。例：

① Mr. Charles H. Franklin

② Messrs. John Henry & Co., Ltd. （以人名為商號名稱）

③ Taiwan Power Company （非以人名為商號名稱）

⑶收信人為個人，而有職銜時，其職銜可寫在人名之後，也可放在人名的下一行。

例如：Mr. Charles H. Franklin, Vice President

或 Mr. Charles H. Franklin

　　Vice President

⑷如寫給商號經理 (Manager)、董事 (Director) 等，而不寫出其姓名時，應在商號名稱上面一行加上 "The Manager" 等字樣。例如：

The Manager (or The President, The Director...)
Taiwan Trading Co., Ltd.
100 Hung Yang Road
Taipei, Taiwan

⑸公司行號名稱：商號名稱應照收信人自用的名稱，不可任意變更。有些商號名稱前面有 "The"，有些沒有。例如中央信託局的英文名稱為 "Central Trust of China" 並不冠 "The"，而華南商業銀行的英文名稱卻冠以 "The" 即 "The Hua Nan Commercial Bank"。

也不可隨意將其名稱予以縮寫或簡寫，例如不可將 "The United States Steel Corporation" 改為 "The U.S. Steel Corp."。

⑹收信人地址寫在商號名稱下一行，通常分為三行，第一行為門牌號碼及街名；第二行為城鎮及郵遞區號（ZIP Code，為 Zone Improvement Program Code 之縮寫）；第三行為國名。門牌前面不加 "No." 或 "#"。

Right	Poor
61 Broadway	No. 61 Broadway
New York, NY 10021	New York, NY 10021
U.S.A.	U.S.A.

五、稱　呼

稱呼為寫信人在進入正文前，對收信人的敬稱，等於中文書信中的「謹啟者」、「敬啟者」、「大鑒」等。性質等於 "How are you?" 的問候語。

1.位置：在 Inside address（或 Attention）下面兩行的地方，並與左緣靠齊。

2.注意事項：

⑴稱呼的第一字母及頭銜必須大寫。例如：

Wrong	Right
Dear sir	Dear Sir
My Dear Sir	My dear Sir
Dear doctor Chang	Dear Doctor Chang

⑵如收信人為個人時，不可將其全名寫出。

Wrong	Right
Dear Mr. John C. Chang	Dear Mr. Chang（張先生台（鈞）鑒）
	Dear John（但正式的商用信不宜用暱稱）

(3)如果收信人為經理 "The Manager" 等,而沒有人名,在稱呼上只能用 "Dear Sir",不能用 "Dear Sirs", 因為這是給經理一個人的。同時 "Gentlemen" 是多數,因此收信人為 "The Manager" 時,就不能用多數的 "Gentlemen" 來稱呼,但也不能用 "Gentleman" 一語。

(4)收信人為商號時, 用下列幾種稱呼 (用多數):

Dear Sirs（英式,對男人或男女組成的商號）

Gentlemen（美式,對男人或男女組成的商號）

Mesdames⎫
⎬（對女人組成的商號, 英美均通用）
Ladies ⎭

(5)Mr., Messrs., Mrs. 和 Miss 不能單獨用作 Salutation, 後面必須有姓。

Wrong	Right
Dear Messrs.:	Gentlemen:
Dear Mr.:	Dear Mr. Chang:
My dear Mrs:	Dear Mrs. Chang:

(6)在 Salutation 一項內頭銜不可縮寫, 但 Mr., Mrs., Prof., 及 Dr. 除外。

Wrong	Right
D'r Sir:	Dear Sir:
Dear S'r:	Dear Sir:
Gents:	Gentlemen:
Mmes:	Mesdames:

(7)稱呼後面的標點有 ":", ",", ":....", 但以前兩者較通用。

(8)商用信中不宜用 "Dear Friend", "My Dear Friend", "Dear Miss", "Dear Customer" 的稱呼。

在大宗的推銷函件、大宗的印刷信件或大宗的通告函, 雖然也有用這種稱呼, 但推銷函件用這種方式, 其效果如何頗令人懷疑。

茲將尊稱與 Salutation 的配合用法列表於下:

收信人		尊　稱	例　示	稱　呼
男性	單數	Mr. (Mister) Dr. (Doctor)	Mr. E. Hemingway Dr. R. P. Watson	Dear Sir Dear Mr....
	多數	Messrs. (Messieurs)	Messrs. Wilson & Co.	Dear Sirs Gentlemen
女性	未婚 單數	Miss	Miss Janet Parker	Dear Miss... Dear Madam
	未婚 多數	Misses	Misses Lucy and Bessy Smith	Dear Ladies Dear Mesdames
	已婚 單數	Mrs. (Mistress)	Mrs. Judy Ford（將 Mrs. 冠以夫姓） Mrs. Maggie Krook（寡婦）	Dear Mrs.... Dear Madam
	已婚 多數	Mmes. (Mesdames)	Mmes. Lucy Cole and Jane Bennet	Dear Ladies Dear Mesdames
	不分已 婚否	Ms. 〔註〕（發音：[miz]）	Ms. Betty Ford	Dear Ms.... Dear Madam

註：Ms. 的多數形不詳，該是 Mses. 吧!

例示：

Messrs. Kahan, Sons & Co.
21, Marylebone Road
London, NW 1
England

Dear Sirs,

Irving Trust Company
1 Wall Street
New York, NY 10021
U.S.A.

Gentlemen:

Mr. Charles H. Chang
Central Trust of China
49 Wuchang Street, Sec. 1
Taipei, Taiwan
Republic of China

Dear Sir:（美式）
Dear Sir,（英式）

Mrs. George Shaffer
California Trading Corp.
15 North Eastern Avenue
Los Angeles, CA 90259
U. S. A.

Dear Madam,

六、正　文（本文）

一封信的正文就是信的主體，也是一封信最重要的部分。其位置在稱呼或事由的下面兩行。正文視內容的繁簡，分為若干段。至於正文的排列型式，請參閱第四節。

發信人寫信的目的，無論告訴人家一件事或請求人家協助，盡在正文中表現。能否達到目的，就要看是否寫得暢通。如果平時多練習，瞭解字的用法、句子的構造、文法的正誤，再研究修辭學，那麼寫起來，就不困難。不過，還有幾點要注意：

1. 要簡潔、明晰，不可拖泥帶水。所用字句不可奧僻，令人讀了難於瞭解，甚至誤會。為便於歸檔或處理，最好一信一事。

2. 正文之中，開頭句及結尾句關係非常重大，如果開頭句及結尾句不精彩，整封信因而黯然失色。關於此，本書第六章有專章論述。

3. 每一種話題 (Topic) 放在同一段 (Paragraph) 中。

4. 每一行的最後一字盡可能排齊，除非不得已宜避免分開最後一個字。如須分開，加連字號 (Hyphen)，並依分寫法的習慣分開，但人名絕不可予以分開。例如：

Un-acceptable　　　　　　　　　　　　　　　　　Remit-tance

Stop-ped　　　　　　　　　　　　　　　　　　　Dis-regard

Example: Beautiful

　　　　　　　　　　　　　　　　　　　　　　　...beauti-

ful..................................

　　　　　　　　　or

　　　　　　　　　　　　　　　　　　　　　　　...beau-

tiful..................................

七、客套結束語

客套結束語是寫信人在信尾客氣的自稱。有如中文信的「謹啟」、「順頌籌祺」、「謹請財安」。其性質等於 "Good-bye"。位置則在正文下面兩行。如用 Full block form 應由左緣開始，其他格式則由信箋的正中央起，向右邊寫。現代商用英文信普通所用的客套結束語有下列三種：Yours truly, Yours very truly, Very truly yours，但

1. 英國常用：

Yours faithfully, 或 Faithfully yours,

而美國則屢用下列用語，以示對收信人的友誼：

Yours cordially, 或 Cordially yours,

2.表示特別尊敬（例如自薦信、下屬對上司、給婦女），可用：

Yours respectfully, 或 Respectfully yours, 或 Yours very respectfully,

3.有親交的商業社交信可用：

Yours sincerely, 或 Yours very sincerely, 或 Sincerely yours,

客套結束語必須與稱呼、正文語氣相稱。茲將配合稱呼的客套結束語列表於下：

Salutation		Complimentary Close
Male	Female	
Sir Sirs	Madam Mesdames	Yours respectfully, Yours very respectfully, Respectfully yours, Yours faithfully, Faithfully yours,
Dear Sir Dear Sirs Gentlemen My dear Sir	Dear Madam Mesdames Ladies My dear Madam	Yours truly, Yours very truly, Very truly yours, Yours faithfully, Faithfully yours, Yours sincerely, Sincerely yours,
Dear Mr. Brown My dear Mr. Brown	Dear Mrs. Brown Dear Miss Brown Dear Ms. Brown My dear Mrs. Brown	Sincerely, Yours sincerely, Sincerely yours, Very sincerely yours, Yours very sincerely, Yours cordially, Yours very cordially, Cordially yours,

八、發信人簽署

簽署包括寫信人的姓名、職銜及簽字，信件寫好之後，必須由發信人簽字，以示負責。手寫簽字往往難於辨認，所以簽名的下面須將發信人姓名和職銜打出。Signature

的位置在 Complimentary close 下面，左邊對齊（Indented form 除外）。例如：

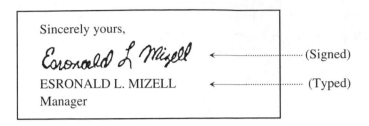

注意事項：

1. 簽字應由發信人親手簽字，不可用橡皮戳。

2. 除非確有特殊情形，不宜由他人代簽，如果由秘書代簽，應在簽字最後一字的
   右下方註明代簽人的 Initial。如：

> Sincerely Yours,
>
> *Megan R. Moore*
>
> MEGAN R. MOORE
> Vice President

3. 如發信人是女性，應將 "Miss"，或 "Mrs." 加註於姓名之前，以便收信人回信時
   易於稱呼。

> Faithfully yours,
>
> *Mary McLaughlin*
>
> (Miss) MARY MCLAUGHLIN
> Assistant Manager

4. 商號名稱因已印在 Letterhead 裡，所以通常不必在簽名的上面再打上商號名稱。
   但比較正式和重要的信函，也打上發信人所屬商號名稱。這時，須將商號名稱
   用大寫打出。

```
Yours very truly,
TAIWAN TRADING COMPANY

Eugene P. Mohan

EUGENE P. MOHAN
General Manager
```

5. 為表明有權代表商號正式簽名起見，往往在商號名稱前加上 "Per Pro" 或 "P. Pro." 或 "P. P." 字樣，這是拉丁文 "Per Procuration"（授權代理）的縮寫。

```
Faithfully yours,
P. P. THE BROWN WATSON & CO.

James B. Johnston

James B. Johnston
Section Chief
```

如無正式代理簽署權，但經商號負責人同意可在其職責範圍內代表商號行文者，應在商號名稱之前，加上 "For" 字樣，或在簽名之前面加打 "By" 字樣，以代替 "P. P."。

九、案　號（參考文號）

現代的大商號，往來函件很多，因此為便於雙方將來查考之用，每一封信都編有案號，註明請收信人在覆信時提及該案號。案號的位置印在 Letterhead 或 Date 下面，靠左緣或右緣。其文字有下列幾種：

1. Please refer to:

2. In reply, please refer to:

3. In reply, please quote:

4. Our Ref. No.

5. Ref.

6. File

7. File No.

8. Letter No.

十、特定受理人（經辦人）

如果收信人對象為商號，而發信人希望某人或某部門特別注意到該信時，可加「經辦人某某」（或某某人查照）一行，這稱為 Attention line。其寫法有：

1. Attention: Mr. William Hunt

2. Attention of Mr. Charles C. Chang

3. Att: Mr. William Hunt

Attention line 通常位於 Inside address 與 Salutation 之間，緊靠左緣，但也有放在中央低於 Salutation 一行或與 Salutation 同一行的。例如：

Irving Trust Company
1 Wall Street
New York, NY 10021
U.S.A.
Attention: Letter of Credit Department

Irving Trust Company
1 Wall Street
New York, NY 10021

Gentlemen: Att: Mr. Ernest Shaw

十一、事由（主旨）

為使收信人商號的收發人員能將信件迅速傳遞到有關部門或有關人員處理，可在信上標明事由或主旨。事由的內容不外兩種：

1.收信人來函的文號或日期或兩者。

2.本信所討論的主題。

事由應放在稱呼及正文之間上下各空兩行。其表現方式有三：

1. 以 "Subject" 字樣開頭，緊靠左緣：

Gentlemen:

Subject: Your order No. 123

We are pleased to inform you⋯⋯⋯⋯⋯⋯⋯⋯⋯⋯⋯⋯⋯⋯⋯⋯⋯⋯⋯⋯

2. 不註明 "Subject" 字樣，但放在中央，加底線：

Gentlemen:

<u>Your order No. 123</u>

We are pleased to inform you⋯⋯⋯⋯⋯⋯⋯⋯⋯⋯⋯⋯⋯⋯⋯⋯⋯⋯⋯⋯

3. 加 "Re"（Reference 的縮寫）字樣，不用底線：

Gentlemen:

Re: Your order No. 123

We are pleased to inform you⋯⋯⋯⋯⋯⋯⋯⋯⋯⋯⋯⋯⋯⋯⋯⋯⋯⋯⋯⋯

第 3 種 "Re" 字樣，有些人認為係陳腐的表現法。

十二、鑑別符號

為了便於查考起見，商用英文信中常將發信人及打字員姓名的第一字母 (Initial) 打在信箋左下面與簽署最後一行平行或下一、二行，緊靠左緣，其表現方式如下：

1. 發信人的 Initial 大寫，打字員的 Initial 小寫：

CYC: ob   CYC/ob   CYC ob

2. 發信人、打字員的 Initial 都大寫：

CYC: OB   CYC/OB   CYC OB

十三、附件符號

如果信中有附件時，應在正文中提及，並在鑑別符號下一行的位置註明，如此一方面可提醒發信部門，以免遺漏；他方面可引起收信人的注意。這種說明有附件的記號稱為 Enclosure mark。其表現方式有多種，茲列若干於下：

Enclosure

Enclosures

Encl.

Encls.

　　用於不重要附件

Enclosure: As stated

Enclosures: a/s

Encl.: As stated

Encls.: a/s

　　同上（As stated 或 a/s 為「附件如文」之意）

Enclosure: 1

Enclosures: 2

Encl.: 1

Encls: 2

　　用於較重要附件

Enclosure: A check #123 for US$100

Enclosures: one cable copy and one catalog

Encl.: one price list

　　用於重要附件

十四、副本抄送單位記號

　　如發信的同時，需將副本抄送有關單位時，以 "C. C."（Carbon Copy 的縮寫）打在 Enclosure 下面的地方，這種副本通常須由發信人簽字。至於其表現方式如下：

　　CC to; CC: ; cc: ; c. c.–; cc–; ccs:（多數時）; Copy to

　　例：

　　　　CC: XYZ Company

　　　　CC to XYZ Company

　　　　Copy to XYZ Company

　　如副本寄送單位多時，為使副本收信單位注意起見，在其名稱前面加 "√"：

　　　　CC: ABC Trading Company

　　　　　XYZ Trading Company

　　　√ Taiwan Marine Insurance Co., Ltd.

十五、附 啟

當信已打好，忽然想加幾句話或補充一件事時，可以在信箋最下方也即在副本抄送單位記號的下面寫上 "P.S." 符號，然後再將要追加的文字加上去。"P.S." 為 Postscript 的縮寫。"P.S." 有如中文信中的「附啟」、「再啟」、「又啟」、「附言」。

注意事項：

1.除非時間關係不便重寫，否則應盡量避免使用附啟。

2.附啟最末一字之後應由發信人加簽（可以 Initial 代替）以示負責。

例：

    CYC: ob

    Encl.: a/s

    CC: New York Corp.

    P.S.: Your offer of August 10 has just been received. We will reply before August 15. cy. 〔cy 為寫信人的草簽（Initial）〕

十六、續 頁

最理想的商用英文信為打字部分的面積約占全頁的四分之三，這樣上下左右都留有適當的空白，型式美觀。如信文過長，不是一張信紙所能容納，則需用續頁。在這種情形，要注意：

1.續頁紙必須用白紙，不宜用印有 Letterhead 的信紙。

2.在各續頁紙的頂端，應將收信人的姓名或商號名稱、頁數、日期分別打出：

例：

ABC Trading Co.	–2–	May 10, 20–

ABC Trading Co. May 10, 2004	Page 2

Page 2   May 10, 20–
ABC Trading Co.

第二節　信封的寫法

一封信寫好後，便要寫信封。信封上記載的文字有：

1. Return address　　　　發信人名稱、地址
2. Mailing address　　　　收信人名稱、地址
3. Mailing direction　　　郵遞指示
4. Others　　　　　　　　其他

信封上的收信人名稱、地址必須正確，以免誤遞，且須與 Inside address 完全相同。所用格式與標點也應一致。

收信人名稱、地址排列行數如不超過四行，則每行之間可採用雙行間隔 (Double spacing)，四行以上則宜採單行間隔 (Single spacing)。

信封的寫法並無標準格式可言，但下面的例子是比較普遍的格式：

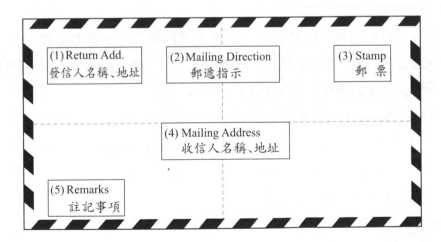

(1)的位置印有發信人商號名稱及地址。如以個人名義發信，則須加打個人姓名，
　　以備無法投遞退回時，可退還個人。除此之外，也有在 Return address 上面加印
　　如下的文字：

　　If undelivered, please return to....（如無法投遞，請退還……。）

If not delivered within...days, please return to....

（如於××日內無法投遞，請退還……。）

⑵的位置為 Mailing direction，可用於指示下列各事項：

Via airmail（航空信）

Express（快信）

Special delivery（限時專送）

⑶的位置為貼郵票處。

⑷的位置為收信人姓名、地址。其位置在信封下半部稍偏右方。其寫法、標點、格式須與 Inside address 一致。須轉交 (In care of, c/o) 時，c/o 放在收信人姓名下一行轉交人姓名前面。

Mailing address 的寫法舉例

例如：

Mr. Charles C. Chang
2516 South Grand Avenue, Apt. 5A
Los Angeles, CA 90018
U.S.A.

---

① 最後一行為國名，即 U.S.A.， U, S, A 後面都須加上省略記號。

② 倒數第二行為：市名、州名、郵遞區號。市名後面必須加 ","，州名縮寫二個大寫字母後面不可加省略記號，即 "."。

③ Room, Suite, Apt. 等番號必須接在 Street address 之後，並排在同一行。

⑸的位置為註記事項，在信封左下方。有的航空信封在這裡事先印上 "By Air Mail"， "Par Avion" 或 "Correo Aereo" 字樣。在此情形，⑸的位置稍向上升。

註記事項可能為下列的一項或多項：

Attention of Mr. William Taylor	專陳威廉泰樂先生
Confidential	機密
Introducing Mr. Charles Chang	介紹張先生
Kindness (or by courtesy) of Mr. A	煩請 A 先生轉交
Photo inside	內有照片（請勿折疊）

Printed matter	印刷品
Private	私函（限由收信人親拆）
Personal	親啟
Registered mail (or registered)	掛號郵件
Sample of no commercial value	無商業價值樣品
Sample of no value	貨樣贈品
Second class airmail	第二類航空郵件
With compliments of...	……敬贈
Strictly confidential	極機密
Urgent	急件

也有將 Return address 寫在信封蓋 (Back flap) 上的。其型式請看下面的例示：

第三節　信紙的折疊法

英文信的折疊方式，雖是微末細節，但因為它是給收信人一個最先的印象，因此不能忽略。

摺信紙的方法，依信封的大小而定。不過總是要摺得美觀，便於郵寄並使收信人易於拆閱。

一、大型信封

使用大型信封時，信紙的摺疊法為先由下端向上端疊 $\frac{1}{3}$，然後再向上疊與上端平齊，成三等分，然後裝入信封。

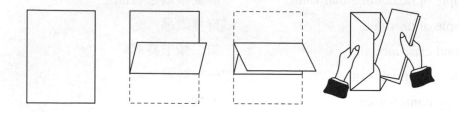

二、小型信封

使用小型信封時，信紙的摺疊法為先由下端向上對折，再從右向左折 $\frac{1}{3}$，然後再由左向右折 $\frac{1}{3}$，再將六折的信紙裝入信封中。

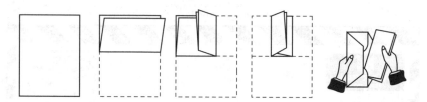

三、開窗信封 (Window envelope)

開窗信封即在信封的中央留一個窗口覆以透明紙。使用這種信封時，折疊信紙必須將收信人姓名、地址折於外面，使其裝入信封後，收信人姓名及地址可在窗口露出。使用開窗信封可免在信封上重打收信人姓名、地址，增加效率，茲舉一例於下：

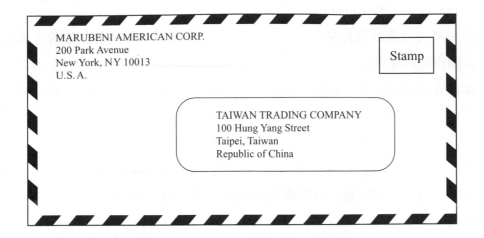

MARUBENI AMERICAN CORP.
200 Park Avenue
New York, NY 10013
U.S.A.

Stamp

TAIWAN TRADING COMPANY
100 Hung Yang Street
Taipei, Taiwan
Republic of China

第四節　商用英文信的格式及標點

一、商用英文信的格式

　　商用英文信的各項構成內容已如前述。至於各部分的編排方式——即格式 (Forms or styles)——以及標點方法，視寫信人的喜愛而定。茲將常見的格式介紹於下：

1. 全齊頭式 (Full block form)：Date, Inside address, Salutation, Body, Complimentary close 及 Signature 等的每一行都一律從左緣開始，排列整齊，不向右縮進。優點：打字不必考慮各段第一行縮進的問題，打字節省時間。缺點：失去平衡，不美觀，不便於閱讀。

2. 半齊頭式 (Semi-block form)：Inside address 及 Salutation 採齊頭式，Date 放在右上方與 Inside address 並列。Body 採縮進式，Complimentary close 與 Signature 移到右下方。

   ①優點：Body 的每段第一行縮進若干字母，便於閱讀，而且 Inside address 與 Salutation 在左上方，Complimentary close 與 Signature 在右下方，看起來顯得平衡。因此這是目前最常用的一種格式。

   ②缺點：因每段第一行須縮進若干字母，打字較費時。

3. 改良齊頭式 (Modified block form)：Inside address, Salutation 及 Body 採齊頭式，每一行都從左緣開始，Date 移到右上方與 Inside address 第一行並列，Compli-

mentary close 與 Signature 移到右下方。

全齊頭式的優點為打字省時，缺點為不平衡，改良齊頭式則將此兩者加以調和，兼顧省時與平衡。

4. 縮進式 (Indented form)：Inside address, Body 的每段開頭，向右縮進 (Indented) 形成梯形。此格式是傳統的型式，但現在用的人較少。優點為各部分區分明顯，缺點為打字效率較差。

5. 懸段式 (Hanging paragraph form)：又稱為 Hanging-indention form，即 Body 的每段第一行從左緣開始，其餘各行則縮進若干字母。其他部分與半齊頭式及改良齊頭式同。

Full block form, Open punctuation

BURDA ENTERPRISES INC.
Exporters-Importers-Manufacturers

E-MAIL: burda@ms1.hinet.net
E-MAIL: burda1@ms22.hinet.net
5TH FL., 26 SEC. 3, JEN-AI ROAD
TAIPEI, TAIWAN, R.O.C.

REFERENCE BANK:
FUBON COMMERCIAL BANK
HUA NAN COMMERCIAL BANK

PHONE : (O2) 2705-9286 (10 LINES)
FAX : (O2) 2701-5235
2701-5236
MODEM : (O2) 2705-9289

Your ref.

Our ref. No. 123

October 15, 20–
Mr. Charles Ford
61 Broadway
New York, NY 10021
U.S.A.

Dear Mr. Ford

This letter is written in full block form, in which no indentions are used. The punctuation is open.

Every line of the heading, inside address, the first line of every paragraph, the complimentary close, as well as the signature, begins at the left marginal stop. Paragraphs are set off by a double line spacing. This form is economical in time and effort for firms doing a large volume of correspondence.

However, full block form is never used in handwritten letters.

Yours very truly

Steve M. McCarthy

Steve M. McCarthy
Sales Production Manager

SMM/csc

　　這種格式在印刷物中常用。書信中很少用，優點為每段都突出，在分項敘述時較易引起注意。

Semi-block form, Mixed punctuation

BURDA ENTERPRISES INC.
Exporters-Importers-Manufacturers

E-MAIL: burda@ms1.hinet.net
E-MAIL: burda1@ms22.hinet.net
5TH FL., 26 SEC. 3, JEN-AI ROAD
TAIPEI, TAIWAN, R.O.C.

REFERENCE BANK:
FUBON COMMERCIAL BANK
HUA NAN COMMERCIAL BANK

PHONE : (O2) 27O5-9286 (1O LINES)
FAX : (O2) 27O1-5235
      27O1-5236
MODEM : (O2) 27O5-9289

Your ref.

Our ref.

Mr. Charles Ford   October 15, 20–
61 Broadway
New York, NY 10021
U.S.A.

Dear Mr. Ford:

    This letter is written in semi-block form with mixed punctuation.

    The heading and the inside address are blocked. However, the first line of each paragraph of the body of the letter is indented five spaces from the left margin. Each succeeding line of the paragraph is carried to the left margin. The lines of the body are single spaced, with double spacing between the paragraphs.

    Because of the length of the heading and the inside address, two line spaces separate the first line of the address from the date line. Spacing at this point is a variable matter, dependent on the length of the letter and other conditioning factors.

    The complimentary closing, the writer's typed signature, and the writer's signature begin at the center of the page. These are in block form, corresponding to the inside address.

Very truly yours,

Steve M. McCarthy

Steve M. McCarthy
Sales Production Manager

SMM: csc

註：上述各種格式的名稱，學者間有若干不同命名，例如：

① Full block form → Block form → Extremely block form

② Semi-block form → Modified block form

③ Modified block form → Block form

Modified block form, Mixed punctuation

BURDA ENTERPRISES INC.
Exporters-Importers-Manufacturers

E-MAIL: burda@ms1.hinet.net
E-MAIL: burda1@ms22.hinet.net
5TH FL., 26 SEC. 3, JEN-AI ROAD
TAIPEI, TAIWAN, R.O.C.

REFERENCE BANK:
FUBON COMMERCIAL BANK
HUA NAN COMMERCIAL BANK

PHONE : (02) 2705-9286 (10 LINES)
FAX : (02) 2701-5235
2701-5236
MODEM : (02) 2705-9289

Your ref.

Our ref.

Mr. Charles Ford   October 15, 20–
61 Broadway
New York, NY 10021
U.S.A.

Dear Mr. Ford:

This letter illustrates the modified block form of letter dress, which has become one of the most widely used methods of arranging letters.

It is called modified block form because it modifies the full block style by moving the complimentary close and signature group to the right to achieve balance. It has the advantages of the full block but also gets a better appearance.

If you desire your letters to be attractive in appearance and economical with regard to stenographic time, I heartily recommend the modified block form as the most suitable for the needs of your office.

Sincerely yours,

Steve M. McCarthy

Steve M. McCarthy
Sales Production Manager

SMM: csc

Indented form, Closed punctuation

BURDA ENTERPRISES INC.
Exporters-Importers-Manufacturers

E-MAIL: burda@ms1.hinet.net
E-MAIL: burda1@ms22.hinet.net
5TH FL., 26 SEC. 3, JEN-AI ROAD
TAIPEI, TAIWAN, R.O.C.

REFERENCE BANK:
FUBON COMMERCIAL BANK
HUA NAN COMMERCIAL BANK

PHONE : (O2) 2705-9286 (10 LINES)
FAX : (O2) 2701-5235
2701-5236
MODEM : (O2) 2705-9289

Your ref.

Our ref.

Mr. Charles Ford,　　　　　　　　　　　　　　　October 15, 20–
　61 Broadway,
　　New York, NY 10021,
　　　U.S.A.

Dear Mr. Ford:

　　In reply to your letter of June 15, I would like to illustrate a traditional form of business letters.

　　This letter is written in the indented form, with closed punctuation.

　　The first line of the inside address begins at the left margin, while each succeeding line begins two to four spaces to the right of the preceding line. The first line of each paragraph of the body of the letter is indented five spaces from the left margin, but each succeeding line is carried to the left margin. The lines of the body are single spaced, with double spacing between the paragraphs.

　　Closed punctuation in the addresses is rarely used by typists today. Open punctuation is now preferred by most letter writers.

　　I hope you will find this illustration helpful.

　　　　　　　　　　　　　　　　　　　　　　Sincerely yours,

　　　　　　　　　　　　　　　　　　　　　　Steve M. McCarthy

　　　　　　　　　　　　　　　　　　　　　　Steve M. McCarthy
SMM: csc　　　　　　　　　　　　　　　　　Sales Production Manager.

Hanging indention form, Mixed punctuation

BURDA ENTERPRISES INC.
Exporters-Importers-Manufacturers

E-MAIL: burda@ms1.hinet.net
E-MAIL: burda1@ms22.hinet.net
5TH FL., 26 SEC. 3, JEN-AI ROAD
TAIPEI, TAIWAN, R.O.C.

REFERENCE BANK:
FUBON COMMERCIAL BANK
HUA NAN COMMERCIAL BANK

PHONE : (02) 2705-9286 (10 LINES)
FAX : (02) 2701-5235
        2701-5236
MODEM : (02) 2705-9289

Your ref.

Our ref.

Mr. Charles Ford   October 15, 20–
61 Broadway
New York, NY 10021
U.S.A.

Dear Mr. Ford:

This letter is written in what is called the hanging indention form, with mixed punctuation.
    The inside address and the signature are in block form.

The body of the letter begins as always two spaces below the salutation. The first line of
    each paragraph begins at the left margin, but each succeeding line is indented five
    spaces. The lines of the body are single spaced, with double spacing between the para-
    graphs.

You will notice that the effect of this form of letter is to emphasize the first word of each
    paragraph. In a carefully planned sales message this device can be used to good advan-
    tage by headlining the salient facts concerning the product sold.

Very truly yours,

Steve M. McCarthy

Steve M. McCarthy
Sales Production Manager

SMM: csc

二、英文信上各部分的標點 (Punctuations)

英文信上的 Date, Inside address, Salutation, Complimentary close 及 Signature 等各行末端所用的標點符號打法有閉式 (Closed punctuation)、開式 (Open punctuation) 及混合式 (Mixed punctuation) 三種。

1. 閉式：Date, Inside address, Salutation, Complimentary close 及 Signature 各行後面加上適當標點符號，如逗點 (,)、句點 (.) 或冒號 (:) 者，稱為 "Closed punctuation"。閉式標點多半配合 Indented form, Semi-block form 及 Modified block form 使用。

2. 開式：前述各項每一行後面都不加任何標點符號者稱為 Open punctuation。開式標點多配合 Full block form 使用。

3. 混合式：前述各項除了 Salutation 後面用逗號 (,) 或冒號 (:) 及 Complimentary close 後面用逗號 (,) 外，其餘都不加標點符號者稱為 Mixed punctuation，現在大都採用混合標點法。

Indented form, Closed punctuation

Full block form, Open punctuation

Semi-block form, Mixed punctuation

②_____   ①_____
_____
_____
③_____ : (,)
    ④_____
_____
_____ .
_____
_____
_____ .
                        ⑤_____ ,
                        ⑥_____
                        _____

Modified block form, Mixed punctuation

②_____   ①_____
_____
_____
③_____ : (,)
④_____
_____
_____ .
_____
_____
_____ .
                        ⑤_____ ,
                        ⑥_____
                        _____

說明:

① Date

② Inside address

③ Salutation

④ Body

⑤ Complimentary close

⑥ Signature

美國各州名的略字

(State abbreviations in U.S.A.)

The state abbreviations (new and old) shown in the following list are officially accepted by the United States Post Office Department.

Alabama	Ala.	AL	Nebraska	Nebr.	NB
Alaska		AK	Nevada	Nev.	NV
Arizona	Ariz.	AZ	New Hampshire	N. H.	NH
Arkansas	Ark.	AR	New Jersey	N. J.	NJ
California	Calif.	CA	New Mexico	N. M.	NM
Canal Zone		CZ	New York	N. Y.	NY
Colorado	Colo.	CO	North Carolina	N. C.	NC
Connecticut	Conn.	CT	North Dakota	N. Dak.	ND
Delaware	Del.	DE	Ohio		OH
Florida	Fla.	FL	Oklahoma	Okla.	OK
Georgia	Ga.	GA	Oregon	Oreg.	OR
Guam		GU	Pennsylvania	Pa.	PA
Hawaii		HI	Puerto Rico		PR
Idaho		ID	Rhode Island	R. I.	RI
Illinois	Ill.	IL	South Carolina	S. C.	SC
Indiana	Ind.	IN	South Dakota	S. Dak.	SD
Iowa		IA	Tennessee	Tenn.	TN
Kansas	Kans.	KS	Texas	Tex.	TX
Kentucky	Ky.	KY	Utah		UT
Louisiana	La.	LA	Vermont	Vt.	VT
Maine		ME	Virginia	Va.	VA
Maryland	Md.	MD	Virgin Islands		VI
Massachusetts	Mass.	MA	Washington	Wash.	WA
Michigan	Mich.	MI	West Virginia	W. Va.	WV
Minnesota	Minn.	MN	Wisconsin	Wis.	WI
Mississippi	Miss.	MS	Wyoming	Wyo.	WY
Missouri	Mo.	MO	District of		
Montana	Mont.	MT	Columbia	D. C.	DC

註： 州名必須按上面二個大寫字母縮寫而不可加上省略記號 (.)，即應寫成 NY，不宜寫成 N. Y.。

第三章
撰寫商用英文信的七要 (Seven C's of English Business Letter Writing)

一封完善的商用英文信應具備 Seven C's 的要件，所謂 Seven C's 指：Completeness, Clearness, Correctness, Concreteness, Conciseness, Courtesy 及 Consideration 而言。茲分別說明於下：

第一節 完備 (Completeness)

所有的信都是為某種目的而寫，那麼為達成此目的，就應將其文意加以完備 (Complete) 的敘述，不可有所遺漏。假如一封信寫得不完備 (Incomplete)，不但不能達成目的，反而可能引起反效果。例如 Dispute（糾紛）、Complaint（訴怨）或 Claim（索賠）等等即往往因為不完備的通信而引起。因此，與遠隔異地的對方通信的 Foreign trade，首先要注意其完整齊備。

在國際貿易中，Terms and conditions of sale（買賣條件）比國內買賣要複雜得多。在通信中，自應將其有關的條件逐一交代清楚，不得遺漏。再如 Particulars of documents（文書內容），Instructions（指示），Specifications（規格書）或 Clauses of contract（契約條款）等等，如其說明稍有欠缺或有一語半字的錯誤，難免需要另行查詢照會，以致浪費時間與金錢；有時，甚至貽誤商機或引起嚴重的糾紛。茲舉例於下：

【例 1】

（不完備）Dear Sirs,

　　　　We are returning the goods you shipped, as it arrived too late.

　　　　　　　　　　　　　　　　　　　　　　　　　　　　　　Yours truly,

這種惜字如金的信，除非賣方標榜 "Return goods welcome"（歡迎退貨）另當別論外，就 Completeness 而言，可謂一文不值。謹慎的商人 (Careful merchant) 起碼將改寫如下：

（完　備）Gentlemen:

We are returning the cotton goods, your invoice No. 105, by s.s. "President" today.

This was received May 2, too late for our Spring Sale. You will find on reference to our order of January 3, that this was to be shipped to reach us not later than March 15.

Yours faithfully,

【例 2】

（不完備）Shipment will be made in due course.

（完　備）Your order for 500 units personal computers will be shipped at the end of this month. You should receive them early next month.

【例 3】

某出口商收到一張訂單，所訂貨品為：

Cellulose tape $\frac{1}{2}'' \times 3$ yds. with plastic dispenser 700 doz.

—ditto—but $\frac{1}{2}'' \times 5$ yds., 1,000 doz.

　　出口商將 "Ditto" 解釋為僅指 "Cellulose tape"，於是裝出附有 Dispenser（切膠帶用具），寬半吋，長三碼的膠帶 700 打，及未附 Dispenser，寬半吋，長五碼的膠帶 1,000 打。但買方收到貨後立即提出索賠。理由為其中 1,000 打未附 Dispenser。這種情形是由於買方所發 Order 不完備所致，並不能完全責備出口商，使用 "Ditto"（略語為 Do.）這種字眼時，宜極謹慎，以免發生誤會。

第二節　清楚 (Clearness)

　　Clearness 與 Lucidity 同義，寫信時應推開窗子說亮話，措詞明白清楚，使閱讀的人能立即明瞭內容，不致引起誤會。商用英文最忌含糊 (Obscurity)、模稜兩可 (Ambiguity)、不明確 (Vagueness)。信的內容應具有一貫性、段落分明、理路清晰，且排列妥當，使其相互關係非常清楚，茲分述於下：

一、避免使用意義不明確的語句

【例 1】

（含糊）*Fluctuations in the freight* after the date of sale shall be for the buyer's account.

"Fluctuations in the freight" 一詞含有運費上漲或下跌之意，因此欠明晰，容易引起爭執。如改寫成下，就清楚明白了。

（清楚）*Any increase* in the freight after the date of sale shall be for the buyer's account.

（出售日以後運費如有上漲，歸買方負擔。）

（清楚）Any change up or down (or Any increase or decrease) in the freight after the date of sale shall be for the buyer's account.

（出售日以後運費如有上漲或下跌，均歸買方負擔。）

【例 2】

（含糊）As to the steamers sailing from Keelung to Sydney, we have *bimonthly* direct services.

"bimonthly" 有「一個月兩次，即每半個月」或「兩個月一次」之意，因此欠明確。

（清楚）We have two direct sailings every month from Keelung to Sydney.（……每個月二次。）

（清楚）We have semi-monthly direct sailing from Keelung to Sydney.（……每半個月一次。）

（清楚）We have a direct sailing from Keelung to Sydney every two months.（……兩個月一次。）

【例 3】

（含糊）We have duly received your order, for which please accept our thanks.

這個句子，文法上並無錯，但在訂單很多的場合，到底是那一訂單，欠明確。因此，必須將訂單號碼、商品名稱、日期等敘明才可。

（清楚）We have duly received your order No. 123 dated March 10 for 1,000 doz. of garments, for which we thank you.

二、修飾詞應放在適當的地方

【例 1】

a. We shall be able to supply 100 cases of the item *only*.

b. We shall be able to supply 100 cases *only* of the item.

a. 為「以此商品為限」，b. 為「以一百箱為限」，由此可見修飾詞的位置的不同，其意思迥異。

【例 2】

（含糊）The goods we received *contrary to our instructions* are packed in wooden cases without iron hoops.

原意是想說「木箱沒有鐵箍條，是違反我們的指示」，但由於 "contrary to our instruction" 放錯地方，變成「違反我們的指示而收到……」。

（清楚）The goods we received are packed in wooden cases without iron hoops *contrary to our instructions.*

三、使用修飾詞應注意其聯貫性 (Coherence)

1.由副詞、副詞子句、關係代名詞所構成的 Clause，應盡可能將 Clause 放在被形容之詞的旁邊，但非修飾文中特定語的 Sentence adverb，應盡可能放在文首或文尾。

（含糊）The proprietor's personality and sincerity have made, *in our opinion*, the firm what it is today.
由於 "in our opinion" 插在及物動詞與其受詞之間，使句子的意義不明，如將它放在文首就清楚了。

（清楚）In our opinion, the proprietor's personality and sincerity have made the firm what it is today.
（我們認為該店主的人格及誠懇使他的店有今日的成就。）

2."Accordingly" 有 "Therefore"（因此）及 "In agreement with what has preceded"（遵照……）兩種意義，如當作後者之意時，應放在句尾。

【例】If you will give us 24 hours' notice, we will arrange accordingly.
（如於 24 小時前通知我們，則我們將遵照指示安排。）

3.句子中避免使用數個關係代名詞。

（含糊）We have received with thanks your letter of May 15, *which* instructs us to purchase for your account 1,000 tons barbed wire, BWG #12, *which* we have today filled at US$190 per M/T.
第二個 which 以 barbed wire 為先行詞，結果變成 to fill barbed wire 意思不通，因為只能說 to fill order，而不能說 to fill goods。

（清楚）We received with thanks your letter of May 15 *instructing* us to purchase for your account 1,000 tons barbed wire, BWG #12, and have today filled *your order* at US$190 per metric ton.

四、忌用生僻字、新撰字、俚語、土語或貿易界不通用的略語

要記住「現代化」與「實際化」聯繫起來，才是正確的原則。不能一味翻新，將尚未通行於一般社會或商界的新撰字，胡亂運用。局部通行的俚語以及一地方的土語，應避免使用。例如「徹底做去」一語，不要說 "To go the whole hog"，應該說 "Do it thoroughly"。商用英文中的略字，習慣上也有一定的限度，使用時應以通用者為限。例如

“M/T” 有 “Metric Ton” 及 “Measurement Ton” 之意。“C. C.” 有二、三十種不同的意義。用生僻字，使讀者不易瞭解。只求詞達意，不必咬文嚼字，過求雅典，反令人難解。例如不要把 “Daily”（每日的）說成 “Diurnal”，“Wealth”（財產）說成 “Pelf”。

五、注意代名詞、關係代名詞與其先行詞的關係位置

（含糊）Mr. Chang wrote to Mr. Ford that he had received his order.

　　　　（張先生寫信給福特先生說他的訂單他已收到了。）

　　　　這裡的 “his” 是誰不明確。因為可以是 “Mr. Chang” 也可以是 “Mr. Ford”。

（清楚）Mr. Chang wrote to Mr. Ford that he had received Mr. Ford's order.

　　　　（張先生寫信給福特先生說福特先生的訂單他已收到了。）

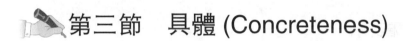

第三節　具體 (Concreteness)

　　所謂 “Concreteness” 就是要言之有物，措詞、文意切中要領，據實直陳，切忌空泛、抽象。

一、避免抽象籠統

　　譬如說貨色好，不要單說 Best, Fine, Highest Grade, Supreme（高級）等抽象的形容詞。一定要將它的優點一一具體的臚列出來。抽象的字眼非但人家不會相信，而且意義上也不明確。

【例 1】

（不具體）Our apples are excellent.

（具　體）Our apples are juicy, crispy and tender.

【例 2】

（不具體）Dear Sirs,

　　　　　　We acknowledge with thanks receipt of your inquiry of May 2.

　　　　　　We can supply you with *the types of Cigarette Lighters you described. If you send us your reply soon, we might be able* to meet the due date you mentioned.

　　　　　　Awaiting your early reply,

　　　　　　　　　　　　　　　　　　　　　　　　　　　Yours very truly,

　　這封信敘述不夠具體。因此，除非 Letter subject（案由）已將品質、規格、數量等

標示，否則收信人會懷疑寫信人對該品質、規格及數量是否已正確了解。

"send us your reply soon" 一句中，"reply" 是抽象的字，並未具體要求怎樣的答覆。"soon" 一字太籠統。"we might be able to..." 有「也許可能，也許不可能」之意，使人覺得不可靠。

（具體）Dear Sirs,

We shall be pleased to supply you with *two dozen "Flamex" Cigarette Lighters according to the specifications* in your letter of May 2.

If you will telegraph us confirmation of the order by May 20, we will send them by air freight not later than May 30.

We look forward to the opportunity of serving you.

Yours very truly,

二、日期、期間應明確表示

Early in July; Late in August; Beginning part of May; Middle part of April; Later part of June; Recent; In due course 等語句均含糊不具體。

【例 1】

（不具體）As requested in your *recent* letter we have *already* sent the samples to you.

這句中 recent, already 等字眼均不具體，應將日期具體寫出。

（具　體）As requested in your letter dated October 12, we sent you the samples by air parcel on October 20.

【例 2】

（不具體）You will receive our reply about it *in due time*.

（不具體）We enclose a copy of our latest catalog which you requested us to send *several days ago*.

in due time（在適當時期），several days ago 等語句，均含糊不具體。

（具　體）You will receive our reply about it within one week.

（具　體）We enclose a copy of our latest catalog, as requested in your letter of November 8.

三、明確寫出文件、單證的案號 (Reference)

（不具體）We have effected shipment under your L/C.

這句對於開狀銀行、信用狀號碼均未具體指出。

（具　體）We have effected shipment under your L/C No. 123 issued by Bank of America.

第四節　簡潔 (Conciseness)

　　寫商用書信應力求簡潔扼要，切忌冗長。須知商場中人，業務倥傯，最重視時間，無功夫閱讀連篇累牘的信。因此，商用書信，應在辭能達意的原則下，使用簡單淺明的辭句。

一、以簡單淺明的文字代替長而艱深的字

長而艱深	簡單淺明	長而艱深	簡單淺明
Apparent	Clear	Initial	First
Approximately	About	Initiate	Begin
Ascertain	Find out	Inquire	Ask
Assist	Help	Modification	Change
Commence	Begin	Obtain	Get
Compensation	Pay	Participate	Take part
Conclusion	End	Perform	Do
Construct	Build	Particularly	Especially
Contribute	Give	Procure	Get
Deliberately	Carefully	Provided	If
Demonstrate	Show	Purchase	Buy
Difficult	Hard	Reimburse	Pay
Discontinue	Stop	Remunerate	Pay
Endeavour	Try	Render	Give
Entitled	Called	Require	Need
Equivalent	Equal	Request	Ask
Expedite	Hasten	Sufficient	Enough
Explicit	Plain	Transmit	Send
Facilitate	Made easy		

二、將數語變為一語

數　語	一　語	數　語	一　語
Similar to	Like	In relation to	To, Toward
At an early date	Early, Soon	In spite of the fact that	Although

At all times	Always	In the amount of	For
Attached hereto	Attached	In the case of	If
At the present time	Now	In the event that	If
At this writing	Now	In the event of	If
Be in a position to	Can	In the meantime	Meanwhile
Both of them	Both	In the near future	Soon
By means of	By, With	In the neighborhood of	Nearly, About, Around
By virtue of	By, With	In the time of	During
By return of mail	Soon, Immediately, Promptly	In this place	Here
Costs the sum of	Costs	In view of	Since, Because
Due to the fact	As, Because	Look forward to	Await
For the reason that	As, Because	On a few occasions	Occasionally
In view of the fact that	Since	On behalf of	For
Depreciate in value	Depreciate	On the part of	For, Among
During the course of	During	Owing to the fact that	Because
Enclosed herewith	Enclose	We would ask that	Please
Final completion	Completion	Past experience	Experience
For the purpose of	For	Previous to	Before
In accordance with	By, With	Prior to	Before
Inasmuch as	As, Since	Seems to be	Seems
In connection with	Of, In, On	There can be no doubt that	Doubtless
In order to	To	Under separate cover	Separately
With a view to	To	With reference to	About
In regard to	To	With regard to	About
		With respect to	About

三、將子句變成片語

【例 1】

（冗長）Mr. Ford, *who is president of Taiwan Trading Company*, said he would leave for London on November 18.

將 "who is president of Taiwan Trading Company" 的 Clause 改為 Phrase，以作為 Mr. Ford 的同位格，就簡潔得多了。

（簡潔）Mr. Ford, *president of Taiwan Trading Company*, said he would leave for London on November 18.

【例 2】

（冗長）Please read the third paragraph of our letter No. 125 of August 5 *so that you will get all the facts*.

將 "so that you will get all the facts" 的 Clause 改為 Phrase 就很簡潔。

（簡潔）Please read the third paragraph of our letter No. 125 of August 5 for all the facts.

四、將冗長句子化為短句子

【例 1】

（冗長）*We take the liberty to approach you with request that you would be kind enough to* introduce to us some exporters of iron scrap in your city.

（簡潔）*Please* introduce to us some exporters of iron scrap in your city.

（簡潔）*Would you please* introduce to us some exporters of iron scrap in your city.

（簡潔）*We shall appreciate* your introducing to us some exporters of iron scrap in your city.

（簡潔）*You would oblige* us by introducing to us some exporters of iron scrap in your city.

【例 2】

（冗長）*We are writing to you with a view to* entering into business relations with you.

（簡潔）We are anxious to start business with you.

【例 3】

（冗長）We are *in receipt of* your letters dated August 30 and September 19 *respectively*.

（簡潔）We have your letters of August 30 and September 19.

五、儘可能使用主動語態 (Active voice) 藉以簡化句子

【例 1】

(Passive): It is noted that the sales volume has been increasing.

(Active): We noted that the sales volume has increased.

【例 2】

(Passive): It is believed that this policy will be beneficial to our customers.

(Active): We believe this policy will benefit our customers.

【例 3】

(Passive): It is suggested that your consideration be given to the recommendations in this report.

(Active): We suggest that you consider the recommendations in this report.

由上述例子，可知一般而言，主動語態較被動語態簡潔而且易於瞭解。

六、善用連接詞 (Connectives)

簡潔並不是說亂將句子縮短，假如將短句子一連串並列，將顯得彆扭。為避免這種情形，應善用連接詞。

（拙劣）Thank you for the samples. You sent them to us on June 20. We are pleased to place an order. It is specified on the enclosed order sheet.

這些句子是夠簡單了，但顯得很拙劣，不自然。

（較佳）Thank you for the samples which you sent to us on June 20. We are pleased to place an order as specified on the enclosed order sheet.

茲將常用的 Connectives 列舉於下：

1.針對前面的句子，附加其他事項時：

Again, Besides, Further, Furthermore, In addition, Incidentally, Likewise, Moreover。

2.與前面的文意對照時：

At the same time, For all that, However, In the meantime, Meanwhile, Nevertheless, Notwithstanding, On the contrary, On the other hand, Still yet。

3.表示結論、結果時：

Accordingly, As a result, Because of this, Consequently, For that reason, In view of the above, Hence, So, Then, Therefore, Thus。

4.強調語氣時：

As a matter of fact, At least, In fact, Indeed, Of course, No doubt, To be sure, Without doubt。

5.概括時：

In brief, In other words, In short, On the whole。

簡潔並不是簡陋，簡潔須兼顧 Clearness, Correctness, Courtesy 及 Completeness 等等。

第五節 正確 (Correctness)

商用書信是科學化的應用文學，正確乃為不可或缺的要素。因為商用書信事關權利、義務、利害關係，如內容不正確，則易引起糾紛。所謂正確性包括下列各要件：

1. 敘述正確 (Correctness of statement)。
2. 數字的精確 (Accuracy of numerical expressions)。
3. 商業術語的正確使用 (Correct use of commercial terms)。
4. 正確的文法 (Grammatical correctness)。

茲分述於下：

一、正確的敘述

1. 勿 Overstatement（誇張）或 Understatement（掩飾）：商業最重信用，騙人只能騙一遭，與其過甚其詞，將來損失信用，倒不如老老實實，既不過分誇張，也不過分含蓄或掩飾。

【例 1】

（誇張）This stove is *absolutely the best* on the market.

這是老王賣瓜，自賣自誇的措詞，既抽象又武斷，不如將其性能、優點具體地向客戶介紹。

（正確）Our model TR 123 stove is designed on modern lives and gives, without any increase in fuel consumption, 25% more heat than the older models. So you will agree that it is the outstanding stove for economy.

【例 2】

（誇張）It is the *lowest* price *available* to you.

假如競爭者報出更便宜的價格，豈不有欺騙對方的嫌疑？

（正確）It is the *lowest* price (that) *we can offer you now*.

（正確）It is a *modest* price (which) *we can quote you now*.

【例 3】

（誇張）We *assure* you that the error will *never* occur *again*.

犯錯乃為人之常情，這種表現法未免過分誇張。

（正確）*We will do all we can* so that we may not *repeat* such an error (or to prevent a repetition of such an error).

下列的用語都是加強語氣的字眼，如亂用，易構成 Overstatement，所以使用時應謹慎。

Completely	Quite
Strictly	Especially
Entirely	Utterly
Exceedingly	Very
Extremely	Very much
Greatly	All
Definitely	Best
Perfectly	Utmost
Only	

2.Understatement 與客氣的措詞：Somewhat, Nearly, As it were, In a measure, It seems to us 等字眼常使文章語氣顯得軟弱無力。如用得不恰當，就易構成 Understatement。如想表示客氣，可使用如下的 Past subjunctive（過去假設語氣）。

We *should* be glad to receive it.

You *might* report to us once a week.

二、數字的正確表示法

商用英文信中涉及數字的情形很多。數字的表現如發生錯誤易造成糾紛。尤其，以「以上」、「以下」表示數量、金額；以「以前」、「以後」表示日期時，應特別注意，以免收信人誤解。茲將其正確表現方法列記於下：

1.提及的數字包括在內者：

2 鎊以上	Stg. £2 or (and) above (over)
50 打以上	50 dozen or up, 50 dozen and upwards
US$100 以下	US$100 or less, US$100 and below
一星期以內	One week or less
30 天以下	Not exceeding a total of 30 days
6 月 30 日以前	On or before June 30
6 月 30 日以後	On or after June 30

從 6 月 30 日起	$\begin{cases} \text{On and from (after) June 30} \\ \text{As from June 30} \end{cases}$

2.從……到……為止（包括最後數字在內）：

40 鎊為止	Up to £40 inclusive
10 月 20 日為止	Up to and including October 20

3.頭尾均包括在內：

從 3 月 1 日到 15 日	From the 1st to the 15th of March both inclusive
從星期一到星期五為止	From Monday through Friday

【例 1】

All offers by cable are open for five days inclusive of the date of dispatch.

（所有報價自拍電日起有效 5 天〔包括拍發日〕。）

【例 2】

This contract will come into effect from and including May 1.

（本契約自 5 月 1 日〔包括 1 日〕生效。）

三、使用正確的商業術語

使用商業術語可節省很多時間與說明。但使用時，不僅對該術語的意義須完全瞭解，而且必須是貿易界通行的術語。

1.慎用容易發生誤會的術語：例如 FOB 後面通常加以裝船港名稱，但在美國，FOB 有六種不同的解釋，因此與美國人交易時，如按船上交貨條件交易應在 FOB 與裝船港之間加 "Vessel" 字樣，例如：FOB Vessel New York（紐約港船上交貨條件）。

又如 CAF 在美國為 CFR 之意，但在法國則解釋做 CIF，即 "A" 為 Assurance（保險）之意。$CIFC_3$ 的 C_3 是指佣金 3%，但此佣金是 Selling commission 還是 Buying commission 呢？而且 3% 的佣金是按 $CIFC_3$ 計算還是按 CIF 計算呢？難免發生歧見。

2.不用意義寬泛的字語：例如，與其用寬泛的 "Economic"（經濟的），不如用 "Money-saving"（省錢的）或 "Time-saving"（省時的），更不如用 "Dollar-saving" 或 "Minute-saving"；與其用寬泛的 "Say" 不如用 "Tell", "Inform", "Assert"（主張說），"State"（陳述）。

3.應該用意義特定的字眼，例如：

⑴「利息」應用 "Interest"，「利潤」應用 "Profit"，不要用普通的 "Advantage"（利益）。

⑵「盈餘」應用 "Surplus"，「帳上的結餘」應用 "Balance"，不要用普通的 "Remainder"（餘數）。

4.應該配合所要說的意思，使用恰當的字語：「償舊債」必須說 "Pay off old debt"，「預付」必須說 "Pay in advance"，「分期攤還」必須說 "Pay by instalments"，「見票即付」必須說 "Pay at sight" 或 "Pay on demand"，「見票後 90 天付款」必須說 "Pay 90 days after sight"，這些動詞所接的介系詞不可用錯。

四、正確的文法

文法是從語言文字的習慣逐漸形成的公認的法則。如果我們作文說話不講文法，別人將不瞭解我們的意思。因此撰寫貿易英文信時，必須講求文法，而且還要顧到貿易英文信中的習慣用法。例如在普通文法沒有錯的字句，有的在貿易英文信上卻已不適用。過分拘泥於刻板的文法，流於呆滯，反將書信的語氣和風格減色；有時候普通文法上認為不妥當的，在貿易英文信中卻很通用。關於此，將在第四章及第五章作進一步的說明。

第六節　謙恭 (Courtesy)

謙恭有禮是商場上的重要法則。接待顧客固然要有禮貌，寫信時也不能缺少禮貌。有禮貌的信很可博得收信人的好感。所以書信中如能適當地應用 "Kindly"，"Please"，"Thank you"，"We have pleasure"，"We regret that" 等用語，必能引起閱信人的好感。但亂用恭敬的詞句，也是不對的。例如在 "Please find enclosed cheque."（敬請查收附上的支票。）中，用 "Please" 就不對。如沒有請求收信人賜惠，或沒有過分麻煩收信人時，在貿易英文信中，"Please" 一詞大可不必濫用。「敬請」一詞，不能像在中文信中那樣隨便加上。

商場上的往來，好像一部轉動的機器，有時難免有摩擦，「禮貌」就是商場中的潤滑油。商用書信中，無論是索賠、訴怨、質問或是拒絕往來，其措詞都應力求客氣。

對方的來信無論如何傲慢，都要心平氣和，有禮貌委婉地作答。

要如何寫出有禮貌的信呢？

一、善用被動式 (Passive) 語氣

同樣一件事的敘述，用主動式表達則常有譴責對方的感覺，如改用被動式，則語氣就緩和得多。

（無禮）You made a very careless mistake.

（無禮）You did not enclose the cheque with your order.

（無禮）For the past two years, you did not give us any order.

這些主動式的句子頗有嚴厲譴責對方之感，顯得欠缺禮貌。如將 "You" 省略，並改以被動式，則語氣委婉，不致刺激收信人。

（有禮）A very careless mistake was made.

（有禮）The cheque was not enclosed with your order.

（有禮）For the past two years, no order has been given to us (or no business has been materialized between us).

二、善用含有客氣語意的詞句

（無禮）We demand immediate payment from you.

（無禮）We are disgusted with your manner of doing business.

（無禮）We must refuse your order.

"demand" 一詞有不客氣地要求之意，語氣太重，如改用 "request" 則顯得客氣，使人看起來較舒服。"disgust" 或 "hate" 等詞，語氣都很重，"do not like" 或 "dislike" 則語氣稍緩和，不致使收信人覺得受不了。"We must..." 頗有威脅之感，"refuse" 也語氣太強。

（有禮）We request your immediate payment.

（有禮）We are not completely satisfied with your manner of doing business.

（有禮）We regret that we are not in a position to accept your order.

三、以肯定 (Positive) 詞句代替否定 (Negative) 詞句

否定詞句常會引起對方不快，如改以肯定語氣，則可博得對方好感：

【例 1】

（否定）We do not believe you will have cause for dissatisfaction.

（肯定）We feel sure that you will be entirely satisfied.

【例 2】

（否定）It is against our policy to sell leftover goods below cost.

（肯定）It is our policy always to have full stocks of goods at fair prices.

【例 3】

（否定）We are discouraged to learn that we were in fault.

（肯定）Thank you for calling attention to the unfortunate incident.

【例 4】

（否定）We cannot understand why you have had trouble with this article.

（肯定）We presume that there must be some reason for your having trouble with this article.

四、善用婉轉句法 (Mitigation)

使用婉轉語法以免刺激對方，此類常用詞句有：

We are afraid	It seems (would seem) to us
We would say	We (would) suggest
We may say	As you are (may be) aware （如你所知）
We might say	As we need hardly point out （不用指出）
We (would) think	

【例 1】

（無禮）It was unwise of you to have done that.

（婉轉）*We would say that* it was unwise of you to have done that.

【例 2】

（無禮）You ought to have done that.

（婉轉）*It seems to us that* you ought to have done that.

【例 3】

（無禮）We cannot comply with your request.

（婉轉）*We are afraid* we cannot comply with your request.

使用婉轉句法時，應注意不可違反 Concreteness。

五、使用 Subjunctive mood

表達希望或意見時，可用含蓄語法，以示有禮貌。

【例】

a. We wish you *would* let us have your reply soon.

b. *Would* you compare our sample with the goods of other firms?

c. We *should* be grateful if you *would* help us with your suggestions.

第七節　體諒 (Consideration)

　　寫信人在寫信時，不能一味地只顧從自己的立場著想，而應設身處地，為對方著想，這就是 "You attitude" 或 "You viewpoint" 或 "The reader's point"。具體地說，提起任何事物，應在可能範圍內少用第一人稱的 "I"，"We"，"My"，"Our" 如何如何，而應多說些 "You"，"Your business"，"Your profit"，"Your needs" 如何如何，要把對方的利益放在讀者眼前。以「收信人的利害關係」為前提，是撰寫商用書信的重要原則。因此，書信中的每一段 (Paragraph) 的開頭宜多用 "You" 而少用 "We"。

"We" expression	"You" expression
a. *We* are pleased to announce that...	a. *You* will be pleased to know that...
b. *We* want you to do it.	b. *You* will no doubt do it.
c. *We* follow the policy because...	c. *You* will benefit from this policy because...
d. As to *our* standing, *we* refer you to Bank of America...	d. As to *our* standing, *you* may address any inquiry to Bank of America...

　　然而少用 "We" 起頭並不是說要勉強地以 "You" 起頭，而故意不用 "We"。假如譴責、責問、訴怨的信也滿篇 "You"，"You"，不但不禮貌，反而易招致惡感。例如下面的句子都是以 "You" 起頭，句子中也一再提起 "You"，然而卻是很拙劣的句子。

　(1)Perhaps *you* did not intend to disregard our request.

　(2)*You* attribute your negligence to the fact that *you* are very busy, but it cannot be believed that *you* must take so long in reply.

　　由上面的例句可知，Consideration 與 Courtesy 是不可分的。

第四章
舊式文體與新式文體 (Old-fash-ioned Styles v. Modern Styles)

第一節　舊派與新派之爭

關於商用英文信的用字，一直存在著舊派與新派無休止的論爭。新派主張商用英文信不應：

1. 使用陳腔濫調的客套。
2. 使用深奧的詞語。
3. 使用詞意模糊的詞語。
4. 攙雜閒語。
5. 以分詞式作為結尾。

然而舊派仍繼續使用一些新派所指責的「舊式用語」(Old-fashioned expressions)，並未向新派低頭。就實際情形來看，銀行、保險公司、船公司以及若干大型的公司或公營事業機構比較保守，仍繼續使用所謂的「舊式用語」，而較小型的公司或民營公司則有不使用「舊式用語」的趨勢。

第二節　舊式用語與新式用語的比較

茲將商用英文信中常見的舊式用語與新式用語作比較，讀者如能仔細比較對照，並瞭解其意義，則在運用時，當能作適當的選擇。

1. Above:
   舊派將 "Above" 當做形容詞，習見常用的有：
   The above statement （上列的聲明）

The above address（上開的地址）

The above date（上載的日期）

新派認為不應將 "Above" 當作形容詞用，而宜以 "Foregoing"，"Preceding" 或 "Above-mentioned"等字來代替，例如：

The foregoing statement

或將 "Above" 正式地當做副詞用，例如：

The address given above

至於 "The above date" 宜改用明確的日期，例如：

Yours of above date

宜改為：

Your letter of May 2

評：其實 "Above" 可當做形容詞用，所以 "The above statement" 的用法，並沒有什麼可議之處。

2.Advise:

舊派將 "Advise" 當做「通知」解釋，例如：

Please advise us of the shipping date.（請通知我們裝運日期。）

新派認為 "Advise" 的正確意義是「勸告」或「提供意見」，只能用於請教人家進告意見的時候。例如：

Please advise us how to proceed.（請告訴我們如何進行。）

而不能用作 "Inform"，"Notify" 或 "Tell" 解釋，因此：

Please advise us of the shipping date.

應改為：

Please inform us of the shipping date. 或

Please notify us of the shipping date.

評：銀行的通知、裝船通知都用 "Advice"，其動詞為 "Advise"，美國出版的 *Webster New International Dictionary* 也明白的解釋成 "To give information or notice to"，因此 "Please advise us...."這種用法並無什麼不妥當。

3.Agreeable to your request; Pursuant to your request; Conformable to your request; As per your request:

這四個片語意義大同小異，舊派將其當作中文信中的「謹遵台命」用。

新派認為這是畸形的用法，尤其 "Agreeable to" 及 "Conformable to" 用字太曖昧，應避免使用。例如：

⑴Pursuant to your request, we are sending you the catalog.

應改為：

We are sending you the catalog you ask for. （我們奉上承索的目錄。）

⑵We have examined our goods as per your request.

應改為：

We have examined our goods as requested in your letter of May 3.

（敝處已按照 5 月 3 日尊函要求把我們的貨物檢查過了。）

評：Pursuant to your request, As per your request 這些片語現在仍然很多人用。

4.Acknowledge receipt of; Acknowledge with pleasure; Acknowledging yours of:

舊派認為這些片語用起來堂皇鄭重，因此在很多書信中，尤其大的公司行號都樂於使用。

新派則認為這些片語，裝腔作勢，不平易，因此，應予摒棄不用。例如：

We acknowledge receipt of yours of (June 3)...

應改為：

We thank you for your letter of (June 3)... 或

We have received your letter of (June 3)...

評：收到比較重要的文件時，用 "acknowledge receipt of" 並沒有什麼不恰當。但以 yours 代替 your letter 語意含混，而且不禮貌。

5.And oblige:

舊派將 "And oblige" 這個詞放在信中的最末尾，接以 "yours truly" 以示謙恭，有如中文書信中的「為荷」、「為禱」、「為感」。

新派認為這個詞，顯得卑屈，毫無意義，而且減弱了結尾的語氣，應予刪除。

例如：

Kindly ship the enclosed order and oblige. （請將訂貨即行運下為感。）

應改為：

Please ship the enclosed order immediately.

評：有道理，贊成新派說法。

6.Along this line; Along these lines:

舊派將此兩片語解做「關於此（方面，類事，行業）」，「依函中所述」。例如：

Our business along this line has been very satisfactory.（我們此一行業非常滿意。）

新派認為這類片語用的太多了，為了免除陳腔濫調，不如以 "Of this kind"，"Of this nature"，"To this effect" 等片語代替。例如：

Our business of this kind has been very satisfactory.

又如：

He said much along these lines.

宜改成：

He said much to this effect.（關於這類事情他談得多了。）

評：同意新派說法。

7.Answering yours of:

舊派書信中常以這個片語作為起頭，例如：

In answering yours of March 5...

其意義為「准……華翰，茲覆……」。新派認為以分詞式開頭的用法是過時的，應改為：

We answer your letter of March 5...

評：贊成新派。

8.As per:

舊派將 "As per" 當做「依照」，「如……」解釋，"Per" 為拉丁文，等於 "By"，"Through"。例如：

As per your letter

As per your instructions

As per offer sheet enclosed

新派認為這種用法太陳腐，而且在同一信內混用兩國文字不妥當。除了法律文字外，不要用，而應改用：

According to your letter

According to your instructions 或 As instructed by you

As described in the attached offer sheet

評："As per" 一詞，在貿易書信中用得很多，在有些場合用起來很方便。但在普通信中不用為宜。

9.Appreciate:

舊派將此詞用作 "Realize" 或 "Understand"（瞭解）解釋。例如：

You appreciate that we had no previous information on the subject.

（我們對此事事前並不曉得，你是瞭解的。）

新派認為 "Appreciate" 的原意為 "To estimate justly"（公正的估計）或 "To value highly"（很重視），不應用作 "Realize" 或 "Understand" 解釋。因此上面的句子應改為：

You understand (or realize) that we had no previous information on the subject.

此外，"Appreciate" 已含有「非常鄭重」之意，所以不要仿舊派另加 "Greatly"，"Highly"，"Very much"，"Deeply" 等副詞。例如：

We appreciate your service highly.（我們很感謝你的服務。）

應改為：

We appreciate your service. 或

We appreciate the favor you have done me.

評：贊成新派。

10.As under:

舊派喜歡將「如下」以 "As under" 表示。但新派認為「如下」應以下列各詞表達：

As follows, As below, As given below, As stated below,

As detailed below, As listed below, As enumerated below,

As described below, As shown below, As outlined below,

As mentioned hereunder, etc. 注意在這些詞後面都要加上 Colon (:)。

評：贊成新派。

11.At all times:

舊派書信中常用這個片語，其意為「總是」。例如：

We shall be pleased to talk with you at all times.

新派認為這是陳舊的用法，應以 "Always" 代替，既簡又賅。例如：

We shall always be pleased to welcome you at my office.

評：雖然言之有理，但英美人所寫的信中仍常用此片語，足見有些人對此片語仍抱有好感，而且兩者語感略有不同。

12. At an earliest possible moment; At an early date; At your earliest convenience; At your convenience:

舊派將這些片語用作「儘速」，「得便務請從速」，「有便即請」，新派認為不簡潔、不確定、陳腐、曖昧 (Vague)。應改用更 Specific 的詞，諸如 "At once", "Soon", "Immediately", "Promptly" 或 "As soon as you can" 或寫出具體的日期。例如：

Please notify us at an early date. 應改為：

Please let us know immediately (or within 10 days).

We should appreciate hearing from you at your earliest convenience. 應改為：

We should appreciate hearing from you immediately.

評：原則上贊同新派。但不能以使用了上述片語就一概視為係舊式。有時，基於文章的語氣，用這些片語，可能反而生動。此外，"at your earliest convenience" 用在對方應做而未做時，是妥當的。否則就有失禮了。如對收信人有所請求時宜將 "earliest" 改成 "early"，至於新派認為不如將日期明確寫出，但規定期限，容易使對方覺得受到束縛引起反感。

13. At hand; Has come to hand; To hand:

舊派將這些片語解作「收到」，「到手」，「在手邊」。例如「收到訂單」寫成：

Your valued order of the 6th inst. is at hand.

Your valued order of the 6th inst. has come to hand.

新派認為這些用語陳舊，不宜用。因為你的回信就等於告訴對方已收到他的信了。因此應改為：

Thank you for your order of June 6. 或

We have received your order of June 6.

評：幾十年前的商用英文信中常出現這些用語，現在已少見了。但用在 "We note from your letter, to hand yesterday, that..." 時，甚方便，不宜貿然視為陳舊。

14. As stated above:

新派認為太古板，不應使用。

評：普通的信固然不應用這種片語，但在貿易文書中，用這種片語有時候仍覺得格調不錯。

一般而言，以 "As mentioned above", "As noted above", "As referred to above" 代替為宜。

15.Attached hereto; Attached herewith; Attached please find; Attached you will find:

舊派將以上片語用於「附上」，「檢附」之意。例如：

Attached herewith is an invoice.

新派認為 "hereto"，"herewith"，"please find" 這些詞是多餘的，應予刪除，以求簡潔。例如：

Attached is an invoice.

評：這些用語雖有陳舊之嫌，但用在貿易書信中，確有其方便之處，不能一概排斥。

16.At the present writing:

這是「在寫這信的時候」之意，舊派很喜歡用這片語。例如：

At the present writing we have none in stock.（目前我們一點存貨也沒有。）

新派認為商用書信在能表達意思的前提下，應盡量運用簡短而有力的字句。"At the present writing" 這個片語本身雖無弊病，而且過去曾有一段時期被認為是一種很時髦的用語，不過在講求效率的現代商業社會，還是直截用 "Now" 一詞來得自然、簡潔、沉著有力。例如上面的句子改為：

We have none in stock now.

評：原則上贊同新派。但偶爾一用，應屬無妨。

17.At this time:

新派認為應以 "Now"，"At present" 代替。因此

We wish to advise that we are out of stock at this time.

應改成：

We are sorry to tell that we are out of stock at present.

評：贊成新派。

18.Awaiting your letter (order, reply); Awaiting your further wishes:

這是「等待你的信（訂單、答覆）」及「等候你再惠顧」的舊式用法。新派認為以分詞式開頭為過時的用法，而且 "Awaiting your further wishes" 的 Expression 也未免迂迴，不如開門見山，改為：

Please give us your further orders.

至於 "Awaiting your letter" 則改成：

Your early reply will be appreciated.

We hope to hear from you soon.

Please let us hear from you.

May we hear from you immediately.

效果較佳。

> 評：現代的美國商用文，雖認為以 "-ing" 型式作為一封信的信尾不好，但只要不亂用，有
> 　　時也具有很好的語感。因此，不能一概摒棄不用。只是不要不管三七二十一，概以
> 　　"Awaiting"，"Trusting"，"Hoping"，"Thanking you" 等分詞型式作為信的結尾。

19. Beg to acknowledge; Beg to inform; Beg to advise; Beg to assure; Beg to call your attention; Beg to confirm; Beg to state; Beg to suggest; Beg to enclose; Beg to say:

"Beg" 為「請求」，「祈求」之意；"Beg to acknowledge" 猶如中文信中的「敬覆者」之意；"Beg to inform"，"Beg to advise" 等相當於中文信中的「敬啟者」，「茲通知」；"Beg to assure" 則等於「茲保證」。"Beg to call your attention" 為「敬請注意」；"Beg to confirm" 為「謹證實」；"Beg to state" 為「謹告」；"Beg to suggest" 為「謹建議」；"Beg to enclose" 為「謹附上」。

以上均在舊式英文信中習見。新派認為這些都是老古董，過分謙卑，除用於下級人員對於上級人員呈文外，不應用於普通商用書信。即使用 "Beg" 一詞也該加上 "leave"，"permission" 等詞。例如：

Beg leave to say（敬告）

I beg permission to go（請准放行）

現在是民主時代，大家平等，像 Beggar 那樣拼命 Beg，未免卑屈，反而使人瞧不起。所以應改成：

We acknowledge (We have received...)

We inform you

We assure you

We call your attention

We confirm

We write（用以代 Beg to state to you）

We suggest you

We enclose you

評：贊成新派。

20.By return mail; By return of mail; By return post; By return of post; By return:

「請即回示」，「收到即覆」之意。例如：

Please answer by return of post.

這些語法，在郵政不發達時，很習見。往昔郵政人員乘著郵車 (Mail coach) 將郵件送到另一個城市，郵政人員抵達後稍作休息，又將該城市待寄出的郵件帶回。所謂 "By return of post" 即指收信人收到信後立即寫覆信交由原班郵車帶回之意。

現在郵件遞送方法與從前不同。因此新派認為這種語法已不合時代，而且與 "At an early date" 同樣是濫調，不夠 Definite，應以 "Immediately"，"Promptly"，"At once" 等詞代替或明確指出日期。例如：

Please answer promptly.

評：原則上贊成新派，但這種語法迄今未被淘汰，可能是戀舊之故吧。

21.Complaint:

「訴苦」，「抱怨」，「責問」之意。例如：

In reply to your complaint of August 2, ...

新派認為覆信中 "Complaint" 一字既不恭敬（好像說「你發什麼牢騷」）也不正確，應改為：

In response to your letter of August 2, ...

或開門見山，說出要說的話。

評：原則上同意，但對那些一天到晚發牢騷的人，又何嘗不可用這個字？

22.Contents noted; Contents have been noted; Contents duly noted; Contents carefully noted:

舊派的客套語氣，其意為「內容均悉」，「內容詳悉」或「敬悉一是」。例如：

Yours of June 25 to hand, contents of which have been carefully noted.

（6 月 25 日函已收到，詳悉一切內容。）

新派認為這些用語在信中是冗句，徒佔篇幅，毫無意義，犯了繁滯重複的毛病。不知道來信內容，如何覆信呢？所以應改為：

We have received your letter of June 25

    a. which we have read carefully.

    b. which we have read with interest.

    c. regarding the shipment by s.s."Peseus".

    d. with regard to the opening of L/C.

    e. in which you ask us to send samples.

    f. from which we note that the goods arrived damaged.

評：贊成新派。

23.The captioned claim; The captioned matter:

信函列有標題（主旨）時，在本文中，常有 "The captioned..." 的用語。其意為「如主旨所列事宜」。美國的銀行或保險公司的文件中常出現這種用語。新派認為此字太古老，在一般貿易書信中不宜採用，而宜改用 "The above claim"，"The claim above referred to"，"This claim"，"The above matter"，"This matter"。

評：原則上贊成新派。但是為什麼銀行、保險公司可以用而貿易書信中卻不宜用呢？

24.Deal:

舊派將此字當作 "Transaction"（交易），"Arrangement"（協定），或 "Agreement"（協議）用。例如：

We made a deal with that company.（敝行與該公司作了一筆交易。）

新派認為將 "Deal" 當作「交易」用是俗語且含意模糊，不夠 Clear，應改為：

We made a transaction with that company.

評：贊成新派。

25.Deem:

舊派將此字用作「以為」、「視為」、「相信」解釋。新派認為太雅了，法律味道太濃，應以 "Think" 或 "Believe" 代替。

評：贊成新派。

26.Due to the fact (that...):

這是「鑒於……」，「由於……」的舊式用法。新派認為這種措詞太迂迴，不如單刀直入，以 "Because" 代替。

評：贊成新派。

27.Duly:

「適當地」之意。例如：

We have duly investigated this matter.（我們已適當地調查了本案。）

Please return one copy duly signed.

新派認為句中加了 "Duly" 一字，即表示調查所作的努力仍有保留，所以語氣顯得不肯定而乏力。因此，應予刪除，改為：

We have investigated this matter.

Please sign and return one copy.

評：除了契約的詞句，宜避免使用此字。

28.Enclosed find; Enclosed please find; Enclosed you will find; Enclosed herewith:

這些用語意指「茲檢附……請察收為荷」，「茲檢附……」。例如：

Enclosed please find a copy of our price list.

新派認為這種措詞未免太滑稽，既然已隨函附上，難道還要發信人請求發現 (Please find)，收信人才會去覓找？不要把收信人當傻瓜吧！因此應改為：

We enclose a copy of our price list. 或

We send you enclosed a copy of our price list.

再者 "Enclose" 後面不應再加上 "herewith" 字樣。

評：贊成新派。

29.Esteemed favor:

"Esteemed" 為「可敬的」、「受尊重的」之意，"Favor" 為「信札」(Letter) 之意。所以舊派將「大札」、「尊函」稱為 "Esteemed favor"，新派認為 "Favor" 為「恩惠」之意。收到人家指責或索債的信，稱其為 "Favor" 已經不恰當，再加上 "Esteemed"，更顯得矯飾過分。業務上的通信也不必用此 Overpolite expression. 因此應改為 "Kind letter" 或只說 "Letter" 就夠了。此外如 "Esteemed letter"，"Esteemed inquiry" 中的 "Esteemed" 也應予刪除。

評：贊成新派，尤其不說 "Letter" 而說 "Favor" 太古板。但 "Favor" 用於 "I shall thank you for any favor you can show me."（如承照拂，不勝感謝。）卻是適當的。

30.Even date:

"Even date" 為「同一日期」(Of the same date)「當日」之意。往昔商用英文中

很習見。例如：

We have received your esteemed favor of even date.

新派認為這是欠明確 (Not specific)，易使人混淆的用語，和 "At present writing"，"At this writing" 一樣應在摒棄之列，而將日期明確寫出。例如：

We have received your letter of August 10.

評：贊成新派。實際上現在已無人使用。

31.Favor:

如前所述，舊派將 "Favor" 當作 "Letter" 解釋，新派認為這是陳腐的用法。須知一封信的本身並不是恩惠，而只是一種加惠的手段，難道討債的信也是恩惠嗎？

所以 "Favor" 一字只有用在下面的場合才算適當：

It will be a great favor if you can let me know...

（如果你能夠給我知道……那將是一大恩惠。）

此外，舊派也將 "Favor" 當作動詞「惠賜」、「給與」解釋。例如：

Please favor us with your reply at your earliest convenience.

新派認為這是 Overpolite expression。應改成：

We are looking forward to your early reply.

Your early reply will be appreciated.

評：原則上贊成新派，但 "Favor" 當作動詞「惠賜」的用法迄今仍常見。

32.For your Information:

「供你參考」之意。很多人喜歡用這種 Expression 以致有陳腔濫調之嫌，所以新派認為應盡量不用，而將所要說的話直接寫出來。

評：贊成新派。但偶爾一用，是無可非議的。"For your reference" 係用於有關具體參考事項、參考物時，因此不可與 "For your information" 混淆。

33.Hand you herewith:

「茲遞交」、「茲檢附」之意。例如：

We beg to hand you herewith our latest catalog.

新派百思不解的是：在信中如何遞交？因為 "Hand you" 是「親手交給你」之意，與事實不符，所以應改為：

Enclosed is our latest catalog.

評：贊成新派。

34.Hasten：

「趕快」、「急忙」之意。例如：

We hasten to inform you some good news.

新派認為這個用語令人有虛飾、沉不住氣的感覺，而且難免令人擔心會不會忙中有錯。如真要快，應用 Cable 或 Telex。所以應改為：

We are pleased to inform you of some good news.

評：新派言之有理。

35.Have before us：

「（來信）在我們面前」也即「我們已收到（你的信）」之意。例如：

We have before us your favor of October 10.

這種 Expression 新派認為是無理取鬧，因為覆信時，來信未必在覆信人面前，而且在不在覆信人面前，對收信人毫無關係。因此，應改為：

We have received your letter of October 10.

評：贊成新派，實際上舊派人也不用了。

36.Hereby advise：

"Hereby" 為「茲」之意，"Hereby advise" 為「茲通知」、「茲奉告」之意。新派認為 "Hereby" 這個字是多餘的，要通知什麼，開門見山地說出就好，何必「茲」不「茲」。

評：在比較正式的文書，尤其官方公文書，仍使用 "Hereby"，以示嚴肅。在普通貿易書信中，能免則免。

37.Have the pleasure to $\begin{cases} \text{inform;} \\ \text{submit;} \\ \text{enclose;} \end{cases}$ Have the pleasure of $\begin{cases} \text{informing;} \\ \text{submitting;} \\ \text{enclosing;} \end{cases}$ Take pleasure in...-ing:

"Have the pleasure of (To)...", "Take pleasure in..." 都是「樂於……」之意，這種客套話，假使能夠適當地應用，可給人好感。不過，新派認為使用太多，則令人覺得虛偽。如向對方索帳、譴責、索賠也用 "Have the pleasure to inform you" 的話，豈不令收信人啼笑皆非。

評：贊成新派，只要符合 "Right word to the right situation" 的原則即可。

38.Hoping for an immediate answer; Hoping to hear from you; Hoping to be favored with a reply:

「希即賜覆」，「希賜覆」之意。舊派商用英文，常將這些措詞作為信尾結束之用。新派認為這種 Expression 充分顯示寫信人的矜持與無能。因為只敢「希望」(Hope) 而已，足見沒有自信心。宜改為：

We are looking forward to your prompt reply. 或

We are looking forward to hearing from you soon.

評：贊同新派。

39.I remain; I am:

舊派常將這些用語作為信尾結束之用，例如：

I remain yours truly (or sincerely, etc.)（你永遠忠實的）也即中文信的「謹上」。

因此 "I remain" 後面不應有逗點 (Comma)。舊派的信，雖然喜歡用此詞，卻在 "I remain" 後面加上逗點，這是畫蛇添足也。而且 "I remain" 係用於第二次以後的通信才能用，初次通信，頂多只能用 "I am yours truly"。新派認為這種客套太陳舊。因此：

Hoping to hear from you in the near future, I remain, 應改為：

I look forward to hearing from you soon.

而 Thanking you for your kind consideration, I am.

應改為：

Your consideration is appreciated.

評：贊成新派。

40.In accordance with; In compliance with:

「依照」，「遵從」之意。例如：

In compliance with your recent request, we are pleased to forward a copy of our latest catalog.

新派認為 "In compliance with" 是 Overformal。貿易書信中不必如此拘謹、誇張做作。因此應改為：

We are glad to send you our latest catalog, which you requested.

（或以 as requested 代替 which 以後的文字）

評：目前仍有很多人用這種措詞，新派人士是否有一點神經過敏？

41.In addition:

「加之」，「又」之意。

新派認為：何不用簡潔的 "Also" 呢？

評：為避免滿篇 "Also"，偶爾用 "In addition" 何嘗不可。

42.Inasmuch as:

「……因此」，「所以」之意。

新派認為應以簡潔的 "Because" 或 "Since" 代替。

評：同意新派。在同類語中，語氣最強的字是 "Because"，其次是 "Since"，再其次是 "As"，"For"。"Because" 是直接說明對於 "Why?" 的「理由」、「原因」。"Since" 一向是站在「既是……就……」的意義去說明事件關係中的自然結果。

43.In answer to same; Same; The same:

舊派將 "Same" 當作代名詞用，等於中文的「該」，「這」或「他，他們，它，它們」(He, She, They, Them, It, This) 等等，其所指的主體可能單數，也可能多數。

例如：

We regret the delay, and hope that same has not caused you inconvenience.

（對於遲延一事，我們很抱歉，希望這一延擱，並未引起你們的不便。）

We have heard from Mr. Ford and have written to same.

（福特君有來信並已覆訖。）

新派認為 "Same" 所代表的主體不明，易引起誤會，除了法律文書外，不應使用，而應用具體的代名詞 "It"，"They"，"Them"，"Him" 等。例如：

We regret the delay, and hope it has not caused you inconvenience.

We have heard from Mr. Ford and have written to him.

評：贊成新派。

44.In due course; In due course of time; In due time:

"In due course" 為「適時」，「屆時」之意；"In due course of time" 為「在相當時間內」之意；"In due time" 為「在適當時間內」之意。例如：

We have received your order and it will be processed in due course of time.

新派認為這些措詞均有含糊不肯定之嫌，且語氣軟弱，應說出具體的日期。例如：

Thank you for your order No. 123. We will ship it by September 30.

評: 美國人士屬於新派，但英國人士認為如對收信人有所請託時不宜將日期明確規定，以免失禮。總之，是否宜用這些措詞，應視情形而定。

45.In re; Re:

「關於……」之意。

舊派在信的「事由」欄（或「主旨」欄）前面往往加上 "Re" 這個字，表示「關於」。例如：

Re: Your order No. 123 of July 5.

又在信中則以 "In re" 表示「關於……」。例如：

In re your request of August 3, would say we can not give you definite reply.

新派認為「事由」前面加上 "Re" 是陳舊的用法。只有在需要寫出「事由」的情形下，逕寫出事由，並在下面畫一橫線，而不用 "Re"。至於 "In re" 更應予摒棄而改用 "Concerning"，"Regarding"，"Relative to"。所以上面兩例應分別改為：

<u>Your order No. 123 of July 5</u>

We are, unfortunately, unable to give you a definite reply concerning your reguest.

評: 贊成新派。但在事由欄前面加 "Re" 的情形迄今仍習見。

46.In reply to your favor; In reply, would state that; In reply, wish to say; In reply, we wish to state that:

都是「謹覆」之意。

新派認為寫貿易書信不是投稿，需按字計酬。因此應力求簡潔，開門見山，不要拖泥帶水，廢話連篇。所以：

"In reply, we wish to state that the samples you requested have been airmailed today."

一句中的 "In reply, we wish to state that" 這一段應予刪除，而直書為：

The samples you requested have been airmailed today.

評: 贊成新派。

47.In response to yurs; In response to your favor:

均與前面的 "In reply to your favor" 同義。新派認為拐彎抹角，應予摒棄。

評: 贊成新派。

48.In the amount of:

為 "For" 之意。例如：

Your check in the amount of US$10,000 has been received.

新派認為寫文章投稿時不妨用 "In the amount of" 的片語。但從事貿易的人還有很多事情要做，如以 "For" 代替，那將感謝不盡，故宜改為：

Your check for US$10,000 has been received.

評：在比較正式的文書裡面，仍可用 "In the amount of"。

49.In receipt of; In possession of:

表示「受領」，「占有」的狀態，也即「收到」之意。例如：

We are in receipt of your favor dated June 5.

新派認為這種 Expression 確具文學風味，可是站在貿易書信的立場，應用簡潔的 "Receive"。即：

We have received your letter of June 5.

評：贊成新派。

50.In the near future:

「在不久的將來」之意。例如：

We shall inform you about this matter in the near future.

（我們將於不久的將來再行奉告。）

新派認為這種表現法抽象、不明確。應明確指出日期，例如：

We shall inform you about this matter next week.

如實在無法確定日期，也應用簡潔的 "Soon" 代替。例如：

We shall inform you about this matter soon.

評：其實老練的人卻喜歡用 "In the near future" 這個片語。因為沒有把握能很快 (Soon) 通知時，用此片語，很可敷衍對方也。

51.Inst.; Ult.; Prox.:

這三個字分別為 "Instant"（本月），"Ultimo"（上月）及 "Proximo"（下月）的縮寫。例如：

Your inquiry of the 4th inst....

Yours of the 9th ult. received.

Your order will be shipped on the 10th prox.

新派認為現代的貿易書信貴在明確，用了這些字，讀者必須在腦海中多打一個轉才能領悟實際的月份。何況 "Ultimo"，"Proximo" 又是拉丁文，犯了用字生僻的毛病。因此應改為：

Your inquiry of June 4....

We have received your letter of May 9.

Your order will be shipped July 10.

評：贊成新派。

52.Kind favor; Kind order:

「親切（或仁慈）的信或訂單」之意。例如：

Thank you for your kind favor (order).

新派認為用此措詞時應小心，因為不是所有的來信都是「親切」的。至於 "kind order" 等於中文的「親切的訂單」，這種措詞也顯得怪彆扭。因此應改為：

Thank you for your letter (order).

評：贊成新派。此外，"Thank you for your (kind enquiry, kind news)." 中的 "kind" 也應予刪除。如一定要這樣說，那麼應以 "Thank you for your kindness in sending this enquiry." 的方式表現。

53.Line:

舊派將此字當作貨品 (Merchandise) 或各種貨品 (Line of goods) 用。例如：

Our salesman, Mr. Jones, will gladly show you our line.

但新派認為 "Line" 此字含意模糊。應改為：

Our salesman, Mr. Jones, will gladly show you our merchandise (or line of goods).

評：贊成新派。此外舊派也將 "Line" 當作 Business 解釋，應予排斥。

54.Kindly advise; Kindly be advised; Kindly inform:

分別為「請賜告」，「敬告」，「請賜知」之意，例如：

Kindly advise (inform) us when our order will be shipped.

Kindly be advised that your order has been shipped.

新派認為以 "Kindly" 代替 "Please" 是多餘的，而且以 "Advise" 代替 "Inform"，"Tell" 也不合理。同時 "Kindly be advised" 是說寫信人的通知是 "Kindly" 並非說收信人的動作是 "Kindly"，因這種措詞很滑稽，所以應改為：

Please tell (inform) us when our order will be shipped.

We are pleased to inform you that your order has been shipped.

評：贊成新派，但新派認為非以 "Inform" 或 "Tell" 代替 "Advise" 不可一節不能苟同。此外還有 "Please kindly look into this matter without delay" 的表現法，同時使用 "Kindly" 和 "Please" 不但不顯得有禮貌，反而有累贅之嫌，應將 "Kindly" 刪除。

55. Meet with your approval:

「得到你的認可（贊成）」之意。例如：

We hope our plan will meet with your approval.

新派認為應改用：

We hope you will approve of our plan.

評：贊成新派。然而，目前仍然有不少人使用此措詞。

56. Of the above date:

「上列日期」、「上述日期」之意，例如：

Yours of the above date....

新派認為當表示日期時，必須將日期明確寫出，不可含糊。因此應改成：

Your letter of June 2....

評：贊成新派。

57. Order has gone forward:

「訂貨已運出」之意。舊式貿易英文中常可見到這種用語，新派認為沒有客戶願所訂之貨運回頭，但他們希望知道的是「何時」(When)、「如何」(How) 運出的。因此，必須詳述運輸方法、何時運出。例如：

Your order No. 123 was shipped on June 10 by s.s. "President" which will arrive at New York about August 1.

評："Order has gone forward" 這種老古董，現在已無人用。

58. Our Mr. Chang:

「我們（本公司）的張先生」之意。例如：

Our Mr. Chang will call on you next Monday, June 2.

新派認為這樣的稱呼不真實，不如直截了當地稱 Mr. Chang 或將其身分加以說明。例如：

Our representative, Mr. Chang, will call on you next Monday, June 2.

評：贊成新派。為表明 Mr. Chang 與本公司的關係，不宜在 Mr. Chang 之前加上 "Our"，而
　　應用同位語的方式來表示。

59.Our records show; According to our records; Our records do not show:

分別為「我們的紀錄顯示」，「依據我們的紀錄」及「我們的紀錄不顯示」之意。

新派認為這種表現法犯了兩個毛病：①夜郎自大，好像自己的紀錄最正確、最
權威。②兜圈子。因此，不如改用：

We find...（我們發現……）

We do not find... 或 We fail to find...（我們未發現……）

評：贊成新派。

60.Permit us to say; Permit us to explain:

分別為「希准我們說」，「希准我們來解釋」之意。這種客套為舊派所樂用。新
派認為在信裡要說什麼，由寫信人自行作主，不必要收信人允許。何況寫信時
已說出在先，又何必請求准許說出於後？豈非揶揄？所以這是多餘的。

評：禮多人不怪，偶爾用一用，又何妨？

61.Please be advised that:

舊派人士迄今仍很喜歡用此措詞以表示「茲奉告」之意。例如：

Please be advised that we shall execute your order in the near future.

新派認為以這種措詞作為敘述某一事的起頭毫無意義，要說什麼直說好了，何
必多費筆墨。因此，宜改為：

We shall fulfil your order by October 2.

評：在貿易書信中宜少用這種措詞，但公文書或銀行、保險界迄今仍樂用此措詞。你能說
　　他們不合時宜嗎？

62.Please favor us with your views (reply) urgently (at once; immediately; promptly; as
soon as possible; at your earliest convenience):

「請速惠示卓見」，「請立即惠示卓見」，「請儘速惠示卓見」之意。新派認為這
種措詞，表面上很客氣，但骨子裡卻在催人家趕辦，好像你的事最要緊。所以，
實際上則很不禮貌，因此應改為：

Please let us have your opinion soon. 或

Please let us have your reply soon (so that we may ship your order before August 10).

評：贊成新派。

63.Participal conclusion:

舊式書信中，差不多每封信都用分詞式作為結尾語。例如：

Trusting that this information may be entirely satisfactory.

Looking forward to hearing from you.

Awaiting the favor of your early reply.

Thanking you for your trouble.（謝謝你的勞神。）

新派認為這是老套，而且這樣的句子既沒有主詞 (Subject) 也沒有述詞 (Predi-cate)，不合文法。應改正如下：

We believe this information will be found satisfactory.

We are looking forward to hearing from you soon.

We thank you for your trouble.（但最好改為：Thank you for the assistance rendered.）

評：原則上贊成新派。但語言是約定俗成的東西，沒有什麼合不合文法的問題。

64.Posted:

舊派將此字當做 "Informed"（熟悉）用。例如：

He is well posted on this matter.（他很熟悉這件事。）

新派認為這是拙劣的用法 (Poor usage)，也是生僻的用法，應以 "Informed" 代替。例如：

He is well informed of this matter.

評：贊成新派。

65.Per:

當做「由」，「經」(By, By means of) 或「每」(For each) 解釋。例如：

per post; per steamer; per arrangement; $5.00 per yard; $1.00 per copy

新派認為不必賣弄拉丁文，尤其將 "Per" 用作 "By" 是誤用。應改為：

By post; By steamer; By arrangement

至於當做 "For each" 用時，雖不能說不好，但用下列方式也佳。

$5.00 a yard; $1.00 a copy

評：其實用 "Per" 也不差。

66. Recent date; Recent favor:

分別為「近日」及「近日的信」之意。例如：

We acknowledge the receipt of your recent favor.

In reply to your letter of recent date, we wish to state the samples you requested have been airmailed today.

新派認為這種不明確指出日期的措詞，係草率而不禮貌。應改為：

We have received your letter of July 3.

The samples you requested in your letter of June 3 have been airmailed today.

評：贊成新派。

67. Referring to; Regarding; Referring to yours of...wish to say that:

「關於……」之意，例如：

Referring to yours of June 6 wish to say that we are pleased to enclose a copy of our latest catalog in accordance with your request.

新派認為以分詞做為開頭的老套應盡量避免，也不要用 "yours" 來代替「來信」。

至於 "wish to say" 這種庸俗的措詞也應予摒棄。因此宜改為：

We enclose a copy of our latest catalog as requested in your letter of June 6.

評：贊成新派。但要特別注意的是：在英文書信中以 "Referring to..." 或 "With reference to..." 為開頭表示「關於……」時，只限於相關的信件、電傳或會談，並且不能用 "Regarding"。例如："Regarding your letter of..." 應改為 "Referring to your letter of..."。

68. Right away; Right off; Right now:

「立即」，「馬上」之意。

新派認為這是美國俗語，應在摒棄之列，而應以 "At once" 或 "Immediately" 代替。

評：贊成新派。

69. Said; The said:

「該」，「上述的」之意。例如：

We have notified them of (the) said arrangement.

新派認為商用書信中大可不必賣弄法律或契約用語，應予摒棄。因此，應改為：

We have notified them of the arrangement.

評：贊成新派。

70.Sorry to say:

「抱歉」，「可惜」之意。例如：

We are sorry to say that we do not have these goods.

（我們沒有這些貨，抱歉之至。）

新派認為應該說成：

We are sorry that we do not have these goods.

否則意思混淆，到底那「所說的話」使你抱歉呢？還是那「事實本身」使你抱歉呢？總之，現代商業書信已進步得像自然科學一般，要講求邏輯及清晰，絕不能如舊式書信那樣含糊不清。

評：贊成新派。

71.Take into consideration:

「加以考慮」之意。例如：

We shall take into consideration of this matter.

新派認為這是陳腐、不簡潔的表現法，不如逕用 Consider。例如：

We shall consider this matter.

評：贊成新派。不過，為求變化，偶爾用這種 Expression，不應指摘其為陳腐、不簡潔。

72.Take pleasure in:

「樂於……」之意。例如：

We take pleasure in announcing our fall line of shoes for women.

（我們樂於宣布我們的冬季女用靴〔已上市〕。）

新派認為 "Take pleasure in" 這個片語是老套。應改用 "Are pleased"；"Are happy" 或 "Are glad" 才顯得親切自然。例如：

We are pleased to announce our fall shoes for women.

（我們很高興宣布我們的冬季女用靴〔已上市〕。）

評：比較正式的文書中，仍習用此用語。

73.Take the liberty of:

「冒昧」之意。例如：

I take the liberty of requesting you to send me a copy of your latest price list.

新派認為 "Take the liberty of" 這個曾經出盡風頭的謙虛措詞應該退休了，還是直截了當地說出吧！例如改為：

Please send me a copy of your latest price list.

評：贊成新派。

74.Thank you in advance; Thank you in anticipation:

「謹先致謝」之意（請託用語）。例如：

Kindly mail us any information you may have for removing oil spot. Thanking you in advance for the favor, we remain,（請寄下有關去油漬的資料，謹先致謝。）

新派認為該謝的時候再致謝不為遲，預先致謝會使收信人覺得非為你做不可。

這種客套語，與其說謙恭，不如說不禮貌，如一定要先致謝，也應用 "Appreciate"（感激）此字，等事成後再說 "Thank you" 也不為遲。所以應改為：

We shall appreciate any information you may have for removing oil spot.

（如能示知去油漬的資料，將不勝感激。）

評：新派言之有理。但目前仍有不少非英語國家的人習用此措詞。

75.Taking this opportunity:

「趁機」、「藉此機會」之意。例如：

Taking this opportunity, I wish to inform yon that I shall visit your country next month.（藉此機會奉告我將於下月訪問貴國。）

新派認為何不直接談正題呢？用此措詞頗有藉機作書之嫌。因此，應改為：

I shall visit your country next month.

評：有時候將正題談完之後，順便告訴對方一些並不很重要的事情，何嘗不可？因此，在適當的情況下，使用此 Expression 應屬無妨。

76.Thank you again:

「謹再致謝」之意。

新派認為感謝一次就夠了，不必再謝。

評：禮多人不怪，再度致謝，又何妨？

77.This is to acknowledge; This is to advise; This is to inform you that; This letter is to advise you that:

分別為「茲收到」、「茲通知」、「茲通知你」、「本函通知你」之意。

新派認為商用書信不應使用這種多餘的修飾文句,而應該把要說的話直接說出。

評：這是適切的忠告，但這種語調在某些場合也可用得上，尤其公文書上為然。

78. The true facts are as stated:

「事實真相如所述」之意。

事實就是真相，因此，新派認為 "True" 這個字是多餘的。應改為：

The facts are as stated.

評： 有道理。「正方形」就是「正方形」，不能說「真」正方形；「垂直」就是「垂直」，不
　　 能說「完全」垂直。

79. Trusting this will; We trust this will:

「相信將會」，「我們相信將會」之意。舊派常以此做為信的結尾，例如：

Trusting this will meet with your approval, we remain,

新派認為 "Trust"（確信）一詞既顯得軟弱無力又陳腐，現在多以 "Believe"，
"Hope" 或 "Be confident" 代替。因此，不如直說：

We hope you will approve of our action.（我們相信你會贊成我們的行動。）

評： 贊成新派。

80. Under separate cover:

「另封」之意。例如：

We are sending you under separate cover a copy of our catalog.

（另寄上本公司貨品目錄一份。）

新派認為： 如要表明另外寄上，必需註明分寄的方法，或不說「另寄上」或只
說「寄上」也可。因此，宜改為：

We are sending you by air parcel post a copy of our catalog.

We are sending you a copy of our catalog.

評： 贊成新派，少用就是。

81. The undersigned:

「下面的簽字人」，即寫信人的自稱。例如：

The undersigned wishes to state that we take extreme pleasure in accepting your esteemed order.

這種表現法顯然想給收信人謙恭的印象而避免使用 "I"，"We"。然而，弄巧成
拙，新派認為不要怕使用 "I"，"We"，而應改為：

I am delighted to accept your order.

評：＂The undersigned＂這種用法，頗易使人不快，應予擯棄。

82.Up to this writing:

「到寫此信時為止」之意。例如：

Up to this writing your order has not been received.

新派認為應單刀直入地說，不必以 ＂Up to this writing＂ 作為開頭，或者以 ＂Up to now＂ 代替。例如：

We have not yet received your order (up to now).

評：到現在為止，還有些人很喜歡用此措詞。平心而論，偶爾用一次，應不致被視為老頑固。

83.Upon investigation:

「經調查」之意，例如：

Upon investigation, we find that...

新派認為①這種措詞暗示不相信收信人，而加以調查，使收信人很不受用，②人家要求調查，那麼未完成調查之前，當然不輕言已經調查，既然已經查出結果，就將結果直說好了。所以應改為：

We find that...

評：如因語氣的關係，需要使用這種措詞的話，似不必吹毛求疵，反對這種用語。類似的表現有 ＂On checking the details＂，＂Upon examination＂ 等。

84.Valued order:

「寶貴的訂單」之意。例如：

We appreciate your valued order given to us.

新派認為與 ＂Esteemed order＂ 一樣 Overpolite，以致有虛偽阿諛之嫌。應改為：

We appreciate your order given to us.

評：贊成新派。

85.We are applying:

「正在申請中」之意。

新派認為 ＂We are applying＂ 有 ＂We are going to apply＂ 或 ＂We are about to apply＂ （將要申請）的含意。而所謂「正在申請中」則是「已提出申請，等待核准」的狀況，換句話說「已開始申請」並不是「將要申請」，所以，應改用下面的 Expression：

We have filed our application to BOFT for permission.

86. We see from your letter:

「從大札得悉」之意。

新派認為這種 Expression 毫無生氣、迂腐，以致將全文的生命摧殘殆盡。

評：這種 Old-fashioned 的用法，應予摒棄。

87. We shall be in a position to:

「我們將能……」之意。

新派認為這種表現法不夠簡潔，應改為：

We shall be able to...

評：贊成新派。但有些人為講求文章的變化（避免一再使用同一用詞）仍常用此 Expression。

88. Will appreciate; Will be glad; Will be pleased:

「將感激」，「將高興」，「將樂於」之意。例如：

We will appreciate your giving us an opportunity to display our merchandise.

We will be glad to discuss this matter more fully with you.

這些表達方式雖常有人用，可是新派認為將 "Will" 當做 "Shall" 用是錯的，應將 "Will" 改為 "Shall"。

評：對！

89. The writer:

「筆者」之意。例如：

The writer wishes to acknowledge receipt of your letter.

The writer appreciates your efforts.

這是為了給收信人有禮貌的印象而避免使用 "I"，"We"。新派認為與前述 "The undersigned" 有同工異曲之妙，由於這種表現法很不自然，以致顯得不親切，該用 "I"，"We" 時就用，不要怕用 "I"，"We"，而勉強用 "The writer"。因此，應改為：

I have received your letter...

I appreciate your efforts.

評：贊成新派。

第五章
貿易英文文法 (Grammar of Business English for Foreign Trade)

關於英文文法 (Grammar) 的問題，範圍很廣，專門討論英文文法的書籍也汗牛充棟，本書無意再作詳細的介紹。在本章裡擬就貿易英文中，有關下面的問題酌予檢討，以便讀者撰寫商用英文信時，不致犯了英文文法上的錯誤。

1. 名詞的單數與多數。
2. 應注意的動詞用法。
3. 應注意的形容詞用法。
4. 應注意的介系詞用法。
5. 縮寫與全寫。
6. 數字與數碼字。
7. 字母的大寫。
8. 易於混淆的商業用語。

第一節　名詞的單數與多數

一、通常限用單數形的名詞

1. Advice：當作「忠告、勸告」解釋時，只有單數形：

   He gave me several pieces of advice on the matter.

   假如只有一個「忠告」，可用 "A piece of advice"。

   當作「報告、通知」解釋時，則有單數、多數兩種。

   According to advice(s) from our agent, business is improving.

2. Correspondence：「通信」，沒有多數形。

　You must answer all correspondence in English.

3.Finish：做工、加工、修飾。

　Their finish is decidedly bad.（它們的做工絕對地差。）

4.Information：「情報、消息」，無多數形。

　We attempt to give as *much information* as we can.

　They gave us *a lot of information*.

5.Equipment：「設備、裝置」之意，無多數形。

　Funds for buildings and *equipment*.

6.Literature (=Printed matter; As circular or advertising matter)：「廣告印刷品」之意，無多數形。

　Enclosed is *some literature* on our refrigerators.

7.Machinery：「機械」，無多數形。

　Machinery is being introduced to save labor.

8.Merchandise：「商品」，無多數形。

　The tariff schedules classify the *merchandise* in three groups.

　（該運價表將商品分為三類。）

9.Dozen：「打」，縮寫為 "Doz."，做數量單位時，無多數形。

　Please pack the goods in ordinary cases of 20 *dozen* each.

　表示「很多」時可用多數的 "Dozens"，例如：

　Dozens of times（很多次）

　cf. To pack the goods in dozens（1 打 1 打地包裝）

10.Gross：「籮」，做數量單位時，無多數形。

　5 gross of shell buttons

11.Hundredweight (c.w.t.)：重量單位，通常無多數形，但也可加 "s"。

　Freight rate is US$150 per 16 gross hundredweight.

12.Staff：解作「全體職員」時係集合名詞，不用多數形。因此，假如說「他是本公司職員」則寫成 "He is a member of our staff." 或 "He is a staff member of our company."，「兩位職員」寫成 "Two members of staff"。

二、商品的集合名詞不可加 "s"

Chinaware	（瓷器類）	Jewelry	（珠寶類）
Confectionery	（糖果類）	Machinery	（機器類）
Crockery	（陶器類）	Perfumery	（香水類）
Cutlery	（刀叉剪類）	Porcelain	（瓷器類）
Drapery	（布疋類）	Pottery	（陶器類）
Earthenware	（陶器類）	Produce	（農產物）
Enamelware	（搪瓷類）	Stationery	（文具類）
Footwear	（靴類）	Underwear	（內衣類）
Furniture	（家具類）	Bambooware	（竹器類）
Glassware	（玻璃器具類）	Knitwear	（針織品）
Hardware	（五金類）	Tableware	（餐具類）
Hosiery	（襪類）	Marbleware	（大理石類）
Ironware	（鐵器類）	Cookware	（烹飪器皿類）

1. 以上名詞，計數時可用：

   A piece of..., Pieces of...,

   An article of..., Articles of...,

   例如： 20 cases of hardware

2. "Ware" 單獨使用時，其多數為 "Wares"。

三、通常限用多數形的名詞

1. Amends: 賠償、賠罪。

   I must make amends for my fault.（我必須為我的過失賠罪。）

2. Arrangements: 準備、安排。

   We have made arrangements for shipment of the goods.

   cf. We have made special arrangements (=agreements) with the company.（這裡是指「約定」）

3. Apologies: 道歉。在下面的情形，習慣上用多數形。

Please accept my apologies.

With sincere apologies for....

4. Auspices：贊助、主辦。

The trade fair was held under the auspices of the government.

（該貿易展覽會由政府主辦或贊助。）

5. Authorities：當局。

The document is required for delivery to the *custom-house authorities*（海關當局）

6. Bankers：銀行。

Our *bankers are* the Bank of Taiwan.

cf. Our *bank is* the Bank of Taiwan.

7. Chemicals：化學藥品，與此相似的尚有：

Breeches	（短褲）	Overalls	（罩衣）
Furnishings	（家具、室內陳設品）	Rags	（破布）
Hards	（麻屑）	Scrapings	（削屑）
Hops	（酒花）	Spirits	（酒類）
Molasses	（糖精）	Trousers	（褲子）
Oats	（燕麥）	Wollens	（毛織品）

8. Circumstances：情況、條件。

The delay was unavoidable under the circumstances.

（在此情況下，無法避免遲延。）

9. Conditions (=Circumstances)：情況。

Conditions were favorable for business	情況對生意有利
Under the present conditions	就現況而言
Under the existing conditions	就目前情況而言
Under difficult conditions	在困難情況下

cf. Please let us know upon what *conditions* you are able to deliver the goods.（這裡的 conditions 為「條件」之意）

We shipped the goods *in good condition*.

（良好情形。注意 condition 係單數）

10.Compliments; Congratulations; Condolences:

"Compliment" 作「問候」解釋時，該用多數。例如：

Kindly give her my compliments.

"Condolence" 作「慰問」解釋時，該用多數。例如：

Present my condolences to you.

但在 "A message of condolence"，"A letter of condolence" 中卻用單數。

"Congratulation" 在祝賀時用多數。例如：

Please accept my congratulations.

但在 "Letter of congratulation" 中則用單數。

11.Contents：內容、裡面的東西。

The contents are unknown to us.（內容不詳。）

12.Customs：關稅 (Customs duty)、海關 (Custom-house)。

Customs are collected at the custom-house.（關稅在海關征收。）

13.Damages：損害賠償金 (=Money claimed by a person to compensate for loss)。

They claimed US$100 damages for breach of contract.

（他們就違約索取損害賠償金 100 美元。）

14.Dues：費用、使用費。

The harbour dues are calculated on the registered tonnage of the vessel.

（港口使用費按註冊噸數計收。）

15.Details：詳情。

The enclosed catalog contains details of our new products.

cf. We investigated the matter *in full detail*.（很詳細地）

16.Difficulties (=Financial embarrassment)：財務上困窘。

We hope to overcome our difficulties before the end of May.

（我們希望在 5 月底以前克服我們的困難。）

cf. We have *a lot of difficulties*.（我們有很多困難。）

17.Exports：輸出品 (Exported goods)，輸出額 (Amount of export)。

The imports and exports together amounted to US$16 billion.

（輸出入總額達 160 億美元。）

18.Engagements (=Financial obligations)：債務。

They found difficulty in meeting their engagements. （他們感到難於履行債務。）

cf. We have *many engagements* to fill. （我們有很多諾言待履行。）

19.Instructions：指示、指令。

Buyer has given seller instructions to ship per U.S. flag vessel.

Shipping instructions	裝運指示
Packing instructions	包裝指示
Instructions from head office	總公司的指令

單數的 "Instruction" 乃指學校的「教導」、「教授」而言。

20.Means：資力、手段、方法。

His means are ample. （他的資力雄厚。）

A means of overcoming the difficulty was found. （發現了克服困難的方法。）

註：用做「資力」時，其動詞為多數形；用做「手段、方法」時，單、多數均可。

21.Papers：文件。

The papers are on the way by mail. （該文件在郵途中。）

22.Particulars：詳細說明、詳情。

Full particulars of the shipment are given below: （貨載詳細說明如下：）

23.Proceeds：款項。

Please remit the proceeds promptly.

24.Terms：條件、用語。

Our terms are cash with orders. （我們的條件是隨訂單付款。）

The agreement will be expressed in legal terms. （協議將以法律用語表示。）

25.Thanks：謝意、感謝。

Please accept our thanks for your kind consideration. （謝謝惠賜考慮。）

26.Works：工廠。

The steel works is (or are) located in Kaohsiung. （那鋼鐵廠在高雄。）

四、Company 等的集合名詞

"Company"，"Corporation"，"Firm"，"Bank"，"Mill"，"Factory" 等名詞是集合名詞，因此：

1.指其組成人員時，應視做多數。

The company have agreed upon a course of action.（公司的人們……。）

The committee are debating the question.（委員們……。）

2.當作一單位組織時，應視做單數。

The company is financially strong.（該公司……。）

The committee is considering the proposition.（該委員會……。）

所以，相應的名詞 "Customer"，"Buyer"，"Agent"，"Shipper"，"Manufacturer"，"Exporter"，"Banker"，代名詞、動詞也應隨之採取單數形或多數形。

3.商號名稱後面的代名詞、動詞究應採單數抑多數呢?

⑴視商號名稱的單數或多數而定。

Sulzer Bros. *are manufacturers* of engines.

H. B. Smith *is* our *agent* for buying wool top.

Francis Becker, Ltd. *is* our *agent* in U.K.

Butterfield & Swire, Ltd. *have* opened *their* office in Kobe.

⑵以 Company, Pty. 或 Corporation 表示公司行號組織性質者，究應採單數還是多數，視文章的上下文而定。

The Union Co., Ltd. is located in Taipei.

The Union Co., Ltd. *are* well-known as *exporters* of machinery.

The Bank of Tokyo, Ltd. *are* our *bankers*.

The Bank of Tokyo, Ltd. *has* been our *bank* for many years.

五、複合語文件名稱的複數形

Bills of lading　　　　Bs/L;　　　　Letters of credit　　　Ls/C

也有以 B/Ls, L/Cs 表示 B/L, L/C 等略語的複數形。表示略語、數字、文字等本身的多數，可以 "'s" 或 "s" 附在末尾。例如：three 2's, three 2s, three C's。

第二節　應注意的動詞用法

一、不能用 "V+to-infinitive" 或 "V+O+to-infinitive" 型的動詞

1. Aim：針對、意欲。

   （誤）The move is aimed to eliminate it.

   （正）The move is aimed at eliminating it.

   （正）The move aims to eliminate it.

   註：此動詞用於 "He aimed a gun at the target." 的句型。故，應注意被動形用法。

2. Demand：要求。

   （誤）They demand us to make immediate payment.

   （正）They demand our immediate payment.

   （正）They demand that we (should) pay immediately.

3. Facilitate：使便利、使容易。

   （誤）This will facilitate us to settle the matter.

   （正）This will facilitate our settlement of the matter.

4. Finish：完成。

   （誤）We must *finish to unload* the goods by 5:00 p.m.

   （正）We must *finish unloading* the goods by 5:00 p.m.

5. Hope：希望。

   （誤）We *hope you* to succeed in it.

   （正）We *hope that* you will succeed in it.

6. Prevent：妨礙。

   （誤）The circumstance *prevented us to* ship the goods in time.

   （正）The circumstances *prevented us from* shipping the goods in time.

7. Propose：提議。

   （誤）We *propose you to inquire* into the matter immediately.

   （正）We *propose that you inquire* into the matter immediately.

8. Reject：拒絕。

（誤）We cannot but *reject to accept* your offer.

（正）We cannot but *reject* your offer. (reject=refuse to accept)

9. Solicit：懇請。

（誤）We *solicit you to give* us a prompt order.

（正）We *solicit your* prompt order.

（正）We *solicit for your giving* us a prompt order.

10. Succeed：成功。

（誤）We have *succeeded to obtain* an order from them.

（正）We have *succeeded in obtaining* an order from them.

11. Suggest：建議。

（誤）We *suggest you* to write your reply.

（正）We *suggest that you* (should) write your reply.

二、用 V+to+O+that clause 型的動詞

1. They *admitted* （承認）*to their customers that* they had made a mistake.

2. We *suggested* （提議）*to them that* it might be better to hold off this business.

三、用 V+ing 型的動詞

1. They *admitted having* made a mistake.

2. We *appreciate your sending* us the sample by airmail.

3. We are anxious to *avoid putting* you to inconvenience.

4. We now *confirm* （確認）having booked your order.

5. We cannot *endure being* disturbed by it.

6. We have *enjoyed seeing* you.

7. Would you *mind telling* us who recommended our service?

8. I *remember having* promised to help you.

9. We must *risk losing* the market.

10. Please *stop sending* the goods.

此外尚有 Commence, Forbid, Forget, Solicit, Suggest 等。

四、幾個特殊動詞的用法

1. Address:

⑴ "Address"（地址）前面的 Preposition 是 "at"。例如："I shall write to him *at* his old address."。

⑵ "Address" 作「向……說話」或「寫信給……」解釋，是 Transitive verb。例如："I am addressing you on an important subject." 不可在 "addressing" 後面加 "to"。

2. Advertise:

"Advertise" 作「廣告」解釋；"Advertise for" 作「用廣告徵求」解釋。例如："They try every means to advertise their goods." 和 "I am going to advertise *for* a clerk to help me."。

3. Answer:

⑴ "Answer a letter" 比 "Answer to a letter" 普通。

⑵ "Answer (to) a letter" 通常和 "Reply to a letter" 相同（注意用了 "Reply" 該有 "to"），但有時 "Answer" 有「答覆……的內容」的意思，例如：覆信裡說 "Your letter has come. As I am very busy, I cannot answer it at present. I shall do so next week."。

4. Claim:

本意為「要求或主張當然的或應歸自己的權利」、「聲言」。在商業上則用作「對於損害的賠償要求」或「索賠」之意。例：

⑴ Every citizen may claim the protection of the law.
（每位公民均可要求法律的保護。）

⑵ He claimed that he was right.（他聲言他是對的。）

⑶ The matter claimed our immediate attention.（這事需要我們立即的注意。）
因此「你索賠了 1 萬元」不能寫成 "You claimed $10,000." 而應寫成 "You submitted (sent in) a claim for $10,000."。因為前者的意思為「你主張了 1 萬元」。「我們被索賠了」不能寫成 "We were claimed..." 而應寫作 "We received a claim against..."。

一般而言，將 Claim 當作「請求賠償」用時，只要將賠償的對象寫出，即不致發生錯誤：

To claim compensation for damage.

To claim against damages.

To claim payment of $10,000 from you for damage against this shipment.

To submit a claim against this cargo in respect of quality.

5. Delight:

"Delight in..." 和 "Delight"+Infinitive 都可以用，例如 "I delight in riding." 和 "I delight to ride."。但用作客氣話裡必須用 Passive voice，後面接 Infinitive，例如 "I shall *be delighted* to meet you on Sunday."，不可作 "I shall delight in seeing you on Sunday." 或 "I shall delight to see you on Sunday."。

6. Demand:

⑴ "Demand" 在用作 Noun 的時候後面接 "for"，例如 "A demand *for* payment"；但在用作 Verb 的時候是 Transitive，後面不可接 "for"，例如 "They demanded payment."，不可作 "They demanded *for* payment."。

⑵ 說「要求某人……」不可用 "Demand" + 指人的字 + Infinitive，例如不可作 "He demanded her to pay." 該作 "He demanded that she (should) pay."。

7. Favour:

⑴ "Favour" 不可有 Double object，例如不可作 "Can you favour me an early reply?"，該在 "me" 後面加 "with"。

⑵ 說自己替他人服務，不可用 "Favour"。

8. Follow:

"As follows" 和 "(As) In the following" 二語不可相混，不可作 "as following" 或 "In the follows"。

9. Inform:

⑴ "Inform" 不可有 Double object，例如不可作 "I shall inform you my new address before long."，該在 "my" 前面加 "of"。

⑵ "Inform" 不可不用 Object 而直接以 "that" 引起的 Clause，例如不可作 "I am sorry to inform that we cannot supply the necessary particulars."，該在 "that" 前面加 "you"。

(3) "Inform" 不可作「報告（事件）」解釋，例如不可作 "She informed her father's death to me.", 該改作 "She informed me of her father's death."。

10. Inquire:

"Inquire" 後面不可接指人的字，例如不可作 "I inquired him about the matter.", 該在 "inquired" 後面加 "of"。

11. Interview:

作「訪問」或「謁見」解釋，不作「接見」解釋，例如 "A Interviewed B" 是「A 謁見 B」不是「A 接見 B」（是 A 去看 B 而不是 B 去看 A）。通常較低的人去看較高的人，寫信給人，如說 "If you wish to interview me..." 或 "Please interview me..." 都是很不客氣的話，該作 "If you allow me to interview...", "Please allow me to interview you...", "Please grant me an interview..." 等。

12. Introduce:

⑴不可有 Double object，例如不可作："Allow me to introduce you my sister Mary.", 該作 "Allow me to introduce my sister Mary *to* you." 或 "Allow me to introduce *to* you my sister Mary."。

⑵寫介紹信，該把收信人做 "to" 的 Object，例如 "Introduce Mary to you", 不作 "Introduce you to Mary"。

13. Invite:

⑴說「請人教授」、「請人診病」、「請人幫忙」等都不可用 "Invite"；"Invite" 通常用在不很勞苦而使人高興做的事，例如演講、跳舞、喝酒等。

⑵寫信給對方通常不用 "Invite", 例如不作 "I invite you to dinner tomorrow." 卻可作 "He invites me to dinner tomorrow."。

14. Know:

⑴ "Know" 作「認識（某人）」解釋，"Know of" 作「知道有（某人）」解釋，例如 "I know *of* him, but do not know him (personally)."。

⑵ "I am glad to learn from your letter（或 the newspaper）that you have passed the examination." 等句裡用 "learn", 通常不用 "know"。

15. Like:

用 "I（或 We）should like to...", 不可用 "I（或 We）like to...", 例如作 "I（或

We) *should* like to call on you on Saturday afternoon.", 不可作 "I (或 We) like to call on you on Saturday afternoon."。用 "I (或 We) *would* like to..." 的人也有, 但照理該用 "should"。

16. Miscarry:

"Miscarry" 作「(信件) 不達到目的地」解釋, 是 Intransitive verb, 不可用在 Passive voice, 例如不可作 "Your previous letter has been miscarried.", 該把 "been" 刪去。

17. Oblige:

⑴ "Oblige" 作「賜恩惠給……」解釋, 表示感激該用 Passive voice。那指被感激的人的字的前面該用 "to", 那指感激的原由的字的前面該用 "for", 例如 "I am obliged *to* you.", "I am obliged *for* your help.", 或 "I am obliged *to* you *for* your help.", 不可作 "I am obliged *to your help*."。倘若 "for" 後面接 Gerund, 那 Gerund 前面不該有 "your", 例如不可作 "I am much obliged to you for *your* giving me so much advice.", 該把 "your" 刪去。注意 "for" 後面的 Gerund 是 "you" 的動作, 不是 "I" 的動作, 例如不可作 "I am much obliged to you for *receiving* so much advice."。

⑵ "Obliged" 祇用於小事, 不用於大事, 例如不可作 "I am much obliged to you for saving me at sea."。

18. Present:

不可有 Double object, 例如不可作 "He presented me a fountain-pen.", 該作 "He presented me *with* a fountain-pen." 或 "He presented a fountain-pen *to* me."。

19. Receive:

⑴ "Receive" 是一時的動作, 所以不可和「(for) 若干時間」用在一起, 例如可作 "I have received your letter." 和 "I received your letter three weeks ago.", 但不可作 "I have received your letter (for) three weeks."。

⑵ 說「從……收到 (款項)」用 "Receive from" 或 "Receive of" 都可以, 例如 "Received from (或 of) Mr. Charles. A. Smith the sum of five dollars."。

⑶ 英美人寫覆信, 往往第一句是 "Your letter of...received.", 把 "is", "has been" 或 "was" 省去。

20. Reply:

"Reply" 在用作 Transitive verb 的時候作「把……作答覆」解釋，例如 "I asked him several times, but he had nothing to reply." 和 "She replied that she did not know him at all.",　不作「答覆……」解釋，例如不可作 "I have several letters to reply." 和 "He replied my question immediately.",　該在 "reply" 和 "replied" 後面加 "to"。（注意把這二字改了 "answer" 和 "answered" 便不必加 "to"，加了不很自然。）

21. Report:

"Report" 不可作「向（人）報告」解釋，例如不可作 "He reported me that he had made another discovery." 或 "I will report you as soon as I have interviewed him.",　在第一句裡該在 "me" 前面加 "to"。在第二句裡該在 "report" 後加上 "to" 或把 "report" 改作 "inform" 或 "notify" 或把 "report you" 改作 "let you know"。"Report" 用在指人的字的前面作「告發」解釋，例如 "We had to report the servant to his master for misconduct."。

22. Request:

⑴ "Request" 在用作 Noun 的時候後面接 "for"，例如 "a request *for* assistance"；但在用作 Verb 的時候是 Transitive，後面不可接 "for"，例如 "He requested assistance." 不可作 "He requested *for* assistance."。

⑵ "Request" 雖然和 "Ask" 意義相仿，但在結構上有一個不同的地方：可以用 "Ask" + 人 + "for" + 物，例如 "We asked him for some information.",　卻不可把 "Requested" 用在這種結構裡，例如不可作 "We requested him for some information.",　該作 "We requested some information from him." 或 "We requested that he give（或 would give，或 should give）some information."。作 "We requested to be informed by him." 也可以。

23. Thank:

⑴在 "Thank" + 指人的字 + "for" + Gerund 的結構裡不該在 Gerund 前面加 Possessive，例如不可作 "Thank you for *your* calling my attention to the fact.",　該把 "your"刪去，也不可作 "We thanked them for *their* showing us the way.",　該把 "their"刪去。"Many thanks" 和 "Thanks" 等於 "Thank you"，所以不可作 "Many thanks for *your* lending me your dictionary." 或 "Thanks for *your* lending me your dictionary.",該把 "your" 刪去。

⑵ "Thank you" 最普通, "I thank you" 頗鄭重,比較少用。"Thank you very much" 和 "Thank you" 相仿。"Thank you so much" 和 "Thank you very so much" 也和 "Thank you" 相仿。"Thanks" 比 "Thank you" 隨便些。"Many thanks" 和 "Thank you" 相仿。"A thousand thanks" 似乎很不自然。"Much thanks" 是古語,很不普通。"Thanks much" 似乎不很自然。"Thanks very much" 和 "Thank you very much" 相仿。(就文法論,"Thanks much" 和 "Thanks very much" 都是不妥的,因為把 Noun "thanks" 和 Verb 相混了,但都有人用。)

24. Write:

⑴ "Write" 後面可以接 Double object,例如 "I will write you the result soon." 和 "They wrote us a report last month."。在商業書信裡也往往祇用 Indirect object 而不用 Direct object,例如 "I will write you soon." 和 "They wrote us last month.",在一般通信裡該在 "Write" 和 "Wrote" 後面加上 "to"。倘若 "Write" 的 Direct object 是一個以 "that" 引起的 Clause,也該用 "to",例如 "They wrote to us that they were coming to see the show."。

⑵ "Write" 可以用作「寫信」解釋,例如 "When I write again, I shall tell you more about her."。說「寫信到家裡」可用 "Write home",例如 "I write home every Sunday."。

⑶ "I(或 We)write to...(Infinitive)" 和 "I am(或 We are)writing to...(Infinitive)" 意思相同。

⑷ "At the present writing"(在寫這信的時候)是商業書信裡的陳腐語,該避免。

五、主詞與動詞的呼應

1. (誤) The *goods is* not suitable this market.

   (正) The *goods are* not suitable *for* this market.

   (正) This *article* (or item, commodity, line) *is* not suitable for this market.

   (正) These *articles* (or items, commodities, lines) *are* not suitable for this market.

2. (誤) *Neither* of the plans *are* acceptable by us.

   (正) *Neither* of the plans *is* acceptable to us.

3.（誤）No definite *news have* been received.

　（正）No definite *news has* been received.

4.（誤）The number of orders *we have received last year were* larger than ever before.

　（正）The number of orders *we received last year was* larger than ever before.

　　cf. A number of inquiries have been received.

5.（誤）There *were a shortshipment* of 5 *dozens* of *hardwares*.

　（正）There *was a shortshipment* of 5 *dozen* of *hardware*.

6.（誤）Two-thirds of the shipment *have* been sent today.

　（正）Two-thirds of the shipment *has* been sent today.

註：數量、金額當作一整體時用單數形動詞，例如 Three hundred dollars was paid for the goods.。
　　分數後面之 "of" 的目的語為單數時，動詞用單數形。"of" 的目的語為多數時，動詞採
　　多數形。如：About one-third of *the onions were* in good condition.

7.（誤）*Each* of the workmen *were* granted an increase.

　（正）*Each* of the workmen *was* granted an increase.

8.（誤）A *few* (several, both) *was* requested to remain there.

　（正）A *few* (several, both) *were* requested to remain there.

第三節　應注意的形容詞用法

一、與不定詞連結的形容詞用法

1. We are *anxious to meet* your wishes.

2. Suppliers are *bound to furnish* buyers with a certificate of origin.

3. We are *delighted to hear* of your success.

4. We are *desirous to know* further details.

　cf. We are *desirous of extending* our export trade.

5. We were *disappointed to know* it.

6. We shall be *grateful to receive* your remittance.

7. We shall be *pleased (or glad, happy) to have* your shipping instructions.

8. We are *ready to do* business on your behalf when you want an agent here.

9. We are *sorry to have* to decline your offer.

10. Please be *sure to appoint* us your agents.

此外尚有 Able, Afraid, Apt, Certain, Eager, Fit, Unable, Unfit, Willing 等。

二、伴隨 That Clause 的形容詞

1. They are *anxious that* you should accept the offer.（渴望）

   cf. We are *anxious for* your success.（渴望）

2. You may be *assured (or Please be assured)* that we will do our best to execute the order.

3. You are *aware that* there is great competition on the part of both Germany and the United States.

4. We are *certain that* no house（商號）there can handle your goods in a better manner than we have done.

5. We are *confident that* we can do a large turnover（銷售量）for you annually.

   cf. We are *confident of* being able to give you complete satisfaction.

6. We are *convinced that* we are offering you machines which cannot be obtained elsewhere at the price.

   cf. We are *convinced of* the fact.

此外尚有 Pleased, Sure, Surprised, Afraid, Delighted, Disappointed, Glad 等。

第四節　應注意的介系詞用法──介系詞的誤用

一、遺漏介系詞

1. （誤）We *assure* you our sincere cooperation.

   （正）We *assure* you *of* our sincere cooperation.

2. （誤）We *apologize* you our mistake.

   （正）We *apologize* to you *for* our mistake.

3. （誤）The buyer *complain* the delayed shipment.

 （正）The buyer *complain of* the delayed shipment.

4. （誤）We *deal* the following goods.

 （正）We *deal in* the following goods.

 Deal in＝Buy and sell (goods)

 Deal with ＝ Do business with (a firm)

5. （誤）Such a thing has never *happened* us before.

 （正）Such a thing has never *happened to* us before.

6. （誤）We *hope* your success.

 （正）We *hope for* your success.

7. （誤）They will *inform* you the details.

 （正）They will *inform* you *of* the details.

8. （誤）We *insist* our claim for the damage.

 （正）We *insist on* our claim for the damage.

9. （誤）Please *favor* me an early reply.

 （正）Please *favor* me with an early reply.

10. （誤）There is enclosed a list of the goods that we are *interested*.

 （正）There is enclosed a list of the goods that we are *interested* in.

二、多餘的介系詞

1. （誤）*In answering to* your inquiry, we quote you as given below:

 （正）*Answering to* your inquiry, we quote you as given below:

 （正）*In answer to* your inquiry, we quote you as given below:

2. （誤）Please *ask for* them about the matter.

 （正）Please *ask* them about the matter.

3. （誤）Your letter *dated on June* 16.

 （正）Your letter *dated June* 16.

4. （誤）Your letter of May 5 *crossed with* ours of May 4.

 （正）Your letter of May 5 *crossed* ours of May 4.

5. （誤）You *promised to us* to do all you could.

（正）You *promised us* to do all you could.

6. （誤）*Regarding to* your proposal we think it will be mutual advantage.

（正）*Regarding* your proposal we think it will be mutual advantage.

三、錯用介系詞

1. （誤）We are *grateful for* you.

（正）We are *grateful to* you.

cf. Acceptable to you	Interesting to you
Agreeable to you	Kind to me
Convenient to me	Necessary to us
Helpful to us	Satisfactory to you
Important to them	Suitable to a person
Useful to you	Valuable to us

2. （誤）We shall be *obliged by* you if you will do so.

（正）We shall be *obliged to* you if you will do so.

3. （誤）We feel much *disappointed by* your failure in duty.

（正）We feel much *disappointed at* your failure in duty.

4. （誤）I went to *condole him for* the death of his mother.

（正）I went to *condole with him on* the death of his mother.

5. （誤）*In receipt of* your reply we shall open an L/C *for your favor.*

（正）*Upon receipt of* your reply we shall open an L/C *in your favor.*

四、應用兩個介系詞但遺漏一個的情形

1. （誤）This fact does not *add* but detracts *from* your merits.

（正）This fact does not *add to* but detracts *from* your merits.

2. （誤）We have no *connection* or knowledge *of* the firm.

（正）We have no *connection with* or knowledge *of* the firm.

3. （誤）We are *manufacturers* and dealers *in* these goods.

（正）We are *manufacturers of* and dealers *in* these goods.

4. （誤）This table shows the value of the principal articles *imported* and exported *from* this country.

（正）This table shows the value of the principal articles *imported into* and exported *from* this country.

五、Prepositional phrase 的誤用

1. （誤）We *look forward to hear* from you soon.

（正）We *look forward to hearing* from you soon.

2. （誤）The damage was *due to handle roughly* by the shipping company.

（正）The damage was *due to rough handling* by the shipping company.

3. （誤）All orders are *subject to confirm* by us.

（正）All orders are *subject to our confirmation*.

第五節　縮寫與全寫 (Abbreviating and Spelling out)

1. 縮寫有兩種：一是「略字」(Abbreviations)；一是「減字」(Contractions)。其分別，應該弄清楚。「略字」是正式的縮寫，後面有句點，例如 "*Company*" 的 "*Co.*"，"*Invoice*"（發票）的 "*Inv.*"。「減字」不過是把某個單字減寫幾個字母的一種簡筆而已；減寫的字母是用 Apostrophe (') 來標出的，例如 "*manufacturing*"（製造）的 "*m'f'g*"，"*association*"（公會）的 "*ass'n*"，"*it is*" 的 "*it's*"。不像「略字」，「減字」的末尾是沒有句點的；不過當為了信箋的行列上的地位關係，可以不用 Apostrophe，而後面加一句點，例如 "mfg."。「略字」有一定的格式，不能亂用。「減字」雖不是正當的格式，可是在打字的時候，逢到受信人或寫信人的名稱地址的場合，為配合行列上的地位起見，用起來非常便利；不過即使在急就的商業信件中，這種「減字」在正文中以不用或少用為是。

2. 人名中的基督教名應完全寫出：

"George", "Charles" 等不能寫為 "Geo.", "Chas."。例外: "William" 可作 "Wm.";
"Thomas" 可作 "Tho." 或 "Thos."。

3.公司商號的名稱,應照它們自己所採定而印在 Letter-head 的格式寫;如果正式
格式不知道,每個字都應該全寫,就是 "and" 與 "Company",也不可寫為 "&" 與
"Co."。

4.除非在表格中,月名應該全寫。表格中的縮寫,應該照它們的標準格式寫: Jan.
(1月), Feb. (2月), Mar. (3月), Apr. (4月), Aug. (8月), Sept. (9月),
Oct. (10月), Nov. (11月), Dec. (12月)。May (5月), June (6月), July
(7月),都沒有縮寫。

5.年份簡寫的時候,應該以 Apostrophe 來標出,如把 '02 代表 2002 年,不過也只
能用在表格中。

6.上午,下午,不論在什麼地方,一律縮寫為 "a.m." 與 "p.m."。

第六節　數字與數碼字

1.一個句子的開端,不宜用數碼字 (Figures),無論那數碼怎樣大,應該把那數字
完全拼出來。

2.簡單而又齊頭的數目、年紀、世紀,不宜用數碼字。例如: "The attendance was
about *three thousand*" (出席者約 3,000 人); "He is *twenty-five* years old"; "The
twentieth century"。

3.在同一句子中,兩種不同的數字,不要用不同的寫法: "*Five yards* of ribbons at
fifty cents a yard" (50 碼絲帶,每碼 5 角) 或 "*5 yds.* of ribbon @ 50¢ a *yd.*" 是正
確的寫法;不能寫作 "*5 yds.* of ribbon at *fifty cents* a yard"。

4.一般而論,一切的「小數」、「分數」、「百分數」、「度數」,以及「距離」、「重量」、
「銀錢」等數,什麼地方用數字,什麼地方用數碼字,一律依照 1. 2.兩規則。

5.如果有兩個不同的數字,在一個複合詞中,把簡短的數字拼出來。例如: "*Five
12 inch* bolts" (5 根 12 吋長的鐵栓); "80 *ten pounds* iron bars" (80 根 10 磅重的
鐵條)。

6.一個數目過大時,應該每三個數碼字,夾一逗點: "5,670"; "2,345,780"。但年

份、書的頁碼、門牌號數、電話號碼等，不必用逗點隔開："The year 1977"；"p. 1850"；"2153 Avenue Foch"（福煦路 2153 號）；"Central 3450"（電話中央 3450 號）。

7. 正文中日份應該加 "st"，"nd"，"rd"，"d" 或 "th" 字樣；但月份恰恰放在日份的前面的時候，單寫數碼字就夠了。例如："Thank you for your order of the *12th*, which we will ship on the *21st*."（謝謝你們 12 日的訂單，我們定於 21 日把貨發奉。）；"We shipped these goods on the *17th* of *May*"；"The goods were shipped *May 17*"。

8. 兩個日期或兩個數碼字應該銜接的時候，不用 "to" 把它們聯接起來，中間只要夾一破折號 (Dash)。例如："Our great sale will be during the week October *1–7*"（本店大廉價將於 10 月 1 日至 7 日舉行一星期）；"The numbers on the cases run *1–100* inclusive"（這些箱子上的號數包括 1 到 100）。不過當用 "*from*" 的時候，不能用破折號，一定要用 "to" 了。例如："Our great sale will be held *from* October 1 *to* October 7"；"The numbers on the cases run *from* 1 *to* 100 inclusive"。

9. 複合的「基數」(Cardinal numbers)，從 21 到 99，中間一定要夾一連號 (Hyphen)，不論單獨用，或當作一個大數目的一部分的場合。例如："*Seventy-five*"；"*Fifty-two thousand*"。

10. 複合的「序數」(Ordinal numbers)（就是講第幾個的數字）也是一樣的應該夾著連號。例如："Sixty-second"。

11. *Half*($\frac{1}{2}$) 與 *Quarter*($\frac{1}{4}$) 等字的複合詞，應該聯以連號。例如 "*Half-dozen*"（半打）；"*Quarter-mile*"（$\frac{1}{4}$ 里）。如 "*One fourth*"($\frac{1}{4}$) 等分數，不必聯以連號，除非用作形容詞的時候。例如："This is *one fourth* of the whole"（這是整個的 $\frac{1}{4}$）；"A *one fourth* interest"（一種 $\frac{1}{4}$ 的利息）。分數如 $\frac{21}{75}$ 等，應照下列那樣寫法："*Twenty-one seventy-fifth*"。

12. 分數、小數的寫法：

$\frac{1}{2}$ 或 0.5: One-half

$\frac{1}{3}$ 或 0.$\overline{3}$: One-third

$\dfrac{3}{8}$ 或 0.375: Three-eights

0.36: Thirty-six hundredths

257.4: Two hundred fifty-seven and four-tenths

0.25 M/T: One-quarter of one metric ton

12.6 g: Twelve and six-tenths grams

0.7 g: Seven-tenths of one gram 或 Seven-tenths grams

第七節　字母的大寫

1. 訂單、帳單等的項目，以及任何貨物舉列的時候，都應該大寫，指出它們的數量的字，不必大寫而用縮寫。例如 "100 *pcs. Velvet*"（100 包絲絨）；"*700 lbs. Portland Cement*"（700 磅波特蘭水泥）。

   主要貨物的名稱，雖然非專門名詞，也可以在正文中大寫，給讀者以醒目的表示。例如："The following dealers in your city can supply you with our new *Electric Fans* of all kinds."（下列貴埠各商家，能夠供給你本廠新出的各種電扇。）

2. 時季的名稱，當用來講到目錄或貨樣的時候，應該大寫，例如："Our *Spring Catalogue* will be out next week."（敝公司的春季目錄將於下星期出版。）

3. 有時候為著某一點起見，可以把商業成語，或特種語句寫成大寫。例如："*Small-Profits-and-Quick-Returns* is the motto of our firm."（「薄利多賣」是本號的格言。）

4. 在廣告以及銷貨函件中，可以把整個字大寫，以引人注目。例如：

   We will deliver THE EVENING STAR to you by special carrier for TWO WEEKS, without any charge whatsover, if you will send us back the enclosed stamped post-card with your name and address.

   （倘使你把你的姓名住址，寫在附奉的黏就郵票的明信片背面，付郵寄下，那麼敝報館會派送報人，免費奉送「明星晚報」兩星期。）

   不過 3. 4. 兩項，不能多用，只能用於最重要的一點、兩點，否則觸目皆是大寫反而變為「不醒目」了。

5. North, South, East, West：表示方向的 North, South 等一般為小寫，但如為固有名

詞的一部分，或指某區域時應該大寫，例：

(1) Cotton is the principal crop in *the South*, it is also grown extensively in *the South-West* and in *Southern California*. (the South：南部各州，the South-West：西南部各州，Southern California：加州南部)

(2) South-West Africa.

(3) North Amercia.

6.文件名稱，尤其有案號時，宜用大寫。

(1) We have quoted our best terms on the enclosed *Price List*.

(2) Please let us have a *Credit Note* for the difference.

(3) We thank you for your *Order No. 165*.

7.公司的部、課等名稱，原則上用小寫，但有些人主張將本公司的部、課名稱大寫，對方的部、課名稱則小寫，例如：

(1) You will be interested in the enclosed brochure just issued by *our Sales Department*.

(2) Thank you for the pamphlet issued by *your sales department*.

8.省略公司名稱、都市名稱，單稱公司時，用大寫：

(1) The *Company* will pay you $100 annuity. (公司將付 100 美元年金。)

(2) You will receive the sum from our *Company*.

(3) Your application to our *City Office* for a prospectus has been handed to me.

9.職稱應該大寫：

Our *General Manager*, Mr. Clifford, has asked me to acknowledge your letter of October 12.

10.船名加引號時，頭一字母大寫，如全部大寫則可省去引號。

We have shipped the goods via the s.s.“*London Maru*”of OSK.

(OSK: Osaka Shosen Kaisha)

We have shipped the goods via the *S.S. TOKYO MARU* of PCL.

(PCL: Phoenix Container Liners Ltd.)

第八節　易於混淆的商業用語

撰寫貿易英文信的人，除應精通貿易實務，懂得銷售術之外，對於商業用語的意義尤應熟悉。實務上，由於誤解商業用語的含義，致引起糾紛者，屢見不鮮。以下將易於混淆的商業用語列舉若干加以說明，尚希讀者注意。

1. Accept 與 Receive：

在普通英文裡，"Accept" 與 "Receive" 的意義並無多大區別。但作為商業英語，卻有很大的出入。例如 "We received your order" 只是說收到訂單，至於是否接受訂貨，並未表明。但如說 "We accepted your order" 則此 "Accept" 為 "Receive with approval" 即不但收到訂單而且表示接受了訂貨之意。"Order" 一經 "Accept"，契約即告成立，賣方即有按訂單配貨的義務。

其他如 "To receive a bill"（收到匯票）與 "To accept a bill"（承兌匯票）意義也大不同。

2. Acknowledge 與 Confirm：

"Acknowledge" 為收到 (Receive) 之意。"To acknowledge a letter" 一詞只敘述收到信的事實 (To state as a fact) 而已，並無接受 (Accept) 之意。但 "To acknowledge an order" 一詞，卻意指 "To accept an order"。

"Confirm" 為確認、追認 (To add firmness, To give formal or decisive assent as necessary to a thing'S validity)，其意與 "Accept" 相同。因此 "To confirm an order" 可解作 "To accept an order"。例如："We shall be glad to have your confirmation of the above amended terms." 中的 "confirmation" 即可解作 "Acceptance"。

3. Address 與 Domicile：

在英文履歷表中，相當於「本籍」的英文為 "Permanent domicile" 或 "Permanent address"。「現址」則稱為 "Present address"。"Domicile" 是一正式 (Formal) 的用語。主要用於法律文件上。此外，履歷表中的 "Permanent domicile" 常以 "Birth of place"（出生地）代替。

"Address" 與 "Domicile" 在商業用語方面，作名詞用時，其意為「票據的付款地」，作動詞用時，其意為「將票據的付款地指定在承兌人住所或營業場所以外的地方」。承兌人在承兌票據時，記入該付款地點於票據，那麼這票據即成為 "Domiciled bill"。例如：

We have the pleasure of returning the bill duly accepted and (*made*) *domiciled at the Bank of New York*, New York City.

（指定紐約銀行為付款地點。）

4. Agency 與 Agent：

"Agency" 為「代理業務或代理行為」之意，"Agent" 為「代理商」之意，例如報紙上 "Agencies wanted" 乃指「願充任代理」之意，也即希望充任代理商的一方刊登 "To want to act as an agent"。至於 "Agents wanted" 則指「徵求代理商」之意，係由希望找代理商的一方（即 Principal）刊登 "To seek agents"。例如：

They hold agencies for well-known firms.

（他們擔任著名廠商的代理。）

We want to undertake exclusive agency for canned goods.

（我們願擔任罐頭貨品的獨家代理。）

Taiwan buying agents are open to accept buying agencies from British importers.

（臺灣的採購代理商願做英國進口商的採購代理。）

5. Agent 與 Representative：

此兩詞有時候都用作「代理商、代理行」的同義詞。例如在 "An exporter may conduct business through appointed representative, branch office or subsidiary company." 中的 "representative" 乃指 "Agent" 之意，但 "Representative" 實際上乃指代表他人為某種行為的人。所以，除了 "Agent" 之外，派往國外代表公司的國外駐在員也稱為 "Representative"，因此，其含義較 "Agent" 為廣。

6. Answer, Reply 及 Response：

"Answer" 與 "Reply" 實際上視為同義詞，已不加區別。但嚴格地說，是有些差別的。依韋氏 (Webster) 字典的解釋，"Answer" 係對來信的回答，但是否對來信所詢各點均予回答，卻不一定。而 "Reply" 則係對來信所詢各點均予回答。又依 P. D. Hugon 的 "Morrow's Word Finder" 的解釋，如對方有所要求，而認為其要求不當時，即以 "Reply" 作答；此外有些學者，認為用 "Reply" 較 Formal，而用 "Answer" 則較 Informal。

總之，兩者雖有上述的區別，但現代商業上已不加區別。但兩者的介系詞的有無應加注意：

In answer to
In reply to
Answering
Replying to
} your letter of...we...

"Response" 有「對於對方的要求回信，誠意地回答，使對方有好感」之意，例如：

In response to your letter of... we...

7. As for 與 As to：

兩者都有「就……而言」、「至於……」之意，但 "As for" 用在人的場合較多，而 "As to" 則用於事或物的情形較多，例：

As for me...

As to the reason...

As to the cause of delay...

8. As of 與 As on：

⑴「自……（某日）起」用 As of, As from。

The new regulations will come into force as of May 4.

This amount has been credited to your account as of yesterday.

⑵「至……（某日）止」用 As at, As on。

Your credit balance as at 30th December was $210.

9. Balance 與 Remainder：

"Balance" 一詞，作為 "Remainder"（剩餘）用是錯的，其正確的意義是「差額」，用於帳戶方面，例如：

There is a balance in my favour of $1,000.

（有 1,000 元的差額在我正方——就是我的帳上，尚存 1,000 元。）

至於「剩餘訂貨」或「剩餘貨物」不能以 "Balance of orders" 或 "Balance of goods" 表示，其正確的表現法為：

The remainder (or rest) of the goods (or orders) will be shipped in the next few days.

（剩餘貨物〔訂貨〕將於幾天內裝運）

然而，事實上卻將 "Balance" 與 "Remainder" 誤作同義詞，OED (Oxford English

Dictionary) 說：將 "Balance" 用作 "Remainder" 之意，是商業用語。這種用法來自美國，因此美國書雖稱這種用法為 Colloquialism，但並不排斥這種用法。

10. Bank 與 Banker：

依 OED 的解釋，Banker is: a. the proprietor or one of the proprietors at a private bank（銀行家），b. the manager or one of the managing body of a joint-stock bank（銀行幹部）。例如：

My friend, Mr. Black, banker（銀行家）of this town, will perhaps support my theory.

該辭典又說，股份有限公司組織的銀行 (Joint-stock banking company) 必須用多數，例如：

Our bankers are Bank of Taiwan, Taipei.

與此類似的還有：A marine insurance company 與 Underwriters; A manufacturing company 與 Manufacturers, Makers; An export (or import) company 與 Exporters (or importers), Export (or Import) traders。此外 Agent, Customer, Seller, Buyer, Shipper, Consignee 等嚴格地說，若非個人，應用多數形較佳。

11. Bid 與 Tender：

招標公告中常有 "*Tenders* are invited for the supply of...", "Sealed *bids* will be received at the office of..."

根據 H. W. Horwill 的 *A Dictionary of Modern American Usages*：在美國 "Bid" 解做 "An offer of a price by a prospective purchaser, Esp. At an auction"，而 "Tender" 則解做 "A statement of the price at which one is prepared to undertake a certain piece of work or to supply certain goods"。但在美國 "Bid", "Tender" 已通用，無差別。

12. Carriage 與 Freight：

據 M. S. Rosenthal 著 *Techniques of International Trade* 的解釋，就國內運輸而言 "Freight" 通常指「貨物」，而不指「運費」，在國際貿易上，則將 "Freight" 解做「運費」，而「貨物」則用 "Cargo" 一詞。

在英國 "Freight" 通常指「海洋運費」而言，「陸上運費」則用 "Carriage" 一詞。由於 "Freight" 可解做「運費」或「貨物」，所以為避免混淆，「運費」宜用 "Freight charge" 或 "Freight rate" 表示，例如：

The special containers have today been returned to you by rail, *carriage forward*.（運費預付）

13. Cheap 與 Inexpensive：

據 *The American College Dictionary* 的解釋，"Cheap" 含有 "Inferior" 之意，因此，"Cheap goods" 含有 "Cheap in price, but poor in quality"（價廉但物不美）之意。至於 "Inexpensive" 為「不貴」之意，含有 "The value is fully equal to the cost" 之意，也就是說「價廉，而物則非不美」是真正的便宜，且不主觀。

因此要表示貨物便宜與其用 "Cheap goods" 不如用 "Inexpensive (or Low-priced) goods" 或 "Goods of better (keener) price"。

14. Copy, Duplicate 及 Transcript：

雖然三詞均譯作「副」本，但辭典上說：

Copy is often without the exact correspondence which belongs to a duplicate.（Copy 往往沒有 Duplicate 那樣正確一致。）

換言之，"Duplicate" 與原本 Exactly alike 而 "Copy" 則不盡然。在文件的場合，"Duplicate" 與原件 (Original) 一樣有簽名，且與原件具有同等效力。反之，"Copy" 則通常並不簽名。

嚴格地說，"Duplicate" 係根據原件正確複製而成，且通常只有一份，所以印刷品或照片複印多份而與原件非 Exactly alike 時，不宜稱其為 "Duplicate"。例如：

If a bill of exchange is lost before it is overdue, the holder may ask the drawer to give him a dulicate.（在票據法，duplicate 譯成複本。）

Please sign the agreement and return the duplicate to us.（這裡 duplicate 可譯成副本）

The invoice is to be made in duplicate.（發票須繕製正副二份。）

"Copy" 又有份數 (Number of copies) 之意。不過很容易使意義混淆不清。例如在 "Please make me two copies of this letter." 一句中，到底總共二份信呢？還是除原件外另須二份副本呢？這裡如寫成 "Please make me two carbon copies（或 photostated copies）of this letter." 意義就清楚了。

"Transcript" 為根據原件用手寫成或打字而成的抄本。

15. Correspondent 與 Correspondence Clerk：

"Correspondent" 雖然通常譯成「通信者」、「通信的人」，但此通信者，究係指發信人或收信人，視情形而定，有些辭典解釋作 "One who writes letters to another" 好像專指發信人。再者，在商用書信的場合，"Correspondent" 可能指收信或發信的商號，也可能是指擔任通信的文書人員，因此，其意義如何需從文章的前後關係來判斷。

⑴當作「擔任通信工作的人」(Correspondence clerk) 解釋時：

He is an applicant for the post of correspondent in this office.

（他是向本公司應徵擔任通信工作的人。）

⑵當作「發信人」解釋時：

Postscripts should be avoided, as far as possible, but unfortunately some correspondents make a habit of employing them.

（信中應盡量避免使用「附啟」，但不幸，有些發信人……。）

⑶當作「收信人」解釋時：

A circular letter is one which is meant to be read by a number of *correspondents*.

（通函是準備發給許多收信人閱讀的信。）

⑷當作「往來行號、銀行的往來銀行」解釋時：

The bank sent the documents to its *correspondent* in New York.

（該銀行將單證寄往其在紐約的往來銀行。）

We have agents and *correspondents* in all the large commercial centers.

（我們在所有大商埠有代理商及往來行號。）

16. Date 與 Day：

"Date" 指某月某日的「日期」，例如文件的日期即為 "Date"，"Day" 則為時間的單位，即「日子」，因此 "Day" 的多數係表示期間（例如 For three days 3 天），而 "Date" 的多數則不表示期間。例如：

The remainder of the items were duly advised on their dates.

"Date" 有時也作期間解釋，但較少用，例如：

You must complete delivery within a specified date.（在規定期間內須完成交貨。）

"Date" 在舊式商用英文中常用作 "The same day" 或 "Today" 之意，例如：

Enclosed we send you a statement of account to date.（奉上迄至今天為止的帳單。）

We appoint you our sole agents for a period of twelve months from date.

（自今天起授任貴公司為敝公司獨家代理，為期 12 個月。）

在詢問今天是什麼日子時，兩者均可用，但意義不同，例如：

⑴ What day it is today?（今天是星期幾?）回答為："Today is Sunday."，不可回答 "It's August 25."

⑵ What date it is today?（今天是幾號?）回答是 "It's August 25."，不可回答 "It's Sunday."

17. Delivery 與 Shipment：

"Delivery" 為將貨物實際地交給對方或將貨物所有權移轉給對方之意，而 "Shipment" 則指「交運」、「裝運」之意。以出口港為交貨條件的 FOB, CFR, CIF 條件，"Delivery" 與 "Shipment" 意義相同，解釋做「裝運」，所以 "Date of delivery" 與 "Date of shipment" 均解釋作「裝運日期」。

以進口地為交貨條件的 Ex Ship, Ex Quay 條件，"Delivery" 與 "Shipment" 意義不同，前者為「交貨」，後者為「裝運」。

18. Discharge 與 Land：

兩者均作「從船上卸貨」解釋，但嚴格言，兩者意義略有不同。"To discharge" 為 "To take a cargo out of vessel"，即從船上將貨卸下之意。至於卸到何處並不一定。可能是卸至岸上，也可能卸至駁船上。所以在卸至駁船的情形，尚須從駁船卸至岸上，也即尚須 "Land"。在此意義下的 "Discharge" 與 "Unload" 同義。

至於 "To land" 則為從船上卸至岸上之意，英文海上保險單中有 "...and until the same be there discharged and safely landed" 字樣，將兩者明確地區別。"Port of discharge" 與 "Port of destination" 未必一致，在轉船的場合，前者係指轉船港。

B/l 中將 "Port of discharge" 解釋為 "Where goods are to be transhipped to on carrier"，而將 "Port of destination" 解釋為 "Where goods are to be delivered to consignee by on-carrier。

19. Discount, Reduction 及 Rebate：

"Rebate" 意指對已付金額按一定比率退還的金額而言，可譯成「回扣」。

在商場上 "Discount"（折扣）可分為⑴ "Trade discount"，⑵ "Cash discount" 及 ⑶ "Quantity discount" 三種。"Trade discount" 為製造廠商、批發商對於零售商的同業折扣，通常用於 "Catalog" 上有一般售價的商品。

"Cash discount" 為對於一定期間內付現時所給予的折讓。例如 "5% cash"（=5% discount to be allowed for prompt payment，即時付現有 5% 折扣）。"30 days $2\frac{1}{2}$% 或 $2\frac{1}{2}$% 30 days"（=When payment is made within 30 days, a discount of $2\frac{1}{2}$% to be deducted.）。

"Quantity discount" 為對於大量訂購時所作的折扣。

"Reduction" 與 "Discount" 不同，"Reduction" 為廣義的減價之意。減價稱為 "To reduce price"，減了的價稱為 "Reduced price"。通常不可說成 "To discount the price" 或 "A discounted price"。

20. Duty, Tax 及 Dues：

稅捐除用上述三詞之外，尚有 "Assessment"，"Customs"，"Excise"，"Impost"，"Levy"，"Toll" 等詞，但在商用方面，仍以使用上述三詞者居多。

"Duty" 為對進出口貨品、消費行為等所課征的間接稅。Customs（關稅），Excise（消費稅）即這裡所指的 "Duty"，但在英國除間接稅之外，尚包括對財產讓與、繼承、證書的認證所課的稅。例如 "Stamp duty" 是。

"Tax" 一般稱為租稅，主要是對財產、所得、交易所課征的稅，例如 "Sales tax"（銷售稅）、"Commodity tax"（貨物稅）是。

"Dues" 為法定的使用費及其他稅捐。例如 "Port dues"（碼頭使用費）、"Tonnage dues"（噸稅）。

21. Force majeure 與 Contingency：

法語中的 "Force majeure" 等於英文的 "Superior force"，契約上都用 "Force majeure" 一詞，中文為「不可抗力」，但不可抗力可分為：

⑴天災 (Act of god)：如洪水、暴風雨、地震等非人力所能控制的自然現象——稱為絕對的不可抗力。

⑵人禍：如罷工、戰爭等非當事人所能規避防止的人為現象——稱為相對的不可抗力。因此，"Force majeure" 有廣狹兩義。*The Concise Oxford Dictionary* 對 "Force majeure" 廣義的解釋為天災人禍，又據 *Pitman's Business Man's Guide* 的說明，在英國 "Force majeure" 往往僅作狹義的天災解釋。

Morris S. Rosenthal 也將 "Force majeure" 狹義的解釋為 "Act of god"，廣義的不

可抗力則主張用 "Contingency"（偶發事故）一詞。

由於見解不同，容易造成糾紛，因此訂約時，宜將構成不可抗力的事件一一列出。

22. Honor 與 Protect：

在票據方面 "To honor a bill"，"To protect a bill" 乃指「票據的承兌或付款」而言。

"To honor" 為「尊重匯票發票人的簽名」(Honoring the signature of the drawer) 之意，所以用作 "To accept a bill" 解釋，又重視自己的承兌 "Honoring one's own acceptance" 所以又作 "To pay a bill" 解釋。在遠期匯票的場合，"Honor a bill" 可解作「承兌」(To accept a bill)，在即期匯票的場合，"Honor a bill" 可解作 "To pay a bill"，兩者合起來，可譯成「兌付」。

"To protect" 為 "To provide funds to meet a bill" 之意，因此，在即期匯票時，為 "To pay a bill" 之意，在遠期匯票時，為 "To accept a bill" 之意。由上述可知，"Honor" 與 "Protect" 同義。

23. Order, Command 及 Indent：

均作「訂單」解釋。"Command" 為有禮貌的誇張用語 (Polite hyperbole)，在舊式商用英文信中常用。例如："We trust you will continue to favor us with your command."。

"Indent" 本來指國內買主委託國內進口商，向國外購貨而言，這種用語起源於澳洲、印度、英國殖民地。受委託的進口商則憑其委託向國外出口商發出訂單。因此進口商進口貨品時，其買主已確定。在此情形下，委託者（即買主）稱為 "Indentor"，受託向國外發出訂單的進口商稱為 "Indent house"，但現在已不作如此解釋。換言之，委託者與受託者不限於在同一國內者。由國外委託購買的場合，也用此用語。更進一步廣義地解作進口商向國外出口商所發訂單 (The order of the foreign importer to the exporter abroad) 或從國外代理或代表收到的訂單 (An order for goods from an agent or correspondent abroad)。因此"Indent"與 "Order" 幾乎已無差別。

一般而言，"Order" 一字用的較多，"Command"，"Indent" 用的較少。

24. Reshipment 與 Transhipment：

"Transhipment" 為「轉船」之意。"Reshipment" 因接頭語 "Re" 有 "Again" 及

"Back" 兩義，所以 "Reshipment" 有兩義：作 "Again" 解釋時，意為「再裝船」或「轉船」；作 "Back" 解釋時，意指「運回」或「將卸船之貨再裝上船」。

由於 "Re" 有兩義，所以 "Reshipment" 究應作何解釋，很容易混淆。茲將商業用語中有 "Re" 接頭語的詞語列舉若干於下：

Re=Once more, Again, Anew, Afresh

Readjust, Rearrange, Reassess, Recount, Redistribute, Reinsure, Renumber, Reorganize, Reproduce, Resettle, Restock (=To supply with fresh stock), Revalue, Rediscount, Reorder.

Re=Back（回復原來狀態）

Reappoint（復職，To Appoint afresh to the same position），Reestablish (=Restore), Re-export (To export same goods which have been imported), Re-import, Reopen, Repurchase（買回），Resell（轉賣）.

以上是 COD 的分類法，但歸類於第二類者；也有些解作 "Again" 的。例如 "Reappoint" 即有「任命其他職務」(To appoint afresh to another position) 之意。又如 "Repurchase" 除可解作「買回」之外，也可解作 "To buy again"。

25. Sample, Pattern 及 Swatch：

"Sample" 為從現貨 (Existing goods) 中抽出一部分，作為貨品全體品質、形狀等的代表者。因此 "Sample" 必須為待售貨物的一部分，是一種現物樣品。COD 將其解釋為 "Small separated part of something illustrating the qualities of the mass it is taken from."。像天然產物等難於憑現物樣品買賣的貨物，其樣品稱為 "Standard" 或 "Type"。附於價目表或 Catalog 的雛形樣品稱為 "Specimen"。在紡織品買賣，其設計、花樣是品質構成要素之一，而能同時表示品質及花樣的紡織品樣品則稱為 "Pattern"。

將紡織品剪下一塊作為樣品者，稱為 "Swatch" 又稱為 "Cutting sample"。

26. Shipment 與 Consignment：

在英國 "To ship" 為「用船運出」(To send away goods on board ship) 之意，在美國，以火車、飛機運出時也用 "Ship" 一語。因此，名詞 "Shipment" 在美國也作廣義的解釋。但為區別起見，乃有 "Railroad shipment"，"Airshipment" 等字眼的產生。

至於 "Consignment" 在英國，猶如美國的 "Shipment"，廣義的用作陸、海、空的運送。

27. Term 與 Condition：

"Term" 當作「條件」解釋時，通常用多數形的 "Terms"，關於付款、費用等有關金錢的「條件」通常用 "Terms" 一詞。例如：

Our terms are cash within three months.（敝方的付款條件為三個月內付現。）

但實際上並不如此嚴格。

The goods are not in accordance with the terms of the contract.

（該貨與契約條件不符。）

"Condition" 解作「條件」之意時，其使用範圍較 "Terms" 大。並且用多數形。

The buyer must observe the conditions of payment.

28. Maker 與 Manufacturer：

一般人常將此兩詞視為同義，實際上卻不能一概而論。在許多場合，兩者並不能交替使用。所謂 "Maker" 多半指偏重使用手工的業者，例如 "Shoe-maker"，"Watch-maker,"，"Basket-maker" 等。此外機械的裝配或製造業者也稱為 "Maker"，例如："Sewing machine maker"，"Tool maker"，"Makers of weaving machinery"，但紡織業者卻不能稱為 "Maker"，茲將不能稱其為 "Maker" 的業者列舉於下：

紡織廠	Mill (Cotton, Woollen, etc.)
印染廠	Printing factory
染色廠	Dye-works
皮革廠	Tannery
鐵工廠	Iron works
奶品廠	Milk plant, Milk factory
釀造廠	Brewery
飲料裝瓶廠	Bottling plant
蘇打廠	Caustic soda factory

第六章
貿易英文信的開頭句與結尾句 (The Openings and Endings of a Letter)

第一節　開頭句

一、開頭句的重要性

在商用書信，開頭句最重要，因為一封信的開頭寫得好不好，會給收信人以深刻的印象。開頭句一定要寫得生動有力，如果第一句話能喚起收信人的興趣或注意，他自然會高興地讀下去，否則他就覺得索然乏味，不願再費時間看下去。因此一切陳腐的開頭句，應盡量避免使用，以免減少收信人的興趣及注意。

二、開頭句的要領

下列幾種提示，可以幫助學者怎樣應用適當的開頭句子。

1. 開門見山，廢去虛偽客套：一封信的開頭，應該快鞭直入，立刻告訴收信人所要知道的事情。有許多慣於運用客套語的人，以為一封信中如果沒有那種開場的客套語，就顯得唐突。其實，現代商界中人士注重實際與效率，對於那些徒然浪費他們寶貴的時間和精力的虛偽客套，常引起反感。尤其我們寫貿易英文信時切勿將中文書信的刻板開頭用語加以套用，只有直接、簡潔、明確的開頭句才能使一封信的文字生動有力，而受收信人的歡迎。例如：

We are very sorry to be obliged to inform you that we forwarded an order for cotton-shirtings on the 3rd inst., and have your acknowledgment of the sixteenth, but the goods have not yet at hand.

（鄙號曾於本月 3 日奉上購買棉布疋訂單乙紙，當蒙覆稱於 16 日收到，然迄今該貨尚未寄到，迫切上陳，曷勝惶悚。）

像這種囉嗦的 "We are very sorry to be obliged to inform you that"（等於中文書信中的「迫切上陳，曷勝惶悚」）等句子，在現代，使人看了頭痛。我們應直截了當地這樣說：

On January 6, we placed an order for cotton-shirtings, which was acknowledged on the sixteenth, but the goods have not yet arrived.

（在 1 月 6 日寄上購買棉布疋訂單乙紙，曾蒙覆信說已於 16 日收到，但那批貨迄今仍未寄到。）

同樣，下列的一些套語，應避免用作一封信的開頭句。

I have the honor of informing you that...

I have the honor to address you that...

We beg to inform you that...

I write in haste to tell you that...

We write in a hurry to tell you that...

I take the liberty to address you that...

I beg respectfully to inform you that...

We beg to intimate that...

We beg to announce you that...

We beg to notify you that...

We take this opportunity to inform you that...

We regret to advise you that...

即使在通函 (Circular letter) 中，有時要用到「敬啟者」那樣的開頭句，在現代貿易英文信中也該簡潔地說 "We inform you that..." 或自然地說 "We have pleasure in announcing that..." 或 "We are pleased to inform you of..."。

在舊式貿易英文信的覆信中，更多陳腐而模糊不清的開頭句子，我們應盡量避免使用，例如：

In reply to yours of 30th ult., we desire to say that we have sent you under separate cover a copy of our latest catalog.

（敬覆上月 30 日尊函，茲啟者，敝公司已將最新之目錄，另套寄奉矣。）

譬如上月是 5 月 30 日，為什麼要說是 30th ult. 呢？為什麼要嚕哩囉嗦說 "We desire to say" 等廢話呢？請和下列用語自然、明晰，而又站在讀者的觀點上說話的開頭句，比較一下：

The catalog you requested in your letter of May 30 has been sent today. On pages 14 and 15 you will find a complete description of the sewing machine in which you are interested.

（您 5 月 30 日來信索閱的目錄，已於今日寄上。在第 14 和 15 頁上，您可以找到您所關切的縫衣機的完全圖樣。）

再看下面的陳腐不堪的句子：

Your favour of June 25th at hand and in reply thereto beg to say that we do not handle the sewing machine you mention.

（6 月 25 日尊函收到，敬覆者，尊函所述縫衣機，敝號並無出售。）

同樣的意思，在現代英文商業信中，應該用下面簡明的寫法：

We regret that we do not handle the sewing machine you mention in your letter of June 25.

（6 月 25 日來信所提起的縫衣機，敝號沒有出售，非常抱歉。）

2. 避免用分詞式開頭句 (Participial openings)：分詞式開頭句用法不僅陳腐，而且顯得了無生氣。例如：

Referring to your order of January 21...

應改為：

Your order dated January 21 for 1,000 sets...

又：

Confirming our cable of today...

應改為：

We confirm we sent you the following cable on May 5.

3. 可能的話，盡量以 "You" 開頭，以收 "You-attitude" 的效果：

We have received your inquiry on...

宜改為：

Your inquiry on...has been received.

又：

We wish to ask you to send us your latest catalog.

宜改為：

Will you please send us your latest catalog?

但不是說每一封信均以 "You" 開頭，事實上很多信都以 "We" 開頭，在一封信中滿篇 "We" 當然不好看，但故意不用 "We" 有時反而顯得不自然。尤其在 Claim letter，更不宜用 "You" 開頭，例如：

You did not reply to our letter about...

應改為：

We have not received your reply to our letter about...

又：

Your letter of May 15 failed to tell us when you will open L/C.

宜改為：

We failed to learn from your letter of May 15 when you will open L/C.

4. 盡量避免以消極或否定句開頭：以消極或否定句開頭，易使人產生緊張或不快，以致影響收信人看信的情緒。例如：

⑴（拙劣）We are discouraged that you have not replied to our letter of July 15...

　（較佳）Did you receive our letter of July 15...?

⑵（拙劣）We have your complaint about...

　（較佳）Thank you for writing us about...

⑶（拙劣）We regret to bring to your attention that you have not yet filled our order...

　（較佳）We are expecting your filling our order...every day...

一般而言，一開始就說抱歉的話似乎不妥，但事實上不得不這樣說的話──尤其向人家道歉、謝罪或表示惋惜時，還是直說，以表示係鄭重聲明的意思。例如我們不宜毫不表示歉意地說：

Yours of the 2nd at hand and in reply would say that we cannot accept any more orders for blankets, as our supply is exhausted.

（頃接 2 日台函，茲覆者，敝號毯氈存貨已罄，不能再行承辦矣。）

我們應該寧可刪掉那些陳套，而把歉意鄭重的放在開頭，明顯的表達出來。

We regret that we are unable to fill your order of the 2nd for blankets, as our supply is exhausted.

（敝號非常抱歉，不能承辦台端於 2 日來信中所定購的羢氈，因為敝號的貨物，已經賣完了。）

5.在覆信開頭句中，應將所要答覆的那一封來信指出，以便收信人能立即瞭解覆信的本意，例如下面的幾種開頭句是值得模仿的。

⑴ We appreciate the information you have so kindly furnished us in your letter of January 10.

（敝號非常感激寶號 1 月 10 日來信中，承蒙惠示的情報。）

⑵ We are unable to give you the information you asked for in your letter of May 10 concerning the financial status of Mr. Lee.

（5 月 10 日來信中，承詢有關李君的財務狀況，敝號無法作答。）

⑶ We shall be glad to send our new price list of automobile tires, which you requested in your letter of May 8, as soon as it is printed.

（5 月 8 日來信索閱的汽車胎新價目表，當印好的時候，敝行當欣然立刻寄奉。）

⑷ Thank you for your candid letter of October 15 in which you called our attention to the fact that the undershirts you purchased on September 10 did not come up to the usual high standard of goods handled by our store.

（謝謝您 10 月 15 日很率直的來信，關照敝店注意，您在 9 月 10 日購買的汗衫，不及敝店平素發售的貨物那樣的品質優美。）

⑸ We regret very much our inability to comply with your request of July 4 for a further discount on your last order, but we can grant you easy terms of payment.

（敝號非常抱歉，不能遵從 7 月 4 日來信寶號的要求，把上次寶號的定貨，再打個折扣，不過敝號能夠接受寬鬆的付款條件。）

⑹ We are very glad, indeed, to learn from your letter of June 20 that you are interested in our line of rugs and have on hand just now a full assortment of the style you describe.

（敝公司不勝榮幸，從 6 月 20 日的來信，知道您有意經營敝公司出品的地毯，敝公司剛巧存有您所開示的那種式樣的全般貨色。）

⑺ Your inquiry of the 14th give us an opportunity which we appreciate, as we are always glad to give any information possible regarding our goods.

（先生 14 日來信的詢問，給敝號一種感佩莫名的機會，因為敝號永遠樂於儘可能的供給關於敝號貨物的一切消息。）

三、開頭句用語

以下按照由「我方主動先寫信」時，以及「覆信」時，兩種情形，分別列舉若干開頭句，以供寫信時參考之用。

1.由我方主動先寫信時：

⑴寄奉、寄送文件、目錄、物品時：

$$
\text{We}
\left\{
\begin{array}{l}
\text{send} \\
\text{are sending} \\
\text{have today sent} \\
\text{are glad to send} \\
\text{will send} \\
\text{are pleased to send} \\
\text{shall be pleased to send}
\end{array}
\right\}
\text{you}
\left\{
\begin{array}{l}
\text{under separate cover.} \\
\text{by separate mail (post).} \\
\text{by another mail (post).} \\
\text{separately.} \\
\text{by air.} \\
\text{by airmail.} \\
\text{by air parcel post.} \\
\text{by airfreight.} \\
\text{by sample post.} \\
\text{by s.s....} \\
\text{by seamail.}
\end{array}
\right.
$$

⑵檢附……時（＊舊式用語）：

$$
\left.
\begin{array}{l}
\text{Enclosed pleased find*} \\
\text{Enclosed you will find*} \\
\text{You will find attached to this letter} \\
\text{We are attaching to this letter} \\
\text{We enclose} \\
\text{You will find enclosed} \\
\text{We send you herewith} \\
\text{There is (are) enclosed} \\
\text{Enclosed is (are)} \\
\text{We are enclosing}
\end{array}
\right\}
\text{...}
$$

(3)通知、證明或介紹：

This is to
$\begin{cases} \text{announce} \\ \text{inform you} \\ \text{certify} \end{cases}$ that...
introduce to you Mr....

This serves to acquaint you that...

It is our pleasure to tell you that...

We have the honor
We are pleased $\Big\}$ to inform you...
We have the pleasure

(4)「請寄……」時：

Please send us
Will you please send us
Kindly favor us with

We shall be $\begin{cases} \text{pleased} \\ \text{glad} \\ \text{happy} \end{cases}$ to receive

We shall be $\begin{cases} \text{obliged} \\ \text{grateful} \end{cases}$ if you will send us

We $\begin{cases} \text{shall} \\ \text{will} \end{cases}$ appreciate it if you will send us

your
$\begin{cases} \text{catalog.} \\ \text{general catalog.} \\ \text{revised catalog.} \\ \text{latest catalog.} \\ \text{estimate.} \\ \text{price list.} \\ \text{proforma invoice.} \\ \text{quotation sheet.} \\ \text{specifications.} \end{cases}$

(5)「從……得悉貴商號名字」時：

We $\begin{cases} \text{hear} \\ \text{learn} \end{cases}$ from... that you are $\begin{cases} \text{the leading importers of...} \\ \text{the leading exporters of...} \\ \text{reliable firm of...} \end{cases}$

Your name has been given by CETRA.

We have had your name and address given to us by Mr....

We are indebted for your name and address to CETRA...

Through the courtesy of Panama Embassy in Taipei, we...

(6)表示追述、補充或確認前寄函電、談話時：

As explained in
We confirm
In adding further to our
Further to
Following (up)

| letter |
| cable |
| telex |
| fax |
| e-mail |
| conversation |

...

(I)
We confirm
We wish to confirm
We hereby confirm
We have to confirm
This is to confirm

→ (II)

our telegram of (date) to you
our telegram sent you on (date)
that we cabled you on (date)
having faxed you on (date)
having advised you by cable on (date)
cabling you on (date)
the exchange of telegrams
the interchange of telexes between us
our recent exchange of telegrams
the exchange of faxes within the last few days

→ (III)

reading:
as follows:
which reads:
to the following effect:
conveying to you that
advising you that
reading as follows:

→ (IV) "(text of cable)"

→ (V)

which doubtless (no doubt) has had (has received) your attention.
which we think is self-explanatory.
to which we await your reply.
as per confirmation attached hereto.
as per copies attached.
confirmation copy of which is attached.
as per attached confirmation copy.

【例】 (I) We hereby confirm

(II) our telegram sent you on May 1

(III) reading as follows:

(IV) "...(text of cable)...,"

(V) which we think is self-explanatory.

⑺表示遺憾（開頭句盡量避免使用，但並不排斥）：

$$\text{We regret to} \begin{cases} \text{inform} \\ \text{advise} \\ \text{notify} \\ \text{remind} \end{cases} \text{you that...}$$

We regret to bring to your notice certain irregularities...

Much to our regret we have to advise you our inability to...

2.覆信的開頭句：

⑴表示「來函敬悉，或來函已收到敬表謝意」時：

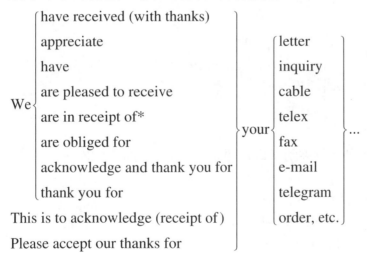

$$\text{We} \begin{cases} \text{have received (with thanks)} \\ \text{appreciate} \\ \text{have} \\ \text{are pleased to receive} \\ \text{are in receipt of*} \\ \text{are obliged for} \\ \text{acknowledge and thank you for} \\ \text{thank you for} \end{cases} \text{your} \begin{cases} \text{letter} \\ \text{inquiry} \\ \text{cable} \\ \text{telex} \\ \text{fax} \\ \text{e-mail} \\ \text{telegram} \\ \text{order, etc.} \end{cases} ...$$

This is to acknowledge (receipt of)

Please accept our thanks for

⑵表示「敬覆者……」時（＊舊式用語）：

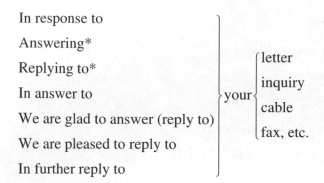

In response to
Answering*
Replying to*
In answer to
We are glad to answer (reply to)
We are pleased to reply to
In further reply to

} your { letter / inquiry / cable / fax, etc.

(3)表示「關於……乙節」時：

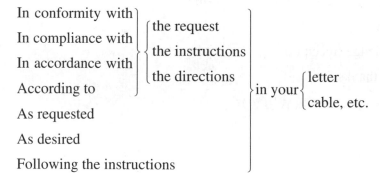

In reference to
With reference to
Referring to*
With further reference to
With regard to
As regards
Concerning
Regarding
In connection with
We refer to
Please refer to
This letter refers to

{ your / our } { letter / inquiry / cable / telex / fax / e-mail / telegram, etc. } ...

(4)表示「遵照實號×月×日來函（電）要求（指示、請求）」時：

In conformity with
In compliance with
In accordance with
According to

} { the request / the instructions / the directions }

As requested
As desired
Following the instructions

in your { letter / cable, etc. }

(5)表示「拜讀×月×日大函，得悉……深感欣慰、遺憾」時：

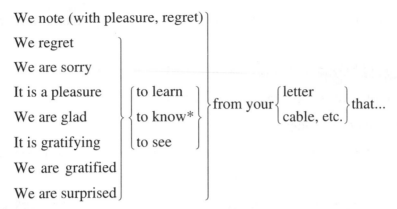

We note (with pleasure, regret)
We regret
We are sorry
It is a pleasure ⎱ to learn
We are glad ⎰ to know* ⎱ from your ⎱ letter / cable, etc. ⎰ that...
It is gratifying   to see
We are gratified
We are surprised

註："to know from..." 的 Expression 最好不用。cf. I know right from wrong.（我能辨別善惡。）

第二節　結尾句

一、結尾句的重要性

　　任何事情的開頭與結尾給予人的印象最深，貿易書信亦然。因此，一封信的結尾句與開頭句一樣，應以簡潔、強而有力的文字表達，使收信人留下深刻良好的印象，從而達成寫信的目的。

　　要一封信前後聯貫，必須先從收信人的觀點，逐漸以合於邏輯的規則推進到寫信人的觀點。因此，開頭句應以收信人的立場開始，結尾句則應站在寫信人的立場結束。換言之，開頭句應著重 "You"，結尾句應著重 "Us"。

二、結尾句的要領

　　以下說明貿易英文信的結尾要領。

1.盡量避免以分詞式結束 (Participial close)：分詞式結尾，不僅顯得軟弱無力，而且俗氣、陳腐，例：

（乏力）Hoping that this information will be helpful to you.

（有力）We sincerely hope that this information will be helpful to you.

（乏力）Trusting that this will be satisfactory to you and hoping that we may be of further assistance to you.

（有力）If we may be of assistance to you in some other way, please feel free to write us.

以下各句都是常見的分詞式結尾句，應盡量少用：

Awaiting your reply, we beg to remain...（鵠候回函。）

Awaiting your further communications, we are...（仍盼德音。）

Awaiting to hear from you at your earliest convenience...（敬俟復命，便祈即復。）

Awaiting your further valued orders...（仍俟源源賜顧為荷。）

Awaiting the opportunity when our services may be of use to you...（俟機圖報。）

Wishing to receive your instructions by letter...（務希函示。）

Begging you to reply by return of mail...（回郵乞賜覆音。）

Soliciting the continuance of the confidence with which we have been hitherto honored...（向承信任，尚祈寵加無已。）

Soliciting your cooperation in this matter and a prompt acknowledgement of the receipt of this letter.（敬祈鼎力合作，並乞接信後，早日俯允裁覆為荷。）

Soliciting the continuance of your kind patronage...（仍希源源惠顧。）

Respectfully soliciting a trial...（請嘗試之。）

Trusting you will excuse our troubling you, and always at your service in similar matters...（瑣瀆清神，務乞恕宥，如有鞭策之處，當無不效勞也。）

Repeating our thanks...（重申謝悃。）

Thanking you in advance for any attention you may show him...

（務希推愛照拂，謹此預申謝悃。）

Thanking you for the confidence which you have shown us...（厚荷信任，特此鳴謝。）

Commending same to your careful attention...（謹此奉托，務希留意。）

Asking your consideration of the matter and with compliments of the season...

（此事務祈卓裁，肅此敬候時祺。）

Assuring you of our best endeavors at all times...（無時不竭誠相報也。）

Regretting the inconvenience you have been caused...（有干未便，殊深抱歉。）

Without having anything further to communicate by this mail...（餘無他陳。）

以分詞式結尾時，有些人喜歡殿以 "We beg to remain", "We remain", "We are",

"We believe" 等詞，以為這樣較客氣，其實都是陳腐的客套。

2.以具體明確的內容結束：結束句子應有具體的內容才能顯得有力。

Trusting that this is satisfactory to you.

We hope to hear from you soon.

上面兩句，既不明確又顯得乏力，下面的就顯得具體有力。

We shall do all in our power to continue to enjoy your confidence in us.

We are interested in hearing that this information fulfills your requirements.

Your inquiry is appreciated and we hope you will follow it with an order.

3.以誠摯的語氣結尾：即用 "Please..." 或 "Kindly..." 的語法，例如：

Please tell us whether we may expect to receive...（懇請惠寄⋯⋯。）

Please let us know about (whether, if)...（懇請賜知⋯⋯。）

Kindly advise us with regard to this matter.（敬請賜知有關此事。）

4.「請求」時以疑問句結尾，可顯得客氣，例如：

Will you please reply us before Friday?

May we ask a favor of you to send us information available?

Won't you let us hear from you promptly?

5.以被動句子結尾：既可避免使用 "We" 又可加強語氣，例如：

Your inquiry is appreciated, and we hope you will follow it with an order.

（感激詢價，並盼接著就寄來訂單。）

Your prompt reply will be appreciated.（迅速回覆，將不勝感激。）

三、結尾句用語

1.表示「請賜覆、請見覆、請惠賜卓見」時：

(I)	We hope to be favored with We hope (or trust) to receive We await (or wait for) We are waiting for We shall be obliged for We shall appreciate We shall be glad to have Will you let us have We look forward (with much interest to)	your a, an	reply, answer. early reply. definite reply. favorable reply. satisfactory reply. good news. further news. comment. opinion. reply by cable, fax. cable reply.

(II)	We trust that we shall hear (further) We trust to hear We hope to hear favorably We shall appreciate hearing We shall be glad to hear Please let us hear Will you let us hear	from you	in this connection. in this respect. in this matter. concerning this matter. about this matter.

以上 (I)，(II) 可分別接下列用語（有「儘速」、「及早」之意）。

(III)	at your earliest convenience. at an early opportunity. as early as convenient. as early as possible. without (the least) delay. on receipt of this letter. (very) soon. at once (promptly, immediately).

2.表示「歉意」時:

⑴ We regret we have not informed you about this matter sooner.

⑵ We are sorry for not having explained our position sooner.

(3) We regret that our reply has been delayed because of the necessity of making a thorough investigation in this matter.

(4) We regret very much that you should have been inconvenienced by this delay.

(5) We are sorry this mistake was not noticed earlier, but trust that this will reach you in time to avoid any misunderstanding.

(6) We hope that this delay will not cause you any inconvenience, and assure you that we shall give your further orders our prompt attention.

(7) We sincerely ask you to accept our apologies for the inconvenience caused to you, and assure you that we will do all we can to see that such an incident will not happen again.

(8) We regret that we are unable to
- meet your requirements.
- be of assistance to you.
- make use of your kind offer.
- avail ourselves of your proposal.
- be of service to you.

3.表示「謝意、惠顧、合作」時：

(1) Please accept our thanks for the trouble you have taken.

(2) We are obliged to you for your kind attention to this matter.

(3) We thank you for your patronage in the past and solicit continuance of the same in the future.

(4) We are greatly obliged for your trial order.

(5) We thank you for the special care you have given to the matter.

(6) We thank you for your generous cooperation.

4.表示「如有機會，必會報答」時：

(1) It would give us a great pleasure to render you a similar service should an opportunity occur.

(2) We shall on a similar occasion be pleased to reciprocate.

(3) We hope to be able to reciprocate your good offices on a similar occasion.

(4) We are always ready to render you such or similar services.

5.懇請惠予多加惠顧:

⑴ We solicit a continuance of your valued favor.

⑵ We hope we may receive your further favor.

⑶ We trust that on further consideration you may decide to give one of these machines a trial.

6.以問候語結尾:

⑴ With the compliments of the season,（歲序更新，特此申賀。）

⑵ With the season's greetings,（謹致時祺。）

⑶ Wishing you a Merry Christmas and a Happy New Year,（謹祝耶誕並賀新年。）

⑷ We take this occasion to offer you our best wishes for a Merry Christmas.
（謹祝耶誕快樂。）

⑸ We send you our cordial Christmas greetings and very best wishes for a Happy New Year.（敬祝耶誕並賀新年快樂。）

⑹ We sincerely hope that in 20– we may continue to serve you as in the past.
（我們誠懇希望在 20– 年能一如過去繼續為您服務。）

⑺ May our business relations in the coming year be more amicable and fruitful.
（願我們來年的業務關係更好、更豐碩。）

⑻ With best wishes and heartiest greetings for Christmas,（衷心敬申耶誕賀忱。）

⑼ With kind remembrances and all good wishes for a Merry Christmas and a bright New Year.（敬賀耶誕快樂與新年燦爛。）

⑽ Heaps of Christmas wishes,（恭祝耶誕。）

⑾ I heartily send all good Christmas wishes to you and yours.
（謹向您及您的家人敬致耶誕賀忱。）

⑿ May this New Year turn out to be the happiest and the best.
（願今年是最幸福與最佳之年。）

⒀ With best New Year wishes,（謹賀新禧。）

⒁ May all your plans and wishes come true in this New Year.
（敬祝您的計畫與願望能在今年實現。）

⒂ May the New Year be a happy year for you.（祝你迎接幸福的新年。）

⒃ With my most hearty greetings and sincere good wishes for you for all the coming year, （謹祝新的歲序將是您及您全家最佳的一年。）

⒄ With all best wishes for a splendid New Year, （敬祝光輝的新年。）

⒅ With the kindest thoughts and the best of good wishes for a Christmas, （敬申耶誕賀忱。）

⒆ With best regards, （特此致意。）

⒇ With kind personal regards,; With kindest regards, （謹致問候。）

(21) We wish you continued success. （謹祝不斷地成功。）

7.表示「希望以後有機會效勞（可奉告）、或將另函奉告等」時：

⑴ We hope that we may be of service to you in some other way.

⑵ We shall write to you further about the matter in a few days.

⑶ We shall keep you informed when conditions improve.

⑷ You will be hearing from us again at a later date.

第七章
進出口貿易步驟 (Steps of Import-Export Trade)

第一節　進出口貿易步驟圖解

任何一筆進出口貿易，從市場調查開始，經業務關係的建立、詢價、報價、訂貨，而至賣方的交貨與買方的付款，完成交易，其間不僅須經過錯綜複雜的手續，而且很多都與貿易英文有關。因此，貿易廠商對其進行的程序，必須先有充分的瞭解，而後才能寫出完好的貿易英文。茲將進出口貿易的進行步驟圖示如下：

建立貿易關係階段

　　1.尋找適當市場：

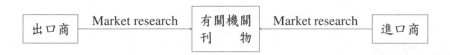

　　2.尋找交易對手：

　　3.發出招攬函、詢價：

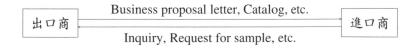

4.徵信:

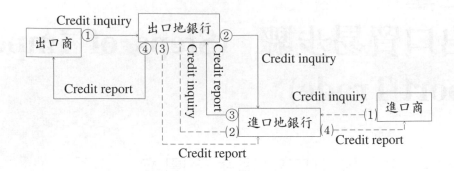

交易磋商與簽立買賣契約階段

5.報價、還價、接受:

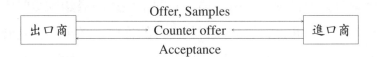

6.訂貨、簽約:

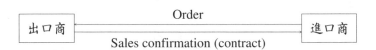

履行買賣契約階段

7.進口簽證、開發信用狀、收受信用狀:

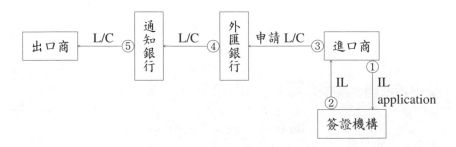

8.備貨:

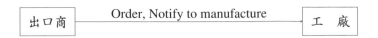

9. 出口簽證：

10. 申請檢驗、公證：

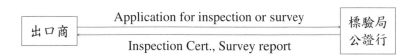

11. 洽訂艙位：

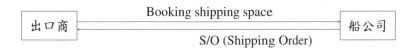

12. 購買保險：

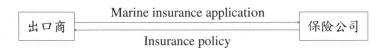

13. 出口報關及裝船：

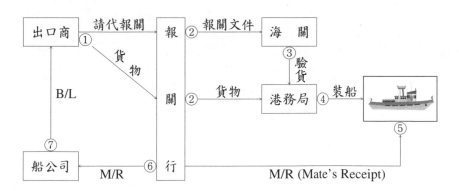

14. 領事簽證、裝船通知等：

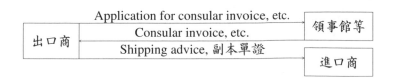

15.出口押匯：

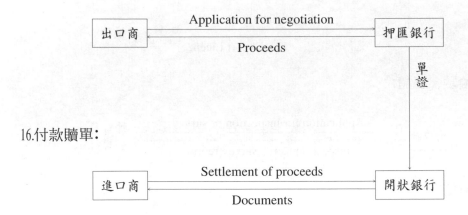

16.付款贖單：

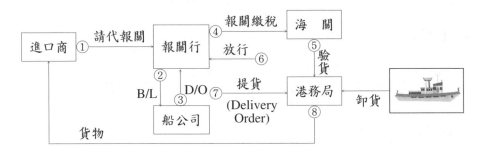

17.進口報關、提貨：

善後階段

18.索賠：

第二節　進出口貿易步驟簡述

1.市場調查：

向	或翻閱	查明
⑴簽證機構	進出口貨品分類審定及管理辦法，中華民國進出口貨品分類表	所經營貨品是否屬於「准許」、「管制」或「禁止」進出口類或「暫停」類，有無配額限制
⑵外國駐本國使領館同業公會、外貿協會	商情報導及有關刊物	那些國家進口或出口所經營貨物、供需量、市價、產銷情形、進出口值、稅率
⑶船公司、航空公司	運價表	航程、運價、班次
⑷國內運輸行		內陸運價
⑸保險公司	保險費率表	保險費率
⑹港務局或港口倉庫公司	港口倉庫及裝卸費率表	倉租及裝卸費率
⑺外匯銀行	結匯收費表、匯率、利率表	結匯費率、匯率、利率
⑻關務署	進出口稅則表	進出口關稅稅率
⑼標檢局、檢疫局、公證行	費率表	檢驗、檢疫、公證費率
⑽報關行	報關業公會收費率表	報關手續費率
⑾駐本國領事館　商會		領事簽證費率　產地證明書費

2.尋找交易對手：

　　詳閱第八章。

3.發出招攬函或詢價：

　　出口商經過上述途徑找到可能的買主 (Prospective buyer) 之後，即可向其發出招攬交易的函電，並寄送價目表、貨品目錄，甚至寄出樣品。

　　反之，進口商根據上述途徑找到可能的賣主之後，即可向其發出詢價函電，索取價目表、貨品目錄或樣品。

4.徵信:

與陌生的國外進出口商初次往來，應先調查其信用情形。最通行的方法為透過本地往來外匯銀行向對方所提供的備詢銀行 (Reference bank) 查詢。

5.報價、還價、接受:

當進口商來函（電）詢價後，出口商即可開列交易條件向進口商發出報價函電，進口商收到報價函電後，如認為價格過高，交貨期太長或其他條件不能接受，當可還價 (Counter offer)，經雙方往返折衝，意見達成一致時，買賣即告成立。

6.訂貨、簽約:

一方報價經他方接受後，買賣契約即告成立，契約一經成立，即簽立買賣契約書，契約書的訂立方式有三:

⑴由賣方繕製售貨確認書 (Sales confirmation) 二份寄交買方加簽後退回賣方一份。

⑵由買方簽發訂單 (Order sheet) 或購買確認書 (Purchase confirmation) 兩份，寄交賣方加簽後退回買方一份。

⑶由買賣雙方共同簽訂買賣契約書 (Contract)，各執一份。

7.進口簽證、申請開發信用狀及收受信用狀:

有些貨品的進口，進口商須先向外匯管理當局（我國為國貿局等簽證機構）申請核發輸入許可證 (Import licence or permit)。取得輸入許可證後，依契約規定以 L/C 為付款方式者，進口商即應向外匯銀行申請開發 L/C 給出口商。

8.備貨:

出口商收到 L/C 經核無誤後即可備貨或通知工廠生產。

9.出口簽證:

在我國，有些貨品的輸出須向國貿局或其他簽證機構申請簽發輸出許可證 (Export licence or permit)，憑以辦理出口報關。

10.檢驗或公證:

依買賣契約規定，貨品須公證者，應請公證行實施檢驗，取得公證報告 (Survey report) 或檢驗證明書 (Inspection certificate)。

有些貨物，依政府規定必須檢驗或檢疫合格才能出口，在此情形，出口商須向標檢局、檢疫局申請檢驗，取得輸出合格證 (Certificate for export)，憑以辦理報關出口。

11.洽訂艙位：

貨物備妥，準備辦理出口簽證時，即可向船公司或航空公司洽訂艙位。

出口商應依契約及 L/C 規定，於交貨期限前選擇開往目的港的適當船隻，以電話或其他方式向船公司洽訂艙位取得下貨單 (Shipping Order, S/O)，以憑裝船之用。

12.購買保險：

按 CIF、C&I 條件交易者，出口商應於裝運之前，向保險公司投保適當保險，取得保險單，以便押匯之用。

13.出口報關及裝船（機）：

凡出口貨物均須向海關申報，稱為出口報關。海關查驗稱「驗關」，通過放行稱「通關」。

為爭取時效，報關多委託報關行代辦。於貨物送進碼頭倉庫之際，將 EL、S/O、Packing list 等文件交給報關行，由其憑以繕打出口報單 (Application for export)，經海關收單、查驗、審核、收費、放行等手續後，貨物即可裝船，然後由船方發給 M/R，憑以換取 B/L。

14.申請領事簽證、發出裝船通知等：

約定出口商須提供領事發票 (Consular invoice) 或產地證明書 (Certificate of origin) 者，出口商應於出口報關時向有關機構申請領事發票或產地證明書。

他方面，出口商於裝船時，應以快速方式通知買方交運貨物數量、船名、預定開航日期 (ETD)、預定到達目的港日期 (ETA) 等，俾進口商做提貨準備。此外，出口商應將提單抄本、發票、裝箱單等副本儘快郵寄進口商，以便萬一經由銀行的正本單證延誤寄達進口地時，可憑這些副本，辦理擔保提貨 (Guarantee on delivery without B/L)。

15.出口押匯：

憑信用狀付款時，出口商於貨物裝船後，即可備齊信用狀規定的單證向銀行申請辦理押匯 (Negotiation) 領取貨款。

按託收條件交易時，出口商可簽發以進口商為付款人的匯票，連同單證送請銀行代收 (Collection)，出口託收有付款交單 (Documents against Payment, D/P) 與承

兌交單 (Documents against Acceptance, D/A) 兩種。

16.付款贖單：

進口商委託銀行開發信用狀時，通常並不需繳足信用狀金額，而是根據進口商的信用情況、進口物資性質，收取部分（如 20% 或 30%）款項，稱為保證金 (Margin money)，待單證寄達時，才由開狀銀行通知進口商前來繳付餘款，進口商繳清餘款即可贖回單證，這個手續稱為「付款贖單」。

至於託收的 D/A、D/P 交易，則其單證經由出口託收銀行 (Export collecting bank) 寄到進口代收銀行 (Import collecting bank)，由其向進口商收取貨款。如果匯票是即期的 (Sight draft)，進口商見票後即須立即付現才能取得單證；如果是遠期匯票 (Usance draft) 的 D/A，進口商只要承兌了匯票，即可取得單證，憑以提貨，俟期限到期，才付款。

17.進口報關及提貨：

進口商自銀行領到單證後，即可委託報關行辦理進口報關提領貨物。於是一筆進出口貿易到此結案。然而事實上卻不然，因為可能還有索賠問題也。

18.索賠：

進口貨物發生量的短缺與毀損，凡可歸責船公司裝載卸貨不當的，應於發覺後，具文檢附事故證明單、公證報告、提單副本、裝箱單及商業發票，要求限期理賠。船公司如不願賠償應敘明理由，發出書面答覆，以供進口商據以向保險公司索賠。

買賣的一方，因不履約或履約不全而引起的貿易索賠，多半起因於①交貨遲延，②品質不符，③數量不足，④包裝不妥，⑤保險不當，⑥不付款，⑦不開信用狀，⑧遲延開發或開發不當信用狀等。如買賣雙方當事人不能以友好方式謀求解決時，即邀請第三人介入協調雙方當事人的意見，或予公平的裁定，俾供當事人遵守，迅求合理的解決。

第八章
尋找交易對象 (Looking for Customers)

第一節　尋找客戶信的寫法

　　進出口商經過市場調查之後，即可就獲得的資料比較分析，選定最有可能的市場作為目標市場，再從這個市場尋找適當的交易對手。尋找交易對手的方法，可分為積極的與消極的兩種。

一、積極的物色方法

　　1.自己直接物色：

　　(1)可在 Internet 上刊登廣告，或從網頁上尋找交易對手。

　　(2)派員常駐國外或出國訪問尋找交易對手。

　　(3)在國外貿易專業雜誌或報紙上刊登廣告。

　　(4)在國內貿易專業雜誌或工商年鑑刊登廣告。

　　(5)參加國際商展 (International trade fair)。

　　(6)利用我國設在國外的貿易中心 (Trade centre) 展出產品。

　　2.委託第三者間接物色：

　　(1)透過外匯銀行介紹。

　　(2)函請本國駐在各國的大使館 (Embassy)、領事館 (Consulate) 及其他駐外政府或半官方單位代為介紹。

　　(3)函請駐本國外國大使館、領事館代為介紹。

　　(4)函請國外進出口公會、商會或有關機構代為介紹。

　　(5)委託國外往來客戶代為介紹。

　　(6)函請親友介紹。

　　(7)向 Data bank 訂閱商情報導資料。

二、消極的物色方法

1. 根據國外發行的新聞雜誌廣告，發函尋找客戶。

2. 根據國外發行的工商名錄、進出口商名錄或電話簿發函聯絡。

3. 根據國內機構所發布的貿易機會發函聯絡。

4. 與來訪國外客戶接談。

5. 與來函國外客戶聯絡。

在上述各種方法中，委託第三者介紹交易對象時，難免要與這些人通信，這種通信的內容，因對象的不同而略有出入，但通常多包括下列幾項：

1. 本公司希望與該國客戶往來，請其介紹客戶。

2. 介紹本公司的營業項目。

3. 介紹本公司的優點——組織、經驗、資本。

4. 提供本公司往來銀行供徵信之用。

請人介紹交易對手的信，可能由出口商採取主動（大多數）撰寫，也可能由進口商採取主動撰寫，茲舉例於下：

No. 1　出口商函請國外商會介紹客戶之一

Gentlemen:

　　We are desirous of extending our connections in your country, and shall be much obliged if you will give us a list of some reliable business houses in New York who are interested in the importation of Chinaware.

　　We are old and well-established exporters of all kinds of Chinese goods, especially of Chinese Typical Chinaware, and therefore, confident to give our customers the fullest satisfaction.

　　To justify our confidence in addressing you, we refer you to the following:

The Bank of Taiwan, Head Office, Taipei.

The Taipei Importers and Exporters Association, Taipei.

Your courtesy will be appreciated, and we earnestly await your reply.

Yours very truly,

【註】

　　1. desirous of extending our connections in your country：「擬拓展本公司在貴國的業務關係」，

"connections" 為 "business connections" 或 "business relations" 之意。

2. be much obliged：「很感激」，"to be much obliged to somebody" 為「感激某人」；"to be obliged to+verb" 為 "have to"，"must" 之意。cf. She was obliged to go back to work.（她不得不回去工作。）

3. reliable business house：可靠的商號；殷實商號。

4. who are interested in：對……有興趣（的商號）。

5. in the importation of：輸入，也可以 "in importing" 替代。

6. chinaware：陶瓷器。

7. old and well-established exporters：歷史悠久且殷實的出口商，因為用 "we" 開頭，所以 "exporters" 用多數，假如用 "I" 開頭，就應用 "exporter"。

8. give our customers...satisfaction：使顧客十分滿意。

9. to justify our confidence in addressing you（為向你證明本公司的信用），也可用 "to justify your confidence in us"。

10. typical：獨特的。

11. refer you to：請你向……查詢。

12. your courtesy will be appreciated.（如承惠辦，則不勝感激。）

"appreciate" 為 "to place a sufficiently high value on" 之意，此字本身已含有「深為感激」之意，所以不必再加上 "greatly"，"deeply"，"highly"，"much" 等副詞，但很多人仍加上這些字，這只好當作「禮多人不怪」吧！

No. 1–1　出口商函請國外商會介紹客戶之二

Dear Sirs,

　　Established in 1950, we have been expanding our business operations around the world as a leading exporter and importer of business machines.

　　Now we are planning to incorporate our business activities in your market as general base in Asia. We will be much obliged for your introduction to a most reliable importer handling business machines.

　　Our line of business includes:

　　　Typewriter, Copying Machines, Calculators, Printing Machines, Cash Register, Addressing Machines, etc.

　　Concerning our financial status and reputation, please direct all enquiries to The Bank of Tokyo, The Bank of Osaka, Osaka or The National Cash Register Co., Ltd. in Osaka.

　　Thank you very much for your cooperation. We hope to hear from you soon.

RH/MY   Yours faithfully,

【註】

   1. a leading exporter and importer：「一主要的進出口商」，這裡 "exporter" 及 "importer" 都用單數。但一般而言，多用複數，即 "exporters and importers"，在此場合 "a leading" 的 "a" 應刪除。

   2. incorporate our business activities in your market：「將貴地市場列入我們營業活動範圍」，即向貴地市場拓展市場 (extend our business activities in your market) 之意。

   3. most reliable：最可靠的。

   4. financial status and reputation：財務狀況及聲譽。

No.2　國外商會覆出口商

Gentlemen:

   We take the pleasure of suggesting the following names of firms who may be interested in the goods referred to in your letter of April 10th.

The ABC Company

100 Broadway, New York City

Messrs. Wilson & Co.

50 Franklin Ave., New York City

   While these concerns are of good repute, we, of course, assume no responsibility for them. They can, no doubt, furnish you with references, and information as to their financial standing may be obtained in the usual way through bankers or commercial agencies.

Yours very truly,

_____ Secretary

【註】

   1. suggesting：通知，也可以 "informing you of" 代替。

   2. firm：商號。"concern" 也是商號，一般而言，"firm" 係指合夥商號 (co-partnership)，而 "concern" 則為各種商號商社的總稱。

   3. be interested in：對……有興趣。

   4. the goods referred to：（來函）所示的商品。

   5. of good repute：聲譽良好。

6. assume no responsibility：不負責任。"responsibility" 主要係指 "moral"（道德上）的責任，債務上的責任則稱為 "liability"。

7. furnish...with：供給；提供。

8. reference：備詢人。

9. financial standing：財務狀況（地位）。也可以 "financial status" 或 "financial conditions" 代替。

10. in the usual way：照例；通常的方法。

11. commercial agencies：徵信所。

No. 3　出口商覆謝

Gentlemen:

　　We have just received your letter of May 20 giving us the names of reliable importers of Chinaware in your city.

　　Please accept our sincere thanks for your information which will afford us a great assistance in forming mutually profitable trade relation with them.

Yours very truly,

【註】

1. 請人幫忙，而別人給予幫忙後，應發出答謝函 (letter of thanks)，注意 "thanks" 是多數。

2. sincere thanks：有誠意的感謝，即「謝忱」。

3. information（無多數形，不加 s）：消息；報告。

4. afford = give。

5. a great assistance：大有助益；很大的幫助。

6. forming mutually...with：與……結成彼此有利的交易關係。

No. 4　出口商函請駐本地外國總領事介紹客戶

Sir:

　　Would you be kind enough to give us the names and addresses of merchants and agents being Panama citizens, and any business organizations established by your countrymen, now having their offices in Taipei or Kaohsiung?

　　If possible, we should like to know what lines of business they are now carrying on.

　　Your early response will be appreciated.

Yours respectfully,

(Signed)

【註】

1. Sir：因為本信是寫給 Consul general，所以用單數，而不用多數的 "Sirs"。

2. Would you be kind enough to give us...?「可否示知……?」，也可以下面句子代替：

    a. We shall be much obliged, if you will give us...

    b. We shall be glad, if you would give us...

3. Panama citizens：巴拿馬公民。

4. business organizations：公司行號。

5. we should like to know：「我們想知道」;「請賜知」。

6. if possible：如屬可能 = If it is possible。

7. lines of business：營業種類，也可用 "lines of goods"（經營商品）。

8. carry on=engage in：「從事於……」。

9. yours respectfully：「謹上」，對官員或官署時用此詞。

No. 5　領事覆函

Dear Sirs:

　　I have received your letter of January 23, 20– requesting the names and addresses of Panama merchants and business organizations having offices in Taipei or Kaohsiung.

　　In reply to your request there is enclosed a list of American firms and individuals having offices in the two cities you named.

Very respectfully yours,

＿＿＿＿＿＿＿＿

(Consul)

【註】

1. in reply to your request：「為覆貴公司請求」;「應貴公司要求，茲覆……」。

2. there is enclosed：隨函附上，也可用下面的句子代替：

    I enclosed...

    Enclosed please find...（舊式）

    Enclosed you will find...（舊式）

3. firms and individuals：商號及個人。

4. you named=you mentioned：你所提及的。

No. 6　進口商函請報社、雜誌社刊登介紹之一

Dear Sirs,

　　We are one of the leading importers and dealers with long experience in international trading in all kinds of building materials, hand tools, hardware, readymade garments, sporting goods, hats & caps, watches and bands, umbrellas, footwear, imitation jewelry, stationery goods, sunglasses, handbags, electrical goods in Nigeria and we are seeking reliable suppliers of these items in your country.

We would appreciate your publishing our name in your newspaper so that any interested exporters and manufacturers in your country can contact us.

Our usual purchasing terms are full letter of credit, with each order, and our bankers are The Bank of America, Lagos, from whom you will be able to obtain all the information you may require in regard to our business integrity and financial standing.

Your help in this matter will be appreciated.

Faithfully yours,

【註】

1. leading importers and dealers：主要的進口商及買賣商，"dealers" 即 "traders"。

2. trading in...：經營……（某種商品）。

3. seeking：尋找；物色。

4. items：項目；商品，可以 "goods"，"commodities"，"merchandise"，"articles" 等詞代替。

5. publishing our name in your newspaper：將我們（本公司）的名字刊登在貴報紙上。類似措詞有：

   placing our name in your publication（出版物）

   putting our name in your magazine（雜誌）

   advertising our name in your bulletin（刊物；公報）

6. interested exporters：有意的出口商；感興趣的出口商。

7. business integrity：業務上的誠實。意指做生意是否誠實。

8. 請報社、雜誌社刊登介紹客戶時，有些需付刊登費用，有些不必付這種費用。一般而言，推廣貿易機構所發行的刊物，都設有「貿易機會」(Trade opportunities) 的專欄，如請其在此專欄刊登，通常都不收費。

No. 7　出口商函請報社、雜誌社刊登介紹之二

Dear Sirs,

We have the pleasure of introducing ourselves to you as one of the most reputable electronics exporters in Taiwan, who has been engaged in this line of business since 1965; particularly, we have been enjoying a good sale of TV sets and Transistor Radios and are now desirous of expanding our market to your district.

We shall, therefore, appreciate it if you will kindly introduce us to the relative importers by announcing in your publication as follows:

"An Export Company of Electronics is now making a business proposal for TV sets and Transistor Radios which are said to have built a high reputation at home and abroad. Contact them by addressing your letter to..."

We solicit your close cooperation with us in this matter.

Faithfully yours,

【註】

1. "we have the pleasure of..." 以新式的 "we are pleased to..." 代替較佳。cf. we have the liberty of...：冒昧地……。

2. introducing ourselves to you as...：謹向您介紹我們（本公司）為……。

3. reputable：信譽良好的。

4. line of business：經營項目。

5. expand our market：拓展本公司市場。

6. relative importers：相關的進口商。

7. announcing in your publication：在貴刊物中發布……。

8. making a business proposal for：提議做……的生意。

9. are said：據說。這種用語有「不負責」之含義。

10. at home and abroad：在國內外。

11. we solicit...：關於此事懇請惠予密切協助，也可以 "your close cooperation with us in this matter will be appreciated..." 代替。

No. 8　我國某銀行對函介顧客的定型覆函 (Form letter)

Dear Sirs:

Your Trade Inquiry dated May 10, 20–

We are pleased to note the trade opportunity directed through us for introduction to interested parties in our country. As our customer files are not precisely sorted to make possible listing of names and addresses of firms likely responding to this particular offer, we feel it better to release the trade information to public attention with the hope that some parties unknown to us may be aroused to react whose credit standing we shall be glad to report upon your future request.

Accordingly, covered under the copies of this letter, your cases (with enclosures, if any) is sent to:

1) China External Trade Development Coucil (CETRA)

333 eighth Floor, Keelung Road, Taipei

2) Central Trust of China, Trading Dept. (CTC)

49 Wu Chang Street, Section 1, Taipei

The former is a non-profit trade promotional institution founded in July, 1970 by government and traders/manufacturers associations serving as trade information center and export marketing consulting agency. The latter is a government agency, one of her authorized functions is to purchase supplies for government organization and public enterprise.

We are sure that CETRA and CTC will write to you (or your customer) by listing names and addresses of recommendable firms and disseminating the trade information through their various publishing media.

It is our wish that our service in this manner will lead to an establishment of trade relationships and that you will give us continued inquiries.

Yours very truly,

【註】

1. trade opportunity：貿易機會。

2. interested parties：有關的人；感興趣的人。

3. release：發布。

4. CETRA：中華民國對外貿易發展協會。

5. CTC：中央信託局。

6. non-profit trade promotional institution：非營利的貿易拓展機構。

7. founded in July, 1970：創設於 1970 年 7 月。

8. government agency：政府機構。

9. supplies：物資。

10. public enterprise：公營事業。cf. state-owned enterprise：國營事業。

11. recommendable firms：值得推介的行號。

12. disseminating：傳播；散布。

13. media：媒體。publishing media 可譯成「出版物」。

No. 9　出口商對於介紹顧客的謝函

Dear Sirs,

We have received with many thanks your letter of May 20, and wish to express our sincere gratitude for your kindness in publicizing our wish in your "Trade Opportunity".

We believe the arrangement you kindly made for us will connect us with some prospective buyers and bring a satisfactory result before long.

We thank you again for your taking trouble and wish to reciprocate your courtesy sometime in the future.

Yours very truly,

【註】

1. wish to express our sincere gratitude：謹致謝忱。

2. publicizing：廣為宣傳，以 "publishing" 代替較妥。

3. will connect us with...：使我們與……取得聯繫。

4. prospective buyers：未來的買主；可能的買主。

5. before long：不久，可以 "soon" 代替。

6. taking trouble：費神。本句不如改為 "We thank you again for your assistance and wish to..."。

7. reciprocate：報答；回報。

第二節　有關尋找交易對象的有用例句

一、開頭句

1. $\left\{\begin{array}{l}\text{As we plan}\\\text{In order}\end{array}\right\}$ to extend our business $\left\{\begin{array}{l}\text{connections in}\\\text{to}\end{array}\right\}$ your

→ $\left\{\begin{array}{l}\text{country}\\\text{city}\\\text{area}\\\text{district}\\\text{market}\end{array}\right\}$ $\left\{\begin{array}{l}\text{would you please send us a list of}\\\text{we request you to introduce to us}\end{array}\right\}$ some reliable $\left\{\begin{array}{l}\text{business houses}\\\text{firms}\end{array}\right\}$ in

→ your $\left\{\begin{array}{l}\text{country}\\\text{city, etc.}\end{array}\right\}$ who are interested in $\left\{\begin{array}{l}\text{importing}\\\text{exporting}\\\text{handling}\end{array}\right\}$ $\left\{\begin{array}{l}\text{Christmas tree light bulbs.}\\\text{household lamps.}\\\text{porcelain.}\end{array}\right\}$

$$
\left\{\begin{array}{l}\text{茲因計畫拓展本公司在}\\ \text{茲為拓展}\end{array}\right\}
貴
\left\{\begin{array}{l}\text{國}\\ \text{市}\\ \text{地方}\\ \text{地區}\\ \text{地市場}\end{array}\right\}
\text{的業務關係}
\left\{\begin{array}{l}\text{盼能提供}\\ \text{擬請推介}\end{array}\right\}
貴
$$

$$
\rightarrow
\left\{\begin{array}{l}\text{國}\\ \text{市}\\ \vdots\end{array}\right\}
\text{若干有意}
\left\{\begin{array}{l}\text{進口}\\ \text{出口}\\ \text{從事}\end{array}\right\}
\left\{\begin{array}{l}\text{耶誕樹裝飾燈泡}\\ \text{家用燈泡}\\ \text{瓷器}\end{array}\right\}
\text{的可靠}
\left\{\begin{array}{l}\text{商家}\\ \text{商號}\end{array}\right\}
。
$$

2.
$$
\left\{\begin{array}{l}\text{We would be very grateful to you}\\ \text{We shall appreciate it}\\ \text{We would be greatly obliged}\end{array}\right\}
\text{if you}
\left\{\begin{array}{l}\text{will}\\ \text{would}\end{array}\right\}
\text{kindly}
$$

$$
\rightarrow \text{introduce to us}
\left\{\begin{array}{l}\text{proper}\\ \text{reliable}\\ \text{relative}\end{array}\right\}
\text{firms with whom we could}
$$

$$
\rightarrow \text{establish a business relationship for the}
\left\{\begin{array}{l}\text{export}\\ \text{import}\\ \text{import and export}\end{array}\right\}
\text{trade business.}
$$

$$
\text{如蒙推介本公司可與他們建立}
\left\{\begin{array}{l}\text{出口}\\ \text{進口}\\ \text{進出口}\end{array}\right\}
\text{貿易業務關係的}
\left\{\begin{array}{l}\text{適當}\\ \text{可靠}\\ \text{相關}\end{array}\right\}
\text{商號}
$$

$$
\rightarrow \text{本公司}
\left\{\begin{array}{l}\text{當感謝你}\\ \text{當感激不盡}\\ \text{當非常感激}\end{array}\right\}
。
$$

3.
$$
\left\{\begin{array}{l}\text{We are desirous of expanding our market to}\\ \text{We are}\left\{\begin{array}{l}\text{seeking}\\ \text{looking for}\end{array}\right\}\text{new business connections in}\end{array}\right\}
\text{your}
\left\{\begin{array}{l}\text{district}\\ \text{area}\\ \text{city}\\ \text{country}\end{array}\right\}
\text{and would}
$$

$\rightarrow$ appreciate your $\begin{Bmatrix} \text{supplying} \\ \text{furnishing} \\ \text{providing} \end{Bmatrix}$ us with the names and addresses

$\rightarrow$ of $\begin{Bmatrix} \text{manufacturers} \\ \text{firms} \\ \text{importers} \\ \text{exporters} \\ \text{concerns} \end{Bmatrix}$ who are interested in $\begin{Bmatrix} \text{garments.} \\ \text{personal computers.} \\ \text{electronics.} \end{Bmatrix}$

二、自我介紹

1. We are one of the $\begin{Bmatrix} \text{most reputable} \\ \text{well organized} \\ \text{experienced} \\ \text{leading} \end{Bmatrix}$ trading firms in Taiwan, who have been

$\rightarrow$ engaged in $\begin{Bmatrix} \text{TV sets} \\ \text{garments} \\ \text{mobile phones} \end{Bmatrix}$ business since 1995.

2. We have close connections with outstanding manufacturers of various kinds of general merchandise, and are very well-placed to supply our customers with high grade

goods at $\begin{Bmatrix} \text{competitive} \\ \text{reasonable} \\ \text{moderate} \end{Bmatrix}$ price.

註：a. outstanding：著名的；b. general merchandise：雜貨；c. well-placed：居於有利的地位；方便。

3. Our firm was $\begin{Bmatrix} \text{established} \\ \text{founded} \end{Bmatrix}$ in 1950 and has enjoyed an excellent reputation for these

$\rightarrow$ $\begin{Bmatrix} \text{lines of product.} \\ \text{items.} \\ \text{lines of business.} \end{Bmatrix}$

4. We are $\begin{Bmatrix} \text{exporters} \\ \text{importers} \end{Bmatrix}$ of ready made textiles for ladies, men and children.

5. We are specially interested in the supply of the following goods.

6. We can supply a wide range of products, especially tools and other hardware items.

　註：a wide range of products：廣泛的產品。

7. Being one of the biggest manufacturers and suppliers of...in Taiwan, we are confident to give full satisfaction to your people.

　註：are confident to give full satisfaction to your people：深信必能令貴國人稱心滿意。

8. As we are long-established exporters of all kinds of porcelain, we have close connections with the leading manufacturers here.

三、提供備詢人

1. To justify your confidence in us, we refer you to the following references:

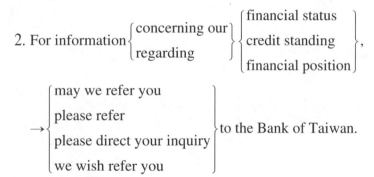

2. For information $\begin{Bmatrix} \text{concerning our} \\ \text{regarding} \end{Bmatrix}$ $\begin{Bmatrix} \text{financial status} \\ \text{credit standing} \\ \text{financial position} \end{Bmatrix}$,

$\rightarrow \begin{Bmatrix} \text{may we refer you} \\ \text{please refer} \\ \text{please direct your inquiry} \\ \text{we wish refer you} \end{Bmatrix}$ to the Bank of Taiwan.

四、請刊登介紹

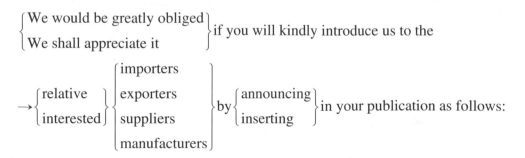

$\begin{Bmatrix} \text{We would be greatly obliged} \\ \text{We shall appreciate it} \end{Bmatrix}$ if you will kindly introduce us to the

$\rightarrow \begin{Bmatrix} \text{relative} \\ \text{interested} \end{Bmatrix}$ $\begin{Bmatrix} \text{importers} \\ \text{exporters} \\ \text{suppliers} \\ \text{manufacturers} \end{Bmatrix}$ by $\begin{Bmatrix} \text{announcing} \\ \text{inserting} \end{Bmatrix}$ in your publication as follows:

五、結尾句

1. We solicit your close cooperation with us in this matter.

2. Your $\begin{Bmatrix} \text{assistance} \\ \text{help} \end{Bmatrix}$ in $\begin{Bmatrix} \text{this} \\ \text{above} \end{Bmatrix}$ matter will be appreciated.

3. Your courtesy and $\begin{Bmatrix} \text{early} \\ \text{prompt} \end{Bmatrix}$ reply will be much appreciated.

　　註："much appreciated" 中的 "much" 是多餘的。

第九章
招攬交易 (Trade Proposal)

第一節　招攬信的寫法

經由各種途徑找到可能的交易對手時，即應作成記錄，以便發出招攬信 (Letter of proposing business)，提議建立業務關係。這種招攬信的內容，一般而言，包括下列各項：

1. 獲悉對方的途徑：如由銀行、外貿協會、進出口商名錄。
2. 表示願意與其建立業務關係：這是招攬信的重心。
3. 自我介紹及經營商品詳情。
4. 交易條件的扼要說明：主要是說明付款的方法。
5. 提供信用備詢人。

如招攬信是由賣方主動發出者，可檢附商品目錄 (Catalog)、價目表 (Price list)，甚至另寄樣品 (Sample)。凡此，均應在信中提及。

一、由出口商發出的招攬信

No.l0　由商會獲悉

Gentlemen:

Your name was given through the...Chamber of Commerce as a firm doing an extensive trade in...（商品）and we take the liberty to write you with the earnest desire of having the opportunity to enter into business relation with you.

For more than twenty years we have been exporting Chinese...（商品）and shipped considerable quantities to your country; but have not had the pleasure of doing any of this business with you. We therefore look forward with much interest to having the privilege of your requirements.

We are well organized with experienced men who have a thorough knowledge of the requirements and the taste of your market. In addition to this, our close connections with many of the leading factories and our investigation into the foreign traffic situation place us in a position to render the fullest satisfaction to our customers.

　　As to the terms of payment, we usually request our customers to open a letter of credit in our favor, under which we will draw at 60 days after sight.

　　In order to show you the excellence of our goods, we are sending you under separate cover our latest descriptive and illustrated catalogue. Our price-list is enclosed into this letter.

　　For any information you may desire in regard to our standing, we are pleased to refer you to the following bank:

(Bank's Name)

　　If you are inclined to entertain our proposal mentioned above, please give us your definite inquiries upon which we shall be pleased to send you our best prices and deliveries.

Yours faithfully,

【註】

1. We take the liberty to write...：「冒昧地寫信」。也可寫成：

    We take the liberty of writing...

    We take liberty in writing...　（注意："liberty" 前面沒有 "the"）

2. to enter into business relations：開始業務聯繫；建立交易關係。

3. considerable = large; no small。

4. look forward to：期待，"to" 是介系詞。因此應接 gerund 或名詞。

5. having the privilege of your requirements：惠請賜顧。

6. a thorough knowledge：精通。

7. the taste：嗜好。

8. to render = to give。

9. terms of payment：付款條件。

10. under which：憑該信用狀之意。

11. under separate cover：另寄（補充說明請看 p. 80）。

12. descriptive and illustrated catalogue：附有說明及插圖的目錄。

13. standing：身分；信用情況。

14. inclined to entertain：俯允；有意接受。

15. mentioned above = above-mentioned。

16. best prices：最克己的價格。

17. deliveries = delivery terms：交貨條件。

No. 11　由外匯銀行獲悉

Dear Sirs,

We have lately requested our Bankers, Bank of Taipei, Taipei, to give us names of reliable firms in London. They have recommended to us your firm in such high terms that we are very anxious to enter into relations with you in our soft goods.

Our firm has been in existence upwards of forty years as Exporters of Silk Piece Goods, in which, we are told, you are specially interested. Our Silks cover many different items, but the under-mentioned are some of our chief exports:

Pongee Silk, Habutae, Crepe de Chine,

Tussore Silk Crepe, etc.

In addition to our specialities we have in recent years been shipping Cotton and Nylon Piece Goods, in which we have done a considerable business with some of the reliable houses on your side up to the present. It is, however, in these Silk Piece Goods that we wish to increase our business with your support and cooperation. The special facilities we have as shippers and our thorough familiarity with the tastes and requirements of your market place us in a favourable position to render you our services to your best advantage. Should you fall in with our proposal and make inquiries for your specific needs, we shall lose no time in sending you samples along with prices and pertinent details.

In reply to inquiries and in making offers we shall quote prices CIF London or FOB Keelung in Pound Sterling, whichever your prefer. Our usual terms are to draw a draft at 60 d/s under a Confirmed Banker's Letter of Credit to cover each order placed with us.

References: For information as to our financial stability, we would refer you to the following:

Bank of Taipei, Taipei

Bank of Taiwan, Taipei

We trust that we may soon have the pleasure of opening up business with your respectable house and of proving that such a connection would yield you a considerable profit, while assuring you of our exceptionally careful and prompt attention to your interests.

Yours faithfully,

【註】

1. in such high terms：如此鄭重地。這裡的 "terms" 為「措詞」(mode of expression) 之意。

   cf. He referred to your work in terms of high praise.（他對你的工作大加讚揚。）

2. enter into relations：開始交易關係。相似的用例有：

   to enter into (business) negotiations with...

to enter into connections (= connexions) with...

to enter into correspondence with...

to open a correspondence with...

to open an account with...

3. soft goods：織物類貨品。

4. in existence：存在，即成立之意。

5. Pongee Silk：府綢紗。

　Crepe de Chine：縐紗。

　Tussore Silk Crepe：山東綢紗。

6. speciality 又寫成 "specialty"，特產品；特製品。

7. shipping：運出，即銷出之意。

8. on your side：在你那邊，即貴地，又可寫成 "at your end"。

9. fall in with our proposal：同意我們的提議。

10. lose no time：馬上。

11. pertinent details：相關的細節。

12. at 60 d/s = $\begin{cases} \text{at sixty days after sight} \\ \text{at sixty days' sight} \end{cases}$：見票後 60 天付款，"d/s" 為 "days after sight" 或

"days' sight" 的縮寫。"at" 表示期限，例如 "at sight" 見票即付。

13. under = according to。

14. cover：抵付；支付；清償。此字在商業上意義很特殊，茲舉例說明其用法：

　a. In order to *cover* our order, we have arranged with The Bank of Tokyo a credit for U.S.$1,000.00 (a credit = a letter of credit).

　b. We enclose a cheque for U.S.$10,000 to *cover* your invoice of the 30th May.

15. yield：產生。

16. 本信最後一段係從收信人利益為著眼，收信人看了之後，一定很高興。這就是 "you attitude" 的寫法。須知，"You attitude" 未必以 "You" 開頭。只要以收信人的利益為前提所寫的信，就符合 "You attitude" 的原則。

No. 12　由刊物獲悉

Gentlemen:

　In a recent issue of the "*Foreign Trade*", we saw your name listed as being interested in making certain purchases in this country.

　We take this opportunity to place our name before you as being a buying, shipping and forwarding agent. If you do not have anyone here to look after your interests in that capacity, we should be glad if you give us your kind consideration.

We inform you that we have been engaged in this business for the past 20 years. We, therefore, feel that because of our past years' experience, we are well qualified to take care of your interests at this end.

Further, as for references, we can give you the names of some concerns in your country and our bankers are The New York State Bank, 20 Wall St., New York, NY 10023.

We look forward to receiving your reply in acknowledgement of this letter and with thanks in advance.

Yours faithfully,

【註】

1. recent issue of the "*Foreign Trade*"：最近的「對外貿易」刊物。

2. being Interested in...this country：有意在本國採購某些貨物。

3. We take this opportunity to place our name...agent.（本公司冒昧向貴公司自我介紹擬代理貴公司在此地之採購、裝船及轉運業務。）

4. we should be glad if you give...consideration：如惠予考慮，不勝感禱。

5. engaged in this business：從事此項業務。（必須用被動語態動詞）

6. we are well qualified to take care of your interests at this end：本公司深信在此方面具有為貴公司提供服務的資格。

7. as for references：至於備詢人。

8. look forward to receiving your reply：盼望能收到貴公司的回音。

9. in Acknowledgement...in Advance：承認收到本信並在此預先致謝。

No. 13　由朋友推介

Gentlemen:

Mr. Charles Chang of Bob AG. Frankfurt, commended your firm to us as one of the largest importers and exporters in Germany.

It so happens that we are not represented in your country at the present time, and we are very much interested in having an active and reputable agent there.

As for ourselves, we are a leading and old established firm of exporters, and we are in a very good position to supply various grades of canned mushroom at competitive prices and for good delivery, and we should be pleased to send you our offers by telex or airmail upon receipt of your specifications covering any grades of mushroom in which you are interested at present.

We also are interested in the import of German cameras and if you are in a position to offer us anything along these lines, we should be pleased to have you send us your quotations, with all perinent information and with covering samples.

We are looking forward to receiving your reply, and hope we shall be able to establish a connection between our two firms which will be pleasant and mutually beneficial.

Yours very truly,

【註】

1. commend：推薦，即 "recommend"。

2. it so happens：適巧；恰好。

3. we are not represented：我們無代表。

4. active：活躍的。

5. as for：至於，後面接人，"as to" 後面接事物。

   例如：As to the reason of delay...

6. old established：歷史悠久的。

7. in a very good position：在很好的位置，意指「很可以」。

8. grades：等級。

9. canned mushroom：罐頭洋菇。

10. good delivery：交貨迅速。

11. specifications：規格。

12. in a position = able：能。

13. mutually beneficial：互惠；雙方均有利。

No. 14　未說明由何處獲悉

Dear Sirs,

As we are given to understand that you are interested in the Taiwan textiles, we take this opportunity of introducing ourselves as a reliable trading firm, established 10 years ago and dealing in the articles ever since with fair record especially with the Southeast Asian countries.

In order to give you an idea we quote some of them without engagement as follows:

Sleeveless shirts, cotton & sythetic fibres mixed,

white or colored, US$15 per doz. CIF your port.

We are sending you under separate cover our catalogues and free samples. Please give us your specific inquiries upon examination of the above as we presume they will be received favorably in your market.

We also handle...(name of goods)...which are selling well in Latin America.

As to our standing, please refer to the following banks:

Bank of Taiwan, Taipei
Central Trust of China, Taipei

We hope to be of service to you and look forward to your comments.

Yours faithfully,

【註】

1. we are given to understand：茲獲悉。

2. take this opportunity of...：藉此機會（向貴公司自我介紹本公司為可靠……）。

3. dealing in：經營。

4. ever since：自……以來。

5. with fair record：獲致良好成績。

6. in order...idea：茲為給予貴公司獲得概念。

7. sleeveless shirts：無袖襯衫。

8. cotton & sythetic fibres mixed：棉與人纖混紡。

9. free sample：免費樣品。cf. The sample is sent gratis.（樣品免費贈送。）

10. presume...favorably：相信必能獲得好評。

11. handle：經銷。

12. sell well：暢銷。

13. as to our standing：關於敝公司的地位，即指關於本公司的 "reputation"。

14. to be of service to you：對你有所助益；能幫助你。

15. your comments：你的教言；你的評語。

二、由進口商發出的招攬信

No. 15 由商會獲悉

Gentlemen:

We are indebted to the Manila Chamber of Commerce for your name and address as one of the respectable concerns, offering Rubber Products of all descriptions.

We are not only buying for our own account, but also booking indent orders for our numerous clients. We are interested in general merchandise, but look specially for the supply of Rubber Products, such as Beltings, Hoses, Tires and Tubes, Athletic Shoes, Shoe Soles, etc.

Being extremely desirous of establishing business relationship with your house, we invite you to offer us the above-mentioned items and send us a range of samples, together with illustrated literature, so that we may scout the market. In indent business, however, we hope you will include in your offer our commission of 3% on CIF cost. Any overprice that we may be able to secure from our buyers is to be understood for our account.

Terms: Irrevocable letter of credit.

You may inquire our business standing and integrity through the China Banking Corporation, and the Kian Lam Finance and Exchange Corporation, both of Manila.

With nothing further by this mail, we look forward to your reply at your earliest convenience.

Sincerely yours,
WRIGHT & CO.
John Smith
Mgr.

JS: A

【註】

1. We are indebted to...：本公司承蒙……。

2. respectable：有聲望的。

3. buying for our own account：自行買進；非代理。

4. indent orders：受託訂購。

5. Beltings：調帶（機械用），Hoses：軟管，Shoe Soles：靴底。

6. illustrated literature：有插圖的說明書。"literature" 又解作「廣告、宣傳用的印刷品」(printed matter)。

7. overprice = overage：溢價。

8. for our account：算我們的收入。

No. 16　由外匯銀行獲悉

Dear Sirs:

We owe your name to The Foreign Department of Bank of Canton, through whom we learned that you are the manufacturers of Textiles, Piecegoods and other General Merchandise, and also that you are Importers & Exporters.

May we introduce ourselves as Importers of all General Merchandise, Exporters of Taiwan Produce, and Manufacturers' Representatives, and Commission Agents.

We have been in business since 1963, and can boast of having vast and wide experience in all the lines we handle.

Our bankers are Bank of Canton, and The Hongkong & Shanghai Banking Corporation of Hong Kong, from whom you will be able to obtain all the information you may require in regard to our business integrity and financial standing.

Will you please let us know your trade terms, etc., and forward samples and other helpful literature, with a view to getting into business in near future.

We hope that this letter will be a forerunner to many years of profitable business to both parties, and look forward to the pleasure of hearing from you.

Yours faithfully,

【註】

1. We owe your name to...：從……得悉貴公司名字。

2. piecegoods：布疋。

3. general merchandise：一般貨品；雜貨。

4. manufacturers' representatives：廠家的代表。

5. commission agents：佣金代理商。

6. can boast of：敢自詡。

7. trade terms：交易條件，這裡主要指付款條件。

8. helpful literature：有助於交易的宣傳品，注意 "literature" 無多數形。

9. forerunner：開端。

No. 17　由商會獲悉

Dear Sirs,

Your name has been listed by the Hong Kong Chamber of Commerce as one of the leading dealers of business machines in your city and a negotiating firm for establishing a base sales point with reliable dealers.

As the enclosed pamphlet shows, we have been specializing in your commodities for over twenty years. Our wide experience in these lines will surely add to your line of business. Moreover, our broad range of business affiliations with many companies in surrounding cities places us in a superior competitive position.

We would like very much to open an account with you, and hope that a mutually agreeable consideration of interests can be arranged.

Enclosed is a sample letter of credit under which a draft may be drawn at sight.

For further information as to our standing, please refer to The Hong Kong & Shanghai Bank, Hong Kong.

We are looking forward to your prompt reply.

Yours faithfully,

【註】

1. a negotiating firm for establishing a base sales point：為建立銷售據點而正在交涉中的商號。

2. specializing in your commodities：專門經營你們這一類商品。

3. in these lines：在這一類項目。

4. add to your line of business：擴大貴公司的營業項目。

5. in a superior competitive position：處於優越的競爭地位。

No. 18　由報紙、刊物獲悉

Dear Sirs,

Your name has been listed in the "*Traders' Express*" published by the China External Trade Development Council as one of the leading exporters of electronic equipment and a firm seeking the possibility of promoting your business in Taiwan.

As the enclosed pamphlet shows, we have been specializing in electronic equipment for over 10 years. Our experience in this line will surely add to your line of business. Furthermore, our broad range of business connections with many firms in Taiwan places us in a very competitive position.

We, therefore, feel that we are well qualified to take care of your interests at this end, and would like very much to open an account with you.

As for references, we can give you the names of some concerns in your country and also our bankers are Bank of Taiwan, Taipei.

We are looking forward to your early reply.

Yours faithfully,

【註】

1. has been listed in：「刊登在……」，也可以下列句子代替：

    a. has been placed in...

    b. has been published in...

2. *Traders' Express*：「貿易快訊」，係由外貿協會出版的日刊。

3. promoting your business：推廣你的業務。

4. this line = this kind of commodity。

5. add to：增加；擴大。

6. to open account with：開立往來帳戶，即建立業務關係。

No. 19　由朋友處獲悉

Gentlemen:

　　Having come to understand from Messrs. Smith & Baker, 100 Montgomery St., San Francisco, of whom, we presume, you are aware, and with whom we are regularly connected, that you are one of the major exporters of Iron and Steel products of all descriptions, we take the liberty of writing to ask if you are in a position to open an account with us.

　　As importers of Scrap Metals, Tin Plates, etc., we have a successful history of thirty years and flatter ourselves that we have efficient sales force and sufficient capital to finance imports on any scale whatever. In recent years our industry has made a remarkable growth in all of its branches, with the results that the demand for these commodities is on the increase, and therefore, we are confident of our being able to introduce your lines in a large quantity in this market and elsewhere. Should you favorably consider our proposal, we shall appreciate hearing from you as early as possible your chief items of export and their quantities available for immediate shipment.

　　As far as payment is concerned, we are prepared to open any type of the letter of credit you desire.

　　As regards our standing and reputation, our Bankers, Bank of Taiwan, Tainan Branch, and The First Bank of Taiwan, Tainan will be pleased to furnish you with any information you may require.

　　We await your prompt reply with keen interest.

Sincerely yours,

【註】

1. flatter ourselves：自誇；自詡。

2. sales force：推銷員陣容；force = staff。

3. on any scale：任何規模。

4. on the increase：增加中。

5. furnish you with: 提供，同義詞尚有：

    a. provide you with

    b. supply you with

No. 20　由工商名錄獲悉

Dear Sirs,

    We owe your name and address to Kelly's Directory, from which we understand that you are general exporters of Taiwan products.

    Therefore, we take the opportunity to introduce ourselves as a reliable firm and inform you that we are very interested in establishing trade relations with your goodselves for the sale of products from your end. we are mainly dealing as commission agents with very wide clientele and therefore it should be understood that if we succeed in securing business, we shall be considered as your "Sole Agents" and you will not deal direct with any of our clients except through our medium.

    We are, at present, very keen on your offer for the following:

Canned Pineapples

Textiles

Chemicals and Fertilizers

    Please let us have your quotations in Sterling currency on basis of $CIFC_5$ Port Sudan.

    We are looking forward to your early news.

Yours faithfully,

【註】

1. clientele [kliːənˊtel]：集合名詞，指顧客而言。

2. sole agents：獨家代理。

3. through our medium：透過我們。

4. very keen on = very interested in：對……很有興趣，eager：渴望。

5. $CIFC_5$：C_5 是指回佣 (return commission) 5% 之意。至於 5% 究係按 CIF 計算或按 $CIFC_5$ 計算並不清楚。因此，為避免糾紛，較謹慎的出口商在報價時多另加說明。例如：The price includes your commission 5% on CIF basis. 但一般而言，多以 FOB 為計算回佣較妥。

第二節 有關招攬交易的有用例句

一、開頭句

1. Your $\left\{\begin{array}{l}\text{good firm}\\\text{name (and address)}\end{array}\right\}$ has (have) been $\left\{\begin{array}{l}\text{recommended to us}\\\text{brought to our attention}\\\text{given to us}\end{array}\right\}$

    → by $\left\{\begin{array}{l}\text{the Hongkong Chamber of Commerce}\\\text{the American consulate at...}\\\text{The Asian Trade Center in Taiwan}\\\text{The China Trade Mission in your city}\\\text{the China External Trade Development Council}\\\text{Messrs. Johnson \& Co.}\\\text{by our bank, City Bank}\\\text{a certain reliable source (某一可靠來源)}\end{array}\right\}$

    → as $\left\{\begin{array}{l}\text{a large exporter}\\\text{large exporters}\\\text{one of the leading exporters}\\\text{one of the greatest importers}\\\text{a supplier}\\\text{a dealer}\\\text{a maker}\\\text{a manufacturer}\\\text{a distributor (經銷商)}\end{array}\right\}$ of...(goods)...

2. We heard from the Chamber of Commerce in your city that you are in the market for...

    註：in the market for...：想買……。

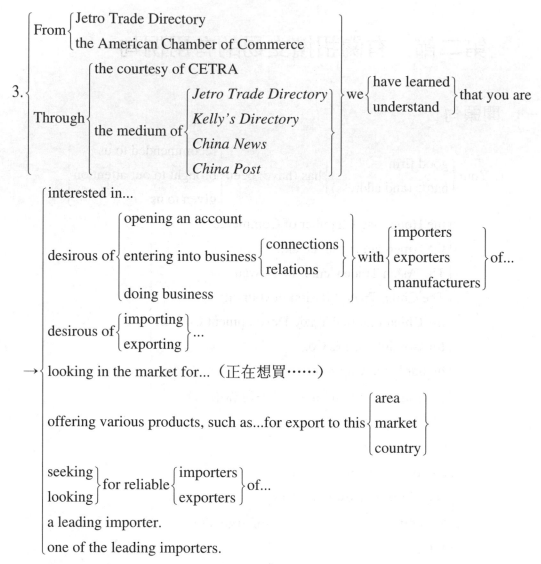

3.
From { Jetro Trade Directory / the American Chamber of Commerce }
Through { the courtesy of CETRA / the medium of { Jetro Trade Directory / Kelly's Directory / China News / China Post } }
we { have learned / understand } that you are

→ {
interested in...

desirous of { opening an account / entering into business { connections / relations } / doing business } with { importers / exporters / manufacturers } of...

desirous of { importing / exporting }...

looking in the market for...（正在想買……）

offering various products, such as...for export to this { area / market / country }

{ seeking / looking } for reliable { importers / exporters } of...

a leading importer.

one of the leading importers.
}

註：a. have learned：聞悉；b. understand：獲知。

二、自我介紹

可參考第八章第二節，這裡再列舉若干供參考。

1. We are well established in manufacturing...(商品)...to which we have devoted years of experience and research. You may rest assured that whatever we offer you excels in workmanship and function.

註：a. devoted：致力於……；b. you may rest assured：保證你。

2. We $\begin{Bmatrix} \text{trust} \\ \text{believe} \end{Bmatrix}$ that our experience in foreign trade and intimate knowledge of international market conditions will entitle us to your confidence.

註：entitle us to your confidence：足以獲得你的信賴。

3. We believe that many years of experience which we have had in this business world and the means at our disposal are sufficient for us to undertake any orders from you.

註：means at our disposal：可供運用的資力。

4. We have a show-room in our office which is very suitable for display of the samples we are requesting you to send us.

三、付款條件

1. As to the terms of payment, we trade on $\begin{Bmatrix} \text{CWO} \\ \text{CAD} \\ \text{D/P} \\ \text{D/A} \\ \text{L/C} \end{Bmatrix}$ basis.

2. Our usual terms are to draw a draft at sight under a banker's irrevocable L/C.

3. We make it our custom to trade on L/C to be opened concurrently with the placing of an order.

註：concurrently with：與……同時。

4. Our terms are shipment within 60 days after receipt of L/C.

四、結尾用語

1. If you are $\begin{Bmatrix} \text{inclined to entertain} \\ \text{interested in} \end{Bmatrix}$ our proposal mentioned above,

→ please $\begin{Bmatrix} \text{give us} \\ \text{let us know} \\ \text{let us have} \end{Bmatrix}$ your definite inquiries upon

→ which we shall be $\begin{Bmatrix} \text{pleased} \\ \text{glad} \end{Bmatrix}$ to $\begin{Bmatrix} \text{send} \\ \text{fax} \\ \text{e-mail} \end{Bmatrix}$ you our best $\begin{Bmatrix} \text{prices and deliveries.} \\ \text{terms and conditions.} \end{Bmatrix}$

2. We $\begin{Bmatrix} \text{trust} \\ \text{hope} \end{Bmatrix}$ that we may $\begin{Bmatrix} \text{soon} \\ \text{in the near future} \end{Bmatrix}$ have the

→ pleasure of $\begin{Bmatrix} \text{doing business} \\ \text{opening up} \\ \text{opening an account} \end{Bmatrix}$ with $\begin{Bmatrix} \text{you} \\ \text{your reputable house} \\ \text{your firm} \end{Bmatrix}$ and proving that

such a connection would yield you a considerable benefit.

3. We are looking forward to $\begin{Bmatrix} \text{receiving your reply} \\ \text{your reply} \end{Bmatrix}$ and hope

→ we shall be able to $\begin{Bmatrix} \text{establish a connection} \\ \text{open an account} \\ \text{enter into business relations} \end{Bmatrix}$ between our two firms which

will be pleasant and mutually beneficial.

第十章
徵信與答覆 (Credit Inquiries and Replies)

第一節　委託徵信

　　接到招攬信後，在未弄清對方身分之前，不宜立即進入交易階段。換言之，在正式進行交易之前，宜先做徵信 (Credit inquiry)，以免日後因對方不誠實而引起糾紛。但徵信費時，因此，在徵信期間，應先覆謝。

No. 21　收到招攬信後，正式建立關係前的覆函

> Gentlemen,
>
> 　　We are grateful to you for your proposal, expressed in your letter of May 10, to open an account with us.
>
> 　　From your letter we are glad to learn that you are interested in Taiwan-made nylon piece goods, and in these we may say that we are specialists. It is our sincere wish to become connected with your establishment, but before we accept your proposal and go into negotiations in a definite way, it is usual with us to take up the reference named in your letter. We shall, therefore, do ourselves the pleasure of communicating with you further as soon as we hear from the bankers you named.
>
> 　　We thank you for your courtesy in making the proposal and hope we may soon be able to do a good business with you.
>
> 　　　　　　　　　　　　　　　　　　　　　　　　　　　　Very truly yours,

【註】

1. establishment：（工廠、商號等）機構，公司，行號。
2. it is usual with us = it is the custom with us = it is customary with us。
3. take up the reference：向備詢人查詢。
4. do ourselves the pleasure：比 "have the pleasure" 鄭重。

如因某種原因，不願意與其建立往來關係也應以委婉的措詞覆函拒絕。下面就是一例：

No. 22 不願與其往來的覆函

Dear Sirs,

Thank you for your letter of May 4 proposing to establish business relations between our two firms.

Much as we are interested in doing business with you, we regret to inform you that we are not in a position to enter into business relations with any firms in your country because we have already had an agency arrangement with Taiwan Trading Co., Ltd. in Taipei. According to our arrangement only through the above firm, can we export our products to Taiwan.

Under the circumstances, we have to refrain from transacting with you until the agency arrangement expires. Your letter has been filed for future reference.

Thank you again for your proposal and your understanding of our position will be appreciated.

Faithfully yours,

【註】

1. much as：雖然很（有興趣）。

2. are not in a position：比 "can not" 委婉。

3. agency arrangement = agency agreement = agency contract：代理契約；代理協議書。

4. only..., can we...：這是加重語氣的措詞。

5. under the circumstances：「在此情形下」，注意 "circumstances" 必須用多數形。

6. refrain from：抑制，強調抑制一種衝動，自動地不做所欲做或所願做的事。與 "abstain from" 同義，但 "abstain from" 係強調以意志力克制自己，也指自動戒除某些所欲的或有害的東西，尤指享樂及飲食。例：He is abstaining from pie. （他在克制自己不吃〔他愛吃的〕餡餅。）

7. transacting：可以 "doing business" 代替。

8. position：立場，也可以 "situation" 代替。

調查對方信用的方法，最常用的有下列三種：

1. 委託本地外匯銀行代為調查。

2. 逕函對方所提供的備詢銀行 (Reference bank) 或備詢商號 (House reference; Trade reference) 查詢。

3.委託徵信所 (Mercantile credit agency) 代為調查。

此外也可委託下列機構代為徵信:

4.委託對方所在地同業調查。

5.委託對方國家的商會或進出口公會調查。

6.委託本國駐外經濟參事處等機構代為調查。

委託徵信時,其所需徵信事項宜具體寫出,通常一封委託徵信函應包括下列各項:

1.被調查商號名稱、地址。

2.調查理由。

3.調查事項:

⑴Character(品性): 指負責人的 Integrity, Reputation(誠實、聲譽),Willingness to meet obligation(履行債務的意願),Attitude toward business(對業務的態度)。

⑵Capacity(才能): 經營者在商場上的經營能力、學經歷。

⑶Capital(資本)。

以上稱為 3C,此外宜包括:

⑷創業日期 (Date of establishment)。

⑸營業項目 (Line of business)。

⑹過去 3 年營運量 (Business volume for the past 3 years)。

⑺過去 3 年損益 (Profit/Loss for the past 3 years)。

⑻員工人數 (Number of employee) 等。

4.表示對所提供消息,保證絕對保密不外洩。

5.表示惠請協助將感激不盡。

No. 23 請求本地外匯銀行代為徵信

委託本地外匯銀行徵信時,可用本國文:

受文者: 臺灣銀行徵信室

主旨: 請代查報 ABC Co., Ltd., 123 Wall Street, New York, NY 10012, U.S.A. 信用狀況。

說明: 1.本公司為貴行進出口及存款多年客戶,開有支票存款第 1234 號戶,每月結匯多筆。
2.茲正與上開紐約新客戶洽商交易中,據悉該商之備詢銀行為 Bank of America, New York。

臺灣貿易公司　謹上

中華民國××年 8 月 10 日

No. 24　逕向對方所提供備詢銀行查詢之一

Dear Sirs,

Anderson Co., Inc., 5000 Market Street, San Francisco, who have recently proposed to do business with us, have referred us to your Bank.

We should feel very much obliged if you would inform us whether you consider them reliable and their financial position strong, and whether their business is being carried on in a satisfactory manner.

In addition to the above, please, if possible, also furnish us with the following information:

1. date of establishment
2. name of responsible officers
3. line of business
4. business volume for the past three years
5. profit/loss for the past three years.

Any information you may give us will be held in absolute confidence and will not involve you in any responsibility.

We apologize for the trouble we are giving you. Any expenses you may incur in this connection will be gladly paid upon being notified.

Faithfully yours,

【註】

1. carry on：經營。

2. furnish...with：提供，同類用語為 "supply...with"，"provide...with"。

3. responsible officers：負責人。

4. incur：發生。

5. upon being notified：一經通知。

No. 25　逕向對方所提供備詢銀行查詢之二

Dear Sirs,

Messrs. H. Butler & Co., Ltd., Holland House, Queen's Road Central, Hongkong, have referred us to your Bank in connection with an account they wish to open with us.

　　We should be greatly obliged if you would inform us what you know about their business integrity, their ability to meet their obligations, and the general reputation they enjoy in your community. If you can add any pertinent matters that you think would be informative to us, we should greatly appreciate them.

　　Any information you may give us will be treated as strictly confidential and reserved to ourselves.

Respectfully yours,

【註】

1. integrity：誠實（尤指履約方面）。

2. to meet their obligations：履行債務，to meet = to fulfil = carry out; obligations = liabilities = engagements（債務；負債）。

3. any information...to ourselves 的相似用語有：

　　a. Any information you may favour us with will be held in strict confidence.

　　b. Your communication on the subject will, you may rest assured, be treated confidentially and discreetly.

　　c. We will make a discreet use of any report you may give us concerning the subject.

　　d. Any information with which you may favour us will be used in absolute confidence.

　　e. Be assured that any data you may give us will be treated with the greatest confidence and discretion.

銀行有代客保密之責，因此逕向對方所提供備詢銀行調查客戶信用時往往得不到結果。所以，還是委託自己的往來銀行徵信較妥。

No. 26　向對方所提供 Trade reference 查詢

Dear Sirs,

　　Your name was referred to us as a reference by ABC Co., Inc., New York, who have recently proposed to do business with us.

　　We shall be grateful if you will furnish us with your opinion on the financial standing and respectability of this firm by filling in the blanks of the attached sheet and returning it to us in the enclosed envelope.

　　We assure you that all information will be kept in strict confidence and will be very glad to reciprocate your courtesy when a similar opportunity arises.

Yours faithfully,

ATTACHED SHEET

Name: ABC Co., Inc.

1. Date of establishment:
2. Line of business:
3. Integrity and ability of its management:
4. Manner of paying obligation:
   ...prompt...medium...slow
5. Your opinion as to this firm:
6. Period of transaction with the firm:
   from...to...
7. Largest amount of one transaction with the firm: US$...

【註】

1. 用 attached sheet 方式，可節省對方時間，方式不錯。

2. respectability：受人尊敬的狀態（指其品格、名譽、社會地位等）。

3. fill in：填寫。"put in" 之意，也可用 "fill out" 一詞。

   cf. fill in an application form：填寫申請書。

4. reciprocate：回報。

5. prompt：迅速。指付款迅速。

6. medium：中等。

7. slow：遲緩。

No. 27　受託銀行轉請國外銀行徵信之一

Request for a report on

XYZ Co., Ltd....

Dear Sirs,

At the request of our valued customer, we would appreciate your furnishing us by airmail with a detailed report on the business, means and credit standing of the above firm.

Any information you supply will be treated in strict confidence and without any responsibility on your part.

We thank you in advance and assure you of our readiness to serve you in a similar manner.

Faithfully yours,

【註】

1. detailed report：詳細報告。

2. means：資力，即財力。

3. We thank you in advance 的措詞宜少用，可將 "in advance" 刪除。

4. readiness：願意。

5. to serve you in a similar manner：為您做同樣（類似）的服務。

No. 28　受託銀行轉請國外銀行徵信之二

Gentlemen:

　　We have been asked by one of our valued clients for current information on the business standing and financial responsibility of American Chemical Imports, 808 Main Street East, Rochester, NY 14603.

　　It occurs to us that the subject may be known to you, and we therefore take liberty of asking you to furnish us with as detailed as possible a report on the subject. We are interested in recent financial figures and would also be glad to receive opinions regarding the subject's reputation and promptness in meeting obligations. A reply by air mail will be helpful.

　　If the subject is not known to you, will you please be good enough to conduct an investigation in our behalf.

　　You may be certain that such information will be conveyed to our client who originated this inquiry, without disclosing your valued name in any manner, and with all the customary disclaimers of responsibility for you, your officers, or informants. We thank you beforehand for your valuable assistance in this matter, and shall welcome an early opportunity of reciprocating your courtesy.

Yours faithfully,

【註】

1. current information：目前的消息。

2. it occurs to us that = we think that。

3. as detailed as possible：儘可能詳細。

4. conduct an investigation in our behalf：替我們調查。

　　in our behalf = on our behalf：替我們；代我們。

5. disclose：透露。

6. disclaimer：否認的聲明。

第二節　徵信報告

No. 29　國外銀行逕向商號提出的徵信報告——Favorable（有利）

Gentlemen:

　　The subjoined is a report on the firm referred to in your letter of March 17. Please note that this information is supplied in strict confidence and without any responsibility on the part of this Bank or any of its officers.

　　We thank you for your offer to reimburse the expense incurred, but the cost was too trifling to bring into account.

Yours very truly,

Anderson & Co., Inc.
5000 Market Street, San Francisco

　　The above-mentioned firm is one of the oldest and best establishments in this city, doing whole-sale and import and export business in general merchandise. They have their branch houses in Portland and Seattle and have been doing a business of considerable volume.

　　The firm has maintained an account with us for the past twenty years and has been one of our best clients. We have often made loans for good amounts, and all these obligations have been met as agreed upon. We are quite satisfied to see that our confidence in the firm has never been misplaced.

　　Mr. W. Anderson, President of the firm and all other directors are everywhere held in the highest esteem, both for their business ability and for their integrity. From our records we do not hesitate to say that the firm may be rated as AI.

S. J.

【註】

1. the subjoined：另附的東西，即「另紙」。

2. reimburse：歸償；歸墊。

3. too trifling to bring into account：微不足道，所以不算帳。

4. made loan for good amounts：貸給相當數額的款項。

5. these obligations...upon：債務全部均依約付清。

6. our confidence...misplaced：我們信任該商號並未錯誤。

7. rated as AI.：信用等級為最佳，AI (éi wʌn) = first rate (class)。

No. 30 國外銀行逕向商號提出的徵信報告——Favorable

Dear Sirs,

We are pleased to supply you with the required information respecting Messrs. H. Butler & Co., Ltd., asked for in your letter of the 12th April.

They have been established for many years as general commission merchants, importers, and exporters. They underwent reorganization in 1940 from the original single proprietorship of Mr. H. Butler into the present joint-stock company with the capital registered at HKD 1,000,000, three-quarters paid up.

So far as we have been able to ascertain, they are regarded in our financial circles as substantial traders with a clean record. They are well-managed, and their financial obligations have all been taken care of very satisfactorily. They have a number of good connections in and out of Hong Kong, the turnover mounting steadily. In our opinion they are rated as AI.

It is hoped that this information, which is given for your confidential use and without liability on our part, will prove helpful to you.

Yours faithfully,

【註】

1. general commission merchants：一般商品（雜貨）的佣金商。

2. reorganization：改組。

3. single proprietorship：獨資商號。

4. joint-stock company：股份公司。

5. the capital registered at = the registered capital of：登記資本。

6. paid up：已繳付。

7. ascertain = find out certainly; get to know：探知；探出真相。

No. 31 銀行逕向查詢者回答——Favorable

This letter is confidential and written without prejudice, as a matter of business courtesy, with the understanding that its source and contents will not be divulged, and that no responsibility therefor is to attach to this Bank or any of its officers or Agents. It contains information and expressions of opinion subject to change without notice, and, while obtained from sources deemed reliable, the accuracy of any statement herein made is not vouched for in any way.

New York, May 30, 20–

A. B. C. Co., Ltd.
Taipei, Taiwan
Gentlemen:

The following is a report on the firm about which you inquired in your letter of May 20. It is understood that this information is supplied in confidence and without responsibility on our part.

Very truly yours,
THE AMERICAN EXCHANGE BANK
W. Lawson
Mgr. Credit Dept.

WL: HT

STRICTLY CONFIDENTIAL
Harris & Co., Inc.
200 Fifth Avenue, New York, NY 10023

The subject is an affiliate of The American Trading Corporation, Inc., one of the oldest and best established concerns in New York City, and commenced operations in 1945 as importers of woolen and silk textiles and exporters of machinery and machine-tools. The subject has many good connections both at home and abroad and is in no small way of business.

The subject has maintained an account with us since July, 1945. Loans we have made at times in sizable amounts have been handled to our satisfaction. A financial statement in our files dated June 30, 2004, reflects a healthy condition.

The management is well known to us, and we have a high regard for their character, integrity, and financial responsibility. We are of opinion that the subject is entitled to the confidence of those with whom they deal.

【註】

1. 銀行對於徵信的回答，往往像本例，在信箋上印好免責條款，相似的條款尚有：

   a. It is understood that the information contained in this letter is given in absolute confidence, and entirely without prejudice or the assumption of responsibility by us.

   b. All persons are informed that any statement on the part of this bank, or any of its officers, as to the responsibility or standing of any person, firm, or corporation, is a mere matter of opinion and given as such, and solely as a matter of courtesy, and for which no responsibility, in any way, is to attach to this bank or any of its officers.

2.without prejudice = without detriment to existing right or claim：不可損害到既得權利或請求權，在這裡意指「回答的人不負責任」。

3.as a matter of business courtesy：做為商業上的禮貌。

4.therefor = for it; for that purpose：「為此」，「為該目的」，不可與 "therefore" 混淆。

5.officers：行員。美國用 "officers"，英國則用 "servants"。

6.subject to change without notice：變動恕不通知。

7.affiliate = affiliate company; subordinate company; subsidiary company：「附屬公司」，「子公司」，與母公司 (parent company) 相對。

8.woolen：英國拼法為 woollen。

9.machinery and machine-tools：機械類及機械工具，"machinery" 為單數集合名詞。

10.at home and abroad = in and out of the country。

11.is in no small way of business = is doing no small amount of business; is in a large way of business; is doing business on a large scale。

12.sizable amount = sizeable amount = considerable amount。

13.financial statement = balance sheet：資產負債表。

14.management：幹部；管理階層人員。係單數集合名詞。

No. 32　銀行對銀行的徵信報告——Favorable

Dear Sirs,　　　　　　　　　　　Credit Report

In answer to your inquiry of October 14, 20– regarding Anderson & Co., London, we are pleased to send you the following information.

This firm was established by Mr. Jones Anderson and his family in 1955. It operates as an importer of ladies' handbags which are distributed through the U.K. in a popular price line.

We have an account of the subject for a good many years and the relationship has been monthly openings of letters of credit in low six figures with deposit balances fluctuating between low five and moderate five figures.

This firm meets all obligations in a prompt manner. The principal is well-known to our officers and is highly regarded by us.

The financial statement as at the end of 1976 indicates that the net worth is GBP 200,000.

We recommend this firm to your customers for their normal business engagements.

Yours very truly,

【註】

1. in a popular price line：以大眾化價格（出售）。

2. fluctuating between...figures：在低五位數與中下五位數之間變動。

   "low five figures" 又可寫成 "low figure proportions" 或 "low five figure category"，均可譯為低五位數。銀行基於業務保密理由，不明示實際餘額，而以位數來代替。茲將其代字的意義列示如下：

   Low …… 1–2

   Moderate …… 3–4

   Medium …… 5–6

   High …… 7–9

   例：low six figures in dollar: $100,000–$200,000; low to moderate five figures: 10,000–40,000; balance in medium Seven figures：餘額為 5,000,000–6,000,000。

3. highly regarded：很受重視。

4. net worth：淨值，即資產減負債的差額。

5. normal business engagements：普通程度的債務。

No. 33　由商號發出的徵信報告——Unfavorable（不利）

Gentlemen,

In response to your inquiry about the firm mentioned in your letter of May 16, we are very sorry to say that our experience with the firm in question has been rather unsatisfactory.

During the past two years, in which the account has been active, we have had some trouble with collections. We have several times been compelled to place the account in the hands of our attorneys in order to protect ourselves.

Under the circumstances, we advise you not to enter into any business relation with the firm question.

Very truly yours,

【註】

1. the firm in question：該商號。

2. the account has been active：有效的帳目；未清的帳目。

3. have trouble with collections：收款遭遇困難。

4. have been compelled：迫不得已。

5. place the account in the hand of our attorneys：將（未清）帳目移請律師處理。

6. protect ourselves：以求自保。

No. 34　無法提供信用資料時

> Your credit inquiry of June 6
on ABC Co., Ltd.
>
> Dear Sirs,
>
> The subject firm does not maintain any account relationship with us, therefore, no information regarding it is available on our file.
>
> Although we have contacted some of our banking friends here trying to obtain information on the subject firm, we failed.
>
> Such being the case, we are not able to furnish you with the credit standing of the subject firm. We regret that we have been unable to assist you in this matter.
>
> Yours very truly,

【註】

1. on our file：在我們檔卷中。也可用 "in our file"。

   cf. We have placed the correspondence on our files.（我們已將該信件歸檔了。）

2. contact：接觸；聯繫。當做動詞用時，係他動詞，因此不能說 "contact with..."。如當做名詞用，則須加 "with"。例：

   to make contact with someone

   to be in contact with someone

3. such being the case：因此；既然是這樣。

4. in this matter：關於這件事。

No. 35　對於提供徵信報告的謝函

> Dear Sirs:
>
> We have received your letter of May 24 giving us a report on the firm we inquired about. We wish to say that your information will certainly be of great assistance to us in deciding the course we have to take.
>
> We express our best thanks for the trouble you have taken, and you may rest assured that your information will be retained in our files in a confidential manner.
>
> Respectfully yours,

TAIWAN TRADING CO., LTD.
C. Y. CHANG
Export Manager

HS: MT

【註】

the course we have to take: 我們應採取的行動方針。

第三節　徵信報告用語釋義

(一) ⎰ Good for ～
　　⎱ Responsible for ～

例：This company is considered Good for its usual business engagements.（此公司對於通常交易所生債務有支付能力。）"Good for" 為 "Able to pay" 之意。

"Responsible for ～" 為「有支付能力」，「有履行義務能力」之意。所以與 "Good for ～" 同義。"Responsibility" 為「支付能力」之意。「有充分的支付能力」以 "Quite good for ～" 或 "Fully responsible for ～" 表示。

"Engagements" 通常為多數，其意為「金錢上的約束」、「債務」。"Business engagements" 為「交易上所生的債務」之意。「履行交易所生債務」以 "To carry out business engagements" 或 "To meet business engagements" 表示。

(二) Usual (Normal, Ordinary, Moderate) engagements

例：This company is good for its usual business engagements.（此公司對通常交易所生債務有支付能力。）"Usual engagements" 為「通常的債務」，意指「該企業通常程度的債務」，因此「通常」的標準，視「企業的交易規模」而定。

"Normal engagements" 為「普通程度的債務」之意，「普通」的標準是指從該企業所屬行業來看，屬於「普通程度的」之意。

"Usual" 為就該企業本身而言是「通常」，而 "Normal" 則就該行業而言，該企業是「普通的」。

"Ordinary engagements" 為「中等程度的債務」之意，即指「不多不少程度的債務」。因此對「金額」有限定的含義。比起 "Usual"，"Normal" 給予人的印象有「異有限制」

的感覺。

　　"Moderate engagements" 為「金額不多的債務」，比 "Ordinary" 消極，與 "Limited amounts"（小額的）之意相仿。因此，如有 "Good for moderate business engagements" 的 Opinion，則與該行號交易時要稍為小心。

㈢Credit information; Credit inquiry

　　"Credit information" 為「有關信用狀況的情報」，"Credit inquiry" 為「關於信用狀況的查詢、徵信」，"Credit report" 為「徵信報告」，"Credit agency" 為「徵信所」，"Credit investigation" 為「信用調查」，「調查信用」為 "To conduct credit investigation(s)"，「查詢信用或做徵信」為 "To make (conduct) a credit inquiry"，「信用情報的蒐集」為 "Assembling of credit information"。

第四節　有關徵信的有用例句

一、請求提供信用資料

1. $\begin{Bmatrix} \text{Please} \\ \text{Kindly} \end{Bmatrix}$ $\begin{Bmatrix} \text{supply} \\ \text{furnish} \\ \text{provide} \end{Bmatrix}$ us with $\begin{Bmatrix} \text{your opinion} \\ \text{a detailed report} \end{Bmatrix}$ $\begin{Bmatrix} \text{on} \\ \text{regarding} \end{Bmatrix}$

→ the $\begin{Bmatrix} \text{financial status and responsibility} \\ \text{financial standing and reputation} \\ \text{financial responsibility} \\ \text{business, reputation, means and credit standing} \end{Bmatrix}$

→ of the $\begin{Bmatrix} \text{above} \\ \text{above-mentioned} \\ \text{following} \end{Bmatrix}$ firm.

請就 $\begin{Bmatrix} \text{上列} \\ \text{上述} \\ \text{下列} \end{Bmatrix}$ 商號的 $\begin{Bmatrix} \text{財務情況及付款能力} \\ \text{財務狀況及聲譽} \\ \text{財力} \\ \text{業務、聲譽、資力、信用狀況} \end{Bmatrix}$ $\begin{Bmatrix} \text{提供} \\ \text{惠示} \end{Bmatrix}$ 卓見。

2. Please get information for us on the financial standing and reputation of...Co., Ltd.

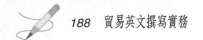

註：get information：蒐集情報；取得情報。

二、將予保密及結尾句

1. $\begin{cases} \text{Any information you may} \\ \text{You can be sure that any information you may} \end{cases} \begin{cases} \text{supply} \\ \text{give us} \end{cases}$

$\rightarrow$ will be kept $\begin{cases} \text{in strict confidence.} \\ \text{confidential.} \\ \text{strictly confidential.} \end{cases}$

2. $\begin{cases} \text{We will be very glad} \\ \text{We assure you of our willingness} \end{cases}$ to $\begin{cases} \text{reciprocate} \\ \text{return your courtesy} \end{cases}$

$\rightarrow \begin{cases} \text{at any time.} \\ \text{when a similar opportunity arises.} \\ \text{if an occasion arises.} \\ \text{if an opportunity arises.} \end{cases}$

註：a. at any time：隨時；b. when a similar opportunity arises：有類似機會時；c. if an occasion arises：如有機會。

3. You are assured of our discretion and we will welcome an opportunity to reciprocate your courtesy.

註：assured of our discretion：保證嚴守機密。

三、情況良好的徵信報告

1. The $\begin{cases} \text{firm} \\ \text{company} \end{cases}$ you inquired about in your letter of...(date)...

$\rightarrow \begin{cases} \text{enjoyed a good reputation in the business circles here.} \\ \text{is considered to have an excellent business reputation here.} \end{cases}$

2. In our opinion, they $\begin{cases} \text{carry on their business satisfactorily.} \\ \text{are considered good for normal engagements.} \end{cases}$

$\left(\text{我們認為} \begin{cases} \text{他們經營得很順利。} \\ \text{可與他們從事正常的交易。} \end{cases} \right)$

3. They are quite punctual in meeting their obligations.

（履行債務很準時。）

4. We have a favorable regard for the company and its management.

（對該公司及其管理階層表示讚許。）

5. We have every confidence to recommend to you the firm you inquire about as one of the most reliable exporters in our district.

（以充分的自信向你推薦……。）

6. Messrs. Smith & Co. is Al in every respect. They always display sound judgement in the conduct of their business, which we suppose, is based on their many years' experience.

（各方面都是第一流的，在業務方面的判斷總是表現得很穩當……。）

7. The company you enquired about was established in 1950. Since then, the president, Mr. A, has gradually leveled up his business standing through the close cooperation of his staff members. His company now commands a large share of trade in the relative market.

　　註：a. level up：提高水準；b. commands a large share：占了很大部分。

8. Our records show that they have never failed to meet our bills since they opened an account with us. The monthly limit of credit we feel we may safely grant them is approximately $... In addition, their sincere attitude toward trade and their extensive business activities merit high esteem.

　　註：a. never failed...bills：從不欠帳；b. monthly limit of credit：每月信用額度；c. grant：給與；d. sincere attitude：誠實的交易態度；e. merit high esteem：值得尊敬。

四、情況令人起疑或欠佳的徵信報告

1. According to the talks in our district, they have recently commenced rather risky speculative business for an amount beyond their capacity.

（據本地傳說，他們最近開始從事超出他們能力所及的金額的相當危險的投機交易。）

2. The reports in circulation indicate that they are in an awkward situation for meeting their obligations.

（傳聞該公司窮於應付債務的履行。）

3. We advise you to proceed with every possible caution in dealing with the firm in question.

（忠告宜特別謹慎。）

4. Reported locally that they are over-trading, do not meet engagements promptly.

（據本地人報告，他們業務擴張過度，不能迅速履行債務。）

5. Said to be involved in a charge of smuggling.

（據說觸犯走私罪。）

6. Maintains a small account with us. Your figures (US$100,000) rather large.

7. In a small way of business. Recommend secured basis.

五、無法提供資料

1. We are sorry we are $\begin{Bmatrix} \text{not able} \\ \text{unable} \end{Bmatrix}$ to give a precise information you ask.

2. We are sorry that we $\begin{Bmatrix} \text{cannot} \\ \text{are unable to} \end{Bmatrix}$ $\begin{Bmatrix} \text{furnish} \\ \text{provide} \end{Bmatrix}$ you with the information you desire.

3. This firm does not maintain an account relationship with us. For this reason, we are not able to comment on the financial responsibility and reputation of the firm.

4. We regret we are unable to give you positive answers about the said concern.

註：positive answers：肯定的答覆。

六、保密、不負責的聲明及結尾

1. $\begin{Bmatrix} \text{This} \\ \text{The above} \end{Bmatrix}$ information is $\begin{Bmatrix} \text{given} \\ \text{furnished} \end{Bmatrix}$ $\begin{Bmatrix} \text{confidentially} \\ \text{in strict confidence} \end{Bmatrix}$ and without responsibility on $\begin{Bmatrix} \text{our part} \\ \text{the part of this bank} \end{Bmatrix}$ and/or its officers.

2. We hope that $\begin{Bmatrix} \text{this} \\ \text{the above} \end{Bmatrix}$ information will be of assistance to you.

3.The information is $\left\{\begin{array}{l}\text{furnished}\\\text{given}\end{array}\right\}\left\{\begin{array}{l}\text{confidentially}\\\text{in strict confidence}\end{array}\right\}$ at your request without any guarantee or responsibility on our part and with the understanding that its sources will not be disclosed in anyway.

註：its sources will not be disclosed in anyway：將不以任何方式洩漏該項資料的來源。

第十一章
一般交易條件協議書 (Agreement on General Terms and Conditions of Business)

國際貿易的進行，常以 Fax 或 E-mail 等電傳方式完成。當進出口商向國外對方發出招攬信後，一接到對方同意建立業務關係函電後，為爭取商機，往往不做徵信，也不約定一般交易條件協議書 (Agreement on general terms and conditions of business)，即以電傳方式進行交易。但電傳通常電文較簡略。在電傳中，所提及的不過是品名、規格、價格、數量、交貨期限、付款條件等。至於其他條件，如檢驗條件、包裝條件及索賠條件等，往往多付之闕如。一旦報價為對方所接受，契約即告成立。雖然雙方為慎重起見，往往另訂書面契約，詳列契約條款，但須知此書面契約，如非賣方製作，即為買方製作，製作者往往將自己有利的條款儘量列入。當送請對方簽署時，如對方無相反意見，自不成問題，但如對方對報價中未言及的條件不同意列入契約中，而雙方又彼此堅持不讓，則糾紛因此而起。在國際貿易中，因這種條件未先約定而引起爭執，最後導致解約或契約不成立的事例，屢見不鮮。

因此謹慎又有意長期繼續與對方交易的進出口商與對方取得連繫後，第一步為調查對方信用，如調查結果認為對方信用良好，值得往來，第二步就與對方洽訂一般交易條件，議訂雙方權義，以作為日後實際交易的基準。尤其對對方的信用調查結果未能十分滿意時，更應訂定相當嚴密的交易條件，約束對方，俾免日後遭受意外的不利或不必要的損失。這種一般交易條件是就雙方所開示的希望交易條件，經往返磋商最後達成時，通常均須作成書面協議書，雙方簽署後，各執一份。協議書簽立後，雙方即可以電傳或書信進行交易。這種電傳或書信，僅記載個別交易的主要條件（品名、品質、數量、價格、交貨期）而已。換言之，憑這種個別交易所簽訂的個別契約需與預先簽立的協議書合併起來才構成一完整的契約。所以在雙方簽有協議書的情形下，買賣雙方進行交易，毋需每次都詳述一般交易條件，這對雙方都省事省錢，頗為方便。

上述協議書通稱為「一般交易條件協議書」或稱為協議書 (Memorandum of agreement) 或 Mutual understanding 或 Basic agreement。其內容隨交易對方的身分、信用、買賣貨物種類以及市場習慣等而有若干的差異，但一般而言，不外包括下列各事項：

1.約定交易的性質：確定交易雙方身分。即契約雙方為賣方與買方的關係。

2.約定今後報價與接受事項。

⑴使用何種貨幣。

⑵使用那一種貿易條件 (Trade terms)。

⑶報價的有效期間。

3.約定每筆交易訂貨事項：約定憑電傳成交後，以書面確認，且一經確認，非經同意不得取消。

4.約定品質、包裝、裝運、保險、付款事項。

⑴品質：①究以裝運時的品質為標準，還是以到貨時的品質為標準，②由何方檢驗品質，如有檢驗費用，應由何方負擔，③品質低劣時，應如何處理。

⑵包裝：約定①包裝方法，②嘜頭。

⑶保險：約定①保險種類，②保險金額，③保險幣類，④理賠地點。

⑷裝運：約定①裝運日期的證明方法，②交貨期限。如契約中採用 Prompt shipment 與 Immediate shipment，應約定此用語是指多長期間，③遲延交貨責任的歸屬，及其善後。

⑸付款：約定①買方採取何種方式付款，②如採用信用狀方式付款，信用狀應於買賣契約簽訂後多少日內開發，③賣方所用匯票究為即期匯票，抑或遠期匯票，如為遠期匯票，貼現利息應由何方負擔。

5.約定索賠及不可抗力事項。

⑴索賠：約定①索賠期限，②索賠通知方法，③索賠證據，④解決方法。

⑵不可抗力：因不可抗力事故發生，而致發生一部分或全部貨物不能於約定期間內交運，依國際商業習慣，賣方不負任何責任。但是否因此解除契約，或待事故結束後再行交運，雙方應事先約定。

6.其他事項的約定。

⑴匯兌風險的協議：契約成立到貨物出口收款一段期間，如匯率變動則必有一方

發生損失，此項匯兌風險究應由何方負擔，須事先加以約定。

⑵運費保險費變動的協議：如以 FAS、FOB 條件交易，運費保險費均由買方負擔，費率發生變動，自應由買方負擔。如以 CFR、C&I 或 CIF 條件交易，其保險費及運費是根據簽約時的費率計算，實際交貨時如發生變動，而此項費率變動風險如擬由買方負擔，則應事先加以約定。

⑶仲裁條款、準據法……。

No. 36. 徵信結果，同意建立業務關係

Gentlemen:

Following our letter of May 17, we are pleased to say that our credit files have now been completed with favorable information from your Bankers, and therefore we are quite willing to accept your proposal made in your letter of May 10 and start business with you in electronic products.

Samples and Prices of Calculators. As requested, we are airmailing you samples of our calculators under Nos. 1/3, particulars of which are given on the Price List No. 50 enclosed. We ask you to note that each of the prices quoted is based on 1,000 sets, which is the minimum quantity of an order we can book. The prices are, of course, without engagement, and we are able to make you a firm offer by cable on receipt of your definite inquiry. We wish you will closely examine the samples, and we are convinced of our goods being found superfine and prices competitive. We are in a position to supply you with lower qualities, which, however, will be unsuitable for your trade.

Our electronic products cover various descriptions. They are all in brisk demand in the U.S. markets. We shall be glad to learn whether you have an opening for TV sets, TV game, Electronic Blocks, etc., in which we have been doing a good business with New York customers. Samples and prices will be forwarded immediately on receiving your inquiries.

Terms and Conditions. We agree to your terms, i.e., draft at 60d/s under an Irrevocable L/C to be opened simultaneously with the placing of an order. In order to preclude any possible misunderstanding which may arise in our transactions, we are enclosing an Agreement on General Terms and Conditions, on which all our future business will be based, for your approval.

We look forward to your further communications and hope that the relations now being established between us will last long and become mutually profitable.

Very truly yours,

【註】

1. without engagement：不受約束，即價格可變更之意。

2. superfine：極精緻的。

3. brisk demand：不斷的需要。 trade is brisk：生意興隆。

4. have an opening for = to be open for = be in the market for。

5. preclude：排除。

No. 37　一般交易條件協議書

AGREEMENT ON GENERAL TERMS AND CONDITIONS

OF BUSINESS AS PRINCIPAL TO PRINCIPAL

THIS AGREEMENT entered into between ABC Co., Ltd., 111 Lin-Sheng N. Road, Taipei, Taiwan, hereinafter referred to as SELLER, and XYZ Co., Ltd., 222 Broadway Street, New York, NY, U.S.A., hereinafter referred to as BUYER, witnesses as follows:

1) Business: Both SELLER and BUYER act as principals and not as agents.

2) Commodities: Commodities in business and their unit to be quoted, are as stated in the attached list.

3) Quotations and Offers: Unless otherwise specified in faxes or letters, all quotations and offers submitted by either party to this Agreement shall be in U.S. dollars on CIF New York basis.

4) Firm Offers: All firm offers shall be subject to a reply within the period stated in respective faxes. When "immediate reply" is used, it means that a reply is to be received within three days from and including the day of the despatch of a firm offer. In either case, however, Sundays and all official Holidays are excepted.

5) Orders: Any business concluded by fax shall be confirmed in writing without delay, and orders thus confirmed shall not be cancelled unless by mutual consent.

6) Payment: Payment to be effected by BUYER by usual negotiable and irrevocable letter of credit, to be opened 30 days before shipment in favor of SELLER, providing for payment of 100% of the invoice value against a full set of shipping documents.

7) Shipment: All commodities sold in accordance with this Agreement shall be shipped within the stipulated time. The date of Bill of Lading is taken as conclusive proof the day of shipment. Unless expressly agreed to, the port of shipment is at SELLER's option.

8) Marine Insurance: All shipments shall be covered ICC (A) for a sum equal to the amount of the invoice plus 10 percent if no other conditions are particularly agreed to all policies shall be made out in U.S. currency and payable in New York.

9) Quality: Quality to be guaranteed equal to description and/or samples, as the case may be.

10) Inspection: Commodities will be inspected in accordance with normal practice of supplier, but if BUYER desires special inspections, all additional charges shall be borne by BUYER.

11) Damage in Transit: SELLER shall ship all commodities in good condition and BUYER shall assume all risks of damage, deterioration or breakage during transportation.

12) Exchange Risks: The price offered in U.S. dollars is based on the prevailing official exchange rate in Taiwan between the U.S. dollar and the New Taiwan dollar. Any devaluation of the U.S. dollar to the New Taiwan dollar at the time of negotiating draft shall be for BUYER's risks and account.

13) Change in Freight and Insurance Rate: Any change in marine freight rate and marine insurance rate is for BUYER's account.

14) Shipment Samples: In case shipment samples be required, SELLER shall forward them to BUYER prior to shipment under the contract of sale.

15) Claims: Claims, if any, shall be submitted by fax within fourteen (14) days after arrival of commodities at destination. Certificates by recognized surveyors shall be sent by mail without delay. All claims which cannot be amicably settled between SELLER and BUYER shall be submitted to arbitration in New York, the arbitration board to consist of two members, one to be nominated by SELLER and one by BUYER, and should they be unable to agree, the decision of an umpire selected by the arbitrators shall be final, and the losing party shall bear the expenses thereto.

16) Force Majeure: SELLER shall not be responsible for the delay of shipment in all cases of force majeure, including mobilization, war, riots, civil commotions, hostilities, blockade, requisition of vessel, prohibition of export, fires, floods, earthquakes, tempests, and any other contingencies which prevent shipment within the stipulated period. In the event of any of the aforesaid causes arising, documents proving its occurrence or existence shall be sent by SELLER to BUYER without delay.

17) Delayed Shipment: In all cases of force majeure provided in the Article No. 16, the period of shipment stipulated shall be extended for a period of twenty one (21) days. In case shipment within the extended period should still be prevented by a continuance of the causes mentioned in the Article No. 16 or the consequences of any of them, it shall be at BUYER's option either to allow the shipment of late goods or to cancel the order by giving SELLER the notice of cancellation by fax.

18) Shipping Notice: Shipment effected under the contract of sale shall be immediately faxed.

19) Packing & Marking: All shipments shall be packed for export and be marked XYZ in Diamond.

In witness whereof, ABC Co., Ltd. have hereunto set their hand on the 1st day of June, 20–, and XYZ Co., Ltd. have hereunto set their hand on the 20th day of June, 20–. This Agreement shall be valid on and from the 1st day of July, 20–, and any of the articles in this Agreement shall not be changed and modified unless by mutual written consent.

SELLER	BUYER
ABC Co., Ltd.	XYZ Co., Ltd.
General Manager	General Manager

〔中譯〕

貨主間一般交易條件協議書

ABC 公司（以下稱為賣方），地址：臺灣臺北市林森北路 111 號，與 XYZ 公司（以下稱為買方），地址：美國紐約市百老匯街 222 號，茲訂立本協議書，（約定）證明下列各事項：

1. 交易型態：當事人雙方均為法律上的本人，而非代理人。

2. 買賣貨物：買賣的貨物以及報價的單位，如附表。

3. 報價：除非傳真或書信中另有規定，本協議書任何一方提出的報價，均按美金計算，而且均以 "CIF NEW YORK" 為條件。

4. 穩固報價：所有穩固報價，均須在個別傳真所載的期限內回覆。在使用「立即回覆」字樣時，應解為：回覆須於穩固報價發出之日起（包括發出之日）3 日內被收到。但星期日及公休日除外不計。

5. 訂貨：憑傳真成交的買賣，應迅速用書面確認，訂貨經確認後，除非經雙方同意，不得取消。

6. 付款方式：買方應以通常的、可讓購、不可撤銷信用狀付款；此項信用狀應以賣方為受益人，於裝運前 30 天開出；規定憑全套貨運單證支付發票金額全額。

7. 裝貨：所有憑本協議書售出的貨物，都必須在約定期間內裝出。提單日期，應視為裝貨日的決定性證據。除非有明確約定，裝貨港由賣方選擇。

8. 海上保險：如未特別約定其他條件，所有裝出的貨物，均應投保 ICC (A) 險，保險金額等於發票金額加一成。所有保險單應載明投保美金，在紐約理賠。

9. 貨物品質：賣方應保證，所裝貨物在品質及狀況方面，與說明及／或樣品相符。

10. 檢驗：貨物將依供應商通常方式實施檢驗，但如買方欲實施特別檢驗時，所有額外費用皆歸買方負擔。

11. 運輸途中的損壞：賣方應裝出情況良好的貨物，買方則須負擔貨物在運輸中損壞、變質或破損的危險。

12. 匯兌風險：以美金報價的價格，乃以臺灣現行美金對新臺幣的官定匯價為準，在押匯時，如美金對新臺幣有任何貶值，則此項風險及損益歸買方負責。

13. 運價、保險費率的變動：任何海運費率及海上保險費率的變動，歸買方負擔。

14. 裝貨樣品：買方要求裝貨樣品時，賣方應於依約裝貨前，將樣品寄給買方。

15. 索賠：如須索賠，應於貨物到達目的地 14 日內，用傳真提出。認可的公證行所簽發的證明書，應速即郵寄。買賣雙方不能友好解決的所有索賠事件，應在紐約交付仲裁，仲裁庭包括兩人，其中一人由賣方指定，另一人由買方指定，如兩人的意見不能一致，則以仲裁人所選評判人的決定為準。而敗方須負擔費用。

16. 不可抗力：在所有不可抗力情形下，賣方對裝運遲延不負責，包括動員、戰爭、騷擾、民變、軍事衝突、封鎖、徵用船隻、禁止出口、火災、洪水、地震、風暴以及在約定期間阻礙裝貨的其他意外事故。萬一上述事故發生，證明事故發生或存在的文件，應由賣方迅速寄給買方。

17. 遲延裝運：在第 16 條所列不可抗力情形下，約定的裝運期限應延長 21 日。如延長期間的裝運，仍因第 16 條所列事故或其影響繼續存在而受到阻礙，則應由買方選擇，或接受後來遲裝的貨物，或用傳真通知賣方取消訂貨。

18. 裝運通知：賣方依買賣契約裝出貨物時，應立即發出傳真通知。

19. 包裝及刷嘜：所有貨儀須施以出口包裝並刷上 xyz 的嘜頭。

為證明上述約定，ABC 公司於 20– 年 6 月 1 日在本協議書簽字，XYZ 公司於 20– 年 6 月 20 日在本協議書簽字，本協議書自 20– 年 7 月 1 日起生效，除非經雙方書面同意，本協議書中任何條款不得變更或修改。

賣方	買方
ABC Co., Ltd.	XYZ Co., Ltd.
總經理	總經理

第十二章
交易條件 (Terms and Conditions of the Transactions)

國際貿易一如國內買賣，一筆交易的成立，買賣雙方必須就其交易的內容有所約定，雙方才能遵照履行。買賣雙方所約定的交易內容稱為交易條件 (Terms and conditions of the transactions)。交易條件的詳略，視貨物種類、買賣習慣以及事實需要而定。但國際買賣每一筆交易通常至少應就商品名稱、品質、價格、數量、包裝、交貨、保險及付款等條件有所規定。這八條件，在報價時，為報價的基本條件，為構成有效報價的基本要素。報價一經有效接受，買賣契約即告成立，而這八條件即轉而成為買賣契約的內容，以下分節說明。

第一節　商品名稱與商品目錄

一、商品名稱

買賣的標的物通常稱為商品，但有時也稱為產品、貨品、貨物、物資、物料或器材等，在英文則有 Commodity, Merchandise, Goods, Product, Produce, Ware, Line, Materials, Supplies, Article, Item 等等稱呼，在保險界及航運界則又稱為 Cargo。

交易上所使用商品名稱，應為國際市場上一般通行者，如以地方性的名稱做為交易商品名稱，在各方面都不方便。

茲將各種有關貨物的名詞列舉於下：

GOODS（貨物、商品、貨品、物資、原料）

Air-borne goods	空運物資	Half-finished goods	半製品
Bargain goods	特價品	Heavy (Light) goods	重（輕）量貨
Canned goods	罐頭品	Household goods	家庭用品
Tinned goods	罐頭品	Luxury goods	奢侈品

Capital goods	資本財	Inflammable goods	易燃品
Coarse (Crude) goods	粗製品	Low-priced goods	廉價品
Clearance goods	出清存貨	Manufactured goods	製成品
Consumer goods	消費品	Measurement goods	體積貨
Consumption goods	消費品	Quality goods	高級品
Contraband goods	走私貨、違禁品	Perishable goods	易腐品
Cotton goods	棉製品	Processed goods	加工品
Customable goods	應課稅品	Piece-goods	布疋
Damaged goods	損壞品	Sporting goods	運動用品
Durable goods	耐用品	Seasonable goods	季節性貨品
Dangerous goods	危險品	Staple goods	重要物資
Earthen-ware goods	瓦器	Strategic goods	戰略物資
Dry goods	布疋（美）	Substitute goods	代替品
Imported goods	進口貨	Sundry goods	雜貨
Fancy goods	精巧品	Miscellaneous goods	雜貨
High-quality goods	高級品	General goods	雜貨
Finished goods	完成品	Woolen goods	毛織品
First-rate goods	一級品	Wet goods	酒類（美）
Second (third) class goods	二（三）級品		

ARTICLE（商品）

Bad article	粗劣品	Coarse (Crude) article	粗製品
Commercial article	商品	Household article	家庭用品
Dutiable goods	課稅品	Inferior article	劣品
Duty-paying article	課稅品	Necessity article	必需品
Fancy article	精巧品	Second hand article	二手貨
Finished article	完成品	Toilet article	化粧品
Unfinished article	未完成品	Useful article	有用品
Gift article	禮品	Hazardous article	危險品

MERCHANDISE（商品，集合名詞）

General merchandise	雜貨	Standard merchandise	標準品
Returned merchandise	退貨	Unclaimed merchandise	貨主不明貨物

COMMODITY（商品、物品、物資）

Daily commodity	日用品	Perishable commodity	易腐品
Essential commodity	基本物資	Staple commodity	重要物資
Marketable commodity	適銷品	Vital commodity	生活用物資

PRODUCT（產品、製品）

Agricultural products	農產物	Intellectual products	智慧產物
Electronic products	電子產品	Foreign products	外國產品
Fishery products	漁產物	Marine products	海產物
Forestry products	林產物	Staple products	重要產品
Industrial products	工業製品		

WARES（手工藝品、物品、製品）

Bamboo-ware	竹器	Iron-ware	鐵製品，鐵器
Brass-ware	銅器	Lacquered ware	漆品
Earthen-ware	陶器	Luxury ware	奢侈品
Enamel-ware	油漆品	Tableware	餐具
Glass-ware	玻璃製品	Wooden-ware	木器
Ceramic-ware	陶磁器	Silver-ware	銀器
Hardware	金屬器具	Alminum-ware	鋁器

LINE（品目、品種）

Best line	超級品	Dry-goods line	布疋

CARGO（貨物，保險界、航運界用語）

General cargo	雜貨	Corrosive cargo	腐蝕性貨物
Fine cargo	精良貨物	Perishable cargo	易腐貨物

Clean cargo	精良貨物	Refrigerating cargo	冷藏貨物
Rough cargo	粗貨	Chilled cargo	冷凍貨物
Dirty cargo	不潔貨	Valuable cargo	貴重物
Liquid cargo	液體貨	Heavy cargo	笨重貨（超重貨）
Dangerous cargo	危險性貨物	Bulky cargo	笨大貨（超大貨）
Explosive cargo	爆炸性貨物	Lengthy cargo	超長貨
Poisonous cargo	有毒性貨物		

二、商品目錄

　　商品目錄又稱為型錄 (Catalog)，是廠商為便於推銷商品，以文字、圖片等說明其所經營商品性能、規格、形狀、重量、尺碼、顏色、包裝方法的印刷物。商品目錄是一種沉默的推銷員 (Silent salesman)，因此廠商在編印時力求內容精彩，外表美觀，以求引人入勝。

　　商品目錄編印目的既為拓展市場，但國外市場區域廣大，各國所用文字不盡相同。因此欲使各國讀者便於閱覽，商品目錄自宜採用各國文字，使閱覽者有親切感。惟英文已成為國際商業語言，所以目前各廠商所編印的商品目錄採用英文者居多。

　　商品目錄的種類有：

Illustrated catalog （有插圖的目錄）	Export catalog （出口貨品目錄）
Latest catalog	Complete catalog （完整目錄）
Revised catalog （修訂目錄）	Supplemental catalog （增補目錄）
New catalog	Spring
Recent catalog	Summer
Catalog of sporting goods	Fall ⎱catalog
General catalog （總目錄）	Winter
Descriptive catalog （有說明的目錄）	New Year's
	Mail order catalog （郵購用目錄）

第二節　品質條件

一、約定品質的方法

　　在洽談買賣時，首先須確定品質。貿易糾紛以品質糾紛為最多，所以對於品質的約定方法應特別小心，約定品質的方法有：

　　1.以樣品為準 (Sale by sample)：

　　　例：Quality: Same as sample submitted by seller on May 10, 20–.

　　　樣品的種類：

　　　①依寄送實務分：

　　　　　Original sample　　　　　　　　　　正份樣品（寄給對方的）
　　　　　Duplicate sample (File sample, Keep sample)　　副份樣品（自己保留的）

　　　②依樣品功能分：

　　　　　Selling sample　　　　　　　　　　推銷用樣品
　　　　　Approval sample　　　　　　　　　核准用樣品
　　　　　Sample for test　　　　　　　　　試驗用樣品
　　　　　Claim sample　　　　　　　　　　索賠用樣品
　　　　　Umpire sample　　　　　　　　　仲裁用樣品

　　　③依提示樣品的人分：

　　　　　Seller's sample　　　　　　　　　賣方樣品
　　　　　Buyer's sample　　　　　　　　　買方樣品
　　　　　Counter sample　　　　　　　　　相對樣品

　　　④依代表部分區分：

　　　　　Quality sample　　　　　　　　　品質樣品
　　　　　Color sample　　　　　　　　　　色彩樣品
　　　　　Pattern (Design) sample　　　　花樣（圖樣）樣品

　　　⑤依代表程度分：

　　　　　Sample of existing goods　　　　現貨樣品
　　　　　Similar sample　　　　　　　　　類似樣品

⑥依取樣時間分：

Advance sample	先行樣品
Shipping sample	裝船樣品
Outturn sample	卸貨樣品

註：

Sample：樣品（主要用於農產品、原料、羊毛、棉花、生絲、雜貨）

Pattern：花樣、款式（主要用於紡織品）

Swatch：樣品（主要用於布料、壁紙等以色樣為主要品質條件者）

2.以標準物為準 (Sale by standard)：

各業公會製有標準樣品 (Standard sample) 者，可約定以標準樣品為品質標準，主要適用於棉花、黃豆、玉米等。

例：如擬以美國二級黃豆為品質標準，可約定：

Quality: U.S. Grade No. 2 yellow soybeans.

至於甚麼是 U.S. Grade No. 2 可不必列明，因為 U.S. Grade No. 2 的標準規格為：

Bushel Wt.: 54 lbs. min.

Moisture: 14% max.

Splits: 20% max.

Damaged total (including heat damaged): 3% max.

Foreign material: 2% max.

Brown, Black/Bicolored: 2% max.

3.以規格為準 (Sale by grade)：

有些商品如水泥、鋼板、鋼筋等已由政府或產業團體或學會制定有關品質的標準規格 (Standard specifications)，這種商品的買賣，其品質可約定以某種規格碼為準。

例：Quality: Conforming to ASTM description C–150–84

　　　　　　　Requirement for Portland Cement type.

　　　　　　　(for cement business)

4.以平均品質或適銷品質為準（Sale on FAQ or GMQ）：

FAQ 為 Fair Average Quality 的略語。FAQ 條件就是指所交商品品質以「裝運時裝運地該季所運出商品的中等平均品質為準」之意 (Fair average quality of the season's shipment at time and place of shipment.)。

GMQ 為 Good Merchantable Quality 的略語，GMQ 條件乃謂保證商品品質在某種商業用途上良好可銷的品質條件。

例：按 FAQ 約定時：

Quality: Brazilian soybeans, 20– new crop, FAQ

按 GMQ 約定時：

The quality of the goods to be of GMQ.（品質須適合商銷。）

5.以牌記為準 (Sale by brand or trade mark)：

即憑牌名或商標約定品質。

例：Quality: Toyota truck

　　　　Model: kp 366

　　　　Year: 20–

　　　　Standard equipment

6.以說明為準 (Sale by specification or catalog)：

即以說明書或型錄說明商品規格、構造、材料、形狀、尺寸、性能等。

例：Specifications as per maker's catalog No. 123.

$$\text{Specifications} \begin{cases} \text{Nitrogen: } 46\% \text{ min.} \\ \text{Uncoated} \\ \text{In granules（粒狀）} \\ \text{（尿素買賣用）} \end{cases}$$

二、有關品質的用語

1.有關樣品的名詞：

Sample card	Sample fair	樣品展覽會
Sample cutting	Sample room	樣品室
Sample book	Sample order	樣品訂單
Sample number	Sample parcel	樣品小包
New (Fresh) sample	Sample post	樣品郵件
Newest sample	Free sample	免費（贈送）樣品

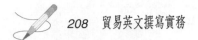

2.有關樣品條件的表現法：

Quality to be {
as per
up to
exactly the same as
identical to
exact to
a match to
in conformity with
conforming to
in accordance with
according to
correspondent with
similar to
about equal to
} the sample sent to you on...(date)

Sample(s) {
which is (are) the nearest to you required. （與你要求的最接近。）

which is (are) most likely to suit your market.（似乎最適合你的市場需要。）

which has (have) been specially prepared for export. （專為外銷而製的。）

representing the bulk. （代表正貨。）

closely resembling to what you want. （與你所需要者近似。）
}

The {
quality
specifications
color
size
material
} must be {
strictly same as
equal to
up to
conforming to
} {
sample
pattern
swatch
} submitted.

The goods must {
agree
comply
} in every respect with our {
specifications.
samples.
patterns.
}

3.有關樣品的動詞:

To maintain the standard
To keep up to the standard　　　保持標準

To be of inferior quality
To be inferior to the sample　　劣等品質
To be below the sample　　　　　比樣品差

To examine
To inspect ⎱a sample　　檢查
　　　　　　　　　　　檢驗 ⎱樣品

To test
To analyze ⎱a sample　　試驗
To assay　　　　　　　　分析
　　　　　　　　　　　分析 ⎱樣品

To sample　　　　　　寄樣品，取樣
To take sample　　　　取樣
To draw sample　　　　從現物取樣

To get
To obtain ⎱a sample　　獲取樣品
To secure

To send
To furnish
To submit ⎱a sample　　寄出
To show　　　　　　　供給
　　　　　　　　　　　提出
　　　　　　　　　　　出示 ⎱樣品

To order from sample　　憑所看樣品訂貨

To sell
To sbuy ⎱on sample　　憑樣 ⎱出售
　　　　　　　　　　　　　 購買

4.有關品質的形容詞:

⑴表示高級品質

Al, O.K., Extra O.K., Extra best, Extra fine, Very best, Very superior, Superior, Superfine, Prime, Fine, Good.

(2)表示中等品質

Second class	
Good fair average	
Middling	
Medium	goods
Common	
Usual	
Ordinary	

(3)表示劣等品質

Inferior	
Low grade	
Bad	goods
Poor	article
Third class	

(4)表示良好或其他品質

Well-conditioned		情況良好的	
Perfect		完整的	
Standard		標準的	
Sound		良好的	
Defective		有瑕疵的	
Damaged		損壞了的	
Deteriorated	goods	變質的	貨品
Faded		褪了色的	
Imitated		仿造的	
Spurious		偽造的	
Meretricious		俗氣的	
Shock-proof		防震的	
Water-proof		防水的	

三、有關品質的有用例句

(一)請送樣品

1. If you have something new in this style, we would thank you to transmit us some patterns.

2. Please send me a pattern in order that I may judge of the quality.

3. It is difficult to judge of a cloth by a narrow cutting, and therefore we expect you to send us fair-sized patterns, with colorings, of your next season goods.

　　註：a. fair-sized patterns：大小合適的花樣；b. colorings：配色。

4. We are on the look-out for the following, and should be obliged if you would send samples of the same.

5. Your offer looks very promising, and we should like to have a complete set of your samples as we cannot do anything without samples.

　　註：promising：很有希望。

(二)送樣品

1. In a few days we will send you some patterns of quality which we consider suitable for your market.

2. We are enclosing a copy of our recent catalog with a few samples which may possibly interest you, and shall be glad to hear from you at any time.

3. If you desire it, we shall be pleased to forward you some samples at once.

4. We have made such a selection as will suit your market, and sent you per sample post.

　　註：sample post：樣品郵件。

5. We send you several samples of papers closely resembling to what you want.

(三)品質必須與樣品一致

1. The goods must be in strict conformity with the shades and patterns supplied.

2. The quality must be $\begin{Bmatrix} \text{quite equal to} \\ \text{same as} \end{Bmatrix}$ the sample submitted.

3. Please see to these goods being in strict conformity with the particulars, any departure from which will be at your own risk, unless expressly authorized by us.

4. The goods must all be of the best quality, as nothing of an inferior kind will suit this market.

5. Should you be unable to find an exact match, kindly send us samples you have to offer so as to enable us to submit them to our customers for approval.

四吹噓品質

1. We take it as a compliment to our workmanship that we have never received a single complaint from our clients.

   註：we take it as a compliment：自我稱讚；我們以此為榮。

2. Our article is the fruit of nearly twenty years' earnest, conscientious work and has had the test of time and wear; this fact, we believe, is the best guarantee of quality and make.

   註：a. fruit：成果；b. earnest：熱心；c. test of time and wear：時間的考驗；d. make：牌子；式樣。

3. For your purpose we think you should find in these machines a marked improvement on the old makes, and they are decidedly superior to anything at present produced.

   註：old makes：老式樣。

4. Our articles are noted not only for their moderate prices, but also for their design in the latest style.

5. We feel confident that a more reliable and good wearing cloth of its kind is not obtainable elsewhere.

   註：wearing：耐用的。

五品質的交涉

1. If it is rather a question of price than of quality, the way out of the difficulty is comparatively easy. We suggest that you stock a lower grade of quite a similar appearance.

   註：the way out of the difficulty：克服困難。

2. We admit there are, as you say, some other lines much lower-priced than ours, but we feel confident that a comparison will reveal to you a wide difference in quality.

   註：a. other lines：別家的貨品；b. reveal：顯出。

3. Our quality has never been sacrified to meet price; we would rather suffer a blame
for high price than produce a low-priced article to the detriment of the quality.

　註：to the detriment of the quality：有損品質。

㈥保證品質

1. We guarantee our machines to wear five years. New machines will be furnished
gratis, if they fail to sustain the guarantee.

　註：a. gratis：免費；b. fail to sustain the guarantee：與保證不符。

2. We give our full guarantee that these are all wool and fast-dyed.

　註：fast-dyed：不褪色。

3. One-year's guarantee goes with our goods, and if they do not come up to our representations, we will replace them free or refund the money.

　註：do not come up to our representations：不符我們所言。

第三節　數量條件

商品品質一經約定，跟著就要約定買賣數量。

一、數量單位

1. Weight　（重量）

Avoirdupois ounce (oz) = 28.35 grams　　　　　　　　　　　　　　　（盎司）

Troy ounce = 31.104 grams

Pound (lb) = 16 oz = 454 grams　　　　　　　　　　　　　　　　　　（磅）

Kilogram (kg) = 1,000 grams = 2.2046 lbs = 35.274 oz　　　　　（公斤）

$$Ton\begin{cases} \text{Long Ton(L/T)=2,240 lbs=British ton=Gross ton} \\ \text{Short Ton (S/T)=2,000 lbs=American ton=Net ton} \\ \text{Metric Ton (M/T)=2,204.6 lbs=1000 kgs=French ton} \end{cases}$$

$$\text{Hundredweight (cwt)}\begin{cases} \text{英制} = 112 \text{ lbs=long cwt} \\ \text{美制} = 100 \text{ lbs=shot cwt} \end{cases}$$　　（匈威特）

2.Number （個數）

Piece (pc, pcs)	個，件	Bale	包
Set	套，組，臺，部	Unit	部
Dozen (doz)	12 個，與數字連用時複	Each (ea)	件，條
	數形式與單數同		
Gross	籮，12 打	Bundle	束
Great gross	大籮，12 籮	Sheet	張
Roll	捲，matting 用	Deca	10 個
Coil	捲，wire 用	Bag	包
Reel	捲，線捲	Case	箱
Ream	令，紙張用	Carton	箱
Pair	雙	Pack	包
Dozen pair	打雙	Head	頭數，複數不加 s

3.Area （面積）

Square feet (sq. ft.)	平方呎
Square yard (sq. yd.)	平方碼
Square meter (m)	平方公尺

4.Length （長度）

Yard (yd.) = 3 feet = 91.4 cm

Foot (ft.) = 12 inches

Centimeter (cm)

Meter (m) = 39.37 inches = 3.28 ft. = 1.09 yds.

5.Capacity （容積）

Gallon $\begin{cases} 英制： 277.42 \text{ cubic inches} \cdots \text{imperial (British) gallon} \\ 美制： 231 \text{ cubic inches} \cdots \text{U.S. gallon} \end{cases}$

Liter = 0.264 gallon

Cubic centimeter (c.c.)

6.Volume　（體積）

Cubic feet (cft., cu. ft.)

Cubic yard (cu. yd.)

Measurement ton $\begin{cases} 40 \text{ cft.} \\ 1 \text{ CBM} \end{cases}$

Cubic meter = CBM = 35.3 cft

二、有關數量條件用語

1. Gross Weight (GW); Tare weight; Net Weight (NW); Net Net Weight (NNW); Legal weight：

"Gross weight"：為毛重，即包括包裝材料在內的重量。

"Tare weight"：為包裝材料的重量。

"Net weight"：Gross weight 減去 Tare weight 為 Net weight（淨重）。

"Net net weight"：純淨重，為毛重除去包裝材料後的重量，也即商品本身的實際重量，所以又稱為 Actual net weight。在很多場合 Net weight 即為 Net net weight。

"Legal weight"：法定重量，為商品重量包含裝飾包裝的重量。

例：10,000 long tons of 2,240 lbs. bagged wheat, gross for net, bags as wheat, such bags to be paid as wheat.

"gross for net"：為以毛重代替淨重之意。按毛重買賣者稱為毛重條件 (Gross weight terms)，按淨重者稱為淨重條件 (Net weight terms)。

2. Shipped quantity terms 與 Landed quantity terms：

"Shipped quantity terms"：為「裝運數量條件」，即賣方交給買方的數量以裝運時的數量為準者。

"Landed quantity terms"：為「卸貨數量條件」，即賣方交給買方的數量以卸貨時的數量為準者。

例：The shipped weight and/or count at the time and place of loading port shall be final.

Landed net weight at port of destination shall be final.

500 metric tons, GSW (gross shipped weight)

3. More or Less, Plus or Minus, Increase or Decrease：

有些貨物因性質關係，技術上不易按約定數量準確交貨，因此常約定可多交或少交若干或要求多交或少交若干，規定這種條件的條款稱為過與不足條款 (More or less clause)，例：

Sellers have the option of shipping 5% more or less of the contracted quantity.

"More or less" 又可以 "Plus or minus" 或 "Increase or decrease" 或 "±" 代替。

三、有關數量的有用例句

1. For any special coloring process, we set a minimum quantity for order. The smallest (minimum) for No. 123 is 25,000 yards.

2. The smallest (minimum) order we can fill for this quality is 3,000 yds. If your requirement is below that, we suggest that you review the enclosed color cards and choose the shade most nearly like it.

   註：a. quality：貨色；b. shade：色度。

3. The above price is based on a minimum order of 1,000 doz.

4. Quantity, unless otherwise arranged, shall be subject to a variation of 5% plus or minus at seller's (buyer's, ship's) option.

5. Any shortage or excess within one percent of B/L weights shall not be taken into consideration.

   註：shall not be taken into consideration：不予考慮；不計。

6. As to weight and/or measurement, an accredited surveyor's certificate shall be final at loading port.

   註：accredited：可信賴的。

7. Public surveyor's certificate of weight at loading port to be final.

8. All landed weight shall be considered final and conclusive.

9. Weighing to be done within six days after delivery in to consignee's craft.

10. Weighing at the seller's works or at the place of despatch shall govern.

第四節　價格條件

買賣雙方來往折衝，多是為了價錢的問題，商人所以熱衷於逐什一之利，也無非想從交易中獲得差價，賺取利潤。

一、交易所使用貨幣種類

在國際貿易，用以表示價格的貨幣不外本國貨幣、對方國貨幣及第三國貨幣，我國目前使用較多的是美金 (US dollars)。

二、價格基礎

在國際貿易做為價格基礎的貿易條件 (Trade terms) 以 FOB，CFR，或 CIF 使用最多，但有時也偶爾以其他條件做為價格基礎。茲據 Incoterms 2000 將其所規定者列舉於下：

1.以出口地為交貨地的貿易條件：

⑴ Ex works (Ex factory, Ex mills...)　　　　　　　　工廠交貨價

⑵ FCA (Free Carrier)　　　　　　　　　　　　　貨交運送人價

⑶ FAS (Free Alongside Ship)　　　　　　　　　　船邊交貨價

⑷ FOB (Free on Board)　　　　　　　出口港船上交貨價，離岸價格

⑸ CFR (Cost and Freight)　　　　　　　　　　　　運費在內價

⑹ CIF (Cost, Insurance, Freight)　　　　運費、保費在內價，到岸價格

⑺ DAF (Delivered at Frontier)　　　　　　　　　　邊境交貨價

⑻ CPT (Carriage Paid to)　　　　　　　　　　　　運費付訖價

⑼ CIP (Carriage and Insurance Paid to)　　　　　　運保費付訖價

2.以進口地為交貨地的貿易條件：

⑴ DES (Delivered Ex Ship)　　　　　　　　　　目的港船上交貨價

⑵ DEQ (Delivered Ex Quay)　　　　　　目的港碼頭交貨價（關稅付訖）

⑶ DDU (Delivered Duty Unpaid)　　　　　　　　　稅前交貨價

⑷ DDP (Delivered Duty Paid)　　　　　　　　　　稅訖交貨價

例:

EUR 34 per case Ex Factory Taipei, delivery during August

（每箱 34 歐元，臺北工廠交貨價，8 月份交貨）

EUR 35 per case FCA CKS Airport, delivery during August

（每箱 35 歐元，中正機場交貨價，8 月份交運）

EUR 38 per case FAS Keelung, delivery during August

（每箱 38 歐元，基隆船邊交貨價，8 月份交運）

EUR 39 per case FOB Keelung, shipment during August

（每箱 39 歐元，基隆船上交貨價，8 月份交運）

EUR 44 per case CFR Hamburg, shipment during August

（每箱 44 歐元，至漢堡運費在內價，8 月份交運）

EUR 45 per case CIF Hamburg, shipment during August

（每箱 45 歐元，至漢堡運費、保費在內價，8 月份交運）

EUR 46 per case DES Hamburg, delivery during August

（每箱 46 歐元，漢堡船上交貨價，8 月份交貨）

EUR 60 per case DEQ Hamburg, delivery during August

（每箱 60 歐元，漢堡碼頭交貨價，8 月份交貨）

EUR 61 per case DDU Hamburg, delivery during August

（每箱 61 歐元，漢堡稅前交貨價，8 月份交貨）

EUR 64 per case DDP Hamburg, delivery during August

（每箱 64 歐元，漢堡稅訖交貨價，8 月份交貨）

貿易條件一方面可表示貨物在運輸中風險移轉時、地，他方面又可表示買賣雙方如何分擔運費、保險費，及其他構成價格因素的費用，所以貿易條件也狹義地稱為價格條件。

國際上解釋貿易條件的規則，除上述 Incoterms 之外，尚有修訂美國對外貿易定義 (Revised American Foreign Trade Definitions, 1990)。因為其解釋與 Incoterms 略有出入，買賣雙方為避免誤會，宜在報價單、訂單或契約中訂明應以何種規則為準，例如約定：Unless otherwise agreed upon, the trade terms in this contract shall be governed and construed under and by the provisions of INCOTERMS 2000.

三、價格變動條款

在國際買賣，自訂約至交貨付款完畢，通常有一段相當長的時間，在此期間如遇到成本、匯率等劇烈變動，其中一方難免將遭受嚴重損失，因此，在物價、匯率變動幅度較大的時期，買賣雙方常約定價格變動條款。

1. 原料成本變動條款：

Seller reserves the right to adjust the contracted price, if prior to delivery, there is any substantial variation in the cost of labor or raw materials.

2. 運費變動條款：

Ocean freight is calculated at the prevailing rate, and increase in the freight rate at the time of shipment shall be for buyer's account.

3. 匯率變動條款：

Exchange risks, if any, for buyer's account.

四、關於價格的用語

1. 各種價格：

Actual price	實價	Unit price	單價
Fixed (Set, Settled) price	標價	Asked price	要價
Import price	進口價格	Bid price	開價
Export price	出口價格	Blanket (Lump) price	總括價格
List price	定價	Reduced price	減價
Market price	市價	Opening price	開盤價
Tag price	掛牌價	Closing price	收盤價
Factory price	出廠價	Ruling price	市價
Buying price	購價	Current price	時價
Purchasing price	購價	Prevailing price	時價
Selling (Sale) price	售價	Base price	基價
Net price	淨價	Fair price	平實價格
Gross price	總價	Firm price	確定價格
Prime cost	成本	High (Low) price	高（低）價
Marked-down price	削碼價	Moderate price	廉價

Cash price	付現價	Official price	公定價格
Credit price	賒帳價	Black market price	黑市價格
Contract price	契約價格	Quoted price	報價
Average price	平均價格	Retail price	零售價格
Flat price	統一價格	Wholesale price	批發價格
Unit price	單價	Price terms	價格條件
Trade price	同業價格	CIFC$_3$ price	包括佣金 3% 的起岸價格
Upset price	開拍價格	Price fluctuation	價格變動
Price current	行情	Price Limit	限價
Price index	物價指數	FOB price	離岸價格

2.表示價錢高、好：

High price	高價	Maximum (Highest) price	最高價格
Extravagant Excessive Exorbitant Unreasonable } price	過高價格	Ridiculous Fabulous Absurd } price	荒唐的價格
		Fancy price	高昂價格
Favorable Handsome Fine (Good) } price	好價格		

3.表示價錢正常：

Moderate (Reasonable) price	中庸(合理)價格	Remunerative price	合算價格
Normal price	普通價格	Right price	合理價格
Satisfactory price	令人滿意的價格	Competitive price	競爭性價格
Usual price	通常價格		

4.表示價錢低廉、便宜、差：

Low price	低價	Losing price	賠錢價
Minimum (Lowest) price	最低價	Rock-bottom price	最低價
Floor price	底價	Bed-rock price	最低價
Unfavorable (Bad) price	價錢差	Half price	半價

Special price	特價
Attractive price	誘人的價格
Popular price	廉價

註：Price 的高低用 "High"，"Low" 形容，不能用 "Dear" 或 "Cheap" 形容，例：The Prices rule high. The goods command a high price. "Dear" 為 "High in price" 之意，"Cheap" 為 "Low in price" 之意，因此我們說：This is dear, but that is cheap. These goods are high in price. The price is too high for me.。

5.與動詞連用：

⑴一般性的

To quote on the goods	就該貨報價
To quote a price for the goods	就該貨報價
To offer a price	報價
To bid a price	要價
To sell above limit	高於限價賣出
To exceed a limit	超過限價
To overcharge	索價過高
To knockdown a price	報價
To raise (advance, increase) a price	漲價
To lower (lessen, bring down) a price	降價
To fix a price	決定價格
To name a price	指定價格
To buy (sell) at limit	限價買進（賣出）
To buy below limit	低於限價購入
To work out a price	算出價格
To make up a CIF price	算出 CIF 價格
To reduce a price	減價

To $\left\{\begin{array}{l} \text{get} \\ \text{make} \\ \text{allow} \end{array}\right\}$ a discount off a price　　　　折價

To shade a price	減一點價錢

⑵表示上漲

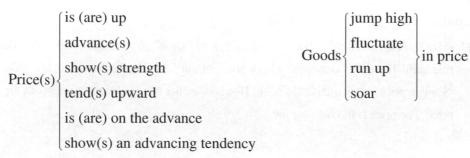

$$
\text{Price(s)} \begin{cases} \text{is (are) up} \\ \text{advance(s)} \\ \text{show(s) strength} \\ \text{tend(s) upward} \\ \text{is (are) on the advance} \\ \text{show(s) an advancing tendency} \end{cases}
$$

$$
\text{Goods} \begin{cases} \text{jump high} \\ \text{fluctuate} \\ \text{run up} \\ \text{soar} \end{cases} \text{in price}
$$

⑶表示下跌

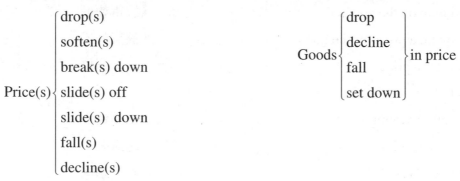

$$
\text{Price(s)} \begin{cases} \text{drop(s)} \\ \text{soften(s)} \\ \text{break(s) down} \\ \text{slide(s) off} \\ \text{slide(s) down} \\ \text{fall(s)} \\ \text{decline(s)} \end{cases}
$$

$$
\text{Goods} \begin{cases} \text{drop} \\ \text{decline} \\ \text{fall} \\ \text{set down} \end{cases} \text{in price}
$$

五、有關價格條件的有用例句

㈠價格的計算

1. The price is net price, without any commission.

2. The price includes 5% commission on FOB basis.

3. US$20 per $\begin{cases} \text{unit} \\ \text{dozen} \\ \text{set} \end{cases}$ CIF New York, including your commission 5% on FOB basis.

4. The price of this offer is calculated at the rate of US$1 to NT$35, when there is any surplus or deficiency in NT$ proceeds at the time of negotiating a bill of exchange, such difference shall be adjusted. Any surplus shall be refunded to buyer, and any deficiency shall be compensated by buyer.

（本報價係按 US$1 = NT$35 匯率計算，押匯時如臺幣收入有多或不足，則應予調整，多餘時應退還賣方，不足時應由買方補足。）

5. No price adjustment shall be allowed after conclusion of contract.

6. Bunker surcharge imposed by the steamship company, if any, shall be for buyer's account.

註: Bunker surcharge imposed by the steamship company.: 船公司所徵收的燃料附加費。

7. Ocean freight is calculated at US$20 per metric ton, any increase in freight rate at the time of shipment shall be for the buyer's risk and account.

8. The goods are sold at US$103 per M/T CIFC$_3$ Kuwait.

9. Please offer us your lowest price for the under-mentioned, FOB Taiwan, and state earliest delivery.

10. The above mentioned price are based on orders of 1,000 dozen and upwards.

㈡價高及上漲

1. In view of the present state of the market, it is evident that prices will rule high for some considerable period.

2. Values all round have an upward tendency, and we believe that you would be well advised to lay in stock sufficient for your requirements during the next season.

3. It is believed that prices will again advance, perhaps higher than the point reached in the last advance.

4. Under the present conditions, when the cost of labor and raw materials may go up any day, it is impossible for us to guarantee the prices for any definite period.

5. Stocks are low, the Japanese crop is late, and there are disturbances in Korea, all which causes may have the effect of bringing about an improvement in prices.

㈢價廉及下跌

1. We are of the opinion that the present quotation US$10 per set FOB Taiwan is a rock-bottom price and a most profitable investment.

2. We are able to make this low prices, because the manufacturers have lowered their prices to us, and rather than keep the difference as extra profit for ourselves, we are going to give our customers the entire benefit of this cut in the prices.

3. Continued weakness in raw (sugar) on the increased competitions from outside sources and almost entire lack of demand for refined were the causes for the decline.

4. Our prices are low because of the peculiarly favorable location of our factory.

5. As we wish to effect a speedy clearance, we are quoting you an extremely cutting price.

6. Owing to the continual fall in prices, buyers are really afraid to place orders for forward shipment, as prices which appear to be attractive one day, are found within a week or so too be high.

㈣討價還價

1. You are no doubt aware that the wool market is very unsettled, and we are therefore obliged to allow ourselves reasonable margin in view of possible disturbances in the present prices.

   註：a. very unsettled：很不穩定；b. reasonable margin：合理的利用；c. disturbances：不正常。

2. Your quotation seems to be exorbitant, in view of the prices prevailing in this market, which stand at something like US$12 per dozen.

3. This price is quite impossible under present conditions, being far above the market here and also the above quotations received from other sources.

4. We are surprised at your remarks, so much so because we have tested this prices in several quarters, only to find that it is at least four dollars below present cost.

5. In view of the prices ruling in this market your prices are prohibitively high, and we commend them to your reconsideration as we believe that there is a mistake on your part in figuring out your quotation.

6. We are surprised that some other exporters in your area are underselling us, so much so because we are the originators of the machine which is protected by patent.

   註：a. underselling us：售價低於我們；b. originators：創作人；發明人。

7. If you can $\begin{cases} \text{make the prices a little easier} \\ \text{shade the prices a little} \end{cases}$, we shall probably be able to see our way to place an order with you.

8. The price is too high to admit of any profit on this transaction.

   註：too high to admit of any profit：太貴以致無利可圖。

9. Your competitors, who made us an offer the other day, charge for the same class of

goods, the quality being exactly the same as yours, only US$ 9 per dozen.

10. The higher prices than our last quotation are accounted for by an increase in ocean freight that has taken place since we last accepted your business.

11. We see no possibility of competing against rival makes unless you can reduce your quotation.

　　註：rival makes：競爭品（牌子）。

12. As there is keen competition for this article, your price will put us out of the market.

　　註：put us out of the market：將我們從市場驅逐出去。

 # 第五節　包裝及刷嘜條件

一、包裝種類

　　進出口貨物因大多需經長途的輾轉運輸，所以其包裝必需適當，才不致在運輸途中遭到損壞。至於何種包裝才適當，因貨物的性質、種類、搬運方法、經過路程、轉運與否、目的地碼頭設備等等情形而異。茲將常見的包裝類別及適用貨品列表於次頁。

外　形	包　裝　名　稱		略　字		適　用　貨　物
	英　　文	中　文	單　數	多　數	
箱　狀	WOODEN CASE	木　箱	C/	C/S	雜貨、儀器、機件
	WOODEN BOX	木　箱	Bx	Bxs	雜貨、儀器、機件
	CHEST	茶　箱	Cst	Csts	茶葉
	CRATE	板條箱	Crt	Crts	玻璃板、機器
	CARTON	紙　箱	Ctn	Ctns	雜貨、塑膠品、奶粉
	SKELTON CASE	漏孔箱	－	－	陶磁、蔬菜、青果
捆　狀	BALE	捆　包	B/	B/S	棉花、布疋
袋　狀	SACK	布　袋	Sx	Sxs	麵粉
	BAG	袋	Bg	Bgs	肥料、水泥等
	GUNNY BAG	麻　袋			玉米、小麥、雜糧
	PAPER BAG	紙　袋			水泥、肥料
	STRAW BAG	草　袋			鹽
桶　狀	BARREL	鼓形桶	Brl	Brls	酒、醬油
	HOGSHED	茶葉桶	Hghd	Hghds	茶葉
	CASK	樽	Csk	Csks	染料
	KEG	小　樽	Kg	Kgs	鐵釘

	TUB	木　桶	–	–	醬油
	DRUM	鐵　桶	Dum	Dums	油、染料
罐　狀	CAN	罐			罐頭食品
	TIN	听			罐頭食品
圓筒狀	CYLINDER	鋼　桶			液體瓦斯、氧氣
	BOMB	鋼　桶			液體瓦斯、氧氣
瓶　裝	BOTTLE	瓶	Bot	Bots	酒類
	DEMIJOHN	大　罎			酸性化學品
	CARBOY	酸　瓶			酸性化學品
	JAR	甕			酒類
	FLASK	細頭鋼瓶			水銀
簍　狀	BASKET	簍			蔬果

常見包裝用語措詞的比較：

POOR	BETTER
1. Press-packed in bales each containing 100 kgs. net	1. Press-packed in bales of 100 kgs. net
2. 200 lbs. to a bale	2. Packed in bales of 200 lbs.
3. B/No. 1–4 contains 20 pcs. each of 60 yds. length	3. B/s Nos. 1–4 each containing 20 pcs. of 60 yds. length
4. Packing: One dozen wrapped in paper, Bale packing	4. Packed in bales of 1 dozen wrapped in paper
5. Packing: 1.25 picul net to a straw bale	5. Packed in straw bales of 1.25 picul net
6. 200 lbs. to a case	6. Packed in cases of 200 lbs.
7. Packing: Packed in strong wooden cases and iron-hooped both ends	7. Packed in strong wooden boxes iron hooped at both ends
8. C/Nos. 1–9 contains 20 pcs. each of passed A	8. Cases Nos. 1–9 packed 20 pcs. A Grade to the case
9. per case N/Net weight 102 lbs.	9. Packed in cases of 102 lbs. net
10. per case 750 yds. (30 pcs.)	10. Packed 750 yds. in 30 pcs. to the case
11. 30 sq. ft. per case	11. 30 sq. ft. to the case
12. Packing: in bulk	12. Packed in bulk
13. Packing: in coil	13. Packed in coils of...
14. 30 kgs. net in burlap bag	14. Packed in burlap bags of 50 kgs. net
15. Each packing in a carton box: 6 in a crate	15. Packed in crates of 6 cartons of 1 piece each

二、嘜　頭

刷嘜 (Marking) 是指在包裝容器上用油墨、油漆或以模板 (Stencil) 加印嘜頭或標

誌 (Mark)，用以有別於其他船貨。所謂嘜頭，廣義的說，是指包裝上所標示的圖形、文字、數字、字母及件號的總稱。

嘜頭的主要功用有四：

⑴在搬運貨物時，易於識別，可避免誤裝誤卸。

⑵貨物包裝刷有嘜頭，賣方製作單證可簡化。就買方而言，容易從單證上明瞭貨物的包裝內容。

⑶可知道貨物的生產國。

⑷嘜頭是以圖形或字母代替文字，多不記明受貨人的全名，所以可使同業不易探知買主，而保持商業秘密。

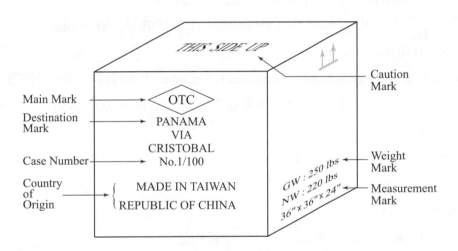

1. Main mark（主嘜頭）：

OTC in diamond		Projecting diamond（突出菱形）
Triangle（三角形）		Three diamond（三菱形）
Circle（圓形）		Square（正方形）
Rectangle（長方形）		Heart（心形）
Hexagon（正六角形）		Triangle crossed（雙三角形）

Cross（十字形）	Oval（橢圓形）
Hourglass（對三角形）	Cross in diamond（菱中十字形）
Downward triangle（倒三角形）	Intersecting parallels（井形）
Star（星形）	T. S. mixed（TS 兩字配合）

2. Port mark or Destination mark：

位於 Main mark 的下面，說明目的港或目的地。

PANAMA
　VIA
CRISTOBAL
　　　PANAMA 為 Destination, CRISTOBAL 為目的港。

3. Case number, Serial number, Consecutive number, Running number（件號，連續號碼）：表示同一批貨物的總件數及本件號碼。"No. 1/100" 表明本批總共有 100 件，本箱為第 1 箱。

4. Country of origin mark：原產地國。

Made in Taiwan

Republic of China

5. Care (Caution) mark 或 Side mark：注意標誌。

茲將常見的注意標誌列出若干於下：

This side up or This end up	此端向上
Handle with care or With care or Care handle	小心搬運
Use no hook or No hook	請勿用鉤
Keep in cool place or Keep cool	放置冷處（保持低溫）
Keep dry	保持乾燥
Keep away from boiler or Stow away from boiler	遠離鍋爐
Inflammable	易燃貨物
Fragile	當心破碎
Explosives	易炸貨物

Glass with care	小心玻璃
Poison	小心中毒
Heave here	此處舉起
Open here	此處開啟
Sling here	此處懸索
Do not drop	小心掉落
To be kept upright	豎立安放
Keep away from heat	隔離熱氣
Perishables	易壞貨物
Guard against wet (damp)	勿使受潮
No smoking	嚴禁煙火
Keep flat (stow level)	注意平放
Not to be thrown down	不可拋擲

6. Weight/Measurement mark：表示貨物毛重、淨重、尺碼。

三、有關包裝、刷嘜條件的有用例句

1. To be packed in

strong wooden case suitable for export

strong wooden case with iron hooped　　　（用鐵條箍住）

zinc-lined case, ...doz each　　　（鋅板襯裡箱）

wooden kegs lined with pitch papers　　　（瀝青紙）

hardwood iron-bound barrels

cardboard box　　　（硬紙板箱）

water-proof canvas　　　（防水帆布）

three-ply kraft paper bags　　　（牛皮紙袋）

new and sound jute bag each contains about 330 lbs.

wooden cases, each weighs about 1,200 lbs. net

export standard packing

seaworthy packing　　　（耐航包裝）

customary export packing　　　（習慣出口包裝）

ordinary export packing

> regular export packing
>
> conventional export packing （慣例出口包裝）
>
> 55 U. S. gallon, 24 gauge new steel drum, each containing 200 kgs. net

2. One piece in a polybag, 12 polybags in a paper box, then 60 boxes packed in an export carton, its measurement about 26 cft.

3. Each piece must be wrapped up in paper and packed in a zinc-lined case.

4. Marking: Every package shall be marked with "OTC" in diamond and the package number.

5. The following shipping marks must be clearly stenciled on each package, and stated on invoices and packing list.

　註：stencil：以模板鏤花。

6. Unless otherwise requested and instructed by the buyer in time, the seller will decide the marking at their discretion.

　註：at their discretion：任由他們決定。

7. The goods must be packed in multi-wall paper bags each containing 50 kgs. net.

8. As we undertake our own packing we can save the money for you on this score.

　註：on this score：在這一點上。

9. All goods must be packed in waterproof cotton.

10. Each piece to be wrapped up in paper and packed in P.V.C. bags.

　註：P.V.C. bags：塑膠袋。

11. Please see that each piece is rolled on a board of usual size and wrapped in kraft paper.

12. These casks should be made of oak, water-tight and iron-bound.

13. Please remember that the goods must be extremely well packed, as the shipments will be transhipped at Hongkong.

14. Mark the cases ⟨ CTC ⟩ and number consecutively.

第六節　保險條件

一、保險投保人

在一筆國際買賣中，貨物的保險究應由買方或賣方負責付保，端視其貿易條件而定。如以 EXW, FCA, FAS, FOB, CFR, CPT 條件成交，則除非另有協議由賣方代為購買保險，否則應由買方自行投保。如以 EXS、DDU、DDP 或 EXQ 條件成交，則貨物應由賣方保險。如以 C&I、CIF 或 CIP 條件成交，則賣方有義務購買保險，但裝船後如發生危險事故，則由買方負責向保險公司索賠。正因如此，在 C&I、CIF 或 CIP 交易時，買賣雙方應就保險條件作適當的約定。

二、保險的種類

1.基本險類
- ICC (A)　　　　　　　　　　　　　　　　　　　A 款險
- ICC (B)　　　　　　　　　　　　　　　　　　　B 款險
- ICC (C)　　　　　　　　　　　　　　　　　　　C 款險

2.附加險
- TPND (Theft, Pilferage and Non-Delivery)　　　竊盜、遺失險
- RFW/D (Rain and Fresh Water Damage)　　　　雨水淡水險
- COOC (Contact with Oil and/or Other Cargoes)　接觸險
- JWOB (Jettison and/or Washing Overboard)　　投棄浪沖險
- Heat & Sweat damage　　　　　　　　　　　　發熱或汗濕險
- Breakage　　　　　　　　　　　　　　　　　　破損險
- Hook hole　　　　　　　　　　　　　　　　　　鉤損險
- Shortage　　　　　　　　　　　　　　　　　　短失險

上述 ICC (A), ICC (B), ICC (C) 均不包括兵險 (War risks) 或罷工暴動險(SR&CC)。因此，如須獲得這種危險的保障，須另加保。

三、保險金額

根據 Incoterms 2000 應按 110% of CIF value 付保。

根據 Warsaw-Oxford Rules 應按 110% of CIF Value 付保。

根據 UCP（信用狀統一慣例）應按 110% of CIF Value 付保。

四、有關保險條件的有用例句

1. Marine insurance to be effected by you on our behalf for gross amount of the invoice plus 10 per cent.

2. We attend to insurance, so please inform us of the name of vessel and the date of sailing, loading port, quantity shipped, value, etc., at the same time of loading by telex.

3. Please advise Messrs. Hall & Co., agents for Lloyd's, of shipment, declaring invoice value of goods, with 10% added thereto.

4. Insurance to be covered by buyer, and any kind of possible loss/damage, such as breakage, shortage, theft, pilferage, etc., after loading shall be covered by insurance by buyer at buyer's option or risks.

5. Insurance to be covered by seller.

6. Insurance to be covered by seller against ICC (B) including TPND. If not arranged otherwise, war risk insurance to be covered by buyer.

7. Premium of war risks is calculated at 0.1%, any increase in premium of war risk subsequent to conclusion of the contract shall be for buyer's account.

   Therefore, L/C must include following clause:

   "If war risk premium is higher than 0.1%, beneficiary is authorized to draw the difference in excess of L/C amount."

8. Age premium, if any, shall be for seller's account.

9. Marine insurance covering ICC (A), War Risk and SR&CC for full CIF value plus 10% shall be effected by the seller and should be covered up to buyer's warehouse in Taipei.

第七節　交貨條件

所謂「交貨」(Delivery) 即賣方將貨物交給買方之意。買賣雙方對於交貨條件所要

約定的有:

　　1.交貨地點 (Place of delivery)。

　　2.交貨時間 (Time of delivery)。

　　3.交貨方法 (Method of delivery)。

一、交貨地點

　　以FOB, CFR, CIF, FCA, CPT, CIP條件交易時，交貨地點為出口港船上，所以所謂「交貨」實際上即為在裝運地點 (Place of shipment) 交出貨物之意。但按DES, DEQ, DDU, DDP條件成交時，因交貨地點在目的港，所以在此情形，"Delivery" 與 "Shipment" 並不同。為避免混淆，在FOB, CFR, CIF, FCA, CPT, CIP等條件時，宜用 "Shipment" 一詞。

二、交貨時間

　　1.即期交貨:

　　(1) Immediate shipment (delivery)　　　隨即裝運（交貨）⎫

　　(2) Prompt shipment (delivery)　　　　即期裝運（交貨）⎬四星期內

　　(3) Shipment as soon as possible　　　儘速裝運　　　　⎭

　　(4) Shipment by first available steamer (ship, boat, vessel)　　有船即裝

　　(5) Shipment by first opportunity　　優先裝運

　　2.定期交貨:

　　(1) Shipment ⎰in⎱ July
　　　　　　　　 ⎱during⎰

　　(2) Shipment ⎧in the beginning⎫ of May　⎧上旬⎫
　　　　　　　　 ⎨in the middle　 ⎬　　　　⎨中旬⎬裝運
　　　　　　　　 ⎩in the end　　　⎭　　　　⎩下旬⎭

　　(3) Shipment ⎰during the first half　⎱ of May　⎰上半月⎱裝運
　　　　　　　　 ⎱during the second half⎰　　　　　⎱下半月⎰

　　(4) Shipment ⎰by　　　　⎱ August 20　　8月20日之前裝運
　　　　　　　　 ⎱on or before⎰

　　(5) Shipment on or about August 20　　8月15日至8月25日之間裝運

　　(6) Shipment in early November　　11月初旬裝運

(7) Shipment within 30 days after receipt of L/C

(8) Shipment within 30 days $\begin{cases} \text{after} \\ \text{on} \\ \text{of} \end{cases}$ receipt of remittance　收到匯款後 30 天內裝運

(9) Shipment must be effected $\begin{cases} \text{on or before...} \\ \text{by end of...} \\ \text{during...} \\ \text{during April and May.} \\ \text{within...days after receipt of L/C.} \\ \text{early in December.} \end{cases}$

3.交貨時間附帶條件：

(1) Shipment during August subject to shipping space available.

(2) Shipment by April 30 subject to L/C reaching seller on or before March 15.

(3) Shipment during February/March subject to approval of $\begin{cases} \text{export licence.} \\ \text{export permit.} \end{cases}$

三、交貨方法

1.可否分批裝運：

(1) Partial shipments (to be) $\begin{cases} \text{allowed.} \\ \text{permitted.} \\ \text{not allowed.} \\ \text{unallowed.} \\ \text{prohibited.} \\ \text{forbidden.} \end{cases}$

(2) Shipment in two approximately equal monthly instalments,

→ $\begin{cases} \text{beginning} \\ \text{commencing} \end{cases} \begin{cases} \text{with} \\ \text{in} \end{cases}$ March.

(3) Ship 1,000 doz. during September and the balance a month later.

(4) Shipment to be spread equally over three months $\begin{cases} \text{beginning} \\ \text{commencing} \end{cases} \begin{cases} \text{in} \\ \text{with} \end{cases}$ May.

(5) Shipment to be effected between June 15 and September, 19— in four equal shipments with interval at least 15 days apart.

2.可否轉運：

$$
\text{Transhipment}
\begin{cases}
\text{allowed.} \\
\text{permitted.} \\
\text{not allowed.} \\
\text{unallowed.} \\
\text{prohibited.} \\
\text{forbidden.}
\end{cases}
$$

3.運輸工具：

(1) Shipment per s.s. "Travestein" sailing around April 15 from Keelung.

(2) Shipment to be
$$
\begin{cases}
\text{effected} \\
\text{made}
\end{cases}
$$
per APL's steamer.

(3) Shipment to be
$$
\begin{cases}
\text{effected} \\
\text{made}
\end{cases}
\text{per}
\begin{cases}
\text{American President Line.} \\
\text{U.S. flag vessel.} \\
\text{container vessel.} \\
\text{airfreight.} \\
\text{air.} \\
\text{air parcel post.}
\end{cases}
$$

四、有關交貨的動詞

To request delivery

To promise delivery

To guarantee delivery

To expedite delivery

$$
\text{To}
\begin{cases}
\text{deliver} \\
\text{make} \\
\text{effect} \\
\text{give} \\
\text{undertake}
\end{cases}
\text{delivery}
$$

To stop delivery

To refuse delivery

To dispatch (despatch)

To send

To take delivery

To accept delivery

To begin delivery

To complete delivery

五、有關交貨條件的有用例句

1. We can generally make delivery of fairly large quantities not later than prompt, but this must not be taken as a promise in so far as the time we shall require for delivery would naturally depend upon the size of your order and the conditions at our mills at the time your order is received.

   註：not later than prompt：三星期以內。

2. According to our contract with our clients we have to deliver it by semi-monthly instalments of 1,000 M/T, failing which we are under penalty of a heavy fine.

   註：failing which...a heavy fine：否則我們將被罰鉅額違約金。

3. We are under contract to deliver the goods by August 1, failing which will mean a serious loss to us.

4. Owing to great pressure at the mills, we are afraid we cannot guarantee delivery within less than three months from this date.

5. Owing to the unprecedented demand for the goods, we are reluctantly compelled to ask for one month delivery, but every effort shall be made an early delivery of your order.

6. We have a large stock of this year's crop, and this, together with our unsurpassed shipping facilities, places us in a position to give delivery at any time on receipt of your order.

   註：unsurpassed：最佳的。

7. Please let us know per return whether you can undertake delivery within 14 days from receipt of order.

   註：undertake delivery：交貨。

8. If you will give us some ideas as to the quantity you are likely to require, we will name the shortest period in which we can promise delivery.

第八節　付款條件

在一筆買賣中，賣方的首要義務為將約定貨物交給買方，而買方的首要義務為依約將貨款付給賣方。由於國際貿易的買賣雙方遠隔異地，交貨與付款不能同時履行，所以對於貨款的清償方式，須視雙方的各種情況而異。

一、付款方式

 1.預付貨款 (Payment in advance)：屬於這種方式的有：

⑴ CWO (Cash with Order)：訂貨時付現，也稱為 CIA (Cash in Advance)。

⑵ Anticipatory credit（預支信用狀）。

 2.裝運付款 (Payment on shipment)：屬於這種方式的有：

⑴ CAD (Cash against Documents)：通常指在出口地交單證時，進口商或其在出口地的代理人即須付款，但也有解釋為後述的 D/P。

⑵ Sight L/C（即期信用狀）詳參閱第廿一章。

 3.延期付款 (Deferred payment)：屬於這種方式的有：

⑴ COD (Cash on Delivery)：買方收到貨物時付款，美國則稱為 "Collect on delivery"。

⑵ D/P (Documents against Payment)：付款交單。即賣方將貨物交運後備齊貨運單證，經由銀行向買方收款，買方付款的同時取得單證，茲圖示如下：

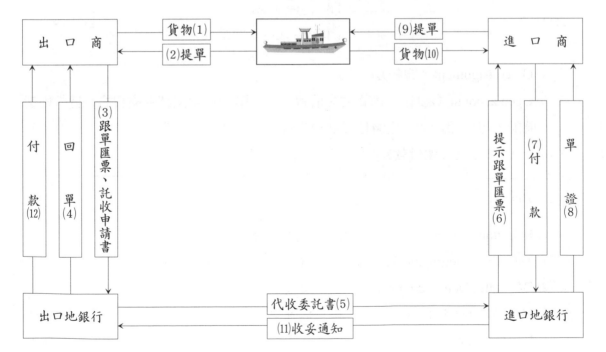

⑶ D/A (Documents against Acceptance)：承兌交單。與 D/P 不同之點在於交單條件，在 D/A 條件下，買方一經承兌賣方所簽發匯票便可取得單證，至於貨款則俟匯票到期時才支付。

茲圖示如下：

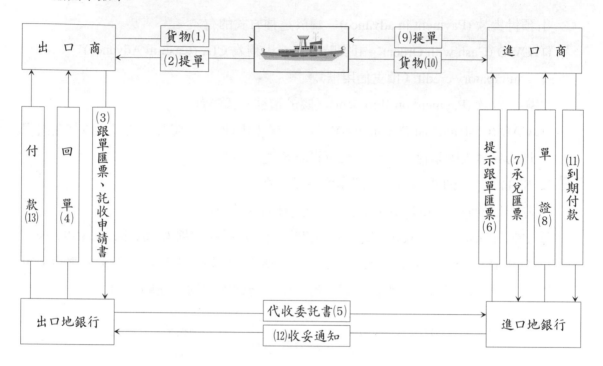

⑷ On consignment（寄售）。

⑸ Open account（記帳）：即賣方先將買方所訂購的貨物陸續運交買方，貨款則暫時記入專戶帳，於一定時日後結帳匯付賣方，實際上即為賒帳。

⑹ Instalment（分期付款）。

二、付款工具

1. Remittance（匯付）：係由買方將貨款匯交賣方的方法。

⑴ T/T, TT (Telegraphic Transfer)　　電匯，又稱 "Cable transfer"

⑵ D/D, DD (Demand Draft)　　票匯

⑶ M/T, MT (Mail Transfer)　　信匯

⑷ Personal check　　私人支票

2. Drawing（發票）：即由賣方向買方簽發匯票收取貨款的方法。

⑴ Clean Bill (C/B)　　　光票（不附單證）

⑵ Documentary bill　　　跟單匯票，"Bill" 為 Bill of exchange 之簡稱。

三、關於付款的用語

1. 名詞：

Settlement	Cash payment
Reimbursement　求償，歸償	Prompt payment
Repayment　還款	Full payment
Refundment　退款	Payment in full
Mode of payment	Delay in payment
Prepayment	Extension of payment
Monthly instalment	Application for payment　請求付款
Part ⎫ Partial ⎭ payment	Request for payment　要求付款
	Demand for payment　催促付款

2. 動詞：

To pay	To defray　支出
To ⎰make⎱ payment 　⎱effect⎰	To repay　償還
To settle	To balance, square　結算
To ⎰make⎱ settlement　清償 　⎱effect⎰	To ⎰reimburse⎱ oneself for　求償 　⎱cover⎰
To pay down　即時支付	To enforce payment　強迫付款
To pay in advance	To withhold payment　拒絕付款
To clear off　償清	To suspend payment　暫停付款
To apply for payment　請求付款	To stop payment　止付
To solicit payment　懇求付款	To defer payment　延付
To request payment	To delay payment　遲付

To demand payment　　　　　　　　To escape payment

To send check in payment $\begin{cases} \text{of an account} \\ \text{for some goods bought} \end{cases}$

四、有關付款條件的有用例句

1. Payment shall be made Cash With Order by means of T/T or M/T.

2. Payment shall be made in advance, for full contract value by any of the following means:

   a. T/T or M/T

   b. Bankers' Draft

   c. Check (Payment shall not be deemed received unless the amount of the check has been collected.)

3. Ten percent of the contract value shall be paid in advance by cash, and ninety percent by sight draft drawn under an irrevocable L/C.

4. Terms of payment: Net cash against documents payable in New York.

5. Terms of payment: Cash on Delivery.

6. Payment against goods shipped on consignment.

7. Payment shall be made by a prime banker's irrevocable & transferable L/C in favor of seller, available by draft at sight for 100% invoice value.

8. Payment shall be made by an irrevocable L/C available by sight draft drawn on a prime bank, when accompanied by the following documents:

   a. Clean ocean B/L

   b. Commercial invoice

   c. Inspection certificate

   d. Marine Insurance Policy

9. L/C must be opened within 10 days after conclusion of contract, otherwise this contract shall be cancelled unconditionally.

10. Payment shall be made by draft drawn under L/C payable 180 days after presentation of documents to the drawee bank, together with an interest of eight percent for

buyer's account.

11. Payment to be made by draft drawn on buyer payable at sight, D/P.

12. Payment shall be made by 180 days' sight bill, D/A.

13. For payment, we shall arrange with the Bank of Taiwan, Taipei for an irrevocable L/C in your favor for the amount of US $10,000.

14. Should you be able to execute the order we would establish credit with approved bankers immediately on receipt of your cable.

15. Terms of payment: O/A, payment to be made within 60 days after B/L date.

16. Terms of payment: payment shall be made by check when the goods have been sold.

17. Terms of payment: payment shall be effected by T/T or M/T immediately after sale of goods.

第十三章
詢價 (Trade Inquiry)

第一節　詢價信的寫法

　　傳統的貿易做法是在初次提議交易 (Trade proposal) 時，先徵信 (Credit inquiry)，次之，訂立一般交易條件協議書，而後才正式進行實際的交易行為。但現今的做法，往往單刀直入，素不相識的買賣雙方，對方一有提議交易，即逕行商談 (Business negotiation)，俟達到某程度的具體化後，才進行徵信，等到交易成立階段，才又將附有一般交易條件的 Sales note 或 Order 寄交對方確認。事實上，在很多場合買賣雙方取得聯繫或獲悉對方之後，不管對方情形如何，出口商即逕向對方進口商推銷其貨品，進口商則逕向出口商寫信探詢有關貨品的種種問題。由進口商主動寫信向賣方提出有關貨品的詢問時，這種詢問稱為 "Inquiry"，但為求與前述 Credit inquiry 有所區別起見，常常稱為詢價 (Trade inquiry) 或業務詢問 (Business inquiry)。詢價又稱探詢，實際上就是買方對某種貨品的查詢。買方的查詢固然多屬於價格方面，但並不以此為限。諸如索取目錄、樣品、往來條件、貨品的有無、種類、數量、交貨期等也在查詢之列。這種 Trade inquiry，有時將其希望購買貨品名稱、品質、數量、交貨期等具體記載，已略具 Buying offer 的雛形，再者，如採行單刀直入的做法，則這種 Trade inquiry 與前述 Trade proposal，在事實上往往很難區別。

　　Trade inquiry 依其性質可分為二：

　　1.查詢某事：這種 Inquiry 並不一定立即進行交易，其內容包括：

　　⑴請寄送某種商品的樣品、目錄或價目表。

　　⑵探詢某種商品的品質、數量、價格、交貨期等等。

　　2.請求報價：這種 Inquiry 實際上就是 "Request for offer"。

　　買方已準備購買某種商品，請賣方就某一種商品報價。

　　寫詢價信內容約有三：

　　1.開頭先將所要詢問的問題提出：如問題簡單，一句話即可，如問題複雜，應逐項清楚列出，不要囉嗦陳述許多無謂的話。

2.其次陳述詢問目的：如採單刀直入做法者，應先自我介紹。

3.有禮貌的結束。但宜避免 "Thank you in advance" 的濫調。

No. 38　查詢錄音機價錢，並索取目錄樣品

Dear Sirs,

Cassette Tape Recorder

Thank you very much for your extensive consideration in establishing business relations with our company. With the prospects of great success we wish to start off with an initial order for 500 sets of your most popular cassette tape recorder, Model CRC–137.

As the demand for inexpensive cassette tape recorders is high, we may expect a successful sale depending on the cost and quality of your machines.

Incidentally we shall be much obliged if you will send us your latest catalog listing your tape recorders and price with samples.

Your prompt reply will be much appreciated.

Yours very truly,

【註】

1. your extensive consideration：深思熟慮；深入的考慮。

2. with the prospects of great success：預期未來往來關係可獲致很大的成就。

3. start off：開始。

4. initial order：初次訂單。cf. trial order：嘗試訂單。

5. the demand for...is high (or active, bullish, brisk)：對……需求殷切。

6. successful sale：銷售良好；暢銷。

7. depending on：決定於。

8. incidentally：附帶地，又可以 "meanwhile"，"in the meantime" 代替。

9. latest catalog：最新的貨品目錄，不可用 "newest catalog"。

No. 39　請寄府綢樣品及報價

Gentlemen:

We appreciate the information you have so kindly furnished us in your letter of January 30.

We deeply regret that business between us has been suspended for some time owing to the slackness of trade. Business in general, however, seems to be picking up, and demands will revive. We feel confident, therefore, that we shall be able to resume dealings in your Silk Piece Goods. As to Tussore Silk, in which you wish to do business, prospects are also hopeful as we have a fairly large outlet for it. We ask you to send us samples of Tussore Silk and quote best rates. On receipt of your samples we shall place them before our prospects, and if it is possible for us to do business at your figures, we shall cable you an offer in the usual way.

We appreciate your keeping us well informed as to the trend of your market and welcome any suggestions you may offer for our import operations.

Very sincerely yours,

【註】

1. slackness = dullness; depression; stagnancy; stagnation：不景氣；蕭條。

2. to be picking up = to be improving：轉好。

3. revive：復甦。

4. tussore silk：府綢。

5. prospects = prospective buyers (or customers)：可能的顧客，英國人常用 "likely buyers" 一詞。

6. outlet = market：銷路；出路。cf. Bombay is an excellent outlet for garments.

7. figure = price。

8. to keep one well informed：使某人消息靈通。informed = posted。

9. trend = condition; state; tendency：趨勢；情勢；狀態。

No. 40　請寄刺繡樣品及報價

Dear Sirs,

We are interested in importing 36″ Nylon Embroideries. We anticipate that a substantial import licence will be granted us, but apart from importing in our own name we would also be offering these goods to other wholesalers and manufacturers.

If, therefore, your quotations are satisfactory, substantial business could follow.

Please send us by airfreight quality sample feelers and shade cards and CIF quotations—the quotations to include a 5% commission to us.

Yours faithfully,

【註】

1. 36″ Nylon embroideries：寬 36 吋的尼龍刺繡。

2. grant：給與。

3. apart from importing in our own name：除我們自行進口之外。

4. quality sample feelers：品質樣品，"feelers" 指為觸摸用的樣品，主要用於織品。

5. shade cards：色度卡（用以表示顏色深淺）。

No. 41　詢購壓克力毛線衫

Acrylic Sweaters

Gentlemen,

We have recently received many inquires from retailing shops in Hamburg area about the subject articles and sure that there would be very brisk demands therefor at our end. We, therefore, are writing to you for your quoting us your most competitive prices on a CIF Hamburg basis for the following:

Commodity: Acrylic Sweaters for men and ladies in different color/pattern assortments.

Quantity: 1,000 doz.

Size assortments: $\dfrac{S}{3} \dfrac{M}{6} \dfrac{L}{3}$

Packing: to be wrapped in polybags and packed in standard export cardboard cartons.

Since this inquiry is an urgent one, please indicate in your quotation the earliest shipment you are able to make for delivery. Competition of these articles is very keen here, therefore, not only your prices should be most competitive but also the quality should be the best.

Meanwhile, please also send us by air parcel one sample each of these garments for our evaluation.

Should the prices you quoted be acceptable and the quality of your samples meets with our approval, we will place orders with you forthwith.

Your prompt response is requested.

Faithfully yours,

【註】

1. acrylic sweaters：壓克力毛線衫。

2. retailing shops：零售店。

3. color/pattern assortments：顏色及式樣的搭配。

4. size assortments：尺寸的搭配。

5. cardboard：硬紙板。

6. evaluation：評估。

7. forthwith：立即。

8. response：答覆。

No. 42　詢購特級蓬萊米

Re: Bonlai Rice

Dear Sirs,

　　We have just received an enquiry from one of our Singapore customers who needs 5,000 M/T's of subject rice and shall appreciate your quoting us your best price immediately.

　　For your information, the quality should be Bonlai Rice, year 2003 autumn crops packed in new PE woven bags of 100 kgs. net each. In the meantime, the rice should be inspected by an independent public surveyor as their quality and weight at the time of loading.

　　The buyers in Singapore will arrange shipping and insurance, therefore, the price to be quoted on an FOB Stowed Kaohsiung basis.

　　As there is critical shortage of rice in Singapore, the goods should be ready for shipment as soon as possible. Please be assured that if your price is acceptable, we will place order with you right away.

　　Your prompt reply is requested.

Yours very truly,

【註】

1. Bonlai Rice：蓬萊米。

2. PE woven bags：PE 編織袋。

3. independent public surveyor：獨立公證行。

4. at the time of loading：裝船時。

5. FOB Stowed：船上交貨包括艙內堆積費在內條件，在大宗物資的交易，如係袋裝者（例如水泥等）在出口港船上的堆積費往往約定由賣方負擔。

6. critical = serious：嚴重。

7. ready for shipment：儘速（早）裝運；馬上裝船。

8. if your price is acceptable：如果價格相宜。

9. right away：馬上。

No. 43　詢購柳安合板

Lauan Plywood

Dear Sirs,

As we are in need of one million sq. ft. of the subject commodity for prompt delivery, would you please quote us your best price therefore as early as practicable. For your reference, we are giving below the details of this enquiry.

Specification: Lauan Plywood, Rotary Cut,
　　　　　　　Type III, 3–ply
　　　　　　　Size: 1/8″×3′×6′

Quantity: One million sq. ft.

Price: Either FOB Taiwan or CIF Bangkok

Shipment: Prompt delivery

Payment: By Irrevocable Sight L/C

It will be appreciated if you will let us have your quotation before the end of this month.

Yours very truly,

【註】

1. sq. ft. = square feet：平方呎，為合板的數量單位，至於單價數量單位通常以 MSF (=1,000 sq. ft.) 表達。

　　例： US$68 per MSF FOB Taiwan port。

2. specification：規格，簡稱為 "Spec."。

3. lauan plywood： 柳安合板，"3–ply lauan plywood" 為三夾板。

4. rotary cut：圓切。

5. size: 1/8″×3′×6′：尺寸：厚度 $\frac{1}{8}$ 吋，寬 3 呎，長 6 呎。

6. sight L/C：即期信用狀，與遠期信用狀 (usance L/C) 相對稱。

No. 44　查詢印刷機器

Gentlemen:

Rossetti Printing Machine

Thank you very much for your letter of May 1 offering a proposal to commence negotiation in the sale of your business machines.

Upon reading the pamphlet introducing your company, we have learned that you are the Sole Agent of Rossetti Printing Machine. This interests us very much.

We are now in demand for your machine, rotary offset with high efficiency, Model ROH-650. We would like a fully detailed pamphlet concerning the above machine. Also, please send us your latest catalog and technical information for the above with price list and possible delivery date as soon as possible.

If your price is reasonable and delivery is superior to other suppliers, we will be pleased to place an order with you.

Yours very truly,

【註】

1. commence negotiation in the sale of：開始洽商有關……的銷售。

2. be in demand for：需求；擬購買。cf. be in demand：銷路好。

3. rotary offset：回轉式凸板印刷機。cf. rotary press：回轉式印刷機。

4. high efficiency：高速。

5. technical information：技術資料（為機械交易不可或缺者）。

6. price is reasonable：價錢公道。

7. be superior to：勝過；比……好。這裡是指交貨比別的供應商迅速之意。

No. 45　詢購油漆並請寄樣品

Dear Sirs,

We are open to consider quotations for the supply of your "Lion" Brand Paint in large quantities and should be prepared to entrust you with our business should your product and prices prove satisfactory.

Your Catalogue No. 50 has been shown to us by one of our friends. Colours of the paint we require are white, black, red, blue, yellow, brown, and green. It will be necessary, however, for us to submit your product to a rigorous test, and should the result be satisfactory, we would place a standing contract for 500 drums in assorted colours monthly. The contract will run for one year. Business of this type, you will agree, is most desirable and well worthy of special discount.

We ask you to inform us immediately of your special prices and send sample tins of all the colours mentioned above.

Yours faithfully,

【註】

1. be open to = be ready to：預備；準備。

2. brand：廠牌；品牌。有時也作「品質」解釋，例如：

This wine is of fine brand.

3. be prepared to entrust you with our business.：擬將我們的業務（生意）交給你做。

4. to submit...test：將你的製品付諸嚴格的試驗。

cf. to submit the plan for approval：提出計畫請求核准。

to submit the matter to arbitration：將該事項付諸仲裁。

5. a standing contract：繼續契約。

cf. a standing order：繼續訂單，意指在一定期間，繼續訂購。

a standing business：繼續的交易。相反詞為 "a casual business"。

6. drum：鐵桶。cf. steel drum：鋼桶。

7. to run：（契約）繼續（有效）。cf. to run out：（契約）終止。

No. 46　詢購魚油脂

Gentlemen:

Your very helpful letter of April 28 and the samples of Sardine Oil Nos. 1/3 have been received with thanks.

We are specially interested in this line, and having heard that you are one of the most reliable refiners and exporters of Fish Oils, we have no doubt that we could not do better than avail ourselves of your valuable services.

We have closely examined the samples which you were good enough to send us and found them suitable for our trade. We, therefore, have the pleasure of sending you an inquiry as detailed below:

1,000 cases Sardine Oil No. 1.

60 lbs. net to be contained in a can; 2 cans to be packed in a case.

Quote your finest pride CIF San Francisco per ton (2,000 lbs.) of net shipped weight.

Loss in weight exceeding 2% to be allowed for by you.

Your terms and conditions are agreeable to us, and should business result, we shall arrange a credit in your favor. As we are likely to place large orders regularly, we must make it clear from the very beginning that a competitive price is essential.

Yours earliest reply with full particulars will be appreciated.

Very truly yours,

【註】

1. sardine oil：沙丁魚油脂。

2. Nos. 1/3: 為 No. 1 至 No. 3 共三種（樣品）。

3. refiner: 煉製廠。

4. could not do better than: 最好。

5. avail ourselves of your valuable services: 利用你寶貴的服務。

    cf. avail yourselves of this good opportunity: 利用這個好機會。

6. net: 這裡指 "net weight"。

7. finest = lowest = best。

8. shipped weight = intake weight: 裝船重量，與此相對的是 "landed weight"（卸貨重量），油類在運輸中易發生漏損，所以特別言明按裝船重量交易。

9. loss in...by you:「重量減少 2% 時，超過部分由你負擔」。to be allowed for by you = for your account: 即 "for seller's account"。

10. should business result: 假如生意有結果，即假如成交之意。

11. credit = letter of credit。"arrange a credit" 為「安排信用狀」也即開發信用狀之意。

12. in your favor: 以你為受益人。

13. make it clear: 說明；說明清楚。

14. from the very beginning: 一開始就。"very" 是一種加強語氣的字眼。

    cf. that's the very thing! 就是那個!

        He is the very man I saw yesterday.（他就是我昨天看到的那個人。）

15. full particulars = full details: 全部詳情。

No. 47　請報美棉價

American Raw Cotton

Gentlemen,

    Would you please quote us by fax 1,200 bales of the subject cotton with the following particulars:

    Grade: Strict Low Middling

    Staple: 1–1/16″

    Pressley: 90,000

    Micronaire: 3.8 NCL

    Please quote the price either on FAS Gulf ports basis or on CIF Keelung basis. We are in urgent need of the cotton and should like these 1,200 bales to be delivered at 300 bales each monthly from July through October, 20–

    We are expecting to receive your earliest reply to this enquiry.

                          Yours faithfully,

【註】

1. raw cotton：原棉。

2. bale：美國原棉每包約 500 lbs.。

3. particulars：細節，這裡可譯成「規格」。

4. Strict Low Middling：原棉等級的一種，簡稱為 "SLM"。美國將美國產白棉 (white cotton) 分為八級，即：

Middling fair	Strict low middling
Strict good middling	Low middling
Good middling	Strict good ordinary
Middling	Good ordinary

5. staple：指「棉花纖維長度」。1–1/16″，1 又 1/16 吋。

6. pressley：拉力（以 78,000 磅以上為好棉）。

7. micronaire：纖度。

8. NCL：為 "no control limits"（無上下限）的縮寫。

9. Gulf port：指位於墨西哥灣的美國港口。

10. in urgent need of：急需。

（以下 No.48–53 是單刀直入法的例子）

No. 48　請就棉織品報價

Dear Sirs,

Having heard from our chief supplier that your company is a leading firm specializing in cotton and rayon goods, we wish to make a spot purchase from you on the following fabrics, each measuring at least 50,000 yards:

1. First class tweed
2. First class black serges

Owing to an influx of orders after opening a new office in Tokyo, our stocks are nearly exhausted. This compels us to compile a large inventory immediately in order to meet the needs of our customers.

We shall appreciate your lowest possible prices on CIF Osaka with immediate delivery.

We are doing business on an Irrevocable L/C by The Bank of Osaka, under which you may draw a draft at sight. For any information concerning our credit, please refer to the above bank.

If your goods are satisfactory in quality and delivery, we will place repeat orders with you in the near future.

Yours faithfully,

【註】

1. spot purchase：立刻購入。

2. fabrics：織布。

3. measuring：數量達……。

4. tweed：花呢。

5. serges：斜紋布料。

6. an influx of orders：訂單湧進，也可以 "rush of orders" 表示。

7. to compile：編列。

8. repeat orders：再訂貨；繼續訂購。

No. 49　請寄腳踏車目錄及通知價格

Dear Sirs,

It has come to our attention through our Chinese friends in Taiwan that you are one of the foremost manufacturers and exporters of Bicycles in Taiwan.

Being in the market for Bicycles and Accessories, we shall be greatly obliged if you will send us a copy of your illustrated catalogue, informing us of your best terms and lowest prices CIF Singapore. As we are in a position to handle large quantities, we hope you will make an effort to submit us really competitive prices.

For any information as to our financial standing, we refer you to:
　　The Hongkong & Shanghai Banking Corporation
　　Singapore

We recommend this matter to your prompt attention.

Yours faithfully,

【註】

1. in the market for = on the lookout for; In want (= need) of：擬購買……。cf. to be on the market：推出市場；上市。to put (place) on the market：上市。to be out of the market：停止出售。

關於 market 的用例：

The market is active (brisk).：市場很活躍；呈現活氣。

The market is dull (weak, heavy, slack, depressed).：市場陷入低潮；沒有活力。

The market is firm (steady).：市場堅挺（堅穩）。

The market is rising (in the upward tendency).：行情看漲中。

The market is falling (in the downward tendency).：行情看跌中。

The market remains unchanged (stationary).：行情堅守原盤。

The market is strong (bullish).：行情看漲。

The market is weak (bearish).：行情看跌。

The market is sagging.：行情平抑。

The market soars up (advances rapidly).：行情猛漲。

The market falls sharply (slumps).：行情猛跌。

2. accessories：附件。

3. illustrated catalogue：插圖貨品目錄。

4. recommend：建議。

No. 50　詢購電扇、電鍋

Dear Sirs:

　　We are much indebted to China External Trade Development Council (CETRA) for the name and address of your company and are pleased to know that you are one of the main manufacturers/exporters of Electric Fans and Rice Cookers in Taiwan.

　　You would be delighted to know that there is a ready market for these electrical appliances to be explored in Indonesia right now and we should like to place substantial orders with you provided the prices you quote are very competitive. We should also like to serve as your Sole Distributor in Indonesia after business relationship between us has been well established. For the time being, we shall appreciate it if you will quote us your rockbottom prices for 500 each of electric fans of different sizes and rice cookers for different numbers of persons at your earliest convenience.

　　As to payment, we will ask our bankers to issue irrevocable letter of credit in your favor for the total value right after orders have been confirmed by us.

Faithfully yours,

【註】

1. electric fans：電扇。

2. electric rice cookers：電鍋。

3. you would be delighted to know：你會很高興知道。

4. ready market：現成的市場。

5. electrical appliances：電器用品。

6. explore：探測，這裡可譯成「推銷」。

7. provided = if。

8. sole distributor：獨家經銷商，又稱為 "exclusive distributor"，與 "sole agent" 不同，"sole agent" 係指獨家代理商，但實務上往往將其混為一談。

9. rockbottom price：最低價格。即 "lowest price"。

10. right after...by us：在確認訂貨之後立即……。

No. 51　詢購罐頭蘆筍

Canned Asparagus

Dear Sirs,

　　We are much indebted to CETRA for the name and address of your company and are pleased to learn that you are one of the leading producers of the subject item in Taiwan.

　　We are now in the market for the subject product and shall appreciate your quoting us therefor either on FOB Taiwan port or on CIF Hamburg basis at your earliest convenience. For the moment, we are in need of approximately 300 cartons each of Tips & Cuts, Center Cuts, and End Cuts to be packed in export cartons of 48 cans each. Since consumers at our end are very demanding about quality, you are requested to quote us of Al quality.

　　As there are many other competing brand scrambling for the consumers' dollars in our market, it is, therefore, imperative that the prices quoted have to be very competitive.

　　We are looking forward to receiving your quotation very soon.

Yours faithfully,

【註】

1. leading producers：主要生產廠商。

    cf. main producers

    　　reputable producers

    　　reliable producers

2. quote us therefor：quote us for that product。

3. tips & cuts：尖端及切片。

4. center cuts：中段切片。

5. end cuts：後段切片。

6. to be packed...cans each：用外銷紙箱包裝，每箱 48 罐。

7. at our end = on our side：在此地。

8. very demanding about = very critical about：對……很苛求；對……很挑剔。

9. competing brand：競爭的牌子。

10. scrambling for：爭取。

11. imperative：必須。

No. 52　詢購手工藝品

Handicrafts

Gentlemen,

You would be pleased to know that during the last ten years we have been one of the leading importers of Japanese handicrafts in the United States. Now, we have decided to expand the sources of supply to Taiwan and Hongkong.

To enable you to quote us the right items, we are listing hereunder the articles in which we are particularly interested:

1. Lanterns, Palace Lantern in particular.
2. Artificial flowers, including leis.
3. Scrolls, reprinted from Chinese paintings.
4. Bamboo Products—such as basket, furniture, screens, mats, bamboo wares, etc.
5. Embroidery and needle work.
6. Ceramics.
7. Chinaware.
8. Hat and hatbodies—paper hats, Tacha hats, hatbodies.
9. Rattan furniture.
10. Insect specimens, especially butterflies.

Since this is the first time we plan to deal in Taiwan handicrafts and because we know little about the quality of them, it is, therefore, absolutely necessary for you to submit samples of all the items you are to offer us for our inspection. Orders will be placed only after we have approved the samples received.

We are informed that you are one of the most reputable exporters of handicrafts in Taiwan, and are sure that you will supply us goods of the best quality.

Your early response to this enquiry will be appreciated.

Yours faithfully,

【註】

1. palace lantern：宮燈。

2. in particular：尤其。

3. artificial flowers：人造花。

4. lei：花環。

5. scrolls：卷軸。

6. reprinted from：從……翻印的。

7. bamboo products：竹製品。

8. screens：簾。

9. mats：席。cf. straw mats：草席。

10. bamboo wares：竹器品。

11. needle work：針織品。

12. ceramics：陶器。

13. chinaware：磁器。

14. hatbodies：帽胚。

15. Tacha hats：大甲帽。

16. specimens：標本。

17. to deal：做生意。

No. 53　請發塗漿物預期發票及樣品

"Jet Size" Sizing Stuff

Dear Sirs,

　　We read with much interest your ad as inserted in the July 2004 issue of Japan Textile News concerning the subject sizing stuff.

　　We want to use the sizing on a trial basis and shall appreciate your sending us Proforma Invoice, in duplicate, for 2,000 kgs., each of No. 66 for Sheeting and Jeans and Nos. 88–98 for Poplin and Broadcloth at your early convenience. Please note the prices you quote should be on FOB Japanese ports basis. Please also send us additional detailed information and a few pounds of samples of these two kinds for our test.

　　We shall appreciate your giving immediate attention to this matter.

　　　　　　　　　　　　　　　　　　　　　　　　　　　　　　Faithfully yours,

【註】

　　1. sizing stuff：塗漿物。用於布疋上漿的物質。

　　2. ad = advertisement。

　　3. proforma invoice：預期發票；估計發票。

4. sheeting：床單布。cf. cotton sheeting：棉布。

5. jeans：斜紋布。

6. poplin：毛葛。

7. broadcloth：寬幅絨布。

No. 54　中央信託局所用詢購定型函 (Form letter)

CENTRAL TRUST OF CHINA

Phone: (02) 1234–5678 (30 lines)
Fax: (02) 234–56789

PURCHASING DEPARTMENT
49, WU CHANG STREET, SEC. 1,
TAIPEI, TAIWAN

IN REPLY PLEASE REFER
TO OUR INVITATION NO.

Date:

ENQUIRY

Re: Our Invitation No.:
Commodity:

Gentlemen,

　We have been requested by　　　　　　　　　　　　　　　　　　to:

☐(1) Furnish price indications for reference in application of foreign exchange.

☐(2) Furnish price lists and technical literature for study.

☐(3) Furnish market price information for reference in setting ceiling price prior to tender opening.

☐(4) Furnish general price information for statistical purpose, etc.

　in accordance with the following conditions:

(a) Description:

(b) Unit/Quantity:

(c) Price: CFR Keelung/Kaohsiung; FOB Japanese port; FAS US port; CIF Taipei via Parcel Post/ Air Freight; by...liner vessel, including packing costs, inland freight, inspection charge and all other expenses.

(d) Insurance: To be effected by CTC, Insurance Department.

(e) Packing: Standard export packing unless otherwise specified.

(f) Shipment date:

(g) Inspection: To be made by independent inspector/manufacturer on specifications quality, quantity, weight, proper packing and marking.

(h) Designated manufacturer/or approved equal:

(i) Other conditions as per our standard foreign purchase invitation form.

(j) Additional conditions:

Please send us the required information in...copies by airmail/telegram as soon as possible/before.

Your very truly,

CENTRAL TRUST OF CHINA

Purchasing Department

cc:

_____

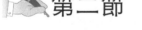 第二節　有關詢價的有用例句

一、開頭句

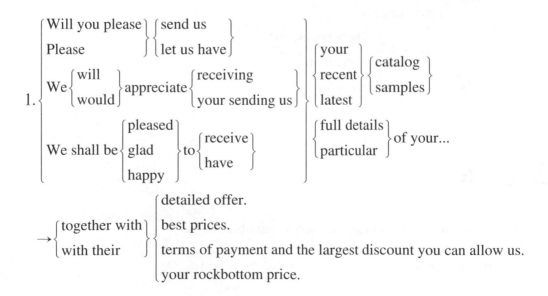

2. $\left\{\begin{array}{l}\text{We are interested in...}\\ \text{We are in the market for...}\\ \text{We are in urgent need for...}\\ \text{We}\left\{\begin{array}{l}\text{have}\\ \text{had}\\ \text{received}\end{array}\right\}\left\{\begin{array}{l}\text{an inquiry}\\ \text{inquiries}\\ \text{many inquiries}\end{array}\right\}\text{from our}\left\{\begin{array}{l}\text{customer}\\ \text{trade connections}\end{array}\right\}\left\{\begin{array}{l}\text{here}\\ \text{in...}\end{array}\right\}\text{for your...}\\ \text{We have}\left\{\begin{array}{l}\text{read}\\ \text{seen}\end{array}\right\}\text{your ad in}\left\{\begin{array}{l}\text{May issue}\\ \text{the Summer issue}\\ \text{a recent issue}\end{array}\right\}\text{of Taiwan Export}\\ \text{Your ad in today's China News interests us}\end{array}\right.$

$\left\{\begin{array}{l}\text{and we will be}\left\{\begin{array}{l}\text{glad}\\ \text{pleased}\end{array}\right\}\text{to}\left\{\begin{array}{l}\text{receive}\\ \text{have}\end{array}\right\}\\ \text{and we would like to}\left\{\begin{array}{l}\text{receive}\\ \text{have}\end{array}\right\}\\ \text{and we would be}\left\{\begin{array}{l}\text{grateful}\\ \text{obliged}\end{array}\right\}\text{if you would send us}\\ \text{and we would}\left\{\begin{array}{l}\text{welcome}\\ \text{appreciate}\end{array}\right\}\end{array}\right.$

$\rightarrow\left\{\begin{array}{l}\text{your}\left\{\begin{array}{l}\text{price list.}\\ \text{your samples.}\end{array}\right.\\ \text{samples with your}\left\{\begin{array}{l}\text{best prices.}\\ \text{best terms.}\\ \text{rockbottom prices.}\\ \text{lowest prices.}\\ \text{most competitive quotations.}\end{array}\right.\end{array}\right.$

二、條　件

1. Will you please inform us (as soon as possible) of the $\left\{\begin{array}{l}\text{prices at which}\\ \text{terms on which}\end{array}\right\}$

you can supply...

2. $\left\{\begin{array}{l}\text{Please}\\\text{Will you please}\end{array}\right\}$ quote us your competitive prices $\left\{\begin{array}{l}\text{for}\\\text{on}\end{array}\right\}$ the following

$\rightarrow\left\{\begin{array}{l}\text{goods.}\\\text{articles.}\\\text{commodities.}\\\text{items.}\\\text{qualities.}\end{array}\right.$

3. $\left\{\begin{array}{l}\text{We would appreciate your sending us}\\\text{Please send us}\\\text{Please let us have}\end{array}\right\}\left\{\begin{array}{l}\text{samples}\\\text{patterns}\\\text{a full range of sample}\end{array}\right\}$

$\rightarrow\left\{\begin{array}{l}\text{you can supply}\\\text{you can deliver}\\\text{can be shipped}\end{array}\right\}$ of... which $\left\{\begin{array}{l}\text{you can supply}\\\text{you can deliver}\\\text{can be shipped}\end{array}\right\}$

$\rightarrow\left\{\begin{array}{l}\text{promptly.}\\\text{immediately.}\\\text{during May.}\\\text{in (within)...after receipt of our order.}\end{array}\right.$

4. Please $\left\{\begin{array}{l}\text{inform us}\\\text{let us know}\end{array}\right\}$ $\left\{\begin{array}{l}\text{how soon you can}\left\{\begin{array}{l}\text{ship}\\\text{deliver}\end{array}\right\}\text{them.}\\\text{how long it will take you to}\left\{\begin{array}{l}\text{fill}\\\text{ship}\\\text{execute}\end{array}\right\}\text{an order.}\end{array}\right.$

註：a. fill an order：配發一批訂貨；b. ship an order：交運一批訂貨；c. execute an order：配交一批訂貨。

5. $\left\{\begin{array}{l}\text{If you can supply us with goods of high quality at competitive prices}\\\text{If your}\left\{\begin{array}{l}\text{prices}\\\text{terms}\end{array}\right\}\text{are}\left\{\begin{array}{l}\text{competitive}\\\text{attractive}\\\text{satisfactory}\\\text{reasonable}\end{array}\right\},\end{array}\right.$

If your quality is { good / right / excellent / superior } and the price { is competitive / is suitable for our market },

If { your / the } goods { are equal to the samples / meet with our approval / are of superior quality },

{ If / Provided / Suppose / So long as } you can { guarantee regular supplies / promise prompt delivery },

we may place a { big / substantial / large / trial } order with you.

we may place { big / substantial / large / bulk （龐大的） / regular } orders with you.

we would consider to sign { an agency / a sole agency / a long-term } contract with you.

your goods should { sell well / sell readily / find a ready sale } in our market.

註：a. attractive：引人感興趣；b. right：合適；c. meet with our approval：合我們的要求；d. guarantee regular supplies：保證經常供應；e. trial order：試驗性訂單；嘗試訂單；f. sell well：暢銷；g. sell readily：容易出售；h. find a ready sale：立即找到銷路。

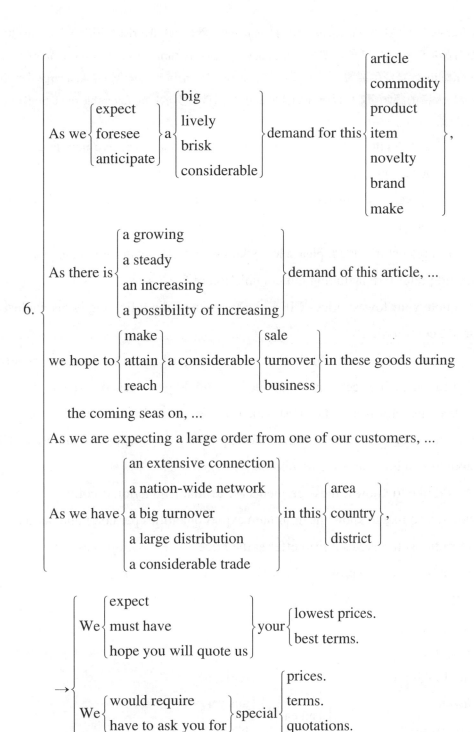

6.

As we $\begin{Bmatrix} \text{expect} \\ \text{foresee} \\ \text{anticipate} \end{Bmatrix}$ a $\begin{Bmatrix} \text{big} \\ \text{lively} \\ \text{brisk} \\ \text{considerable} \end{Bmatrix}$ demand for this $\begin{Bmatrix} \text{article} \\ \text{commodity} \\ \text{product} \\ \text{item} \\ \text{novelty} \\ \text{brand} \\ \text{make} \end{Bmatrix}$,

As there is $\begin{Bmatrix} \text{a growing} \\ \text{a steady} \\ \text{an increasing} \\ \text{a possibility of increasing} \end{Bmatrix}$ demand of this article, ...

we hope to $\begin{Bmatrix} \text{make} \\ \text{attain} \\ \text{reach} \end{Bmatrix}$ a considerable $\begin{Bmatrix} \text{sale} \\ \text{turnover} \\ \text{business} \end{Bmatrix}$ in these goods during

the coming seas on, ...

As we are expecting a large order from one of our customers, ...

As we have $\begin{Bmatrix} \text{an extensive connection} \\ \text{a nation-wide network} \\ \text{a big turnover} \\ \text{a large distribution} \\ \text{a considerable trade} \end{Bmatrix}$ in this $\begin{Bmatrix} \text{area} \\ \text{country} \\ \text{district} \end{Bmatrix}$,

→

We $\begin{Bmatrix} \text{expect} \\ \text{must have} \\ \text{hope you will quote us} \end{Bmatrix}$ your $\begin{Bmatrix} \text{lowest prices.} \\ \text{best terms.} \end{Bmatrix}$

We $\begin{Bmatrix} \text{would require} \\ \text{have to ask you for} \end{Bmatrix}$ special $\begin{Bmatrix} \text{prices.} \\ \text{terms.} \\ \text{quotations.} \\ \text{discounts.} \end{Bmatrix}$

We look forward to a favorable quotation from you.

註：a. foresee：預見；b. anticipate：預料；c. lively：熱烈的；d. brisk：興隆的；e. novelty：新奇品；f. make：牌子；型式；g. steady：堅穩；h. turnover：營業量；i. an extensive connection：廣泛的聯繫；j. a nation-wide network：全國性的聯絡網；k. a large distribution：大量的經銷量；l. a considerable trade：相當的生意；m. favorable quotation：有利的報價。

7. We would appreciate being placed on your mailing list to receive new and revised copies of your catalogs and new product announcements.

8. We would appreciate your letting us know what discount you can grant if we give you a large order.

9. If there is any cost involved, please let us know so that we can reimburse you.

10. We would prefer CIF including our commission of 5%.

11. Please quote your lowest prices CIF Kobe for each of the following items, inclusive of our 5% commission.

12. Your price should be on a CIF basis, and include packing in tin-lined water-proof wooden cases, each piece wrapped in oilcloth, and 30 pcs. packed in one case.

13. Please keep us informed of the latest quotations for the following items.

14. Before you proceed to make sample pieces, please give us approximate prices CIF Singapore including our 5% commission.

15. We should like to know if these garments are available in nylon tricot.

16. We would also like to know the minimum export quantities per color and per design.

17. Do your utmost to give us a firm offer at the price and keep it open as long as possible, as the buyer would probably increase his offer considerably.

三、結尾用語

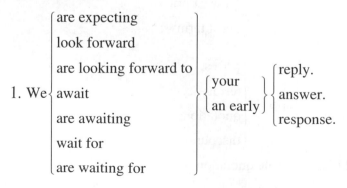

1. We
{
are expecting
look forward
are looking forward to
await
are awaiting
wait for
are waiting for
}
{
your
an early
}
{
reply.
answer.
response.
}

2. We hope to $\begin{Bmatrix} \text{receive your reply} \\ \text{hear from you} \end{Bmatrix}$ $\begin{Bmatrix} \text{soon.} \\ \text{by air.} \\ \text{by e-mail} \\ \text{by fax.} \\ \text{by cable.} \\ \text{at your earliest convenience.} \end{Bmatrix}$

3. Your $\begin{Bmatrix} \text{prompt} \\ \text{immediate} \end{Bmatrix}$ attention to this matter $\begin{Bmatrix} \text{would} \\ \text{will} \end{Bmatrix}$ be appreciated.

4. $\begin{Bmatrix} \text{An} \\ \text{Your} \end{Bmatrix}$ early reply $\begin{Bmatrix} \text{by fax} \\ \text{by cable} \end{Bmatrix}$ $\begin{Bmatrix} \text{will} \\ \text{would} \end{Bmatrix}$ be appreciated.

5. Your $\begin{Bmatrix} \text{cooperation} \\ \text{kind consideration} \end{Bmatrix}$ in this matter $\begin{Bmatrix} \text{will} \\ \text{would} \end{Bmatrix}$ be appreciated.

第十四章
答覆詢價 (Reply to Inquiry)

第一節　答覆詢價信的寫法

賣方收到詢價信之後，應儘速答覆，答覆的內容要正確不含糊，並設法抓住買方的買意，以免貽誤商機。

答覆詢價信時，其內容應注意下列各點：

1. 對來信表示謝意。

2. 針對所詢問題作確切的答覆。

3. 檢送所索資料，諸如商品目錄、價目表及樣品等，或說明另行寄送。

4. 適當地說明市況，激發對方早日訂購的意願。

5. 如果無對方擬購的貨品，可藉機推薦代替品，或介紹其他貨品。

6. 以期待能收到訂單做結束。

No. 55　對卡式錄音帶詢價的答覆 (See No. 38)

Dear Sirs,

Cassette Tape Recorder

Thank you very much for your inquiry of July 26 concerning your purchase of 500 sets of our popular cassette tape recorder, Model CRT−153.

In compliance with your request, we have enclosed our latest tape recorder catalog with our price list No. 250. Among our best selling products, we recommend our Model CRT−508 because it has won the Chinese Government Good Design Award, and has been enjoying excellent sales ever since.

If satisfactory, we will send 400 sets from the production line. Owing to a rush of orders, this will be our maximum quantity for the present.

Together with the printed materials requested, we have sent samples, including several other models, by special delivery.

We are very pleased to have concluded business relations with you and hope that our appreciation will continue for a long time to come.

Yours very faithfully,

【註】

1. inquiry of July 26 concerning：7月26日有關……的詢價。

2. in compliance...request：「遵從」,「依照」之意, 新派人士認為這是 Over-formal 的措詞, 貿易書信中不必如此拘謹, 可以 "as requested" 代替。

3. Chinese Government Good Design Award：中華民國政府優良設計獎。

4. ever since：從那時候起到現在; 此後一直。

5. from the production line：從生產線。

6. rush of orders：訂單湧入。

7. together with：可以 "along with" 代替。

8. printed materials：印刷品, 包括目錄、技術資料、說明書等。

9. special delivery：限時專送。

10. conclude business relations：建立業務關係。

11. to come：未來的; 將來的。"for a long time to come" 為「未來長時間」, 即「今後長久」之意。

No. 56　對府綢詢價的答覆 (See No. 39)

Gentlemen:

Tussore Silk

Your letter No. 100 of February 18 has given us much pleasure as it serves as a means of stimulating business between us.

Samples and Prices. We have sent you by airmail a pattern book of our Tussore Silk. The book contains all the patterns that are typical of the goods, which, we trust, are best adaptable for your selling purposes. From the Price List inclosed you will observe that our prices are exceptionally low. This sacrifice is entirely due to our recognition of the necessity for price cutting in order to counter competition.

Market. The market here is recovering somewhat from the long period of depression under which it has labored, and there is a decided improvement in the general tone. Inquiries, too, are more frequent and more often lead to business than has of late been the case. Manufacturers in most branches are getting busier, and we may look forward before long to an upward tendency in prices all round. We believe, therefore, that the present is a very favorable opportunity to make bull purchases.

We wish to assure you that we always adhere to our policy of providing high-grade products at competitive prices. Perhaps you would be good enough to entrust us with a trial order, which will assuredly redound to your advantage.

Your inquiry will be promptly answered, and more detailed information furnished upon request.

Very sincerely yours,

【註】

1. a pattern book = a book of patterns：樣品簿（款式簿）。

2. sacrifice = sacrifice price：犧牲價，即 low price 之意。

3. to counter competition：應付競爭。

4. recovering：（從不景氣）恢復，又稱為 "rally"，如從上漲跌回來稱為 "react"。

5. depression：不景氣。

6. labored：受……不利的影響；為……所支配，與 Under 連結。

7. decided improvement：決定性的改善。指市況轉好，價格上漲。

8. the general tone：一般景況。這是市場用語的一種。

9. more frequent：更頻繁。

10. lead to business：成交。

11. of late = recently。

12. branches = trades：行業。

13. getting busier：越來越忙。生意好，自然忙。

14. all round：一齊；全盤。

    cf. a. Prices in silks are soaring *all round* with the approach of the new season's demands.

        b. There being a good deal of business doing in silks, prices *all round* are sure to appreciate very soon.

        c. An *all round* decline was registered in the Yokohama silk market.

15. bull purchases = to buy in expectation of a rise：預期價格會上漲而買進。

16. adhere to our policy：固守我們的政策。

17. entrust us with a trial order：惠賜試驗性訂單。

18. redound to your advantage：增加你的利潤。

19. upon request：一經要求。

No. 57　對刺繡品詢價的答覆 (See No. 40)

36″ Nylon Embroideries

Dear Sirs,

We thank you for your letter of 4th August, 20–, asking us to quote you on 36″ Nylon Embroideries. In response we have forwarded today our samples and shade cards by Air Parcel.

Our quotation, subject to your reply being received here before 15th September, are as follows:

Sample Nos.	Description		Unit Price
A–100	Cutwork	36″×25 yds.	US$0.50 per yd.
B–105	Stitchwork	36″×25 yds.	US$0.55 per yd.
C–110	Drawnwork	36″×25 yds.	US$0.56 per yd.

The above quotations are CIF Wellington, including a 5% commission to you, and for a minimum order of 50,000 yards.

We can ship the goods within a month after we receive your order, provided the shipping space is available.

For payment we shall draw on you against an irrevocable letter of credit.

We hope that our samples and terms prove to be satisfactory and that you will favor us with an initial order which will lead to cordial business relations for our mutual advantage.

Yours faithfully,

【註】

1. cutwork：空花繡，尼龍製刺繡品有三種，除 cutwork 之外尚有 stitchwork（針繡）及 drawnwork（抽絲刺繡）。

2. provided...is available：假如有艙位的話。

3. draw on：向……開出（匯票）。

4. favor...with：惠賜。

5. lead to cordial business relations：導致友善的業務關係。

6. mutual advantage：相互的利益，共同的利益，對雙方均有利。

No. 58　對印刷機詢價的答覆 (See No. 44)

Dear Sirs,

Thank you for your letter of...expressing that you are interesting in our Rossetti Printing Machine.

We have already dispatched the detailed pamphlet, latest catalog and technical information one copy each for the Machine with price list you requested by special air delivery.

As our market is now somewhat dull and prices generally low, you are very fortunate in making purchase at this time. European buyers, however, seem to be picking up in activity, so we advise you to buy the goods before the recovery reaches a peak.

Accordingly, the price will be kept effective only three weeks from the date of this letter and we wish to receive your order by end of May 10. Delivery can be made within 30 days after receipt of your L/C if it reaches us by end of May 30.

We are looking forward to your immediate reply.

Yours very truly,

【註】

1. dispatch = send。

2. market is somewhat dull = market is somewhat inactive：市況呆滯。

3. picking up in activity：逐漸活躍起來。

4. before the recovery reaches a peak：行情回到顛峰之前。

5. accordingly = therefore。

6. keep effective：繼續有效。

No. 59　對詢購油漆及樣品的答覆 (See No. 45)

Dear Sirs,

We are obliged for your letter of the 3rd May and are glad to learn that you are particularly interested in our "Lion" Brand Paint and intend placing a standing contract should our product and prices be found satisfactory.

We are pleased to say that the standing contract you are going to make deserves our special consideration, and we are accordingly prepared to allow you a special discount of 5 per cent off our catalogue prices, in addition to the usual quantity discount of 3 per cent. This special discount will involve our selling practically at cost, and we trust you will recognize that this concession must be regarded as exceptional.

You will readily admit that this concession is made on the stipulation that you bind yourselves to place with unfailing regularity an order monthly, running for one year, for 500 drums of our "Lion" Brand Paint in assorted colours, or the special concession is to be revoked if your monthly order falls below this figure, i.e., you are to pay the full prices for the quantity you do take. This condition does not, of course, prejudice your right to reject any goods that are not up to samples, or are unsuitable or shipped later than the time stipulated in each contract.

As requested, we have sent our sample tins for your test. We also enclose our Catalogue No. 51, in which you will find that our regular prices for "Lion" Brand Paint are the same as described in Catalogue No. 50.

We trust that the result of your test of our product and our special concession in the prices will induce you to enter into the contract.

Yours faithfully,

【註】

1. deserve our special consideration：值得我們的特別考慮。
2. discount of 5 per cent off our catalogue price：按目錄價格折扣 5%。
3. this special discount will involve...cost：這個特別折扣難免使我們按成本出售。"involve" 為……「難免」，「使……必需」之意。
4. concession = discount。
5. readily = promptly。
6. on the stipulation that：以……為條件。
7. bind yourselves：拘束你；使你負有義務。
8. with unfailing regularity：正確無誤地。
9. running for one year = continuous for one year。
10. in assorted colours：以各種顏色。
11. revoke：取消（約定）；解除（契約）。
12. falls below...：低於……。
13. full prices = regular prices：正規的價格。此外又有「可獲充分利益的價格」之意。
14. do take："do" 為加強語氣的助詞。
15. prejudice your right：損及貴方權利。
16. not up to samples：達不到樣品的標準。
17. sample tins：樣品罐。
18. induce = cause：促使（某人做某事，後接不定詞）。
    cf. What induced you to do such a thing?（什麼誘導你做這樣的事?）
19. enter into contract：締約。
20. 這封信雖然寫得相當長，但是寫得很具體、詳細，使收信人覺得寫信人的認真、精細、週到。

No. 60　對詢購腳踏車的答覆 (See No. 49)

Dear Sirs,

We are very glad to learn from your letter of the 28th March that you are interested in our Bicycles. As requested, we take pleasure in sending you separately a copy of our latest Catalogue No. 15, together with our revised Price List No. 20 showing prices and rates of quantity discount for our supplies.

The Catalogue contains illustrations and descriptions of our entire line of Bicycles. We have already had the pleasure of supplying our overseas customers with our products in enormous quantities and presume that you must be well aware of the superiority of our products. It is no exaggeration to say that they leave nothing to be desired in both construction and efficiency; indeed, they have reached the highest degree of perfection. Please remember that we have been in the manufacture of Bicycles for 40 years, and in that time we have built up a large trade by guaranteeing all that we manufacture. Our reputation is much too valuable for us to risk selling anything but the very best.

The Catalogue sets out our terms in detail, and as our prices have been cut to a point where our margin of profit is almost insignificant, you will understand that these terms must be strictly observed.

We thank you for the inquiry you have kindly made and assure you of our prompt and special attention to your interests at all times.

Yours faithfully,

【註】

1. separately = under separate cover。

2. rates of quantity discount：數量折扣率，即對於一定數量以上者，予以折扣的百分率。

3. overseas customers：海外顧客。

4. leave nothing to be desired：一點缺點也沒有。

5. in the manufacture of：製造。

6. built up：建立。

7. our reputation is much too valuable for us to risk...：我們的名譽是如此的寶貴以致不能冒險（出售好品質之外的貨品）。

8. set out：表示；記載。

9. margin of profit = profit margin：為利潤額之意，即成本與售價的差距。

No. 61　覆告無法供貨

Dear Sirs,

PVC Resins for Wire Coating

We thank you for your letter of May 10 concerning PVC Resins for Wire Coating.

We appreciate your interest in our products, but regret to inform you that due to the current critical shortage we are not in a position to offer this material.

The present outlook for PVC supplies in this country is not good and it appears that this shortage will continue throughout 2003.

Towards the end of 2004, we expect to have a new large PVC plant on stream and hope at that time we will be in a position to serve our foreign customers.

We thank you again for your interest in our products.

Yours faithfully,

【註】

1. concerning：同義詞有 "in relation to"，"relating to"，"regarding"，"in reference to"。

2. PVC resins for wire coating：做電線護皮用塑膠樹脂。

3. due to current critical shortage：由於目前嚴重缺貨。

4. not in a position to offer：無法供應。

5. present outlook：目前的展望。

6. a new large PVC plant on stream：一座新的大塑膠工廠開始生產。

7. will be in a position to serve our customers：將能對顧客服務（意指可供貨）。

No. 62 對於詢購貨品無法供應，另推薦代替品

Gentlemen:

We refer to your letter of August 15 with regard to "Sunshine" brand recording tape.

Unfortunately, Sunshine Industries, Ltd., has already tied up with an agent on recording tapes on your side, and consequently we are unable to offer this line. We have, however, made a contact with another supplier, Moonshine Industrial Co., Ltd., who manufactures a practically complete line of recording tapes, and the quality of their product is kept up with, even better than, that made by Sunshine Industries, Ltd.

The enclosed price list will give you our best CIF Singapore prices which we believe are highly competitive.

We realized that it might be pretty difficult for you to push an unknown brand, but if you think there is a possibility of promoting this line, we shall be happy to send you sample.

Yours faithfully,

【註】

1. be tied up with：與提攜，名詞為 "tie-up"。"be tied up" 也用於船運方面，指因碼頭工人罷

工被套牢無法進退。

2. push an unknown brand = promote the sale of unknown brand：拓銷未出名牌子的東西。

No. 63　對於詢購貨品無法供應，另介紹公司

Dear Mr. ×××，

　　Thank you for your letter of October 12. Unfortunately, we don't have any parts for your AM–34, and we don't know when any will come in. I'm afraid we aren't of much help at this time. You might try the XYZ Company in Tainan. They try to carry a complete stock of Daisy Tractor parts.

Sincerely yours,

【註】

措詞非常 Personal，這種寫法常見於美國商用英文書信教科書。用於美國國內商業，似無不妥，但就國際貿易的書信而言，似太 Personal。

 # 第二節　有關答覆詢價的有用例句

一、開頭句

1. $\left.\begin{array}{l}\text{Many thanks}\\\text{Thank you}\\\text{We thank you}\end{array}\right\}$ for your $\left.\begin{array}{l}\text{enquiry}\\\text{fax}\\\text{e-mail}\\\text{letter}\end{array}\right\}$ of (date) $\left\{\begin{array}{l}\text{for}\\\text{about}\end{array}\right\}$ ...

2. $\left.\begin{array}{l}\text{We acknowledge with thanks}\\\text{We appreciate}\end{array}\right\}$ your $\left.\begin{array}{l}\text{letter}\\\text{fax}\\\text{e-mail}\\\text{inquiry}\end{array}\right\}$ of (date) $\left\{\begin{array}{l}\text{for}\\\text{about}\end{array}\right\}$ ...

3. $\left.\begin{array}{l}\text{As requested}\left\{\begin{array}{l}\text{by you}\\\text{in your letter of...}\end{array}\right\}\\\left.\begin{array}{l}\text{In answer}\\\text{In reply}\\\text{With reference}\end{array}\right\}\text{to your}\left\{\begin{array}{l}\text{letter}\\\text{fax}\\\text{e-mail}\\\text{enquiry}\end{array}\right\}\text{of...}\end{array}\right\}$, we $\left\{\begin{array}{l}\text{are glad to enclose}\\\text{are happy to enclose}\\\text{enclose}\\\text{are enclosing}\\\text{are pleased to enclose}\\\text{are sending (you)}\end{array}\right.$

→ a copy of our $\begin{Bmatrix} \text{illustrated} \\ \text{revised} \\ \text{Spring} \\ \text{new} \end{Bmatrix}$ catalog

→ {
...in which you will find a detailed description of our line of product.

...in which you will find $\begin{Bmatrix} \text{many} \\ \text{a number of} \end{Bmatrix}$ items that will interest you.

...wherein you will find the brief descriptions of our products.

$\begin{Bmatrix} \text{...including} \\ \text{...containing} \end{Bmatrix}$ information you need.

...of the plastic toys suitable for your business.

...with full particulars of our products.

...which we recommend to your *careful perusal* （詳核）.

...containing quantities for large orders taken from our *existing stock* （庫存）.

...including complete descriptions of our goods.

...of sporting goods, on page 24, you will find marked a machine that may interest you.

...containing a description of the construction and working of this gas-engine, as well as particulars of the various standard sizes.

...and hope you will be able to spare a few minutes to glance through its pages, wherein you will find brief descriptions of many articles which should interest you.

...with a few samples which may possibly interest you, and shall be glad to hear from you at any time.

...and price list, in the hope that you may find something to suit you.

...with lowest prices.
}

二、條　件

1. All $\begin{Bmatrix} \text{details} \\ \text{particulars} \end{Bmatrix}$ are $\begin{Bmatrix} \text{indicated} \\ \text{given} \\ \text{shown} \end{Bmatrix}$ in our $\begin{Bmatrix} \text{catalog.} \\ \text{price list.} \end{Bmatrix}$

2. We $\begin{Bmatrix} \text{can} \\ \text{are ready to} \end{Bmatrix}$ $\begin{Bmatrix} \text{despatch} \\ \text{deliver} \\ \text{supply} \\ \text{ship} \end{Bmatrix}$ any quantity of...

$\rightarrow$ $\begin{Bmatrix} \text{from stock.} \\ \text{immediately.} \\ \text{at once.} \\ \text{within...} \begin{Bmatrix} \text{days} \\ \text{weeks} \\ \text{months} \end{Bmatrix} \begin{Bmatrix} \text{of} \\ \text{from} \\ \text{after} \end{Bmatrix} \text{receipt of order.} \end{Bmatrix}$

註：a. are ready to：隨時都能；b. from stock：由現貨中。

3. We $\begin{Bmatrix} \text{are pleased} \\ \text{happy} \\ \text{glad} \end{Bmatrix}$ to quote you as follows and can promise $\begin{Bmatrix} \text{dispatch} \\ \text{delivery} \\ \text{shipment} \\ \text{supply} \end{Bmatrix}$ within 30

days on receipt of your L/C.

4. $\begin{Bmatrix} \text{Our prices} \\ \text{All prices} \end{Bmatrix}$ are subject to $\begin{Bmatrix} \text{alteration.} \\ \text{change without notice.} \\ \text{market fluctuations.} \end{Bmatrix}$

註：a. subject to change without notice：隨時變更，恕不另行通知；

b. subject to market fluctuations：隨市況波動。

5. $\begin{Bmatrix} \text{For quantity of} \\ \text{For order of} \\ \text{In case of an order for} \end{Bmatrix}$ 10,000 sets and $\begin{Bmatrix} \text{more} \\ \text{over} \end{Bmatrix}$

$\rightarrow$ we $\begin{Bmatrix} \text{will} \begin{Bmatrix} \text{allow} \\ \text{give} \end{Bmatrix} \text{you a special discount of 2\%.} \\ \text{can offer a discount of 10\% on} \begin{Bmatrix} \text{list} \\ \text{catalog} \end{Bmatrix} \text{prices.} \end{Bmatrix}$

6. $\begin{Bmatrix} \text{Regarding} \\ \text{As regards} \end{Bmatrix}$ payment, you are requested to $\begin{Bmatrix} \text{open} \\ \text{issue} \\ \text{establish} \end{Bmatrix}$ through a $\begin{Bmatrix} \text{first class bank} \\ \text{prime bank} \end{Bmatrix}$

an irrevocable L/C in our favor for the full invoice value within 30 days after confirmation of sale.

　　註：as regards 中的 "regards" 是 transitive verb，所以 "s" 不可省去，後面不可加 "to"。

7. Our payment terms are
$$\begin{cases} \text{cash with order (CWO).} \\ \text{cash in advance (CIA).} \\ \text{cash against documents (CAD).} \\ \text{by sight L/C.} \\ \text{by 90 d/s bill, D/P.} \\ \text{by 90 d/s bill, D/A.} \\ \text{by sight bill, D/P.} \\ \text{O/A, payment to be made within 60 days after B/C date.} \end{cases}$$

8. We believe our new model illustrated in our catalog will make outmoded other products of the similar type.

9. The catalog inclosed gives you a small knowledge of the vast range of the articles we are now handling. If you have requirement for any other sorts of new style, please give us an opportunity to demonstrate our ability of making a creative design.

10. The pamphlet we have sent you is compiled as descriptively and systematically as possible, so that you need only point out the item numbers and quantity when inquiring.

三、結尾句

1.
$$\begin{cases} \text{As our} \begin{cases} \text{stock is} \\ \text{stocks are} \end{cases} \text{running} \begin{cases} \text{low} \\ \text{out} \\ \text{short} \end{cases} \\ \text{As prices are rising} \\ \text{As we are booking heavy orders every day（每天接受大量訂單）} \\ \text{In view of the} \begin{cases} \text{heavy} \\ \text{great} \end{cases} \text{demand for this article} \\ \text{As we fill all orders in strict rotation（嚴格依序發貨）} \end{cases},$$

$$\rightarrow \text{we would advise you to} \begin{Bmatrix} \text{order} \\ \text{place an order} \end{Bmatrix} \begin{cases} \text{soon.} \\ \text{without delay（不可延誤）.} \\ \text{without loss of time（不可猶豫）.} \\ \text{by return.} \\ \text{immediately.} \\ \text{at once.} \end{cases}$$

　　註：a. running low：逐漸減少；b. running out：即將售完；c. running short：逐漸短缺。

2. We hope that you will find the article(s) you want in our catalog and are looking forward to receiving your order.

3. Please let us know if our $\begin{Bmatrix} \text{offer} \\ \text{quotation} \end{Bmatrix}$ does not contain what you $\begin{Bmatrix} \text{want} \\ \text{require} \end{Bmatrix}$ in order to

send you further samples.

4. May we $\begin{cases} \text{hear from you soon?} \\ \text{have the pleasure of hearing from you soon?} \end{cases}$

5. We $\begin{Bmatrix} \text{look forward to} \\ \text{hope to have} \end{Bmatrix}$ the $\begin{Bmatrix} \text{pleasure} \\ \text{opportunity} \end{Bmatrix}$ of $\begin{cases} \text{serving you.} \\ \text{receiving your order.} \end{cases}$

6. We look forward to your answer and a pleasant business $\begin{cases} \text{cooperation.} \\ \text{association.} \\ \text{relationship.} \end{cases}$

$$\left(\text{企望覆信並祝業務} \begin{Bmatrix} \text{合作} \\ \text{提攜} \\ \text{關係} \end{Bmatrix} \text{順遂。} \right)$$

7. We assure you that it is always a pleasure to serve you.

8. We stand ready to be at your service and await your order, which shall have our best and quickest attention.

9. Any order that you may place with us will have our prompt and careful attention.

10. We $\begin{Bmatrix} \text{hope} \\ \text{shall be pleased} \end{Bmatrix}$ to receive your order.

11. Always with pleasure at your service.

12. We welcome you as a customer and hope that this will lead

→ to $\begin{Bmatrix} \text{permanent} \\ \text{lasting} \end{Bmatrix}$ business $\begin{Bmatrix} \text{relations} \\ \text{cooperation} \end{Bmatrix}$ of $\begin{Bmatrix} \text{mutual advantage.} \\ \text{with mutual benefits.} \end{Bmatrix}$

$\begin{pmatrix} 永久性的 \\ 永恆的 \end{pmatrix}$　　　　$\begin{pmatrix} 互利的 \\ 互惠的 \end{pmatrix}$

13. We $\begin{Bmatrix} \text{hope} \\ \text{are sure} \end{Bmatrix}$ that these samples will $\begin{Bmatrix} \text{prove satisfactory to you} \\ \text{meet with your approval} \\ \text{meet your requirement} \end{Bmatrix}$

→ and $\begin{Bmatrix} \text{hope to receive your order.} \\ \text{that we may be favored with your order.} \end{Bmatrix}$

註：a. will prove satisfactory to you：會使你滿意；b. will meet with your approval：能獲得你
　　的讚許；c. will meet your requirement：符合你的需要。

14. We strongly advise you to $\begin{Bmatrix} \text{take advantage} \\ \text{avail yourselves} \end{Bmatrix}$ of this exceptional opportunity.

註：a. take advantage of：利用；b. avail yourselves of：掌握；利用。

15. Please $\begin{Bmatrix} \text{send us} \\ \text{let us have} \end{Bmatrix}$ your $\begin{Bmatrix} \text{order} \\ \text{comments} \end{Bmatrix}$ by $\begin{Bmatrix} \text{cable.} \\ \text{return mail.} \\ \text{e-mail.} \\ \text{fax.} \end{Bmatrix}$

第十五章
報價 (Offer)

第一節　報價信的寫法

一、報價與接受的意義

當賣方答覆買方有關商品的詢價時，其答覆往往就是一封報價函電。國際貿易所謂「報價」(Offer)，即我民法上所稱的要約。

所謂報價即買賣當事人的一方就某一商品向對方提出一定條件，表示願依這些條件與對方成立買賣契約的意思表示。報價的一方稱為「報價人」(Offeror)，被報價的對象為被報價人 (Offeree)。報價必須有被報價人的接受 (Acceptance)，契約才能成立。所謂「接受」在我民法上稱為「承諾」，即被報價人願依報價人所開條件訂立契約的意思表示。

二、報價的種類

1. Selling offer 與 Buying offer：通常 Offer 多指 Selling offer（售貨報價）而言。所謂 Selling offer 乃賣方將擬售商品名稱、品質、數量、價格、包裝、裝運、保險及付款等條件等以函電等通知買方表示願按這些條件將商品賣給對方的意思表示。所謂 Buying offer（購方報價）則係買方將希望購買的商品名稱、品質、數量、價格、包裝、裝運、保險及付款等條件等以函電等通知賣方，表示願依這些條件向對方購進商品的意思表示。

2. Firm offer 與 Non-firm offer：Firm offer（穩固報價）即報價中載明接受期限，在期限內不變更所報各項條件，Offeree 只要在期限內接受，契約即告成立。Non-firm offer（非穩固報價）則指報價中所載條件不明確或有所保留的報價，Non-firm offer 又可分為：

　(1) Offer without engagement（不受約束報價）：即報價人可隨時改變條件的報價，

這種報價實際上只是 Invitation-to-offer 的一種。

⑵ Offer subject to confirmation（確認後有效報價）：即 Offeree 的接受報價，必須經原報人的確認，契約才能成立。

⑶ Offer subject to prior sale
Offer subject to being unsold （有權先售報價）：即 Offeree 的接受報價，以商品未售出才算有效的報價。

3. Counter offer（還價）：如 Offeree 就 Offeror 所報價的條件，加以變更接受，則等於 Offeree 就 Offeror 的報價提出還價。Firm offer 一經 Offeree 的 Counter offer，即失去效力。

三、報價的內容

報價的主要內容包括下列各項條件：

1. Kind of offer	（報價種類）	6. Packing	（包裝）
2. Commodity	（商品名稱）	7. Insurance	（保險）
3. Quality	（品質）	8. Shipment	（裝運）
4. Quantity	（數量）	9. Payment	（付款）
5. Price	（價格）	10. Others	（其他）

四、報價實例

報價可以信函，報價單 (Offer sheet) 或傳真 (Fax) 方式進行，也可以 E-mail 方式進行。在後兩者的情形，除了報價時效很短者外，往往於發出 Fax 或 E-mail 後，立即再以信函確認。撰寫確認報價的函件，應先說明發出 Fax 或 E-mail 的日期和內容，然後將有關各項再予說明或補充。

No. 64　對於蓬萊米詢價的 Firm offer (See No. 42)

Dear Sirs,

Bonlai Rice

We have received your letter of June 6, 20– asking us to offer 5,000 metric tons of the subject rice for shipment to Singapore and appreciate very much your interest in our product.

In compliance with your request, we hereby offer you as follows:

1. Commodity & Description: Bonlai Rice year 2003 Autumn crops.

2. Quantity: Five Thousand (5,000) metric tons.

3. Price: US$250 per metric ton FOB Stowed Kaohsiung.

4. Packing: To be packed in new PE woven bag of 100 kgs. net each.

5. Insurance: Buyer's care.

6. Shipment: Within 30 days after receipt of L/C from Kaohsiung to Singapore, by vessel. Partial shipments to be permitted.

7. Payment: By irrevocable L/C in our favor through a prime bank available by draft(s) at sight.

8. Validity: This offer is good until June 30, 20–, our time.

Your attention is drawn to the fact that we have not much ready stock on hand. Therefore, it is imperative that, in order to enable us to effect early shipment, your L/C should be opened within 10 days after our offer is accepted by you.

We look forward to your prompt reply.

Faithfully yours,

【註】

1. in compliance with your request = to comply with your request = as requested。

2. buyer's care：由買方照顧，即由買方負責投保之意。

3. validity = expiry：有效期限。

4. to be good until：有效到……。

5. our time：以我方時間為準。

6. your attention is drawn to：請注意……。

7. ready stock on hand：手頭存貨。

No. 65　對於柳安合板詢價的 Firm offer (See No. 43)

Lauan Plywood

Dear Sirs:

We thank you for your letter of May 30, 20– enquiring about the captioned building material and take pleasure to enclose our Offer No. TT–110 for your consideration.

May we draw your attention to the facts that the price we quoted is on C&I Bangkok basis instead of on either FOB Taiwan or CIF Bangkok basis and that our offer will be valid until June 30, 20–. As the demands for plywood have been very heavy recently, your early decision and reply are requested.

We are waiting for your prompt and favorable reply.

Faithfully yours,

【註】

1. re = Reference：案由。新派的人認為不宜用此字。

2. captioned：上開。新派人士認為這種字不該用，而應改為 "subject" 或 "above" 或 "above-mentioned" 等字樣，其實這是迷信。

3. C&I：保險費在內價，即等於 FOB 加上保險費的價格。當運價變動不定時，常用此條件代替 CIF。

No. 65–1　報價單

OFFER SHEET

To: Bangkok Trading Co., Ltd.　　　　　　　　　　　　　　　　　June 5, 20–
　　P.O. Box 123
　　Bangkok

No. TT–110

Dear Sirs:

We take pleasure in offering you the following commodity at the price and on the terms and conditions set forth below:

Payment: Against 100% confirmed, irrevocable and transferable Letter of Credit in our favor.

Insurance: ICC (A) plus war for 110% of invoice value.

Shipment: During August, 20– subject to your L/C reaches us by end of July, 20–.

Packing: Export Standard Packing.

Validity: June 30, 20–. Our Time.

Remark: Mill's inspection to be final.

Item	Commodity Description & Specifications	Quantity	Unit Price	Total Amount
TT–12	Lauan Plywood, Rotary cut, Type III, 1/8″, 3–ply, Grading as per JPIC Standard. Size: 1/8″×3′×6′	1,000,000 sq. ft.	C&I Bangkok US$60/MSF	US$60,000.00

Yours very truly,

【註】

　1. our time：以我們的時間為準，意指買方的接受報價函電須在有效期間內到達賣方才有效。

　2. mill's inspection to be final：品質以工廠檢驗為準。

　3. JPIC = Japan Plywood Inspection Council：日本合板檢驗協會。

　4. MSF：1,000 sq. ft.。

No. 66　對詢購府綢的 Cable confirmation of firm offer (See No. 39)

Gentlemen:

　　We thank you for your telegram of February 28 and now confirm our cable dispatched to you to offer firm the undermentioned goods, subject to your acceptance in our hands by March 7:

Goods: 300 pcs. No. 450 Tussore Silk Crepe,28″×40 yds. per pc.

Price: @ $1.20 per yd. CIF New York.

Shipment: Per steamer during April via Panama.

Packing: 20 cs. each in a tin-lined case.

Terms: As usual.

　　For your information we may say that it would be advantageous for you to buy the goods during the present prevalence of this low price, as large orders for Silk Goods are expected to come in at this time of the year. We have no doubt that these orders will affect the market to a considerable extent and that any delay in purchasing will make you pay higher prices.

　　Your specific inquiries for any of our lines would be appreciated. We assure you that we will make every effort to meet your requirements.

Sincerely yours,

【註】

　1. $1.20 讀成 "one dollar twenty" 但 $0.50 則讀成 "fifty cents"。

2. per steamer...via Panama.：於 4 月份由經由巴拿馬運河的船運出。

3. a tin-lined case：錫箔襯裡箱。

　　cf. a zinc-lined case：鋅箔襯裡箱。

　　　　an iron-hooped case：加箍鐵皮箱。

　　　　a wire-hooped case：加箍鐵線箱。

4. terms：指付款條件而言。

5. it would be...low price. = it would be to your interests to buy the goods while the price is so greatly in your favor (while the price rules low).。

6. will affect...extent = will have a great affect on the market。

No. 67　對於詢購毛線衫的 Offer subject to prior sale (See No. 41)

Dear Sirs,

We acknowledge with thanks the receipt of your letter of May 12 and as requested we are pleased to offer you the following items subject to their being unsold upon receipt of your reply.

Commodity: Acrylic Sweaters for men and ladies in different color/pattern assortments.

Quality: as per samples submitted to you on May 18 by air parcel.

Quantity: 1,000 doz., Size assortments: $\frac{S}{3} \frac{M}{6} \frac{L}{3}$

Price: US$14 per doz. CIF Hamburg.

Packing: To be wrapped in polybags and packed in standard export cardboard cartons.

Insurance: To cover ICC (A) plus war for 110% of invoice value.

Shipment: September/October, 20–

Payment: by Al bankers' irrevocable and without recourse L/C available by draft(s) at sight. Such L/C must reach us one month prior to shipment.

This low price can be quoted because we have employed a new mass production system since last Spring. However, we are now receiving many inquiries from other areas we have recently contacted. The price will surely rise when the present stock is exhausted with the approach of the season. Therefore, it is advisable that you take advantage of this rare opportunity by accepting this offer as soon as possible.

We look forward to your early acceptance soon.

Very truly yours,

【註】

1. Offer...subject to their being unsold...reply.：在接獲回答時尚未售出方屬有效的報價。

2. $\dfrac{S\ M\ L}{3\ 6\ 3}$：意指每一打的尺碼搭配為小號 (S) 3 件，中號 (M) 6 件，大號 (L) 3 件。

3. without recourse L/C：無追索權信用狀。與 with recourse L/C 相對。

憑有追索權信用狀開出的匯票，萬一遭到拒付 (unpay) 時，持票人可向背書人請求償還票款。反之，憑無追索權信用狀開出的匯票，如遭到拒付時，持票人不能向背書人請求償還票款。換言之，前者有追索權，後者則無追索權。

就法律觀點而言，匯票為票據的一種，其製作需依票據法的規定。這裡所稱票據法，係指行為地的票據法而言，也即指簽發匯票人所在地的票據法而言。依我國票據法第 29 條及第 39 條規定，匯票上有「無追索權」字樣者，其記載應屬無效。

4. employ：啟用。

5. new mass production system：新式大量生產系統。

6. when...is exhausted：當目前存貨售罄時。

7. take advantage of this rare opportunity：把握此一難得的機會。

8. acceptance：接受；承諾。

No. 68　對沙丁魚油脂詢價的 Offer subject to market changes (See No. 46)

Gentlemen:

We wish to thank you for your inquiry of May 10 for our Sardine Oil, on which we are pleased to quote you as follows:

Article: 1,000 cases Sardine Oil No. 1.

Price: @ $80.00 per ton of 2,000# CIF San Francisco for net shipped weight.

Shipment: During July, 20–

Packing: 60 # net to be contained in a can; 2 cans to be packed in a case.

Terms: Draft at 60 d/s under an Irrevocable Credit.

N.B. The price quoted is subject to market fluctuations. Loss in weight exceeding 2% to be allowed for by us.

You will observe that our price is exceedingly low. We would remind you that this low price reflects in no way on the quality of the goods and is rendered possible only by our system of mass-production and our policy of eliminating the usual middlemen.

We trust you will give us a trial order, which will be highly valued and executed to your satisfaction in every particular.

Very truly yours,

【註】

1. 2,000# = 2,000 lbs.：# 放在數字後面時表示「磅」。

2. during July：7 月間（裝運），shipment during July = July shipment：意指賣方得在 7 月 1 日至 7 月 31 日裝船，但在此期間不能再分批裝運。

3. subject to market fluctuations = subject to the fluctuations of the market; subject to market change：隨市價波動；隨市價漲跌，因此本報價並非穩固報價。

4. mass-production：大量生產。

5. to eliminate the usual middlemen：排斥普通的中間商。middlemen = intermediary：中間商。

No. 69　對詢購原棉的 Firm offer (See No. 47)

Dear Sirs,

Subj. American Raw Cotton, SLM

Thank you for your letter of June 1, 20– informing us that you are in urgent need of the subject cotton. To meet your requirements, we confirm we have faxed you today to offer the following goods, subject to your acceptance arrives here by end of June, 20–.

(decoded)

Commodity: American Raw Cotton; Type: TASK; Grade: Strict & Low Middling; Staple: 1–1/16″; Pressley: 90,000

Spec.: Micronaire: 3.8 NCL

Quantity: 1,200 bales of approximately 500 lbs. each.

Price: US¢52 per lb. FAS Gulf ports.

Shipment: August, 20– through November, 20– at 300 bales each monthly.

Payment: By irrevocable L/C in our favor available by draft(s) at sight one month prior to each monthly shipment.

We have to inform you that we are unable to comply with your wish so far as shipment is concerned because the 20– new crop would not be available until August and trust that you would agree to our proposed shipping schedules. Meantime, we are awaiting your early acceptance in order to catch up with our shipping schedules.

Yours sincerely,

【註】

1. Subj.：為 Subject 的略字，即案由之意。

2. subject to...here：以你的接受（函電）在（某日以前）到達此地才有效的條件，因規定了接受期限，所以本報價為穩固報價。

3. new crop：新作物，"crop" 主要指土地生產物而言。

4. proposed shipping schedules：所提議的裝運預定日期。

5. catch up with：趕上。

No. 70　對詢購罐頭蘆筍的 Firm offer (See No. 51)

Dear Sirs,

Canned Asparagus, Al Grade

We thank you very much for your letter, Ref. INQ/89/123, of May 3, 20– inquiring about the subject item and take pleasure in quoting you as follows:

ITEM	QUANTITY (carton)	FOBC$_3$ in EUR	CIFC$_2$ Hamburg in EUR
		(Unit price per carton of 48 cans)	
Tips & Cuts	300	30.00	35.00
	48 cans×12 oz.nw		
Center Cuts	300	27.50	32.50
	48 cans×12 oz.nw		
End Cuts	300	24.00	29.00
	48 cans×12 oz.nw		

N.B. 1. Packing: Standard export cardboard cartons.

2. Insurance: ICC (B), plus TPND and war risk for 110% of CIF value.

3. Shipment: To be shipped in one lot during August, 20–.

4. Payment: By a prime banker's irrevocable L/C to be opened in our favor 30 days before shipment.

5. Validity: 15 days from the date hereof.

Please note that the prices we quoted above are the rockbottom ones and that our products would compete favorably with other brands in your market because of their superior quality.

As our stocks have been running low and the demands therefor are brisk right now, we advise you to make decision as soon as possible.

Very truly yours,

【註】

1. FOBC$_3$：船上交貨價含佣金 3%。"C" 代表 commission 而且此項佣金係指回佣 (return com-

mission)，即將來需退還給對方的佣金。

2. EUR：歐元。

3. carton of 48 cans：每箱裝 48 罐。

4. N.B. = Nota Bene（拉丁語）= note well（注意）。

5. cardboard cartons：（硬）紙板箱。

6. in one lot：以一批（裝出）。

7. 15 days...hereof：hereof 指 "of this offer"。

8. stocks...running low：存貨逐漸減少。

No. 71　對詢購電扇、電鍋的 Offer subject to confirmation (See No. 50)

Gentlemen:

Your letter of March 2nd has been received and we are very much pleased to be informed that you are interested in our products and are plan to explore the Indonesian market for them. Your efforts in this respect is much appreciated.

In compliance with your request, we are happy to offer you the following items under the terms and conditions as follows:

Electric Fans	FOB net/unit in US$	CIF net/unit Jarkata in US$
Table Fan 12″	8.00	9.00
14″	9.00	10.00
16″	9.50	10.00
18″	10.00	11.50
Floor Fan 22″	25.00	26.70
24″	30.00	31.80
26″	35.00	37.00
Electric Rice Cookers		
for 6 persons	6.00	7.00
for 8 persons	7.50	8.50
for 10 persons	9.00	9.00
for 16 persons	12.00	14.00

Packing: Standard export wooden case.

Insurance: ICC (A) plus war for 110% of CIF value.

Shipment: Within 60 days after receipt of L/C.

Payment: By irrevocable L/C to be opened in our favor available by draft(s) at sight. Such L/C must be opened within 20 days after confirmation of sale.

Validity: All orders are subject to our final confirmation.

Min. order: 500 units for each item.

Meanwhile, we should like to assure you that the goods to be delivered to you are all of the best quality obtainable elsewhere.

Please let us have your early reply should our offer are acceptable.

Yours faithfully,

【註】

1. explore the Indonesian market：開拓印尼市場。

2. in this respect：在這一方面，也可以 "in this regard"（關於此事）代替。

3. FOB net/unit：每部船上交貨淨價，"net" 指淨價，也即不包括佣金之意。

4. electric rice cooker：煮飯用電鍋。

5. for 6 persons：6 人用（的電鍋）。

6. standard export：標準出口包裝用的（木箱）。

7. min. order：最低訂購量，也有以 "minimum quantity" 表示。

8. we should like to assure you：我們願向你保證，也可以 "you may be rest assured" 代替。

No. 72　對詢購塗漿物發出 Proforma invoice (See No. 53)

Dear Sirs,

"Jet Size" Sizing Stuff

Many thanks for your inquiry about the subject stuff as stated in your letter of May 12, 20– and we are pleased to enclose herewith our Proforma Invoice No. 123 for your consideration.

We are producers of the stuff for many years and our stuff has been very popular among the end-users in Japan as well as in the Southeast Asian countries. Under separate cover and by sea parcel, we have despatched to you sample bags of 50 kgs. each of these two qualities along with brochures explaining the merits of our products.

We are looking forward to receiving your comments on the samples and your early order.

Yours faithfully,

Encl.: a/s

【註】

1. enclosed herewith：檢附，"herewith" 是多餘的，應予刪除。

2. proforma invoice：預期發票，又寫成 "pro-forma invoice"，"proforma" 為「假定的」、「假設的」之意。

3. popular：受歡迎。

4. end-users：最終用戶；直接用戶。

5. sea parcel = surface parcel：海運包裹。

6. sample bags：樣品包。

7. along with = together with。

8. merits：優點。

No. 72–1　預期發票 (Proforma invoice)

PROFORMA INVOICE NO. 123　　　　　　　　　May 20, 20–

To: Taiwan Trading Co., Ltd.
　　Taipei, Taiwan

MARKS & NOS. Pakg.	COMMODITY DESCRIPTION	QUANTITY (kg.)	UNIT PRICE	TOTAL AMOUNT
	"Jet Size" Sizing Stuff No. 66 for Sheeting & jeans	2,000	FOB US$0.50/kg.	KOBE US$1,000.00
	No. 88–98 for Poplin & Broadcloths	2,000 TOTAL	US$0.60/kg. FOB KOBE	1,200.00 US$2,200.00

Remarks: PACKING: In PE bags, each containing 50 kgs. net.
　　　　　INSURANCE: Buyer's care.
　　　　　SHIPMENT: Within 30 days after receipt of L/C.
　　　　　PAYMENT: By irrevocable Sight L/C which must be opened within 30 days after confirmation of sale.
　　　　　VALIDITY: This offer is valid until June 20, 20–.

【註】

　　PE bag = polyethylene bag：一種塑膠袋。

No. 73　對詢購棉織品的 Firm offer (See No. 48)

Dear Sirs,

　　Thank you very much for your inquiry of June 18 requesting us to quote the prices for our first class tweed and black serge.

Concerning the above, we have just cabled you the following firm offer:

5,000 YDS SAMPLE NO. 235 BLACK SERGE @ STG. £35 AND NO. 685 TWEED @ STG. £38 PER YD CIF KEELUNG SHIPMENT AUGUST SUBJECT REPLY RECEIVED HERE JUNE 20.

We are sure you will find our prices very reasonable. The market here is enjoying an upward trend, and we have no further stock available to offer at the same price.

We advise you not to overlook this opportunity and hope to receive your prompt order by cable.

Yours faithfully,

【註】

1. price for: The price of a thing：用於價格已經確定時。

　例：What's the price of this book? （這書的定價多少？）

　the price for a thing：用於賣方所要求的價錢。

　"price for" 為 price charged for; price paid for; price to be charged for 之意。

　例：The price for a smaller quantity will be slightly higher.

　the price on a thing：用於詢問價格。"price on" 為 price set (put) on 之意。

　例：What is his price on this lot? (= What is the price he puts on this lot?)

2. market is enjoying an upward trend：市況顯示上昇趨向。

3. no further stock available to offer：已無存貨可供應。

No. 74　對詢購照相機的 Firm offer

Gentlemen:

Thank you for your inquiry of June 25, against which we have quoted the price as shown in the following firm offer cabled today subject to your reply here by the 15th.

Commodity: Model FAU–150 Full-automatic tube camera equipped with automatic exposure meter & range meter.

Quality: Specified in the attached specification sheets.

Quantity: 500 sets.

Price: US$250.00 CIF Keelung per unit.

Payment: By Bankers' irrevocable L/C payable by draft at sight.

Shipment: Before July 30, 20–.

> With the approach of the leisure season, the demand for high quality cameras is increasing. The compact styling and superb quality will certainly present excellent sales as shown by our production ration.
>
> We trust you will take advantage of this seasonal opportunity and favor us with an early reply.
>
> Very truly yours,

【註】

1. full-automatic tube camera：全自動 35 mm 照相機。

2. automatic exposure meter & range meter：自動曝光表及距離表。

3. with the approach of the leisure season：隨休閒季節的到臨。

4. production ration：生產實績。

5. take advantage of：利用。

第二節　有關報價的有用例句

一、開頭句

1. $\begin{Bmatrix} \text{Thank you} \\ \text{Many thanks} \\ \text{We thank you} \end{Bmatrix}$ for your $\begin{Bmatrix} \text{inquiry} \\ \text{letter} \end{Bmatrix}$ of (date) $\begin{Bmatrix} \text{about} \\ \text{on} \\ \text{concerning} \end{Bmatrix}$ ...and (we) are $\begin{Bmatrix} \text{pleased} \\ \text{happy} \\ \text{glad} \end{Bmatrix}$

to offer them as follows:

2. $\begin{Bmatrix} \text{In response} \\ \text{In reply} \\ \text{With reference} \end{Bmatrix}$ to your inquiry of (date) $\begin{Bmatrix} \text{about} \\ \text{on} \\ \text{concerning} \end{Bmatrix}$ ...we have the pleasure of

offering you on the following terms and conditions.

3. $\begin{Bmatrix} \text{In reply} \\ \text{In response} \\ \text{With reference} \end{Bmatrix}$ to your inquiry of..., we are $\begin{Bmatrix} \text{pleased} \\ \text{glad} \\ \text{happy} \end{Bmatrix}$ to

→ $\begin{Bmatrix} \text{offer} \\ \text{make an offer to} \\ \text{submit our offer to} \end{Bmatrix}$ you $\begin{Bmatrix} \text{on} \begin{Bmatrix} \text{the following} \\ \text{the below-mentioned} \end{Bmatrix} \text{terms/conditions.} \\ \text{as follows:} \end{Bmatrix}$

4. In compliance with your request of..., we are pleased to $\begin{Bmatrix} \text{submit} \\ \text{make} \end{Bmatrix}$ our offer as follows:

5. In response to your inquiry of..., we have $\begin{Bmatrix} \text{cabled} \\ \text{faxed} \end{Bmatrix}$ you our offer as shown in the enclosed confirmation.

6. Thank you for your inquiry of...concerning...as requested, we are $\begin{Bmatrix} \text{pleased} \\ \text{glad} \\ \text{happy} \end{Bmatrix}$ to offer them as follows:

二、條 件

1. $\begin{Bmatrix} \text{This} \\ \text{The} \\ \text{The above} \end{Bmatrix}$ offer is $\begin{Bmatrix} \text{firm} \\ \text{valid} \\ \text{good} \\ \text{effective} \\ \text{open} \\ \text{in force} \\ \text{available} \\ \text{to remain in force} \end{Bmatrix}$ $\begin{Bmatrix} \text{for...days.} \\ \text{until Nov. 10.} \\ \text{till Nov. 10.} \\ \text{until further notice.} \end{Bmatrix}$

2. All $\begin{Bmatrix} \text{offers} \\ \text{quotations} \end{Bmatrix}$ and sales are subject to the terms and conditions printed

$\rightarrow \begin{Bmatrix} \text{on the reverse side hereof.} \\ \text{on the back hereof.} \end{Bmatrix}$

3. This is a combined offer on all or none basis.

（此為聯合報價，所列貨品必需全部接受或全部不接受。）

4. We renew our offer of June 20 $\begin{Bmatrix} \text{on the same terms and conditions.} \\ \text{subject to the following modifications.} \end{Bmatrix}$

$\left(我們 \begin{Bmatrix} 基於原來條件 \\ 依照下列變更 \end{Bmatrix} 更新 6 月 20 日的報價。 \right)$

5. $\begin{Bmatrix} \text{This} \\ \text{Our} \\ \text{The} \end{Bmatrix} \begin{Bmatrix} \text{offer} \\ \text{quotation} \end{Bmatrix}$ is subject to $\begin{Bmatrix} \text{prior sale.} \\ \text{being unsold on receipt of your acceptance (reply).} \\ \text{your reply} \begin{Bmatrix} \text{received} \\ \text{reaching} \end{Bmatrix} \text{here} \begin{Bmatrix} \text{by...} \\ \text{not later than...} \\ \text{on or before...} \end{Bmatrix} \end{Bmatrix}$

$\left(此報價 \begin{Bmatrix} 有權先售。 \\ 收到接受時未售才有效。 \\ 以你的答覆 \begin{Bmatrix} 在……以前 \\ 不能晚於 \\ 在……或以前 \end{Bmatrix} \begin{Bmatrix} 收到 \\ 到達這裡 \end{Bmatrix} 才有效。 \end{Bmatrix} \right)$

6. This offer is subject to shipping space available.

7. $\begin{Bmatrix} \text{Due to} \\ \text{Because of} \end{Bmatrix}$ the sensitive market situation here, we are

→ $\begin{Bmatrix} \text{unable} \\ \text{not in a position} \end{Bmatrix}$ to $\begin{Bmatrix} \text{submit our offer} \\ \text{make our offer} \\ \text{offer} \end{Bmatrix}$ that might be outstanding for over 15 days.

$\left(\begin{Bmatrix} 由於 \\ 因為 \end{Bmatrix} 本地市場的敏感性，我們無法 \begin{Bmatrix} 提出 \\ 報出 \end{Bmatrix} 有效至 15 天以上的價。 \right)$

8. We $\begin{Bmatrix} \text{are able to} \\ \text{can} \\ \text{are in a position to} \end{Bmatrix}$ compete with any other firms $\begin{Bmatrix} \text{concerning} \\ \text{regarding} \end{Bmatrix}$ the quality and

price of the merchandise.

9. We $\begin{Bmatrix} \text{may} \\ \text{are prepared to} \end{Bmatrix}$ grant you a 3% discount $\begin{Bmatrix} \text{provided} \\ \text{on condition} \end{Bmatrix}$ that you $\begin{Bmatrix} \text{take} \\ \text{order} \\ \text{purchase} \\ \text{buy} \end{Bmatrix}$

more than 10,000 dozen.

10. We are now pleased to submit to you a firm offer for Taiwan jades as required by you some time ago, subject to your reply here by noon our time on Monday, June 25.

11. As stated above, the supply being extremely limited, this offer is made subject to the goods being unsold on receipt of your $\begin{cases} \text{acceptance.} \\ \text{order.} \end{cases}$

12. This offer is CIF any port on the Pacific Coast subject to our confirmation and for your reply here on June 30, our time.

13. The best firm offer we can make is FOB Keelung subject to immediate reply.

14. We are busily working on your offer, in the hope that we can bring about business.

    註：a. working：處理；b. bring about：促成；完成；成交。

15. These goods being in great demand we cannot hold this offer open; we are therefore obliged to make the offer subject to prior sale.

16. As the goods now under offer are commanding a ready market, there is no knowing when they will be sold out. You will, therefore, understand that we hold this offer good only for acceptance by Friday, May 25, our time, and after that, it is subject to the goods being unsold.

    註：commanding a ready market：擁有現成的市場，意指暢銷，可隨出銷出。

第十六章
接受 (Acceptance)

第一節　接受信的寫法

接到報價之後，應迅速檢討報價的內容，以便決定接受，或另提出自己的條件，或因條件相差太遠根本不接受。無論那種情形，均應以迅速答覆為原則，不可置之不理。因此對於報價的答覆可分為三種：

1. Complete acceptance（全部接受）：即對於對方報價所開的全部條件無條件接受。如前所述，Acceptance 意指被報價人願依報價人所開條件成立契約的意思表示。接受的內容除接受的意思表示外，尚應將商品名稱、品質規格、數量、價格、裝運、包裝、保險及付款條件等內容予以重述。Acceptance 一如 Offer，往往以電傳完成，所以發出電傳之後，通常尚須發出 Confirmation，並附上 Purchase order 或 Order sheet。

2. Conditional acceptance（附條件的接受）：即將對方的報價條件加以變更或追加新條件而接受。這種接受是不完全接受 (Imcomplete acceptance)，所以實際上是「反報價」也即還價 (Counter offer)，還價本質上不僅為拒絕原報價，而且也是被報價人對原報價人的新報價。因此，如原報價人接受「還價」，契約即告成立。在實際交易中，很少一報價即被接受。大部分情形是經過一連串的討價還價，然後契約才告成立。

3. Non-acceptance（不接受）：也即拒絕對方的報價 (Offer declined) 之意。

    拒絕報價時，其拒絕信應包括：

    (1)謝謝其報價。

    (2)表示不能接受的遺憾。

    (3)解釋不能接受的理由。

    (4)暗示將來還有機會往來。

No. 75 覆告蓬萊米報價已接受 (See No. 64)

Dear Sirs:

Re: Bonlai Rice

We thank you very much for your letter of June 12, 20– offering us 5,000 M/T of the subject rice and wish to confirm this order with the following particulars:

Commodity
        &
Specifications: Bonlai Rice, year 2003 Autumn crops.

Quantity: 5,000 M/T's.

Price: US$250.00 per M/T FOB Stowed Kaohsiung.

Packing: To be packed in new PE woven bag of 100 kgs. net each.

Insurance: Our care.

For your information, we have already applied for the issuing of an L/C in your favor to Bank of Canton, Singapore and trust you would receive it in a few days. It will be appreciated if you will effect shipment of these 5,000 M/T's in about two equal lots by direct steamer as soon as you receive our L/C. In the meantime, although the price contracted is on an FOB stowed basis, we shall appreciate your arranging with shipping companies on our behalf for delivery of the goods with ocean freight to collect.

Yours faithfully,

【註】

1. applied for...to Bank of Canton：向廣東銀行申請開發 L/C，"for" 後面接申請的東西，"to" 後面接「人、機構等」。

2. effect shipment：裝運（船），也可以 "make shipment" 代替。

3. in about two equal lots：分約二等批，意指分兩批裝，兩批數量大約相等。

4. arrange with shipping companies on our behalf...：代我們與船公司安排交貨事宜，指由賣方代洽船，"on our behalf" 可以 "on behalf of ourselves" 或 "in our behalf" 代替。

5. freight to collect：運費到付。freight prepaid 為運費預付。

No. 76　接受柳安合板報價並發出購貨訂單 (See No. 65)

Dear Sirs:

Lauan Plywood

We acknowledge with thanks the receipt of your letter of June 5, 20– together with one copy of your offer sheet No. TT–110 for one million sq. ft. of the captioned commodity.

After careful perusal of the terms and conditions of your offer, we have found them quite acceptable, therefore, we are enclosing our Purchase Order No. PO–123 in duplicate for your signature. Please sign and return one signed copy thereof for our files.

As we are in urgent need of this materials, you are requested to effect shipment during August, 20– as promised in your offer. Meanwhile, we shall apply for the opening of an L/C in your favor within one week after we have received the copy of our purchase order duly signed by you.

We are looking forward to receiving your confirmation soon.

Very truly yours,

Encl.: a/s

【註】

1. careful perusal：仔細的審閱；詳核。

2. one signed copy thereof = one signed copy of our purchase order。

3. for our files：俾供我們存檔。

4. duly signed：簽妥字。

No. 76–1　確認購買柳安合板並發出購貨訂單

PURCHASE ORDER

Taiwan Trading Co., Ltd.　　　　　　　　　　　　　　　　　　　　June 20, 20–

Taipei, Taiwan, R.O.C.　　　　　　　　　　　　　　　　　　Order No. PO–123

Dear Sirs,

　　We confirm having purchased from you the following commodity on the following terms and conditions:

1. Commodity: Lauan Plywood, Rotary Cut.

2. Specifications: Type III, 1/8″, 3-ply, Grading conforming to JPIC standard.
　　　　　　　Size: 1/8″×3′×6′

3. Quantity: 1,000,000 sq. ft.

4. Price: US$60 per MSF C&I Bangkok.

5. Packing: Export standard packing.

6. Insurance: ICC (A)plus war for 110% of Invoice value.

7. Shipment: To be shipped in August, 20— in one lot.

8. Payment: By irrevocable and transferable L/C in your favor to reach you by end of July, 20–.

9. Inspection: Mill's inspection at mill to be final.

Bangkok Trading Co., Ltd.

【註】

1. transferable L/C：可轉讓信用狀，以前又稱為 "assignable L/C"，現在最好不要用。

2. mill：工廠，木材廠，麵粉廠，紙廠等的工廠多用 "mill" 這個字。

No. 77　接受壓克力毛線衫報價並發出訂單 (See No. 67)

Dear Sirs,

Acrylic Sweaters

Your letter of May 25, 20– offering the subject garments has been received, and we are pleased to inform you that your terms and conditions are acceptable to us.

By means of this letter, we are pleased to place our order with you for the garments of the following particulars:

Acrylic Sweaters for men:

Quantity: 500 doz.

Colors and size assortments per dozen:

Size　　Color	S	M	L
Red	1	2	1
Green	1	2	1
Yellow	1	2	1

Acrylic Sweaters for ladies:

Quantity: 500 doz.

Color and size assortments per dozen:

Size　　Color	S	M	L
Navy	1	2	1
Blue	1	2	1

Purple	1	2	1

As the season is coming, you are requested to effect shipment the total quantity of this order in one lot not later than October 5. Meanwhile, please send us your Proforma Invoice immediately and we will open our L/C for the total value in due course upon receipt of it.

Your prompt confirmation is awaited.

Yours faithfully,

【註】

1. by means of this letter：憑此信。

2. color and size assortment：顏色及尺寸的搭配。

3. in due course：在適當時期；不久以後。

No. 78　買方訂購蘆筍函 (See No. 70)

Gentlemen:

Canned Asparagus, Al Grade

We have received your letter of May 15, 20– submitting your offer for three different qualities of the subject item.

Since the prices you quoted are quite reasonable, we hereby place the following order with you:

Item	Quantity
Tips & Cuts	300 cartons, 48 cans×12 oz., N.W./CTN
Center Cuts	ditto
End Cuts	ditto

So far as prices are concerned, we prefer $CIFC_2$ Hamburg to $FOBC_3$. It is understood that the prices for the former are EUR35, EUR32.50, and EUR29 per carton for these three qualities respectively.

As requested in your letter under reply, we will issue an L/C through our bankers, Dresdner Bank, Hamburg, in your favor around June 30, in order to enable you to make delivery during August, 20–.

Please let us have your confirmation at your earliest convenience.

Very truly yours,

【註】

1. quite reasonable：（價錢）很公道。

2. ditto：「同上」之意。

3. so far as...concerned：就……而論，也可以 "as far as...concerned" 代替。

4. prefer...to...：prefer 常與 "to" 連在一起，意指「喜歡……而不喜歡……」，例如：I prefer tea to coffee.（我喜歡茶而不喜歡咖啡。）

5. it is understood：諒解；同意；心照不宣。

6. around June 30：6 月 30 日左右，"around" 也可以 "about" 代替，不過比較不確切。

No. 79　賣方寄出蘆筍售貨確認書函 (See No. 78)

Dear Sirs,

Re: Canned Asparagus, A1 Grade

Thank you very much for your letter of May 25, 20– placing an order with us for 300 cartons each of three qualities of the captioned canned food and we take pleasure to inform you that we have already started processing this order. Please be assured that we will effect shipment of the total quantity during August provided your L/C is received prior to June 30, as indicated in your letter under reply.

When the goods are ready for shipment, we will fax you upon shipping space has been arranged. Enclosed please find our Sales Confirmation in duplicate for your signature and we shall appreciate your returning to us one duly signed for our file.

We thank you again for your interest in our product.

Yours faithfully,

【註】

1. take pleasure to inform = take pleasure in informing。

2. for our file：可以 "for our records" 代替。

No. 79-1　售貨確認書

TAIWAN TRADING COMPANY, LTD.

Taipei, Taiwan

Sales Confirmation

X.Y.Z. A.G.　　　　　　　　　　　　　　　　　　　　　　　　　　June 1, 20–

P.O. Box 123　　　　　　　　　　　　　　　　　　　　　　　　　Sale No. 89/123

Hamburg

Dear Sirs,

We confirm that we have sold to you the following goods on the following terms and conditions:

1. Commodity: Canned Asparagus, A1 Grade of Tips & Cuts, Center Cuts, and End Cuts.

2. Quantity: 300 cartons each of Tips & Cuts, Center Cuts, and End Cuts. Each carton contains 48 cans×12 oz., N.W.

3. Price: Tips & Cuts: EUR35 per carton $CIFC_2$ Hamburg. Center Cuts: EUR32.50 per carton $CIFC_2$ Hamburg. End Cuts: EUR29 per carton $CIFC_2$ Hamburg.

4. Packing: Standard export cardboard packing.

5. Insurance: ICC (B) plus TPND and War for 110% of invoice value.

6. Shipment: During August, 20–.

7. Payment: By a prime bank's irrevocable L/C which must be opened in our favor 30 days prior to shipment.

Accepted on...(date)

by:

Yours faithfully,

Taiwan Trading Company, Ltd.

No. 80　訂購電扇、電鍋 (See No. 71)

Dear Sirs,

Electric Fans and Rice Cookers

We acknowledge with thanks the receipt of your offer dated March 10 and take pleasure in placing our order with you for the following:

Electric Fans	Quantity
Table fan 14″	500 units
ditto 16″	500 units
Floor fan 20″	500 units
ditto 26″	500 units

Electric Rice Cookers	
for 8 persons	500 units
for 10 persons	500 units
for 16 persons	500 units

Since this is the first time we explore the Indonesian markets for your products, it is, therefore, absolutely essential for you to see to it that all the goods delivered are the best quality and that all of them are thoroughly inspected prior to shipment.

We will arrange with our bankers for establishing an L/C as soon as we receive your confirmation of sale.

Yours faithfully,

【註】

1. to see to it = to take care of it：注意；照拂。

2. prior to shipment = before shipment。

3. arrange with...for...：與（我們的銀行）協商（開 L/C）。

4. confirmation of sale = sales confirmation。

No. 81　訂購塗漿物 (See No. 72)

Gentlemen:

"Jet Size" Sizing Stuff

We have received your letter of May 20, 20– together with your Proforma Invoice No. 123 for the subject stuff.

We hereby place an order with you for 2,000 kgs. each of Nos. 66 and 88–98 Sizing Stuffs at US$0.50 and US$0.60 per kg. FOB Kobe respectively. We have submitted our application for import license this morning and will inform you the relative number as soon as our application has been approved by the Government Authorities concerned. Of course, we shall request our bankers to open L/C immediately upon obtaining of the import license.

As we need the stuffs quite urgently, you are requested to make delivery within 30 days after receipt of our L/C.

Faithfully yours,

【註】

Government Authorities concerned：政府有關當局。

No. 82　確認接受縐紗報價，發出購貨確認書 (See No. 66)

Dear Sirs,

We have pleasure in confirming our telex, accepting your firm offer of the 28th March on 200 pieces No. 450 Crepe de Chine, and now send you herewith our Purchase Note No. 150, which we trust you will find in order.

In order to cover the amount of this purchase, we have arranged with Lloyds Bank, Ltd., London, for a Confirmed Letter of Credit in your favour. This credit will be telexed to your bank as usual. This business being very important on our part, we ask you to give it your best attention so as to be able to satisfy us in every respect.

Tussore Crepe No. 380. We have just been approached by our reliable friends in Liverpool to make them an offer on this commodity, and accordingly we have just telexed, as per copy enclosed, asking you to telex us a firm offer CIF on 500 pieces for May shipment. We hope you will do everything you possibly can to make us your best offer, as we think it is a good chance to dispose of a large parcel of this commodity. We look forward to your telex offer, so that we may close the business to our mutual advantage.

Faithfully yours,

H. WHITEHALL & CO., LTD.

W. Meyer

President

WM: FS

Enclos. 3 Copies of telexes

    1 Purchase Note

    P.S. Your telex offer of the 30th March on 200 pieces Pongee Silk No. 190 has just reached us. We regret that we find it unworkable at your figure, but we will make every effort to cable a feasible offer at an early date.

W. M.

【註】

    1. crepe de chine: 縐紗的一種。

    2. purchase note = bought note = purchase confirmation。

    3. in order...purchase = in settlement for our purchase: 為清償購貨價款。

    4. parcel: 貨物,包成一定數量的貨物。

    5. to close: 成交;締結(契約)。

    6. unworkable: 難於處理,意指價高無法接受。

    7. feasible offer: 可行的報價。feasible = practicable,這裡的 "offer" 為 "buying offer"。

No. 83　對尼龍內衣報價的還價

Dear Sirs,

    We thank you for your letter of July 20 and for the samples of nylon underwear you kindly sent us.

    We appreciate the good quality of these garments, but unfortunately your prices appear to be on the high side for garments of this quality. To accept the prices you quoted would leave us little profit on our sales since this is an area in which the principal demand is for articles in the medium price range.

    We like the quality of your goods and also the way in which you have handled our enquiry and would welcome the opportunity to do business with you. May we suggest that you could perhaps make some allowance on your quoted prices that would help to introduce your goods to our customers. If you cannot do so, then we must regretfully decline your offer as it stands.

Yours faithfully,

【註】

1. the samples of nylon underwear you kindly sent us：承蒙惠寄的尼龍內衣樣品。本句中的關係代名詞因係受格並係限定使用法，故省略。請參閱下例：The offer (which) you made is too high.（貴公司所作的報價太高。）These are the best terms (which) we can grant.（此乃本公司能給予的最優惠條件。）

2. appear to be on the high side for garments of this quality：與同品質的成衣相較價格似嫌偏高。

3. to accept the prices you quoted would leave us little profit on our sales：如接受貴公司所開之價，將使本公司的銷售幾乎無利可圖。本句的 little 乃強調所剩無幾之意，如用 a little 則強調尚有幾分利潤。

4. articles in the medium price range：中等售價的貨品。

5. the way in which you have handled our enquiry：貴公司對本公司詢價的處理方式。

6. you could perhaps make some allowance on your quoted prices：可否對所作的報價略予讓步。此句中的 could 是一種請求的禮貌說法，例：Could you do me a favor? 另 "allowance" 也可譯為折扣。

7. then, we must regretfully decline your offer：否則祇能謝拒貴公司的報價了。"decline" 比 "refuse" 要客氣。

8. as it stands：照現狀；照現在的報價。

No. 84　接受尼龍內衣還價 (See No. 83)

Dear Sirs,

　　We are sorry to learn from your letter of August 5 that you find our prices too high. We do our best to keep prices as low as possible without sacrificing the quality and to this end, are constantly enquiring into new methods of manufacture.

　　Considering the quality of the goods we quoted, we do not feel that the prices are at all excessive, but bearing in mind the special character of your trade, we have decided to offer you a special discount of 3% on a first order for US$12,000. We make this allowance because we should like to do business with you if possible, but we must stress that it is the furthest we can go to help you. We hope this revised offer will now enable you to place an order.

Sincerely yours,

【註】

1. without sacrificing the quality：不犧牲品質水準。

2. to this end, are constantly enquiring into new methods of manufacture：為達到此目標，不斷探求生產製造的新方法。

3. bearing in mind the special character of your trade：考慮到與貴公司交易的特別性質。

4. we must stress that it is the furthest we can go to help you：我們必須強調這是本公司能給予貴公司最大的優待。

5. this revised offer：此經過修正的報價。

No. 85　對胚布報價的還價

> Dear Sirs,
>
> We acknowledge your letter of the 1st May, together with Price Lists for your White and Grey Shirtings, and are grateful for the supply of samples, which are quite up to our expectations.
>
> We have studied your quotations with interest, and though your products have impressed us very favourably, we regret that business cannot be considered at the prices stated, and therefore keener prices are necessary. As we intimated to you in our last letter, our requirements for whites and greys are fairly heavy, and it is hardly necessary to remind you of the benefit likely to accrue to you from competitive prices.
>
> It is our intention to place our orders with you, and we are prepared to take 100 pieces for No. 10 and 200 pieces for No. 12 respectively by way of trial, so that we trust you will make every effort to revise your prices.
>
> Your earliest reply will oblige.
>
> Yours faithfully,

【註】

1. white and grey shirting：素色及原色胚布，"grey" 為英國寫法，美國則拼成 "gray"。

2. keener prices = lower prices。

3. whites = white shirting; greys = grey shirting。

4. by way of trial = as a trial; for a trial。

5. make every effort：儘量。

6. to revise：修正；改正，這裡指修正價格，也即減價之意。

7. will oblige：這種措詞嫌舊，宜改為 "will be appreciated"。

No. 86　對於胚布還價不予同意 (See No. 85)

Dear Sirs,

We note with regret from your letter of the 9th May that our prices are not low enough to meet your needs.

The high quality of our goods, which has caused them to gain your approval, cannot be maintained at lower prices. Reduction has already been made for a large order which you intend to place with us. Our prices barely cover the cost of production, and although you are so good as to give us an order for our Nos. 10 and 12, we deeply regret that we are unable to make any further shading of prices.

We hope that you will consider the advisability of placing your order with us immediately.

Yours faithfully,

【註】

1. to meet your needs：滿足你的要求。needs = requirements。

2. reduction：減價。

3. to cover：抵付；足數。

　　例：The profits realized at this season do not cover the working expenses.

4. shading = cutting：減價。

　　to make...prices = to cut the prices finer (make the prices easier)。

No. 87　對染料報價的還價

Gentlemen:

We appreciate your offer of February 24 on your various Dyestuffs, but regret that we are far from accepting it.

You have pointed out that prices are unfortunately high owing to the paucity of stocks on the market, and that the commodities are available only through sources where you have to pay comparatively high premiums. We are open for some of your offerings, but are surprised to receive your offer of these high prices. We are reported that there has been a decline in the prices of raw materials, and we can see no justification for the raising of your prices.

You have probably underestimated the importance of our firm. We find, by offers recently made to us by other firms, that we have been paying you some 5% more than the prices obtainable from your competitors. Seeing that we have dealt with you so long, we prefer to continue buying from your house, but you must help us to do so by bringing your prices down, at least, to the same level as those of your neighbors.

While thanking you for your offer, we sincerely ask for your cooperation to bring about a good business mutually profitable.

Very truly yours,

【註】

1. premiums：額外價錢。

2. to be open for = to have an opening for; be in the market for...（擬購入）。

3. offerings = goods offered：出售的貨品。

4. to underestimate...our firm：低估本公司。"to underestimate" 為輕視、低估之意，相反詞為 "to overestimate"。

5. neighbor = competitor; rival。

No. 88　對水泥報價的還價

Gentlemen:

Thank you for your offer on Portland cement.

We immediately contacted our prospective clients but so far have been unable to interest them. They state that they have received from other source in Taiwan a much more attractive offer which is around 10% below our price.

After intensive negotiations, however, we have been successful in obtaining from them a counter offer of US$27 per M/T CFR Singapore. Hence we have despatched to you cable today which reads as follows:

RYT TWENTYTHIRD COUNTER OFFER DOLLARS 27/MT CANDF STRONGLY RE-QUEST ACCEPTANCE ADDITIONAL BUSINESS ANTICIPATED WRITING SINGTRACO

Will you please therefore contact your supplier again to see if they can improve their price. Please bear in mind, in this connection, that cement is one of the items that our clients regularly need, and that once we were successful in securing a foothold in this line, we could expect to place additional business with them.

We await your favorable reply.

Sincerely yours,

【註】

1. source = source of supply。

2. hence = for this reason。

3.電文譯文：

"With reference to your cable of the 23rd, we counter offer at US$27 per M/T CFR Singapore. We strongly request you to accept this counter offer in view of the high possibility that additional business can be expected. We shall writing in details on this matter."

RYT = Refer to your telegram。

CANDF = C&F = CFR。也有用 "CNF" 者。

SINGTRACO：發報人的電報掛號。

4. improve one's price = lower price or reduce price：改善價錢，不用 "lower"，"reduce" 而用 "improve" 相當好，也可用 "shade price" 表示。cf. Can't you shade your price?

5. secure a foothold：獲得據點。

6. place additional business = place additional order：追加訂單。

No. 89　對男西裝衣料的討價

Gentlemen:

　　We have received both your quotation of March 25 and the samples of MEN'S SUITINGS, and thank you for these.

　　While appreciating the good quality of your suitings we find the price of these materials rather high for the market we wish to supply. We have also to point out that very good suitings are now available in Eastern countries from several European manufacturers, and all of these are at prices from 10% to 15% below yours.

　　We should like to place orders with you, but must ask you to consider whether you can make us a more favorable offer. As our order would be worth around $40,000, you may think it worth while to make concession.

Yours truly,

【註】

1. men's suitings：男子用衣料。

　cf. gentlemen's suitings：男用衣料。

　suit：（衣服等）一套。

　a suit of clothes：一套衣服。

　a woman's suit：女子的一套衣服（包括上衣和裙子）。

　a man's suit：男子的一套衣服（包括外套、背心和褲子）。

　a two pieces suit：一套二件的衣服。

2. point out：指出。

3. worth while：值得的，此詞用在動詞 to be 之後（在本例文 it 後面省略了 "is"）。如用在名詞之前，應為 "worthwhile"。

4. make concession：讓價。

No. 90　不同意討價

Dear Sirs:

　　Many thanks for your letter of May 18, in which you ask us for a keener price for our Pattern 102.

　　Much as we should like to help you in the market you mention in your letter, we do not think there is room for a reduction in our quotation as we have already cut our price in anticipation of a substantial order. At 8 cent per yard this cloth competes well with any other product of its quality on the home or foreign markets.

　　We are willing, however, to offer you a discount of 5% on future orders of value $1,000 or over, and this may help you to develop your market.

　　　　　　　　　　　　　　　　　　　　　　　　　　　　　Yours faithfully,

【註】

　　there is no room：無餘地。

No. 91　對討價同意減價若干

Dear Sirs:

　　We have given your letter of September 23 very careful consideration.

　　As we have done business with each other so pleasantly for many years, we should like to comply with your request for lower prices.

　　However, our own overhead has increased sharply in recent months and we cannot reduce price 15% without lowering our standard of quality...and that we are not prepared to do.

　　We suggest an overall reduction of 5%, through the line.

　　We hope you will consider this satisfactory, and that we can continue our long and friendly association.

　　　　　　　　　　　　　　　　　　　　　　　　　　　　　Yours truly,

【註】

    1. pleasantly：愉快地。

    2. overhead：營業開支。

No. 92　對嗶嘰布報價的還價

Dear Sirs,

    We have received your offer of June 15 on 50,000 yds., black serge @ stg. £35 and tweed @ stg. £38 CIF Keelung, against which we have cabled you our counter offer as follows:

    YOURS 15TH ACCEPTABLE IF JULY SHIPMENT

    Your offer meets our requirement both in price and quality. However, we need the goods at latest by the end of August. Therefore, a July shipment is the latest. Please cable your acceptance by June 25.

    We trust that you will be able to accommodate us and give your prompt acceptance by return.

    Thank you for your kind cooperation.

Yours faithfully,

【註】

    1. @ stg. £= at sterling pound（英鎊）。

    2. a July shipment = a shipment during July。

    3. accommodate：給方便；幫助。

      to accommodate a friend（幫助朋友）。

    4.還價不限於價格。凡對於有關買賣條件加以變更者，均視為還價 (counter offer)。

No. 93　對雨傘還價不同意

Dear Sirs,

    We are pleased to know from your letter of July 16 that you will place an order with us for our umbrellas if the price can be more competitive.

    Much as we wish to accept your suggestion, we have to inform you something true that we offered you our rockbottom price which was 5% lower than that we offered to Messrs. Ford Bros. Ltd. in your city for delivery in October. We would like enclose a copy of the L/C we received from them for your reference.

    Such being the case, we suggest that you accept the price and send us your order promptly.

You may be sure that we will quote you the best price as soon as our cost is lowered and hope this can be done when you pass us your fresh order next time.

We appreciate your cooperation and awaiting your order.

Very truly yours,

【註】

1. "know from your letter" 這種措詞不佳，應改為 "learn from your letter"。

2. much as we wish: 雖然我們很想……但……。

3. such being the case: 情形是如此；在此情形下。可以 "under the circumstances" 代替。

4. fresh order: 新訂貨。

No. 94　對全自動照相機報價的拒絕 (See No. 74)

Gentlemen:

Full-Automatic Tube Camera

Thank you for cabling us an offer for your high quality tube camera. However, we have to point out the following:

Although you emphasized the wonderful production record shown for this set, we have on hand a very powerful competitive camera of German make which is also commanding a fine reputation over here.

Moreover, the German make is lower in price by US$20 to US$50 and the design and style are excellent.

Our retail stores report that yours will be competing with other best selling cameras if the price is reduced by US$30. But we think this is too large a discount for you, so we would like to refrain from placing an order with you.

Please accept the decline of your offer in the light of present market considerations and assist us with a more marketable price in the future.

We hope this will meet your prompt attention and induce a proper countermeasure.

Very truly yours,

【註】

1. commanding a fine reputation: 博得好評。

2. retail stores: 零售店。

3. would like to refrain：不（下訂單）；放棄（下訂單）。

4. in the light of present market considerations：鑒於目前市況的考慮。

5. marketable prices：易於銷售的價格。

6. induce a proper countermeasure：請你採取適當的對策。

No. 95　對美棉報價的拒絕 (See No. 69)

Dear Sirs,

American Raw Cotton, SLM

Thank you for your offer of the subject cotton as transmitted by your letter of June 10, 20–, the contents of which have been duly noted.

We regret to say that, owing to the facts that the price you quoted is a little higher than your competitors' and that the shipment you scheduled lags one month behind our need, we are not in a position to accept your offer. None the less, we appreciate very much your kind response to our inquiry.

You may be rest assured that we will write to you again when we are in need of cotton in future.

Yours faithfully,

【註】

1. the contents of which have been duly noted：來函的內容敬悉。新派人士認為這種措詞應予摒棄。

2. we regret to say：可以 "we regret to reply" 代替。

3. your competitors' = your competitors' prices。

4. lags one...need：較我們所需的時間落後。

5. none the less：依然；仍然。

第二節　有關接受的有用例句

一、接受報價

1. 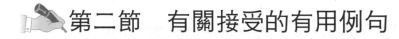 ditions as follows:

2. Thank you for your $\begin{cases} \text{fax offer} \\ \text{quotation} \\ \text{offer} \end{cases}$ of...(date), which we accept on the terms and condi-

tions quoted as follows:

3. We are pleased to confirm acceptance of your offer of...(date)...on 10,000 sets TV sets Model TR−123 at US\$130 per set FOB Keelung, shipment to be effected within 60 days after receipt of L/C.

4. Thank you for your $\begin{cases} \text{offer} \\ \text{quotation} \end{cases}$ of...(date)..., which we

→ accept subject to $\begin{cases} \text{your confirmation reaches} \\ \text{your confirmation received by} \end{cases}$ us before...(date)...
$\begin{cases} \text{the approval of import licence.} \\ \text{modification of the following terms.} \end{cases}$

5. We accept your offer of...(date)...on the terms as follows:

→ ...

→ Please $\begin{cases} \text{cable} \\ \text{fax} \\ \text{e-mail} \end{cases}$ immediately if it is not in order.

（如有問題，或如不同意，請速來電。）

6. We accept your offer of July 10 on Garments but price do better, if possible.

（如可能，請酌減價錢。）

二、還 價

1. In view of the prevailing prices in this market, your $\begin{cases} \text{offer} \\ \text{quotation} \end{cases}$ is a little expensive.

→ We have just $\begin{cases} \text{cabled} \\ \text{faxed} \\ \text{telexed} \end{cases}$ you a counter offer $\begin{cases} \text{asking} \\ \text{requesting} \end{cases}$ for a large discount. Un-

less approved we regret that business agreement will not be acceptable.

2. $\begin{Bmatrix} \text{If} \\ \text{In case} \end{Bmatrix}$ you can reduce your price to US\$10 per dozen, it is possible for us to place an order for considerable quantity of this article.

3. We counter offer US\$10 per dozen subject to your confirmation $\begin{Bmatrix} \text{reaches us} \\ \text{arrives here} \end{Bmatrix}$ on or before July 10.

（本公司還價每打 10 元，但以貴公司確認在 7 月 10 日以前達到我方為條件。）

4. We $\begin{Bmatrix} \text{cannot} \\ \text{are not in a position to} \end{Bmatrix}$ accept your offer of...(date)..., but we $\begin{Bmatrix} \text{submit our} \\ \text{make you} \end{Bmatrix}$ counter offer as follows:

5. Please reduce your price as much as possible without any change of other terms.

（在其他條件不變下，請儘量減價。）

6. Please confirm whether you can shorten the delivery time of your offer of... (date)...

7. We are working, please hold your offer good until...

（我們正在辦理中，請將報價保留至×月×日。）

三、對還價的答覆

1. We are glad to confirm acceptance of your counter offer of August 10 on our note-book computer Model BX−123.

2. The prevailing prices in this market are nearly 10% higher than yours and therefore we regret our inability to accept the counter offer mentioned in your telex.

3. Your request to extend the validity of our offer to the end of this month is acceptable to us.

4. Owing to rising cost of raw materials and wages, we are not in a position to reduce our prices any more.

四、拒絕接受報價

1. Unfortunately we cannot accept your offer. Your prices are prohibitive. （太貴了）

2. We have the impression that your price is higher than that in the average market

which is now showing a decline.

（我們以為你的價格比目前趨跌的平均市價要高。）

3. We feel that your quotation is not proper because the price for such material is on the decline at present.

4. Unfortunately we are not in a position to accept your offer because another supplier in your market offered us the similar article at a price 3% lower.

第十七章
推銷與追查 (Sales Promotion and Follow-up)

第一節 推銷信的寫法

　　推銷信 (Sales letters) 就是賣方向買方主動洽銷生意的信。廣義地說，賣方所發出的每一封信都可視為推銷信，但這裡所要談的是指賣方向已有往來的客戶所發出的狹義的推銷信而言。出口商推銷貨品的方法，或發出推銷信或印發通函 (Circular letters)、商品目錄 (Catalog)、價目表 (Price list)、樣品 (Sample)、市況報告 (Market report)，藉以引起客戶的注意，由而達成推銷的目的。

　　國際性的推銷或以書信或以電傳（包括 Internet）推銷貨物。這種函電的寫作技巧不能忽視，其主要秘訣是要寫成不是為你本身的利益而向他推銷，而是完全為對方能獲利而勸對方進貨。你的貨品性質、價格、其他條件均必須要適合對方的利益。因此，在撰寫推銷信時，應該從這一方面下筆，以引起他的注意，由而促成他的下決心。一般而言，一封完善的推銷信，其內容應以達成下列四項為要：

　　1.引起注意 (Attention)。
　　2.發生興趣 (Interest)。
　　3.喚起欲望 (Desire)。
　　4.決定行動 (Action)。
　　上述四項可簡稱為 "AIDA"。茲略作說明於下。

一、引起注意

　　推銷信的第一段必須寫得引人入勝 (Attract the reader's attention)。收信人未見得一定會拆閱來信，也未必一定會繼續讀下去。信封的型式、信的格式，都可以使收信人決定這封信是否值得一讀。如果收信人要讀這封信，那麼就應該寫得可引起他的注意，使他願意一口氣讀完。所以推銷信的第一段，必須寫得動人。

信文的第一段，在性質上，應以讀者的立場（即 You attitude），用積極的語氣及直接的措詞。在寫法上，可用詢問式、命令式、說明式、假定式或故事式，並力求 Courteous，及 Impressive。

1. You attitude：就是要從讀者立場，說明讀者可以有怎樣的利益，例如：

"As an importer of novelties, you are certainly seeking for more new products to meet your customers' growing requirement. Where are you looking for?"

2. Positive tone（積極的語氣）：與其說：

"You wouldn't import any tropical fish food, would you?" 不如說：

"You want to import tropical fish food, don't you?"

3. Direct expression（直接的表示）：使收信人一看就知道信的內容，例如：

"We have recently developed some new styles of ladies' handbags made of plastics for our customers in Germany. They may be of your interest."

Sales letter 的第一句，如用詢問式，則所問的應該集中讀者的注意，並適合所說的貨物。例如：

"Do you encounter the problem of locating suitable machinery-suppliers and/or mechanical man? If you do, why not let us solve your problem?"

說明式可以採用合適的格言或俗語。例如：

"A penny saved is a penny earned."（省一文就得一文。）

假定式，第一為使讀者相信一種好的標準；再說明你的貨品等於或勝於這個標準。例如：

"If you wish to import cars that will bring you profit, import the 'Ford'."

故事式是利用人類喜歡聽故事的心理，先講故事，再說明你要推銷的貨品。

二、發生興趣

客戶對於你所要推銷的貨品情況不明白，當然就無法發生興趣。所以推銷信在引起了收信人的注意之後，跟著就應敘述貨品的用處、品質、式樣、優點、價格、付款條件、交貨時間等。為求敘述能引起其興趣，寫信的時候，必須從讀者的觀點出發，研究收信人的需要和環境，使信的內容能夠迎合客戶的心理。譬如收信人喜歡價廉的貨物，推銷信就應著重於價格的低廉；如喜歡 Stocklots（存貨），就要著重於 Stocklots 一事以引起他的興趣 (Arousing the reader's interest)。例如：

"We have a great number of inexpensive ladies' garment of various styles, designs, and colors in our stock."

三、喚起欲望

有時收信人對於推銷的貨物，雖則發生興趣，但並無購買的欲望，要喚起收信人的購買欲望 (Create buying desire)，可以用理由或例證，使他確信貨物易於出售、易於獲利、服務好，以及其他優待條件。例如：

"In addition to the trade discount stated, we would allow you 5% discount for order exceeding 1,000 doz."

四、決定行動

喚起了收信人欲望之後，應進一步促其立即決定購買。促使收信人決定購買的行動，可以用存貨不多，優待期有限，或價格有上漲趨勢等等詞句，由而促其採取行動 (Induce action or urge him for taking action)。例如：

1. On June 1, the price of this item will be positively be raised from US$...to US$... per dozen FOB Taiwan. To take advantage of the old price, you must send your order now.

2. Remember too, our supply in stock is limited. Send your order to us before...(date).

No. 96　推銷原子筆

Dear Sirs,

Ball Pens

We have the pleasure of forwarding you, under separate cover, some samples of the ball pens made by us.

We have to draw your attention to our prices mentioned in our price list enclosed that we can offer our products at from 5% to 10% lower than those of other brands.

Notwithstanding the lower prices, you may rest assured that our products are of the first-class quality and of the best workmanship, as can be proved by our samples.

With these advantages you can develop market for our products without difficulty, and so we trust we shall be able to receive your orders soon.

Yours very truly,

【註】

1. ball pens：原子筆，又寫成 "ball point pens"，"ball-point pens" 或 "ball-point"。

2. at from 5% to 10% lower than those of other brands：比其他廠牌便宜 5% 至 10%，"those of" 為 "the prices of" 之意。

3. develop market：開拓市場。

No. 97　推銷紙張

Gentlemen:

As you know, the best quality at the lowest price is the road to success in selling any goods, especially the paper in such a time of keen competition.

We would rather talk quality than price because no other concern manufactures better paper than ours, but we cannot help talking price because no other concern charges as little for them as we do.

We have special machinery designed by ourselves—machinery that may be used by no other concern. This enables us to manufacture better paper at a minimum expense.

We have sent a complete set of samples of our papers to you by parcel post, all particulars of which will be found on the enclosed price-list.

Will you not favor us with an order as a trial?

Yours very truly,

【註】

1. the road to success：成功之路。

2. keen competition：劇烈的競爭。

3. would rather...than：寧……而不願……。

4. concern：行號。

5. cannot help talking price：不能不談價格。

6. a complete set of samples：各種（紙張的）樣品全份。

7. Will you...as a trial?（可否賜一訂單試一試?）

No. 98　推銷華美新奇品

> Dear Sirs,
>
> ### Fancy Novelties
>
> As an importer of novelties, you are certainly seeking for more new products to meet your customers' requirements. Where are you going to look for?
>
> Look for us please. Right here is the Fancy Enterprises Co., Ltd. You can get many newly-developed items of novelties that you may be of interest.
>
> Being one of the leading manufacturers and exporters of novelties, we are always manufacture various fancy novelties to fulfil our customers' needs. we have done outstanding services for our regular customers, so we can do the same for you too.
>
> For your reference, we are pleased to enclose a list of novelties that are manufactured by us. You can find from the list that we have wide range of products. Please don't hesitate to let us know if you find any specific item(s) interesting. We will, upon request, promptly send you detailed information including catalog, prices, and even samples for your evaluation.
>
> We are ready to serve you, why not write us now!
>
> Sincerely yours,

【註】

1. fancy novelties：華美新奇品。

2. newly-developed items：新開發的項目，即新產品。

3. you may be of interest：也許你感興趣。

4. wide range of products：產品種類多。

5. please don't hesitate to = please feel free to：請不必客氣。

6. why not write us now!：請現在就寫信來吧!

No. 99　推銷成衣

> Dear Sirs,
>
> ### Garments
>
> You may be interested, we think, in our newly-developed styles of garments for our overseas customers.
>
> Enclosed is a copy of price list together with a sheet of catalog for your reference. You will ob-

serve from the attached catalog that the new products are of marvelous design. Exotic in style, they look beautiful and appeal to the interest of young women. We believe that these new styles will come in fashion in your market before long.

As you are considered by us one of our best customers, we are very happy to introduce in first priority these new products to you, and we are sure your importing them will bring you handsome profit.

For your study and reference, we have sent you a pack of samples by air parcel post today. It should be reach you about ten days. We are sure you will speak highly of the styles and quality as soon as you evaluate these samples.

We will reserve this first priority for you until October 5, so don't miss the good opportunity, and give us your trial order immediately.

【註】

1. marvelous：稀奇的；奇異的。

2. exotic：異國情調的；外國產的。也可寫成 "exotical"。

3. appeal to the interest of...：投合……的興趣；引起……的興趣。

　　cf. appeal to force：用武力解決；訴諸武力。appeal to reason：訴諸於理。

4. come in fashion：流行。也寫成 "come into fashion"，"come into mode"，不流行為 "go out of fashion"，或 "go out of mode"。

5. handsome profit：厚利。

6. speak highly：讚賞。

No. 100　推銷油漆

Gentlemen:

New "Sealex" Paint

You may be interested in the new "Sealex" paint we have just introduced to the trade. A sample has been sent to you today by parcel post.

"Sealex" is the result of many months of careful research. It is made from a special formula and owes its superiority over other exterior paints to its remarkable ability to allow for the movement of those paint-peeling cracks just visible to the naked eye. This quality to expand with the cracks comes from a very special combination of granite, mica and resin that provides a rich, thick coating twice the thickness of that of the average finish, thus giving long-term protection.

"Sealex" is available in twenty-one basic colors and, as you will see from the enclosed list, prices are surprisingly low. We are nevertheless allowing a special 5% discount to you if you place orders before the end of the current month.

So why not take advantage of the opportunity now and send us an immediate order.

Yours faithfully,

【註】

1. have just introduced to the trade：甫推出上市。

2. the result of many months of careful research：經過很多月份（長久時間）精心研究的結果。

3. made from a special formula：由特別處方製成。

4. superiority over：優於。

5. to allow for the movement...the naked eye：意指使油漆脫落的痕跡不致看得出來。

6. paint-peeling cracks：油漆脫落的痕跡（裂縫）。

7. just visible to the naked eye：肉眼即可看見。

8. granite：花崗石。

9. mica：雲母。

10. resin：樹脂。

11. basic colors：基本色彩。

12. so why...opportunity：所以，為何不利用機會……。

No. 101　推銷牛肉精

Dear Sirs,

New Beef Extract

A sample of our new beef extract, Vimbeff, has been sent to you today by air parcel post, which we hope will reach you in perfect condition.

You will find that it possesses many unique features which definitely place it ahead of its many competing brands. It dissolves readily in water, leaving no trace of sediment. As a result, its digestion presents no difficulty, rendering it particularly valuable in cases of convalescence and general debility. Other points are stressed in the leaflet enclosed.

The many inquiries we have received prove that the public are fully aware of the merits of our new product. There will accordingly be no prolonged holding of stocks, with the loss and difficulties such a practice entails.

After paying due regard to the amount of turnovers, you will agree that the 15 percent trade discount allowed is decidedly generous, particularly as settlement is to be effected on 90 days' D/A basis.

We look forward to adding you to our other importers who are reaping substantial profits from the sale of Vimbeff.

Yours faithfully,

【註】

1. extract：濃縮物；精華。

2. in perfect condition：情況完好。

3. unique feature：獨特的優點。

4. place it ahead...brands：使之列於同類貨品的前茅。

5. dissolve：溶解。

6. leaving...of sediment：毫無渣滓。sediment 為渣滓或沈澱物。

7. digestion：消化。

8. rendering：致……使……。

9. convalescence：（病）復元。

10. debility：虛弱。

11. stress：強調；表明。

12. no prolonged holding of stocks：存貨將不致留存很久。

13. with the loss...entails：（以致存放過久）招致損失與困難。entail 為引起，招致之意。

14. decidedly generous：絕對的優厚。

15. settlement...basis：按 90 天 D/A 結帳。

16. we look forward...importers：我們深盼你們將成為經手我們貨品的進口商之一。

No. 102　推銷吸塵器

Gentlemen:

Vaccum Cleaner

We are sure that you would be interested in the new "Klean-Kwick" Vaccum Cleaner which is to be placed on the market soon. Most of the good points of the earlier types have been incorporated into this machine which possesses, besides, several novel features which have been perfected by years of scientific research.

You will find that a special contrivance enables it to run on slightly more than half the current required by machines of equal capacity. Further, most of the working parts are readily interchangeable and, in the event of their being damaged, they are thus easy to replace. Such replacement, however, will not be the result of ordinary wear and tear, as only toughened steel is used in the manufacture of moving parts.

The essential advantages it offer will make it a quick-selling line, and we are ready also to co-operate with you, by launching a global advertising compaign. Moreover, we are ready to assist to the extent of half the cost of any national advertising.

Bearing in mind the rapid turnover which is likely to result, you will agree that the 10 percent commission we are prepared to offer you is extremely generous.

You will find enclosed leaflet and circulars describing the Klean-Kwick, and we look forward to your agreeing to handle our product as the sole agent in your country.

Yours faithfully,

【註】

1. vacuum cleaner：吸塵器。
2. place on the market：上市。
3. most...incorporated into this machine：所有早先各型貨物的優點，大部分均已為本吸塵器所兼有。
4. novel feature：新特色。
5. contrivance：機械裝置，special contrivance：特殊設計。
6. current：電流。
7. equal capacity：同等效能。
8. working parts：活動部分零件。
9. interchangeable：彼此更換。
10. wear and tear：磨損；損耗。
11. toughened steel：堅強的鋼料。
12. moving parts：活動部分零件。
13. quick-selling：銷得快；暢銷。
14. launching：發動。
15. leaflet：摺疊傳單。
16. circulars：傳單。

No. 103　推銷新肥皂及肥皂粉

Dear Sirs,

New Soap and Soap Powder

We think you will be interested in the new soap and the soap powder we have just introduced to the trade. Generous samples of both have been dispatched to you by parcel post today.

These products are the results of months of careful research, and are likely to revolutionize all the methods in use at present. A trial will convince you of their merits, and we submit them to your strictest criticism with confidence.

From the price list enclosed you will see that the prices are surprisingly low, and we would remind you that it is only our system of mass production that enables us to maintain these economical prices without sacrifice of quality.

Special terms are allowed to customers who place trial orders before the end of the current month.

We accordingly look forward to your advices at your earliest convenience.

Yours faithfully,

【註】

1. generous samples：大量的樣品。

2. result of months of careful research：窮許多月悉心研究所得的結果。

3. revolutionize all the methods in use at present：革新了目前應用中的一切方法。

4. a trial will...merits：一經試用，必能深信其有各種優點。

5. submit...confidence：以極大的信心，供君使用，願接受最嚴厲的批評。

6. system of mass production：大量生產體系。

7. economical prices：經濟的價格；廉價。

8. without sacrifice of quality：無損於原來品質。

9. Special terms are allowed...current month.（如主顧於本月底訂購試用，尚可予以特別優惠條件。）

No. 104　推銷鞋類品

Dear Sirs,

Footwear

You might, we think, find our offer interesting when you read through this letter.

It is our pleasure to introduce you that we have recently developed various kinds of footwear such as sandals, slippers, rubber shoes, boots, canvas shoes, etc. You can tell from the attached sample photos that they look beautiful and attractive enough. They are of unique designs.

You will agree that these items are saleable in your market. Please let us know your requirements if you find any item(s) interesting. We will happy to send you free samples with rock-bottom prices for your approval upon request.

Could you give us a prompt response?

Yours sincerely,

【註】

1. footwear: 鞋類品（注意無多數）。

2. sandals: 涼鞋。

3. slippers: 拖鞋。

4. boots: 長統靴。

5. canvas shoes: 帆布鞋。

6. tell from: 從⋯⋯看得出。

7. saleable: 可銷；有銷路的，又寫成 "salable"。

No. 105　推銷新機械

Dear Sirs,

We regret to note that you have not entrusted us with your orders for some time, and we hope that you have no reason to be dissatisfied with the execution of your past orders.

You may be interested to hear that our factory has recently been extended and new machinery introduced. We are confident that these changes will lead to an improvement in both the quality and the quantity of our output.

From the catalog enclosed you will observe that our terms and prices are even more favorable than before.

We should be pleased to supply you with samples, and feel sure that they would convince you that our claims are not without foundation.

We look forward to a renewal of your custom.

Sincerely yours,

【註】

1. entrusted us with your orders: 惠賜訂貨。

2. we hope that you have no reason...past orders: 希望貴公司並無緣由不滿我們以往所供貨物的情形。

3. extended = expanded：擴充。

4. new machinery introduced：使用新機器。

5. will lead to an improvement：（品質⋯⋯）可獲改進。

6. output：出品。

7. feel sure= be sure：確信。

8. they would...foundation：這些貨品必將令貴公司深信本公司所宣稱者，悉非無根據。

　　"without foundation" 為「無根據」之意。

9. We look forward...custom!（倘荷再惠顧，實為企盼!）

No. 106　推銷花呢

Dear Sirs,

Low-priced Tweeds

　　We think that you will be interested to know that we have recently manufactured some low-priced tweeds of assorted colors.

　　We are going to offer these products to our regular customers, at a considerably low prices. They have a great market in East and South African Areas.

　　Orders can be supplied from stock but only limited quantities are available. When sold out no repeated order will be accepted.

　　Price list together with colorful catalog are enclosed for your reference, and we shall be pleased to send patterns on request.

　　We assure you that no exceptional opportunity would be available if you miss it this time.

Yours faithfully,

【註】

1. low-priced tweeds：低廉的呢絨（花呢）。

2. interested to know：樂聞。

3. assorted colors：各種顏色。

4. regular customers：經常交易的客戶。

5. have a great market in：暢銷於⋯⋯；在⋯⋯有很大的市場。

6. orders can be supplied from stock：如蒙訂購，可交現貨。

7. only limited quantities are available：數量有限。

8. sold out：售完。

9. no repeated order will be accepted：不再接受訂貨。

10. on request：承索（即寄）；索取（即寄）。

No. 107　推銷蜜餞荔枝

Gentlemen:

Laichee Preserve

We have today sent you by separate post, a sample of our new speciality, "LAICHEE PRESER-VE"，and hope it will reach you safely. This new preserve has met with a very gratifying reception from the trade, and we have no doubt that, if you can see your way to stock it, you will find a very large demand in your area.

The preserve is packed in 1–lb. crystal bottle, corked and capsuled with an attractive label, and may be used advantageously in window advertising. The sample sent is a facsimile of our 1–lb. bottle.

We hope to be favoured with an order, after you have tasted the quality. Price list is enclosed for your consideration.

Yours truly,

【註】

1. Laichee preserve：蜜餞荔枝，又寫成 "preserved Laichee"。

2. new speciality：新產品。

3. met with a very gratifying reception：深受歡迎。

4. if you can see your way to stock it：如能購儲若干。

5. find a very large demand in your area：暢銷於貴方市場。

6. packed in 1–lb. crystal bottle：玻璃瓶裝潢，每瓶 1 磅裝。

7. corked and capsuled with an attractive label：瓶蓋用木塞，並以錫紙護封，瓶上貼有美麗悅目的招紙。

8. The sample sent...bottle.（寄送的貨樣即為每瓶 1 磅裝之貨。）

   facsimile [fæk´simili]：精確的複製。

No. 108　推銷捕蠅紙

Gentlemen:

Sticky Fly-net Fly Paper

This article is sold under a double guarantee...that of the manufacturer and our own.

It will keep in any climate.

It will not granulate nor leak.

It is equal to any high-grade fly paper manufactured.

It will catch more flies than any other high-grade fly paper.

It will sell and give your customers satisfaction.

It will please your old customers and attract new ones.

Its merit will make the sales and you will make the profit.

Compare our prices with those you are paying for other high-grade fly paper.

Prices either FOB Keelung or Kaohsiung.

5 c/s	US$2.90 per c/s
10 c/s	2.80 per c/s
25 c/s	2.75 per c/s
50 c/s or over	2.60 per c/s

Minimum shipment from either port 25 c/s, unless ordered with other goods, when any quantity can be shipped.

Yours sincerely,

【註】

1. sticky fly-net fly paper：膠黏捕蠅紙。

2. double guarantee：雙重保證。

3. It will keep in any climate. （無論何種天氣，均保持一樣品質。）

4. It will not granulate nor leak. （無起顆粒或裂罅之弊。）

5. It is equal to any.... （足與各種上等捕蠅紙相較。）

6. It will please...new ones. （能因舊主顧的讚美而吸引新客戶。）

7. Its merits...profit. （如此功效，必能暢銷而獲厚利。）

8. Minimum...can be shipped. （無論從何港裝船，起碼須購 25 箱，如與別種貨物同購，則不論多少均可裝運。）

No. 109　推銷耶誕禮品

Dear Sirs,

Christmas Gifts Items

Have you been looking for some new items, fresh and fancy, to add to your Christmas gifts items of this year? Here is just what could make your dream come true: the Hollow Glass Ornaments made by us after months of careful designing and repeated experimenting. They are so cute and loveable that they would surely win the hearts of children as well as adults.

The Hollow Glass Ornaments are made in the shape of either dolls or animals, adorned with wigs and hats made of bright colored flowers cloth. The dolls include Snow White, Cinderella, Peter Pan etc. and the animals poodles, hounds, puppies ,rabbits, bears, elephants, etc. The color of glass varies in different shades of crimson, yellow, brown, purple, green, blue, grey and white, all transparent. The adornments to each figure are made out under special attention and direction of our artist-designers.

The color illustrations on the enclosed catalogs will show you how eye-catching they are. You would feel that you could hardly wait to see them in real shape. Besides, we believe you will find the price very attractive and reasonable.

It is sincerely hoped that you will take advantage of this opportunity and favor us with your order without loss of time.

Yours very truly,

【註】

1. fresh and fancy：新鮮又華美的。

2. make your dream come true：使你的夢想得以實現。

3. hollow：空心的；中空的。

4. cute and loveable：玲瓏可愛的。

5. win the hearts of：獲得（小孩及大人的）心。

6. ornaments：裝飾。

7. adorned with wigs and hats：戴上假髮和帽子。

8. Snow White：白雪公主；Cinderella：灰姑娘；Peter Pan：潘彼得；poodles：鬈毛狗；hounds：獵犬；puppies：小狗；crimson：深紅色；each figure：每一具型；artist-designers：藝術設計家。

9. eye-catching：引人注目的。

10. you could hardly wait...：你幾乎等不及去；覺得非馬上去（看個真東西）不可。

11. without loss of time：刻不容緩。

No. 110　推銷尼龍織棉緞

Gentlemen,

After the year-end sales, alert dealers are setting their eyes on spring sales. Believing that you are also preparing for the next season, we have today cabled you the following offer as per enclosed copy.

10,000 yds., sample No. 161, Rayon Brocade Satin

@ 50 ¢ per yd. FOB Keelung

Shipment February/March

With the turn of the year, prices of all rayon goods are expected to make a sudden advance. If you lose this opportunity, you may not be able to get the same article even at a higher price.

We trust, therefore, that this offer will meet your immediate acceptance to mark the happy start of this year.

Yours truly,

【註】

1. alert：細心的；機敏的。

2. set one's eyes on：注意於……；注視……。

3. mark the happy...year：以誌今年愉快的開始。

No. 111　推銷照相機

Dear Sirs,

Our New Model: Magic 20

We take pleasure in informing you that we have recently developed a new model called "Magic 20" whose high efficiency has been proven by a scrupulous test of its mechanism and functions as clearly explained in our illustrated catalog inclosed.

We believe that considering the improvements it offers, you will find our new model not only far better in quality than those we offered before but also a very good seller at the competitive price of US$100 per unit FOB vessel San Francisco including the cost of accessaries. Like all other cameras we handle, "Magic 20" is also accompanied with two-year guarantee for our after-care services.

If you have interest in dealing with us in "Magic 20" please inform us of your requirements.

We assure you of our usual best service to you.

Sincerely yours,

【註】

1. scrupulous：審慎的；小心翼翼的。

2. mechanism：機械裝置。

3. functions：功能。

4. good seller：好銷的東西。cf. best seller：暢銷的東西。

5. accessaries：附屬品。

6. after-care services：售後服務，又稱為 "after-sale services"。

第二節　追查信的寫法

賣方向顧客 (Customers) 發出推銷信或循買方的要求報了價，甚至寄出樣品後，自然期待能收到覆信或訂單。假如寄出了信或樣品之後，如石沈大海，毫無音訊時，賣方自宜發出追查信 (Follow-up letters) 查問。追查信的內容大致如下：

1.提醒顧客曾向其發過信、寄過目錄、價目表或樣品。

2.重敘貨品的優點，價格的低廉，願意為其效勞。

3.提出若干理由（例如存貨不多、特價期限快屆滿、價格將提高等），促其趕快採取行動。

4.誠懇要求顧客答覆。

No. 112　寄出樣品後無音訊再發出追查信

Dear Sirs,

It was way back in August last year when we wrote you and sent you a sample of our products under separate cover. Now Spring is here. The trees are budding and the flowers are blooming, but still no word from you.

Did you ever write a man about a matter you felt sure would interest him and then wait and wonder why he didn't reply?

That is what we are doing now.

We presume that our letter could have escaped your attention due to the pressure of your work. So, this time we are going to make it easy for you to answer.

Just look below. Take your pencil and write "yes" or "no" opposite the questions that apply in your particular case. Then return this letter in the enclosed postage-paid envelope.

Do this now. It will pay you, because each of our customers has been successful.

Yours sincerely,

The questions are:

(1) Are our prices in line?
(2) Are our products suitable?
(3) Are you in the market for these products?

【註】

1. It was way back in August last year. （那是去年 8 月的事了。）

2. the trees are budding：樹木皆青。

3. the flowers are blooming：百花爭豔。

4. still no word from you：還是沒有收到你的音訊。

5. Did you...reply?（閣下是否曾經自己充滿了信心寫一封信給人而老是在等著，在猜測對方為什麼不答覆呢?）

6. That is...now. （我們現在的情形正是如此。）

7. our letter could...of your work：可能因為閣下貴人事忙而給疏忽了。

8. Just look below. （請看一下下面。）

9. opposite the questions：在問題的對面（填上 "yes" 或 "no"）。

10. postage-paid envelope：已貼妥回郵郵票的信封。（這封信，顯然是用於國內的 follow-up letter，國際貿易用的信，不流行也無法如此做。因為出口商不會有進口國的郵票，所以 "postage-paid" 可改用 "addressed"。）

11. it will pay you：這對貴公司有利；對貴公司划得來。

12. Are our prices in line?（我們的價錢對不對?）

13. Are our products suitable?（我們的產品合適不合適?）

14. Are you in the market for...?〔你們是否想買（這些貨品)?〕

出口商將樣品與函件寄給進口商後，毫無音訊時，自然很不舒服，對付這種有去無回的客戶，辦法有三：

1. 去函臭罵一頓。例如，指責其為 Sample collector，如奈及利亞的商人來函時，多自稱其為 Leading importer，但寄去樣品之後，往往如「肉包子打狗，有去無回」。這個時候，可以 "You are a leading impostor."（一流的騙子）譏稱之。

2. 算了，自認倒霉。

3. 再接再厲。

第一法，於事無補；第二法是沒出息的做法；第三法才是最好的辦法。

No. 113　寄出雨傘報價單後無反應，發出追查信

Dear Sirs,

　Now that our offer of January 15 for 1,000 dozen of ladies' nylon umbrellas expired yesterday, we wish to know the reason for its failure in interesting your customers so that we may instruct our manufacturers to make better offers.

Since we made offer, we have heard that Hongkong sources are underselling us. But to convince our manufacturers, will you please write to us stating the fact specifically?

This is also a test case for our manufacturers. As you may have seen in the various catalogs we sent you, they are very good manufacturers. So, please let us have a second chance by pointing out how to improve our offer.

Sincerely yours,

【註】

1. offer...expired: 指報價時效已逾期，原報價是 firm offer 定有時效。

2. Hongkong sources: 香港方面。

3. 第二段可改用下列句子:

In quality and delivery we have full confidence, but we are afraid that our prices were not attractive enough, possibly in the face of Hongkong competition. Will you, therefore, let us have your comments on our offer, especially about competition, if any, and the price level workable to you?

4. "So, please...improve our offer." 可以下面句子代替:

But they are not the only source of supply. We have connections with many other leading manufacturers. So, we can choose any of them upon hearing from you.

本來出口商與工廠應同心協力推銷貨品才對。假如改用上面的句子，似乎有使人覺得出口商淪於 Broker 的地位，同時好像原來的工廠不一定是好工廠。所以，非不得已，不用上面的句子為宜。

No. 114 寄出太陽眼鏡價目表及照片後無音訊，再發出追查信

Dear Sirs,

In response to your inquiry of October 2, we forwarded you our price lists along with sample photo of sun glasses on October 15 for your evaluation.

However, so far we have not yet heard from you. We presume that you have been too busy to give us a reply or the above-mentioned mail has gone astray. Whatever it may be, we take pleasure in enclosing again 2 sample photos and 1 price list for your study.

With the increasing demand of our overseas customers for our quality products, our stock is running low and there is an indication the prices will go up.

So, your prompt order will pay you.

Yours very truly,

【註】

1. along with = together with。

2. sun glasses：太陽眼鏡。

3. so far：到現在為止，也可以 "up to this writing" 或 "up to date"，"up to now" 代替，但少用這種舊式的措詞較妥。

4. too busy to give us a reply：太忙以致無法給我們答覆。

5. go astray：迷失（郵件）；誤遞。

   "the mail has gone astray" 可寫作 "the mail has been lost"（郵件遺失）。

6. quality products：品質優良的產品。

7. will pay you：划得來；合算。

8. 最後一句是 "you attitude" 的表達法，值得模仿。

No. 115　寄出樣品、目錄、價目表後無音訊，再發出追查信

Dear Sirs,

Suppose you have written a letter to your girl friend for a date and you don't get any reply, you will feel pretty bad, won't you?

The situation between you and our company is quite similar to that. We wrote and sent samples, catalog and price list to you on May 10 and have been expecting your speedy and favorable response. Much to our disappointment, since then we have not received a word from you.

There must be some important reason why you have not replied us. And if it is important enough to cause you to stay away from us, it is certainly important enough for us to make an effort to find the reason.

Please write and tell us frankly whether our products meet your requirements? Whether our prices workable in your market? or do you have any other comments on our products? Whatever it may be, please let us know, and we will be happy to do what we can do to fulfil your needs.

May we hear from you soon?

Yours sincerely,

【註】

1. date：約會。

2. feel pretty bad：覺得很不適意；覺得很不舒服。

　　feel bad about...：對……覺得遺憾。

3. much to our disappointment：使我們非常失望。

4. stay away from：離開（我們）；疏離。cf. stay away from school：缺課。

5. make an effort：努力。make efforts 努力。make every effort：費盡心血；盡力。

6. workable：可行，意指價格適合否？

No. 116　寄出男用拖鞋報價單、樣品後無音訊，再發出追查信

Dear Sirs,

Gents' slippers

　　Upon receipt of your request of November 3, we mailed you immediately a letter together with a price list and catalog on November 10, and forwarded you one pair sample each of GM–123, GM–234, and GM–345 for your evaluation by air parcel post on November 12. We think you might have already received them.

　　Do our products meet your requirements? Do you find our prices competitive? Do you have good news for us?

　　We are keenly interested in assisting your continued business development and we trust that you will never hesitate to contact us to explore areas of mutual interest.

　　As we mentioned to you in the last mail, we have a very warm feeling towards your firm and sincerely hope that we can stimulate a more active relationship.

　　We wish you every good fortune in the coming Christmas season.

Sincerely yours,

【註】

　　1. gents' slippers：男用的拖鞋。Gents' 為 Gentlemen's 的簡寫。

　　2. explore：探索；開拓。

　　3. mutual interest：互利。cf. mutual benefit：互惠。

　　4. wish you every good fortune：祝您好運！

　　5. in the coming Christmas season：即將來臨的耶誕節。

No. 117　寄出電晶體收音機樣品後無反應，再發出追查信

Transistor Radios

Dear Sirs:

　　In response to your inquiry of April 15, we sent you our price list together with two sample transistor radios per air parcel post. Since then we have heard nothing from your side and are beginning to wonder whether you have received these samples or not. Please let us know by return mail, because, if you did not receive them we will send again.

Perhaps the package has escaped your attention due to the pressure of your work. If this is true, we wish you would do us a favor and spend about ten minutes right now, examining the quality of these unique radio sets and the price list. Then you will find the radio sets we have selected for you will be just the right models for your market, and the high quality of these items with the best price will convince you to place an order with us.

With the rapid growing demand for our quality products, our factory has a very tight schedule of production now. However, as one of our best regular customers, we would like to give you the first priority of executing your order and we promise that your order will be shipped within 30 days upon receipt of your L/C.

If you need our service, would you care to place your order now? A fax order will be more helpful.

Yours faithfully,

【註】

1. pressure of your work：你的工作壓力。

2. right now = right away = right off：立刻；馬上。

3. convince you：使你確信；使你信服。

4. quality products：高品質的產品。

5. tight schedule：（生產）檔期緊湊。

6. first priority：第一優先。

7. fax order：傳真訂貨。

No. 118　寄出樣品後，如石沈大海，再發出追查信

Dear Sirs,

As requested in your letter of October 2, we sent you samples of our products together with our quotation sheet. That was months ago. Since then, we have written you a couple of times, but not a word from you.

No news is good news, so the saying goes. It is certainly not true in our case. Normally, business executives would blow their tops and exclaim: "Hell!"

This is not the way we do business. When people ignore you, there must be a reason. We would appreciate your telling us why. If we are doing something wrong, we want to correct it.

May we hear from you soon?

Yours sincerely,

【註】

1. that was months ago：那是數月前的事了。

2. a couple of times：幾次。

3. no news is good news：沒有消息就是好消息，這句話是 1632 年 Donald Lupton 所說的，但實際上是由意大利的諺語翻成英文的，它的本意是說：如果你沒有接到某人的信息，那麼他必定是安然無恙的。

4. so the saying goes：如諺（俗）語所說。

5. business executives：業務主管。

6. blow one's tops = lose one's temper：發脾氣。

7. exclaim：大聲說。

8. hell!：混蛋！他媽的！

9. ignore：不理（你）。

No. 119　寄出針織衣類樣品後，再發出追查信

Dear Sirs,

Knit-wear

It has been almost two months since we sent ten styles of knit-wear samples to you, and since then we have written to you several times, but we have not received any reply from you.

It seems to us that you ignored our letters. It is unfair that you treat us like that as you have always been considered as one of our good customers. You may know that we are very cooperative with you based on the fact that we sent generous samples to you without charging you even a penny of cost.

To tell you frankly, we have received from other good customers in Saudi Arabia some orders for the same styles that we offered you. From this fact, we quite believe you should have obtained some orders on the ground that it seems impossible that all the styles we offered should be dropped out completely by your clients.

May we expect orders from you or at least hear from you soon?

Yours truly,

【註】

1. knit-wear：針織衣服。cf. knit goods：針織品。

2. without charging you even a penny of cost：連一文錢都沒有向你收。

3. on the ground：基於……的理由。cf. on the ground of：以……為由。

4. drop out：從……撒手。

No. 120　寄出樣品後無回音，再發出追查信

Dear Sirs:

On April 15 we received a request from you for some prices and samples.

We replied to your letter on April 17, and when we did not hear from you, we wrote again to you on May 20.

At the same time we got the necessary information, after which we again wrote you on May 30, asking that you acknowledge receipt of this information, and let us know if you were interested.

To date, we have not had the courtesy of hearing from you.

Don't you think that when you ask for certain information, and a firm goes to the trouble of getting it for you, you should at least repay them with a reply even to say that you are not interested?

Yours very truly,

【註】

1. to date：到今天為止……。

2. we have not had the courtesy of...：我們還沒有得到你們的賜示。

3. repay：報答；付還。

　　cf. a. to repay a kindness：報答恩惠。

　　　　b. It repays your labor well.（你的力氣不會白費。）

　　　　c. He repaid the money he had borrowed.（他把借的錢還掉了。）

4. 最後一段的譯文：你想當你向人家要某項資料，人家花了很多功夫才找到寄給你，你是不是至少該給人一個答覆，即使說你不感興趣？

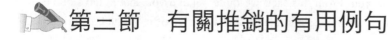

第三節　有關推銷的有用例句

一、開頭句

1. $\begin{cases} \text{We take the liberty of sending} \\ \text{We are very pleased to send} \end{cases}$ you a copy of our latest catalogue and price-list.

2. $\left\{\begin{array}{l}\text{As you have placed many orders with us in the past} \\ \text{As you are one of our oldest and most regular customers}\end{array}\right\}$, we have decided to make you a special offer of...

3. We feel you will be interested in the new...which we are shortly to

$\rightarrow \left\{\begin{array}{l}\text{place on the market.} \\ \text{introduce to the trade.}\end{array}\right.$

4. We are sorry to note that we have been without an order from you for over five months...

5. With reference to the offer we made you on..., we regret that we have not yet had an order from you.

6. Looking through our records we note with regret that we have not had the pleasure of an order from you since last December.

二、結尾句

1. We hope you will take full advantage of this exceptional offer.

2. We are most anxious to serve you and hope to hear from you very soon.

3. We are offering you an article of the highest quality at very reasonable price and hope you will take the opportunity to try it.

4. We feel sure you will find a ready sale for this excellent material and that your customers will be well satisfied with it.

   註：to find a ready sale：發覺好銷。

5. We should be pleased to welcome you to our showroom at any time to give you a demonstration.

6. We are allowing a special 5% discount to you if you place orders before the end of May and look forward to receiving one from you.

7. We look forward to the pleasure of your renewed custom.

   註：renewed custom：再惠顧。

第十八章
訂貨 (Orders)

第一節　訂貨信的寫法

在國際貿易，交易的成立方式有二：

其一為由出口商 Offer，中間經過討價還價，終於 Accept 後，由買方發出訂單 (Order)，或由賣方發出售貨確認書 (Sales confirmation)。

其二為由進口商根據出口商先前寄出的樣品、商品目錄、價目表、估價單，主動向出口商發出訂單，並經出口商接受 (Acknowledge) 或確認 (Confirm)。關於第一種情形已在第十四、十五兩章中介紹。茲將第二種情形予以介紹。

進口商根據出口商原先寄出的樣品、商品目錄或價目表、估價單等主動向出口商發出訂單時，如該訂單經出口商接受 (Acknowledge) 或確認，買賣契約即告成立。當然，這種情形，通常大都進口商前此已與出口商有成交經驗，且對於出口商貨品品質、價格、交貨期等很清楚時才有一拍即合的可能。否則，在此之前，必有函電的往來磋商。

進口商下訂單時或用事先印妥的訂單 (Order sheet) 附以簡單的伴書 (Covering letter)，或以書信方式下訂單——訂貨信 (Order letter)。不論用那一種型式，訂單的內容，與前述報價單 (Offer sheet) 相似，應詳載交易的條件。換言之，每份訂單都應包含下列有關事項：

1. 商品名稱及品質（說明樣品編號、商品目錄編號、品種、價目表編號及其日期等）及訂單編號。
2. 擬購數量。
3. 擬購價格。
4. 包裝方法、嘜頭等。
5. 保險條件。
6. 交貨時間。
7. 付款條件。
8. 其他事項。

由進口商主動發出的訂單，本質上是 Buying offer，訂單一經出口商接受，契約即告成立，所以內容應力求詳細完整。

No. 121　初次訂購 (Initial order)

Dear Sirs,

Thank you for your samples and price list of 10th September. We are pleased to find that your materials appear to be of fine quality.

As a trial order we are delighted to give you a small order for 100 pcs., white shirting S/235. Please note that the goods are to be supplied in accordance with your samples.

The particulars are detailed in the enclosed Order Sheet No. 123. We are in a hurry to obtain the goods, so please cable your acceptance, upon receipt of which we will open an irrevocable L/C through Bank of Taiwan.

If this initial order turns out satisfactory, we shall be able to give you a large order in the near future.

Encl.: a/s　　　　　　　　　　　　　　　　　　　　　　　　　　Yours faithfully,

【註】

1. trial order: 試驗訂貨；嘗試性訂貨。

2. small order: 少量訂貨。trial order 一般而言，均是 small order。

3. in accordance with: 與……一致。

4. the particulars are detailed in...: 細節詳載於……。

5. initial order: 初次訂貨。

6. large order: 大量訂貨。注意: a large order 又有「棘手之事」之意。

No. 121-1　訂貨單 (Order)

ORDER SHEET

Dear Sirs,　　　　　　　　　　　　　　　　　　　　　　　　23rd September, 20–

We have the pleasure of placing the following order with you.

Quantity	Description	Unit Price	Amount
100 pcs. (2 cases)	Cotton White Shirting Sample No. 253 Exactly as shown in the sample.	CIF Osaka in US Dollars @ 12.00/pc.	US$1,200.00

Packing: 10 pcs. in Hessian bale.
　　　　　5 bales in one wooden case.

Marks: with number 1 and 2 under port mark stating the country of origin.

△ OTK
OSAKA
C/#1–2

Insurance: ICC (A) for full invoice amount plus 10%.

Shipment: During October, 20–.

Terms: Draft at 30 d/s under an irrevocable L/C.

Remarks: Certificate of quality inspection & shipment sample to be sent by airmail prior to shipment.

Osaka Trading Co., Ltd.

【註】

1. exactly as shown in the sample：與樣品完全一致。這種品質條件，猶如 "exactly same as sample" 或 "exactly identical to sample" 的條件，對出口商很不利，尤其紡織品的交易不宜以這種嚴格的條件做為品質條件，否則很容易引起糾紛。

2. @12.00/pc.：意指每疋 12 元。

3. shipment samples：裝船樣品。

No. 122　訂購罐頭洋菇

Dear Sirs,

Canned Mushroom

We are pleased to receive your price list of January 1, 20– on the subject goods and hereby place our order with you as per enclosed Purchase Order No. 123 in duplicate.

To cover this order, we will request our bankers to open an L/C in your favor immediately upon receipt of your confirmation of this order. Please arrange to ship the first two items by the first available vessel sailing to Melbourne direct after receipt of the L/C.

We shall appreciate your forwarding us the sample can each of these qualities by airfreight immediately for our evaluation of the quality and meantime, confirm your acceptance of this order by signing and returning the duplicate copy of our Purchase Order.

Yours very truly.

【註】

1. hereby：藉此；特此；茲。

2. to cover this order：為支付本訂貨價款。

3. sample can：樣品罐。

No. 122-1 購貨單

PURCHASE ORDER

No. 123   January 30, 20–

United Export Corp., Taipei

Dear Sirs,

     We are pleased to place an order with you for the following goods on the terms/conditions set forth below:

Commodity
      &
Description: Canned Mushroom of the following four qualities:
          A: 6×68 oz. Button   C: 24×8 oz. Slice
          B: 24×16 oz. Whole   D: 24×4 oz. Pieces& Stem

Quantity: A: 500 cartons   C: 400 cartons
          B: ditto   D: ditto

Price: FOB Taiwan per carton in US Dollars
        A: 16.20   C: 8.25
        B: 14.70   D: 5.00

Packing: to be packed in export standard fiber cartons.

Insurance: Buyer's care.

Shipment: For item A&B: ASAP; For items C&D: within one month after receipt of L/C.

Payment: At sight draft under Irrevocable L/C, assignable and divisible.

Remark: Sample cans of each quality to be airfreighted to us for examination.

                                           Yours truly,

Accepted by:

【註】

1. button：環狀。

2. whole：整片。

3. slice：切片。

4. ASAP：as soon as possible 的縮寫，大多用於 telexese。

No. 123　對電傳訂貨的確認書

Gentlemen:

Porcelain Wares

Pleasure is taken in inclosing our formal Confirmation of Order No. 20 of our recent purchase from you of Porcelain Wares, which has been ensued from the exchange of cables, yours of July 26 and 29 and ours of July 28. We trust that you will find particulars of our order all in the right.

In order to cover this order we have arranged with the National Bank of Commerce, New York, an Irrevocable L/C to the value of U.S. $2,500.00, available till September 30. This credit will be air-mailed to your bankers.

Please give this first account your most careful attention and ship the goods with all speed.

Very truly yours,
THE ATLANTIC TRADING COMPANY
H. W. Dickson
Manager

HWD: 40
Inclos. 1 Confirmation of Order
　　　3 Copies of Cables

【註】

1. pleasure is taken = we take pleasure。

2. formal confirmation of order：正式訂貨確認書。

3. to cover：抵付；支付。

4. to the value of = to the amount of; to the extent of：金額以……為度，也可以 "for" 代之。

5. available = valid：有效。

6. account = business：交易。

7. with all speed = with all (possible) dispatch：迅速；盡快。

No. 124　訂購電晶體收音機電傳確認書

Dear Sirs,

We are glad to confirm our cable order dispatched this morning as follows:

LT

TAITRA

ARRANGE SHIPMENT BX 157 600 TRC 418 800 BOTH DURING AUGUST PV 123 BY SEPTEMBER

ELECTROLEADER

Article	Model	Quality	Shipment
Transister Radio	BX–157	600 sets	August
Tape Recorder	TRC–418	800 sets	August
Portable TV	PV–123	1200 sets	September

In your stock news we found these items available by the date mentioned above. Please arrange shipment accordingly.

We will cable the shipping instructions and the credit immediately upon receiving your acceptance by cable.

We hope that you will promptly accept our order and make your usual careful execution.

Yours faithfully,

【註】

1. cable order dispatched this morning：今晨以電傳發出的訂單。

2. stock news：庫存消息（指 maker 對經銷商、代理商所發出的可供銷售庫存表）。

3. these items available by the date：到今日為止可以購到的項目。

4. arrange shipment：安排船運。

5. shipping instructions：裝船指示。"instructions"解做「指示」時，必須多數。

6. usual careful execution：如往例一樣，注意處理訂單，即指裝船之意。

No. 125　訂購鋼條

June 15, 20–
Per Your telex 6/3, 20–
Our telex 6/12, 20–

Order

Dear Sirs:

We have the pleasure to place you herewith our order for the undermentioned goods on the terms and conditions stated as follows:

DESCRIPTION	QUANTITY	PRICE	VALUE
HOT ROLLED STEEL BARS	10,000 lbs	$6.00 per 100 lbs	$600.00

SPECIFICATION AISI–C1020			
ROUND 1″×18′			
ROUND 3/4″×18′	10,000 lbs	6.00 per 100 lbs	600.00
SQUARE 3/4″×3/4″×18′	15,000 lbs	6.00 per 100 lbs	900.00
FLAT 3/4″×3″×18′	20,000 lbs	6.00 per 100 lbs	1,200.00
			$3,300.00

SAY U.S. DOLLARS THREE THOUSAND THREE HUNDRED ONLY

REMARKS:

1. Prices herein are understood to be on CIF Keelung basis.
2. Payment is to be made through Bank of Taiwan under an Irrevocable L/C.
3. Insurance is to be effected by the seller, covering ICC (C) for 10% above invoice value.
4. Shipment is to be made not later than Aug. 31, 20– and partial shipment is not acceptable.
5. No substitutes should be supplied without our approval.
6. The acknowledgement of this order is required.

Taiwan Trading Co., Ltd.

【註】

1. hot rolled steel bars：熱軋鋼條。

2. round, square, flat：圓型，方型，扁平型。

3. substitutes：替代品。

No. 126　確認訂購絲綢，並附上正式契約書

Dear Sirs,

150 pieces Silk Crepe

　　We acknowledge your telegram of the 21st July, accepting our order of the 19th July for 150 pieces No. 550 Silk Crepe. In confirmation of this order we enclose our formal Contract No. 156, which we trust you will find correct.

　　We have arranged with Lloyds Bank, Ltd., London, an L/C to be opened for this order. We ask you to note specially that as per our contract for this business you must ship the goods to Marseilles/ London on an Optional B/L. Kindly pay your close attention to the secure packing of the goods, so that they may be guarded against the possibility of being pilfered in transit or upon arrival here. The goods must be up to the sample; inferior goods will not be accepted.

You must be aware of the shade assortment required for these 150 pieces as it has already been given in our cable. For caution's sake, however, we again specify the assortment hereunder:

Plain White...	40 pieces	Green...	25 pieces
Light Sky...	30 pieces	Navy...	15 pieces
Mouse...	10 pieces	Dark Blue...	10 pieces
Saxe...	10 pieces	Nigger...	10 pieces

We would ask you to ship this order within the stipulated time and suggest that bulk orders will entirely depend on the outturn.

Yours faithfully,

【註】

1. Marseilles/London：馬賽或倫敦，在選擇卸貨港交易時用。

2. optional B/L：任意港卸貨提單。

3. to pilfer：拔貨，名詞為 "pilferage"。

4. in transit：運輸中。

5. shade assortment：顏色的搭配。

6. for caution shake：為謹慎起見。

7. navy：深藍色。

8. saxe：藍色。

9. nigger：黑褐色。

10. bulk orders = large orders。

11. depend on：視⋯⋯而定。也寫成 "depend upon"。

12. outturn：（處理訂貨的）結果。

No. 127　確認訂購海灘用橡皮球

Gentlemen:

Enclosed you will find copies of cables, yours Nos. 14 and 15 and ours Nos. 11 and 12.

From the exchange of cables as well as your letters Nos. 25 and 26 and ours Nos. 200 and 212, it is our understanding that we have purchased from you, as a sample order, 100 dozen Rubber Beach-ball, 10" and three-colored, at $2.30 per dozen FOB Keelung.

We have today instructed the Chemical Bank & Trust Co., New York City, to cable a letter of credit in the amount of $270.00 in your favor.

We have to express our thanks for your having accepted this small order. However, we hope you realize that we are doing this in order to show our prospects the quality of the merchandise. We shall, therefore, be able to place our orders for larger quantities once this initial shipment has been made and the basis of operating established.

Please book space by the fastest route. We will pay the freight and insure the cargo at this end.

We believe that we can do a very considerable business in this item. We note from your letter that you are willing to accept our proposal and appoint us as exclusive agents for New York, provided that we can place orders with you for fairly large quantities. We would like to go into details regarding the agency and look forward to your further information.

However, the most important thing at this time is to get this sample order on the water. We trust you will do everything possible to expedite shipment.

Cordially yours,

【註】

1. sample order：樣品訂單，即指做樣品用的小量訂貨，與此相對的是 "bulk order"（大量訂單）。

2. rubber beachball：海水浴場等遊戲用的海灘用橡皮球。

3. once this...has been established：「一旦這初次貨裝出，且活動（業務）基礎確立」，意指一旦這初次訂貨裝來，且確立了銷售基礎的話。once = when once。

4. book space：洽訂艙位。

5. at this end = at our end。

6. get...on the water：把……運出。

7. to expedite = to hasten; to quicken：加速。

第二節　有關訂貨的有用例句

一、用　語

1. 名詞：

Firm order	穩固訂單	Sample order	樣品訂貨
Indent	委購訂單	First order	第一次訂貨
Open order	未撤銷前有效的訂單	Initial order	初次訂貨

Repeat order	繼續訂貨	Opening order	首次訂貨
Supplementary order	追加訂貨	Back order	積壓訂單（指訂貨的遲未照發）
Additional order	追加訂貨	Formal order	正式訂貨
Verbal order	口頭訂貨	Limited order	限價訂貨
Written order	書面訂貨	Order note	訂單
Cable (Telegraphic) order	電傳訂貨	Order sheet	訂單
Telex order	電傳交換訂貨	Order form	訂單格式
Fax order	傳真訂貨	Order blank	空白訂單
Trial order	嘗試性訂貨	Suspension of order	擱置訂貨
Goods on order	訂購的貨物	The balance of an order	尚未配發的殘餘訂貨
(Goods ordered but not yet supplied)			
Goods on an order	訂單所載貨物		

2.形容詞：

Large
Heavy
Extensive
Substantial
Important
Bulk

} order　鉅額訂單

Small order　小額訂單

3.動詞：

① To solicit an order　懇請訂貨

　To desire an order　希望訂貨

② To invite an order　邀請訂貨

　To arrange an order with　與……洽商訂貨

　To fix up an order with...　同上

③ To collect an order　蒐集訂單

　To obtain an order　獲得訂單

To secure an order 　　　同上

④ To look for an order 　　　尋找訂貨

To seek an order 　　　尋找訂貨

cf. "We look for large orders to come from our Indian customers." 中的 "look for" 為 "expect" 之意。

To look out for an order 　　　注意訂單的獲得

⑤ To induce a sample order 　　　勸誘樣品訂貨

⑥ To order...from... 　　　向……訂購……

$$\text{To} \begin{cases} \text{give} \\ \text{hand} \\ \text{send} \\ \text{grant} \end{cases} \text{an order to...}$$ 　　　給……以訂單

$$\text{To} \begin{cases} \text{submit} \\ \text{put in} \\ \text{pass} \end{cases} \text{an order to...}$$ 　　　向……提出訂單

To pass an order thru... 　　　經由……訂貨

To pass an order with... 　　　不給……訂單（將訂單給……以外的人）

⑦ To place an order with... 　　　向……訂貨

⑧ To mail an order to... 　　　向……以信函訂貨

⑨ To transit an order to... 　　　將訂單轉給……

⑩ To repeat an order 　　　續發訂單；再訂貨

$$\text{To} \begin{cases} \text{make up} \\ \text{make out} \\ \text{write out} \end{cases} \text{an order}$$ 　　　繕製訂單

⑫ To confirm an order 　　　確認訂貨

$$\text{To} \begin{cases} \text{receive} \\ \text{take} \end{cases} \text{an order}$$ 　　　收到訂貨

⑭ To accept an order 　　　接受訂貨

⑮ To be favored with an order 　　　惠賜訂單

⑯ To $\begin{cases} \text{book} \\ \text{enter} \end{cases}$ an order　　　　將訂單予以登記（表示接受）

⑰ To $\begin{cases} \text{modify an order} \\ \text{make alterations in order} \end{cases}$　　變更訂單內容

⑱ To increase an order　　增加訂貨

　　To reduce an order　　減少訂貨

⑲ To duplicate an order　　加倍訂貨

⑳ To delay an order　　延期訂貨

　　To postpone an order　　同上

二、關於訂貨的有關例句

㈠開頭句

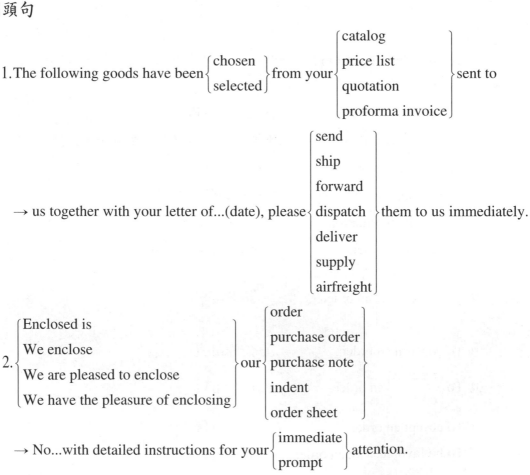

1. The following goods have been $\begin{cases} \text{chosen} \\ \text{selected} \end{cases}$ from your $\begin{cases} \text{catalog} \\ \text{price list} \\ \text{quotation} \\ \text{proforma invoice} \end{cases}$ sent to

→ us together with your letter of...(date), please $\begin{cases} \text{send} \\ \text{ship} \\ \text{forward} \\ \text{dispatch} \\ \text{deliver} \\ \text{supply} \\ \text{airfreight} \end{cases}$ them to us immediately.

2. $\begin{cases} \text{Enclosed is} \\ \text{We enclose} \\ \text{We are pleased to enclose} \\ \text{We have the pleasure of enclosing} \end{cases}$ our $\begin{cases} \text{order} \\ \text{purchase order} \\ \text{purchase note} \\ \text{indent} \\ \text{order sheet} \end{cases}$

→ No...with detailed instructions for your $\begin{cases} \text{immediate} \\ \text{prompt} \end{cases}$ attention.

3. Please $\begin{Bmatrix} \text{deliver} \\ \text{ship} \\ \text{supply} \\ \text{send} \end{Bmatrix}$ the $\begin{Bmatrix} \text{following} \\ \text{below-mentioned} \end{Bmatrix}$ $\begin{Bmatrix} \text{goods} \\ \text{articles} \\ \text{items} \\ \text{qualities} \end{Bmatrix}$ listed in your $\begin{Bmatrix} \text{catalog.} \\ \text{price list.} \\ \text{estimate.} \\ \text{quotation.} \end{Bmatrix}$

4. Please $\begin{Bmatrix} \text{book} \\ \text{fill} \\ \text{execute} \end{Bmatrix}$ the following order according to your $\begin{Bmatrix} \text{catalog} \\ \text{quotation} \\ \text{price list} \\ \text{estimate} \end{Bmatrix}$ of...(date) and samples submitted.

$\left(\right.$ 請按你們×月×日的 $\begin{Bmatrix} \text{目錄} \\ \text{報價單} \\ \text{價目表} \\ \text{估價單} \end{Bmatrix}$ 以及提供的貨樣 $\begin{Bmatrix} \text{接受下列訂貨。} \\ \text{配發下列訂貨。} \\ \text{執行下列訂貨。} \end{Bmatrix}$ $\left.\right)$

5. Thank you very much for your letter of May 5 with catalog and price list. We have chosen three qualities and take pleasure in enclosing our order No. 1234.

6. Your samples of umbrellas received favorable reaction from our customers, and we are pleased to enclose our purchase order for 1,000 dozen.

7. Confirming our exchange of cables, we are enclosing the following order: Order No. 1234 for 1,000 doz. Nylon Umbrellas US$12/doz. CIF Colon.

8. $\begin{Bmatrix} \text{We thank you for} \\ \text{Thank you for} \\ \text{We appreciate} \\ \text{We acknowledge with thanks} \end{Bmatrix}$ your $\begin{Bmatrix} \text{offer} \\ \text{quotation} \\ \text{letter} \end{Bmatrix}$ of...with $\begin{Bmatrix} \text{samples.} \\ \text{catalog.} \\ \text{price list.} \end{Bmatrix}$

9. Please send us by $\begin{Bmatrix} \text{post} \\ \text{air} \\ \text{the first available steamer} \end{Bmatrix}$ this order as $\begin{Bmatrix} \text{per particulars given} \\ \text{below:} \\ \text{specified below:} \end{Bmatrix}$

　　註：first available steamer：第一艘開駛船隻；可利用的第一艘船；近便的船。

10. We are enclosing $\begin{Bmatrix} \text{our order sheet} \\ \text{our purchase note} \\ \text{our purchase order} \end{Bmatrix}$ in confirmation of our $\begin{Bmatrix} \text{telex} \\ \text{cable} \end{Bmatrix}$ dispatched to

you on...(date).

(二)條件（除這裡所列舉者外，讀者尚可參閱第十二章交易條件的有關部分）

1. Please choose nearest substitute for any article out of stock.

   註：nearest substitute：最近似的代替品。

2. This order must be $\begin{cases} \text{filled} \\ \text{executed} \\ \text{delivered} \\ \text{shipped} \end{cases}$ $\begin{cases} \text{immediately} \\ \text{without delay} \\ \text{within...days} \end{cases}$ after receipt of L/C, otherwise it will

   be cancelled.

3. Please $\begin{cases} \text{execute} \\ \text{deliver} \\ \text{ship} \\ \text{pack} \\ \text{mark} \end{cases}$ in strict accordance with our instructions.

4. $\begin{cases} \text{Full} \\ \text{Detailed} \end{cases}$ instructions regarding marks and numbers will follow.

5. Please purchase and ship on our account not later than May 10, 1,000 dozen cotton handkerchiefs of pattern No. 145.

6. The quality of the order must be the same as that of our sample.

7. The material should be of the finest quality and exactly the same as the sample submitted last week.

8. Please note that your goods should meet our requirements.

9. If you do not have them in stock please send us substitutes of the nearest quality.

10. Prompt shipment is $\begin{cases} \text{very important.} \\ \text{essential.} \\ \text{will be appreciated.} \end{cases}$

11. The shipment should be made not later than...

12. If the goods are not delivered before June 1, we will have to cancel the order.

13. When the goods are ready for shipment please let us know. We will then send you shipping instructions and other relative information.

14.Upon receipt of your reply we will open an L/C by cable.

15.For payment we have arranged with the ×× Bank for a confirmed irrevocable letter of credit in your favour for the amount of US$...

16.The L/C for this order has been opened through Bank of Taiwan and we hope you will receive it before long.

17. $\begin{Bmatrix} \text{Please inform us} \\ \text{Please let us know} \\ \text{It will help us to know} \end{Bmatrix}$ by return when we may expect delivery of the goods ordered.

㈢結尾句

1. Please confirm $\begin{Bmatrix} \text{receipt} \\ \text{acknowledgement} \\ \text{acceptance} \end{Bmatrix}$ of this order.

2. Your $\begin{Bmatrix} \text{prompt} \\ \text{early} \\ \text{careful} \end{Bmatrix}$ attention to this order and $\begin{Bmatrix} \text{confirmation} \\ \text{acknowledgement} \end{Bmatrix}$ will be appreciated.

3. We $\begin{Bmatrix} \text{expect} \\ \text{rely on} \end{Bmatrix}$ your careful execution of this order.

4. If this order is satisfactorily $\begin{Bmatrix} \text{filled} \\ \text{executed} \end{Bmatrix}$, we $\begin{Bmatrix} \text{hope to} \\ \text{will} \\ \text{may be able to} \end{Bmatrix}$

$\rightarrow$ place $\begin{Bmatrix} \text{further} \\ \text{large} \\ \text{bulk} \\ \text{regular} \\ \text{substantial} \\ \text{repeat} \end{Bmatrix}$ orders with you.

5. We shall appreciate your careful attention to our instructions.

6. Please inform us whether you can accept this order on these terms.

7. We shall be grateful for prompt delivery as the goods are needed urgently.

第十九章
答覆訂貨 (Reply to Order)

第一節　答覆訂貨信的寫法

　　出口商收到訂單後，應盡速答覆，以免進口商對於進貨或轉售計畫發生困擾。如因存貨關係，無法接受訂貨，也應立即函（電）覆致歉，或推介代替品。

　　對於訂購函電的答覆，可大分為三種：

　　其一為可照訂單全部接受訂購。

　　其二為部分訂貨不能接受或暫時不能供貨。

　　其三為謝絕訂貨。

一、可照訂單全部接受訂購時

　　答覆時應將交易條件內容詳述。其內容大致如下：

　1.對訂購表示謝意。

　2.表示接受訂購。

　3.引述訂單號碼、日期。

　4.重述訂單內容（包括商品名稱、品質規格、數量、價格、包裝刷嘜、保險、裝運、付款等條件）。

　5.其他特別事項 (Special instructions)。

　6.結尾：結尾句須把握下列幾個原則：

　⑴希望繼續惠顧。

　⑵保證履行諾言。

　⑶趁機介紹新產品。

　　原則上，進口商以電傳訂貨 (Teletransmission order)，則出口商應以電傳答覆，隨後以 Sales note, Sales confirmation 或書信確認。

　　訂貨是交易的正式開始，俗語說得好，「好的開始，等於成功了一半」(Well begun

is half done)，能否使進口商滿意，由而繼續惠顧，端視出口商如何處理訂單而定。所以出口商應站在進口商的立場，處處為進口商設想，為進口商做週到的服務。

No. 128　接受訂購罐頭洋菇 (See No. 122)

Dear Sirs,

Subj.: Canned Mushroom

We thank you for your letter of January 30, 20– along with your Purchase Order No. 123 for four different qualities of the subject item.

In acceptance of your order, we are returning herewith the duplicate copy of your purchase order duly signed.

The first two items are now ready, and we are now arranging shipping space for the first available vessel sailing to Melbourne. Therefore, please have the L/C opened at your earliest convenience in order to enable us to effect prompt shipment. As regards the other two qualities, we will ship them within one month on receipt of your L/C.

As requested, we have today airfreighted one sample can each of these qualities for your evaluation. We are sure that the quality would meet your approval.

We take this opportunity to thank you for your initial order and look forward to receiving more orders from you in the near future.

Faithfully yours,

【註】

　1. arrange shipping space：安排艙位，也可以 "book space" 代替。

　2. "receive more orders" 也可以 "favor us with continuous orders" 代替。

No.129　確認接受訂購棉胚布

Dear Sirs,

Thank you very much for your order of July 2 for 50,000 yds. Grey Cotton Twill. We are pleased to confirm our acceptance as shown in the enclosed Sales Note together with the copy of our cable dispatched today.

We have arranged for immediate shipment as requested. Please open your L/C by telegram by the end of July upon accepting our requirement.

The first available ship is scheduled for departure on August 10. We are confident of handling your order to your complete satisfaction, provided your L/C reaches us by the end of July.

We hope that this initial order will be just the beginning of many and that we may have many years of pleasant business relations together.

Yours faithfully,

【註】

1. confirm our acceptance：確認本公司接受（訂貨）。

2. sales note：售貨單，實務上名稱很多，諸如 "sales confirmation"，"sales contract"，"confirmation of sale"，"proforma invoice" 都有人使用。

3. arrange for immediate shipment：安排立即裝船。

4. is scheduled for departure...：預定於……啟航。

5. are confident of...satisfaction：對於照拂貴公司的訂貨有自信令貴公司完全滿意。

6. 最後一段的文意為：我們希望本批訂單只是一個開頭，並望有長年、愉快的業務關係。

No. 129–1　售貨確認書

SALES NOTE

Dear Sirs,
July 10, 20–

We are pleased to confirm our sale of the following goods on the terms and conditions set forth below:

Commodity: Grey Cotton Twill.
Quality: As per sample No. 123.
Quantity: 50,000 yds.
　　　　　118 yds. and up 85% min.
　　　　　40 yds. and up 15% max.
Price: US¢80 per yard $CIFC_8$ Jeddah.
Packing: 10 pcs. in polybag and 50 pcs. in wooden case.

Marking:

Jeddah
C/No. 1 and up
Made in Taiwan
Don't Use Hooks
Insurance: To cover ICC (B) plus war for 110% of invoice value.
Shipment: During August, 20–.

Payment: Draft at sight under an Irrevocable L/C which must reach us by end of July, 20–.

N.B. 1. Certificate of Quality Inspection & Shipment samples to be sent prior to shipment.

2. Unless otherwise specified in this Sales Note, all matters not mentioned here are subject to the agreement of the general terms and conditions of business concluded between both parties.

【註】

1. grey cotton twill：斜紋棉胚布。

2. 118 yds. and up 85% min.：指每疋 (pc) 長度 118 碼及以上者不得少於 85%， min. 為 minimum 之意。40 yds. and up 15% max.：指每疋 (pc) 長度 40 碼及以上者不得多於 15%，max. 為 maximum 之意。

3. Don't Use Hooks：不得用手鉤。

4. subject to：遵守；以……為準。

5. agreement of the general terms and conditions of business：一般交易條件協議書，請參閱第十一章。

No. 130 接受訂購電晶體收音機 (See No. 124)

Dear Sirs,

Please accept our sincere appreciation for your letter of August 5, 20– placing an order for 10,000 sets of transistor radios.

We are returning herewith a copy of your order duly signed by us and assure you that the quality of the goods to be delivered will be conforming to the sample and shipment will be made before November 10.

To achieve this end, your instructions have been immediately passed on to our makers for execution.

We look forward to receiving your L/C before the end of September as stipulated in your order.

Yours faithfully,

【註】

1. to achieve this end：為達成此目的。

2. pass on to = pass along to：(將東西) 轉給；(將指示) 轉達或交付。

例：He passed the jewel on to his wife.

No. 131　對初次訂貨的答謝信之一

Dear Mr. Ling:

It was just fine of you to send us that nice order. Thanks a lot.

For the confidence you have placed in our products, we are very grateful. In return, we shall leave no stone unturned to justify a continuance of that confidence.

You'll always find our organization happy and ready to give you the fullest measure of assistance. Don't feel you'll be putting us to any trouble, because besides selling quality goods, it is our job to create satisfied customers.

We will not forget you after this, your first order. No sir! We could not have carried on for fifty-two years—successfully weathering every business upheaval—were that our policy. This order will be the beginning of a long and pleasant business relationship. Anyway, that's what we shall try to make it.

So kindly think of us the next time you need Our Goods, and if we can help you in any way, please write us.

Our one desire is to serve you faithfully.

【註】

1. just fine：真好。

2. thanks a lot：多謝。

3. the confidence you have placed in our products：你對我們產品的信心。

4. in return：以為報答。

5. leave no stone unturned：盡力。

6. to justify a continuance of that confidence：使（你們）繼續信任（我們的產品）。

7. give the fullest measure of assistance：予以全力協助。

8. don't...trouble：不要覺得會給我們任何麻煩。

9. create satisfied customers：產生滿意的主顧。

10. we will not...this.：我們不會在你們第一次訂貨之後就忘了你們。

11. No Sir!（不會的!）

12. carried on：經營。

13. weathering every business upheaval：渡過各種商業上的動亂。

14. try to make it：努力做到。

No. 132　對初次訂貨的答謝信之二

Dear Mr. ××× :

　　For your first-time order, I want to thank you on behalf of the company. This is your initial order and I know of no greater kick a sales manager can get than the beginning of business with a new customer.

　　The signing of an order is an expression of confidence, and I want you to know that we are fully aware of the responsibility we have for maintaining that confidence.

　　The most important thing to the buyer of any product is the character of the supplying organization; its resources; its facilities; its reputation; and its standards of service.

　　You have at your disposal every facility of our company. We would like to be useful to you beyond the mere necessities of business transaction.

　　We appreciate your business. We want to continue to deserve it.

Sincerely,

【註】

1. I know of no greater kick a sales manager can get than the beginning of business with a new customer.（作為一個銷售經理，沒有什麼能比與一個新客戶開始做生意更為高興的了。）

2. the signing of an order：簽發訂單。

3. supplying organization：供應機構，指賣方。

4. its resources; its facilities：它的來源；它的設備。

5. standards of service：服務的標準。

6. You have at your disposal every facility of our company.（我們公司的設施都任由你來支配。）

7. beyond...：除了……。

No. 133　對於再次訂貨致謝

Dear Mr. Ford,

　　Thank you very much for the order. We appreciate your continued support, and we are pleased to again have the opportunity of cooperation with you. We believe these goods will be of assistance to you in increasing your sales in a satisfactory manner.

　　We are constantly producing unique ranges of modern merchandise and a considerable proportion of the goods, and all the packages, are designed or styled by us.

We won't feel that the sale is "closed" until you receive lasting satisfaction from your purchase —until you honestly feel that you have received full value for every dollar spent!

It is our earnest hope that behind this personal service our business relations may long continue.

Very truly yours,

【註】

1. unique ranges of modern merchandise: 獨特的新式貨品。

2. We won't feel...spent! 本段的意思是說: 我們不認為賣出是「完結」，我們要你們對貨品滿意，要你們真正的覺得貨真價實，每塊錢都花得有價值才算是完結。

3. It is our earnest...long continue. 本段意思為: 我們懇切的希望由於我們忠誠的服務，我們的生意關係會長久的繼續下去。

二、部分訂貨不能接受，或暫時無法供貨

部分訂貨不能接受的原因很多，如庫存賣完，或該貨已停止生產等等。在此種情形，出口商可採下列步驟:

1. 徵求允許先行裝運可供應的部分，並說明另一部分延後交運的理由以及預計何時可出貨。

2. 徵求允許以同等品代替不能供應的部分。

3. 如訂購貨物已停止生產，說明原因，並推介代替品。

因此撰寫這類信時，其內容大致如下:

①表示謝意。

②重述訂單內容、訂單號碼、日期。

③說明那些貨品何時可以交運。

④對於不能供應部分表示歉意，並說明理由。

⑤說明不能供應部分須延至何時才能供貨，或推介代替品。

No. 134　部分不能供應，另推介代替品

Dear Sirs,

Thank you for your order of 18th September requesting a Rotary Printing Press Model PM-600, PM-800 and PM-1600. We are pleased to book all except the first one which is now under mechani-

cal redesigning.

As you requested us to ship them by the end of November, we have just cabled you two alternatives concerning Model PM–600 as shown in the enclosed cable:

1. As an excellent substitute for PM–600 we recommend 630. This is our latest model and much superior in printing speed, 90 revolutions per minute. Considering your inconvenience, we will make a special price discount to stg. £1,500. If this is acceptable please cable us and we will dispatch all items in November.

2. We may ship PM–800 and PM–1600 during November and PM–600 during January next year upon finishing the redesigning, providing you prefer to import PM–600.

We are very sorry for the inconvenience. However, please note our wish to offer an alternative, especially our special discount price.

We hope this will meet your immediate acceptance so that we can execute the order in a most satifactory manner.

Yours faithfully,

【註】

1. mechanical redesigning：機械重設計。

2. alternative：代替辦法；代替品 (alternative product)。

No. 135　無庫存不能立即供應

Dear Sirs,

We thank you for your order for 100 dozen yards of curtain fabric to patterns submitted, but as our stocks have been cleared and the cloth has to be manufactured, we are not in a position to effect delivery in less than a fortnight. It must also be understood that this period will be exceeded if the makers have none of the patterns selected on hand; there is, however, little danger of this.

If your demand is so pressing as to necessitate immediate delivery, we are prepared to supply a material in some respects inferior but satisfactory for most ordinary purposes, as the samples enclosed will show. We suggest, however, that it would pay you to suspend the execution of your order unless such a course is for special reasons inadvisable.

Perhaps you would be good enough to confirm your order subject to these conditions.

Yours faithfully,

【註】

1. curtain fabric：窗簾布。

2. patterns submitted：附來的式樣。curtain fabric to patterns submitted：照所附樣式的窗簾布。

3. little danger of this：無此疑慮；無此危險。

4. it would pay you to...：這樣對你比較划得來。

5. suspend the execution of：擱置配貨。

6. unless such a course is for special reasons inadvisable：除非因特別理由這樣的做法不適當。

      course：做法；方法；行為。

No. 136 因訂單太多，須等一段時間才能供貨

Dear Sirs,

    We thank for your order of July 6, 20– for 10,000 Can Openers but regret that we are unable to execute the order from our stocks, owing to the heavy demand recently for these gadgets.

    Our manufacturers have, however, undertaken to replenish our stocks in one month and we trust that it will not be inconvenient for you to allow us this extension.

    Should you be willing to meet our wishes, we should be grateful if you would confirm your order on the revised conditions.

                          Yours faithfully,

【註】

1. can opener（美）＝ tin opener（英）：罐頭刀；開罐頭用具。

    cf. We eat what we can, and can what we can not.（我們盡量吃，吃不完的予以罐裝。）

2. gadgets：小器具。

3. undertake to replenish：承擔（保證）補充。

No. 137 因工廠失火，暫時無法供應

Dear Sirs,

    Many thanks for your order of September 10, 20– for 1,000 suits in the sizes specified and we trust that your requirements are not so urgent as to render an immediate delivery essential.

    Owing to a recent fire at our factory stocks are considerably destroyed and the high-grade materials necessary for the suits are not now available. If, however, you require the suits for immediate delivery, we can make them in materials nearly as good, but as the original material is likely to be extremely popular this season, we would advise you to wait about four weeks by which time we shall be in a position to supply it.

    We should be grateful to receive your comments very soon.

                          Yours sincerely,

【註】

1. suit：一套衣服。1,000 suits：1,000 套衣服。

2. not so urgent as to render...essential.：不致需貨如此急，以致須立即交貨。"render" 一字很有用處，其用法宜多練習。

3. not available：沒有。

三、謝絕訂貨 (Declining orders)

就出口商而言，當然希望能夠滿足所有來自進口商的訂貨。但有時也會遇到不能不婉謝訂貨的情形。例如已無存貨、已不再產製、訂貨已過多或已逾接受訂購期限等等，均使得出口商不得不拒絕訂貨。在撰寫這種覆函時，必須先表示謝意，然後，說明不能接受訂貨的理由，以獲得進口商的諒解。假如別處能供貨，可告訴進口商向何處訂購，如此不但可獲得進口商的感激，同時還可以給進口商留下好印象，以備將來繼續惠顧。

此外，如價錢已變動，比原來發出的 Price list, Proforma invoice 或 Catalog 所載價格要高的時候，不宜逕予拒絕，而應說明漲價的理由，並希其按調整後的價格訂購。

No. 138　已不再生產，推介新產品

Dear Sirs,

　　Thank you for your letter of May 12 enclosing your order for 1,000 sets of "OK" Brand transistor radios.

　　We are sorry we can no longer supply this model, because it is rather old model and the demand for it has fallen to such an extent that we have ceased to manufacture them.

　　Instead, we should like to recommend our new "YES" Brand transistor radios, which, we think, is more attractive in design and practical for use. The large number of repeat orders we regularly receive from our overseas customers is clear evidence of the wide spread popularity of this brand. At the low price of only US$2.30 each with a leather case and a strap CIF Singapore is much cheaper than "OK" Brand. The sample with description is being forward to you by air parcel today.

　　We are looking forward to receiving your comments or order soon.

Yours faithfully,

【註】

1. practical for use：實用。

2. clear evidence：明白的證據。

3. leather case：皮套。

No. 139　檔期已滿，無法再接訂單，另推介廠家

Gentlemen:

Your order No. 123 dated June 10 for Natural Color Seagrass Door Mats is greatly appreciated, but we regret to have to disappoint you.

Since we have already received so many orders that if we accept your order now, we could not deliver the mats before December 25, which is more than four months later than the time you specified. This situation prevails despite the fact we have doubled our production facilities since last winter.

Please accept our apologies for our inability to accept your order. However, we hope you will be able to obtain the mats elsewhere so that you will not be inconvenienced by our temporary inability to serve you.

On your behalf we have contacted XYZ Enterprises, Inc., and we are informed they are quite willing to supply you. Probably you will hear from them before long.

Yours truly,

【註】

1. natural color：本色的。

2. seagrass：海草。

3. door mat = door-mat：門前擦去靴底污泥的墊子。

4. double our production facilities：增加生產設備一倍。

5. before long：不久；也可以 "soon" 代替。

6. 無法供應時，一般人都未必願意另推介廠家，主要是因同行是冤家之故吧！但本寫信人卻胸襟寬大，為客戶另推介廠家，因此必定會給客戶良好的印象。

No. 140　已停止生產，推薦新產品

Dear Sirs,　　　　　　　　　Subj.: Zepher Silk

We thank you for your order of November 14 for the subject silk but regret that we cannot supply this material, as its manufacture has no longer been continued. It was only when the entire absence of inquiries led us to believe that it had been dropped quite out of favor that we decided to take this step.

We are pleased to hear that it has met with your approval and venture to think that you will find our new "GOSSAMER" brand even more satisfactory.

The new cloth is considerably fine, with a dull lustre that is most attractive; the popularity it enjoys among the leading manufacturers is proof enough of its value.

You will find enclosed a price list and a full range of patterns. It seems to us that a trial order would make you share our confidence.

We look forward to receiving your favorable response.

Yours faithfully,

【註】

1. had been dropped quite out of favor = become unpopular：已不受歡迎。

2. venture to think：膽敢以為……。

   cf. venture to suggest that...：冒昧地提議……。

      venture a guess：大膽地做猜測。

3. dull：暗晦的。

4. lustre = luster：光澤。

5. a full range of patterns：各種式樣。

6. share our confidence：相信我們（所說的）。

No. 141　已漲價無法按舊價供應

Dear Sirs,

We wish to thank you for your Order No. 90 of the 20th July for our various Leather Goods.

It is a great regret, however, that your present order can not be filled at our old prices. As foreshadowed in our last letter of the 5th July, leather and all other materials have risen so considerably in prices of late that we have found ourselves compelled to cancel all prices ruling hitherto and raise our prices for all our products.

We are enclosing our new Price List No. 30. The advance, as you will see, is not very serious, and we suggest that you accept our new prices, for it is more than probable that we shall have to intimate you a further advance before long, so that you need have no fear of any relapse in prices.

In order to save time, we have to ask you to wire us "Proceed" if you decide to confirm your present order at our new prices, so that we can ship all the items immediately.

Yours faithfully,

【註】

1. leather goods：皮貨。

2. to rise = to advance; increase; appreciate; improve; soar; move upwards; go up：上漲；相反詞 為：to fall; decline; decrease; depreciate; drop; move downwards; go down。

3. to raise：提高（價格），是他動詞不可與 "to rise" 混淆；相反詞為 "to reduce"。

4. you need...prices：不必耽心價格會回到原來的狀態，"relapse" 為「回到原來狀態」之意， 這裡指「跌回原價格」。

5. proceed：請進行，這裡指「請配貨」之意。

No. 142　原料、工資漲價，無法照原價供應

Dear Sirs,

<u>Re: 50% Nylon 50% Polyester Mixed Knitted Fabrics</u>

We are pleased to receive your letter of the 4th Sept. together with initial order No. 72/2963 for the captioned article.

Please be advised that we can not supply you this item at price of US$0.95 per yard because recently the price of raw materials and wages have been greatly raised in our country. The best price we can offer presently is US$1.07 per yard $CIFC_5$ Port Louis. We can accept 300 yds. per color instead of 600 yds., but the minimum quantity should be more than 2,400 yards. we further enclose herewith revised quotation No. 610–3013 and the color samples for your selection.

We are looking forward to hearing from you soon.

Yours truly,

【註】

1. 50% nylon 50% polyester mixed knitted fabrics：50% 尼龍 50% 多元酯混合針織布。

2. the best price：最便宜的價錢。

No. 143　存貨已售完，無法供應

Dear Sirs,

We have just received with thanks your order No. 123 of October 2 for 1,000 dozen No. 350 Rayon Socks in assorted sizes for immediate shipment.

Very much to our regret, however, we have to inform you of our inability to accept your order as the goods have now been sold out. A reference to our letter of the 23rd September will remind you that our offer was not firm and that we stressed the importance of an early reply, owing to the fact that bidding was likely to be keen. The response has exceeded our expectations, and we have felt ourselves obliged to close with the offer of a customer prepared to take the goods.

>   We solicit your order for our similar makes, of which you are well aware. If, however, you admit of no substitute, we can give you a special order to our mill to turn out the goods in question and will ship them not later than the 10th November.
>
>   We should be most grateful if you would confirm your order on these conditions.
>
>   Yours very truly,

【註】

    1. rayon socks：嫘縈絲襪。

    2. in assorted sizes：各種尺寸。

    3. a reference...remind you = if you refer to..., you will see that...。

    4. to stress...reply：強調迅速答覆乃為必要。

    5. bidding：買進的出價。

    6. to close...the goods：必須接受擬購入本貨品的某一客戶的訂貨。to close with = to accept。

    7. similar makes：類似製品。

    8. to turn out = to manufacture; produce。

第二節　有關答覆訂貨的有用例句

一、訂貨用語

㈠關於 Execution of order 的用語

1. To $\begin{cases} \text{execute} \\ \text{fill} \\ \text{fulfil} \end{cases}$ an order   執行（處理）訂貨

2. To carry out an order   執行（處理）訂貨

3. To begin the execution of an order   開始處理訂貨

4. To hurry on the execution of an order   趕快處理訂貨

5. To $\begin{cases} \text{expedite} \\ \text{hasten} \end{cases}$ the execution of an order   加緊處理訂貨

6. To delay the execution of an order   延遲處理訂貨

7. To work overtime with an order　　　加班處理訂貨

8. To complete an order　　　完成所訂貨品

9. To get an order ready for shipment　　備妥所訂貨品待運

10. To ship an order　　　運出所訂貨品

11. To $\begin{Bmatrix} \text{dispatch} \\ \text{forward} \end{Bmatrix}$ an order　　發出所訂之貨

(二)名　詞

1. Counter order　　　替換訂單

2. $\begin{Bmatrix} \text{Cancellation} \\ \text{Revokation} \end{Bmatrix}$ of order　　取消訂單

3. Withdrawal of order　　　取消訂單

4. Suspension of order　　　擱置訂單

(三)動　詞

1. To $\begin{Bmatrix} \text{stop} \\ \text{suspend} \\ \text{hold up} \end{Bmatrix}$ an order

停止訂貨

擱置訂貨

擱置訂貨

2. To postpone an order　　　延期訂貨

3. To $\begin{Bmatrix} \text{cancel} \\ \text{annul} \\ \text{revoke} \\ \text{countermand} \end{Bmatrix}$ an order　　取消訂貨

4. To withdraw an order　　　撤回訂貨

5. To hold an order in abeyance　　暫時擱置訂單

（"hold" 可以 "leave" 代替）

6. To hold back an order　　　壓下（扣留）訂單

7. To proceed with an order　　　繼續處理所訂貨品

(To go on with...after interruption)

8. To decline an order　　　拒絕訂貨

9. To accept an order　　　接受訂貨

二、答覆訂貨的開頭句

1. $\begin{Bmatrix} \text{Thank you} \\ \text{Many thanks} \end{Bmatrix}$ for your $\begin{Bmatrix} \text{order} \\ \text{purchase order} \\ \text{indent} \end{Bmatrix}$ No....of (date)..., which we confirm here-

   with as follows:

2. It is a pleasure to receive a $\begin{Bmatrix} \text{trial} \\ \text{sizable} \\ \text{generous} \end{Bmatrix}$ order for...

3. We are pleased to confirm that we have $\begin{Bmatrix} \text{acknowledged} \\ \text{accepted} \\ \text{received} \end{Bmatrix}$ your $\begin{Bmatrix} \text{telex} \\ \text{cable} \end{Bmatrix}$ order

   of...(date)...for...(goods).

4. Your order No....of...(date)..., for which we thank you, has been booked as instructed

   → and we $\begin{Bmatrix} \text{enclose} \\ \text{are enclosing} \\ \text{are pleased to enclose} \end{Bmatrix}$ our confirmation in duplicate. Please sign and re-

   turn us the original.

5. $\begin{Bmatrix} \text{We acknowledge} \\ \text{Thank you for} \end{Bmatrix}$ your letter of...(date)...enclosing your order No....for...(goods),

   which we have passed on to our shipping department for earliest possible shipment.

6. It was a pleasure to receive your order of October 2 for...(goods), which was imme-

   diately taken in hand.

7. We appreciate your order of...(date)...for...(goods).

8. We are very much obliged for your order of...(date).

三、答覆訂貨的文中用語

1. We have instructed our $\begin{Bmatrix} \text{manufacturer} \\ \text{maker} \\ \text{factory} \end{Bmatrix}$ to start manufacture at once, and you may

$\rightarrow \begin{Bmatrix} be \\ rest \end{Bmatrix}$ assured that the goods will be $\begin{Bmatrix} delivered \\ shipped \end{Bmatrix} \begin{Bmatrix} within \\ before \end{Bmatrix}$ the specified shipment date.

2. You may $\begin{Bmatrix} be\ sure \\ be\ assured \\ rest\ assured \end{Bmatrix}$ that your instructions will be carefully $\begin{Bmatrix} followed. \\ adhered\ to. \\ observed. \end{Bmatrix}$

$\left(我們會小心地 \begin{Bmatrix} 遵照 \\ 忠於 \\ 遵守 \end{Bmatrix} 你的指示，請放心。 \right)$

3. This order will $\begin{Bmatrix} have \\ receive \end{Bmatrix}$ our best attention and we assure you that we shall do our

$\rightarrow \begin{Bmatrix} best \\ utmost \end{Bmatrix}$ to $\begin{Bmatrix} fill \\ execute \end{Bmatrix}$ it within the $\begin{Bmatrix} specified \\ prescribed \end{Bmatrix} \begin{Bmatrix} time. \\ date. \end{Bmatrix}$

4. We will ship the goods as early as possible, and will advise you immediately the shipment is made. Enclosed is our Sales Confirmation No....in duplicate, the original of which please sign and return.

　　註：這裡的 "immediately" 是 conjunction，其意為 "as soon as"。

5. We shall give this order our best and prompt attention and hope to be able to manufacture within the stipulated time.

6. We accept your limit at a sacrifice and hope you will compensate us by ordering more from us in the near future. （我們以虧本來接受你限定的價格……。）

四、無法接受訂貨

1. We $\begin{Bmatrix} are\ sorry \\ regret \end{Bmatrix}$ that we are quite unable to $\begin{Bmatrix} fill \\ execute \\ carry\ out \end{Bmatrix}$ your order at the present

$\rightarrow$ moment, due to $\begin{Bmatrix} heavy\ orders. \\ rush\ of\ orders. \\ recent\ rise\ in\ demand. \\ heavy\ demand. \end{Bmatrix}$

2. We have too many orders on hand and could not effect shipment before

$$\rightarrow \begin{cases} \text{September.} \\ \text{the date specified.} \end{cases}$$

3. We $\begin{cases} \text{regret} \\ \text{are sorry} \end{cases}$ that item No....is $\begin{cases} \text{sold out （售完）} \\ \text{out of stock （缺貨）} \\ \text{out of production} \\ \text{out of make （停製）} \end{cases}$ at present.

4. Although we have tried every source of supply we know, we have been unable to get the size you asked for.

5. We are sorry to inform you that we no longer $\begin{cases} \text{produce} \\ \text{manufacture} \\ \text{stock} \\ \text{carry （有）} \\ \text{handle} \end{cases}$ this $\begin{cases} \text{design.} \\ \text{article.} \\ \text{size.} \\ \text{model.} \\ \text{material.} \end{cases}$

6. Although we are unable to fill your order now, we shall be pleased to contact you again, as soon as circumstances allow. （……情況一允許，我們將再與你連絡。）

7. We have been compelled to raise our prices by 10%

$$\rightarrow \begin{cases} \text{under the pressure} \\ \text{of} \\ \text{owing to} \end{cases} \begin{cases} \text{increased labor costs.} \\ \text{the rise in raw material prices.} \\ \text{heavier import duties on raw materials.} \end{cases}$$

8. We are sorry to inform you that these goods cannot be bought now at your limit.

9. Unfortunately it is impossible to accept the price of...（……無法接受……的價格。）

五、建議訂購替代品

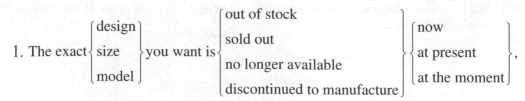

1. The exact $\begin{cases} \text{design} \\ \text{size} \\ \text{model} \end{cases}$ you want is $\begin{cases} \text{out of stock} \\ \text{sold out} \\ \text{no longer available} \\ \text{discontinued to manufacture} \end{cases} \begin{cases} \text{now} \\ \text{at present} \\ \text{at the moment} \end{cases}$,

however, we are sending you by separate mail a sample similar to it, which we think will be a very good substitute for it.

2. Please $\begin{Bmatrix} \text{select} \\ \text{choose} \end{Bmatrix}$ $\begin{Bmatrix} \text{a suitable substitute} \\ \text{another quality} \end{Bmatrix}$ from the enclosed $\begin{Bmatrix} \text{patterns.} \\ \text{samples.} \\ \text{catalogs.} \end{Bmatrix}$

3. The quality of the substitute is even better than the one you want, while the price is same.

4. Quality of item 1 is equally $\begin{Bmatrix} \text{attractive.} \\ \text{hardwearing.（耐穿的）} \\ \text{serviceable.（可用的）} \\ \text{water-repellent.（不透水的）} \end{Bmatrix}$

六、條 件

1. Upon receipt of your $\begin{Bmatrix} \text{L/C} \\ \text{check} \\ \text{remittance} \end{Bmatrix}$, we will $\begin{Bmatrix} \text{ship} \\ \text{deliver} \\ \text{despatch} \end{Bmatrix}$ the goods immediately.

2. Your order No....will be ready for shipment by end of this month. Please let us have your instructions regarding shipping marks together with an L/C to cover this order.

3. In order to process this order, however, we have to ask you to open L/C immediately.

4. We will make every effort to effect shipment to meet your requirements.

七、結尾句

1. We hope that this order will be just the $\begin{Bmatrix} \text{beginning} \\ \text{first} \end{Bmatrix}$ of many and that we may have many years of pleasant business relations together.

（希望本批訂單只是一個開頭，並企望有長年、愉快的業務關係。）

2. You may $\begin{Bmatrix} \text{rest assured} \\ \text{be sure} \end{Bmatrix}$ that we will do our best to $\begin{Bmatrix} \text{fill this order} \\ \text{satisfy you} \end{Bmatrix}$ and we hope that

→ this will be the $\begin{Bmatrix} \text{beginning} \\ \text{first} \end{Bmatrix}$ of a long and $\begin{Bmatrix} \text{happy} \\ \text{pleasant} \end{Bmatrix}$ $\begin{Bmatrix} \text{co-operation.} \\ \text{association.} \end{Bmatrix}$

3. We will be pleased to receive your $\begin{Bmatrix} \text{repeat} \\ \text{continued} \\ \text{further} \end{Bmatrix}$ orders, which will always have our

→ $\begin{Bmatrix} \text{best} \\ \text{prompt} \\ \text{careful} \end{Bmatrix}$ attention.

4. $\begin{Bmatrix} \text{Thank you} \\ \text{Many thanks} \end{Bmatrix}$ for this order and we hope to serve you again in the near future.

5. $\begin{Bmatrix} \text{Please be assured} \\ \text{You may be assured} \end{Bmatrix}$ that we shall $\begin{Bmatrix} \text{do our best} \\ \text{spare no effort} \\ \text{make every effort} \\ \text{do everything possible} \\ \text{do what we can} \end{Bmatrix}$ to satisfy your

→ $\begin{Bmatrix} \text{wishes.} \\ \text{requirements.} \\ \text{customers.} \end{Bmatrix}$

$\begin{Bmatrix} 請相信 \\ 請放心 \end{Bmatrix}$ 我們會 $\begin{Bmatrix} 盡力 \\ 不遺餘力 \\ 盡一切力量 \\ 儘可能 \\ 盡我們能力 \end{Bmatrix}$ ……

6. We sincerely regret that we are unable to serve you $\begin{Bmatrix} \text{at this time.} \\ \text{in this instance.} \end{Bmatrix}$

註：in this instance：在此刻；在這種情況下。

7. As soon as conditions become normal again, we will be pleased $\begin{Bmatrix} \text{to write you.} \\ \text{to contact you.} \end{Bmatrix}$

8. We trust that you will understand that it is not lack of co-operation and good-will but sheer necessity which makes it impossible for us to meet your wishes in this case.

（我們相信你當會了解此次我們無法達到你的願望，並不是由於欠缺合作與善意，而是完全由於不得已的苦衷。）

9. We thank you very much for this business and trust that its success will induce you to favor us with further orders.

10. We believe this treatment of your order will be entirely satisfactory, and hope to be favored with further requirements.

（我們相信這樣處理你的訂單將使你完全滿意，並希望繼續惠顧。）

第二十章
買賣契約書 (Sales Contract)

第一節　買賣契約書的簽立方法

任何一筆國際買賣，經過上述有關各章所述詢價、報價、往返還價，最後經接受；或訂貨經確認，買賣契約即告成立。

就契約本身而言，報價與接受，或訂貨與確認，其所用的函件、電傳，均為契約成立的證據 (Evidence)。至於買賣契約成立後，買方所發出的訂單，或賣方所發出的售貨確認書，甚至買賣雙方另訂立的買賣契約書 (Sales contract) 只不過是契約的證據而已。無 Order, Sales confirmation 或 Sales contract 固不妨礙契約的成立，但有了這些，則顯得更為週到。在實務上，買賣雙方對於交易的意思表示，常以電傳為之，而電傳中有關交易條件，通常僅限於交易商品、品質、數量、價格、交貨期及付款條件等而已。對於包裝、刷嘜、保險、檢驗、索賠、仲裁、不可抗力以及其他關係雙方權利義務的各種條件，往往略而不提。因此為免日後發生糾紛，或解決糾紛時有所依據，進出口雙方均宜在成立契約後，另以書面互為確認，或另簽訂契約書，以證實或補充契約內容。

實務上，買賣契約書的簽立方法，可分為二：

一、以書面確認方式代替買賣契約書

交易成立之後，由當事人的一方，將交易內容製成確認書寄交對方。這種確認書由賣方發出的稱為 "Sales confirmation"（售貨確認書）或 "Sales note"（售貨單）。由買方發出的，稱為 "Order"，"Sheet"，"Purchase order" 或 "Purchase confirmation"。

二、正式簽立買賣契約書

交易成立後，也可由當事人的一方將交易內容製成正式契約書，然後由雙方共同簽署。這種契約書由出口商草擬時，往往稱為 "Sales contract" 或 "Export contract"。如由進口商草擬時往往稱為 "Purchase contract" 或 "Import contract"。

　　確認書與契約書不同，前者，僅為單方的確認，而後者則由雙方共同簽署。然而，當一方提出確認書時，仍可寄出兩份，請對方簽字後寄回一份，如此則確認書不但具有與契約書同等的效力，而且也已具備契約書的效力。

　　較單純或金額較小的交易多採用確認方式簽約，但交易性質較複雜或金額較大的則往往採取契約書方式以昭慎重。

　　關於確認書，在前幾章已陸續敘及，本章擬舉一買賣契約書為例，說明其內容及有關用語。

　　買賣契約書的內容，因交易性質、交易貨品等的不同而異，但較完整的契約應包括下列各項：

1. Title（契約名稱）。
2. Preamble（前文）。
3. Body $\begin{cases} (1) \text{ basic terms and conditions。} \\ (2) \text{ general terms and conditions。} \end{cases}$
4. Witness clause（結尾）。

第二節　買賣契約書實例

No. 144　買賣契約書

CONTRACT

　　This contract is made this 15th day of July, 20– by ABC Corporation (hereinafter referred to as "SELLERS"), a Chinese corporation having their principal office at 19 Wu Chang St., Sec. 1, Taipei, Taiwan, Republic of China, who agree to sell, and XYZ Corporation (hereinafter referred to as "BUYERS"), a New York corporation having their principal office at 30 Wall St., New York, NY, USA, who agree to buy the following goods on the terms and conditions as below:

1. COMMODITY: Ladies double folding umbrellas.

2. QUALITY: 2 section shaft, iron and chrome plated shaft, unichrome plated ribs, siliconed coated waterproof plain nylon cover with same nylon cloth sack.
   size: $18\frac{1}{2}''\times10$ ribs
   as per sample submitted to BUYERS on June 30, 20–.

3. QUANTITY: 10,000 (Ten thousand) dozen only.

4. UNIT PRICE: US$14 per dozen CIF New York.

Total amount: US$140,000 (Say US Dollars one hundred forty thousand only) CIF New York.

5. PACKING: One dozen to a box, 10 boxes to a carton.

6. SHIPPING MARK:

NEW YORK
No. 1 & Up

7. SHIPMENT: To be shipped on or before December 31, 20– subject to acceptable L/C reaches SELLERS before the end of October, 20–, and partial shipments allowed, transhipment allowed.

8. PAYMENT: By a prime banker's irrevocable sight L/C in SELLERS' favor, for 100% value of goods.

9. INSURANCE: SELLERS shall arrange marine insurance covering ICC (B) plus TPND and war risk for 110% of the invoice value and provide for claim, if any, payable in New York in US currency.

10. INSPECTION: Goods is to be inspected by an independent inspector and whose certificate inspection of quality and quantity is to be final.

11. FLUCTUATIONS OF FREIGHT, INSURANCE PREMIUM, CURRENCY, ETC.:

(1) It is agreed that the prices mentioned herein are all based upon the present IMF parity rate of NT$35 to one US Dollar. In case, there is any change in such rate at the time of negotiating drafts, the prices shall be adjusted and settled according to the corresponding change so as not to decrease SELLERS' proceeds in NT Dollars.

(2) The prices mentioned herein are all based upon the current rate of freight and/or war and marine insurance premium. Any increase in freight and/or insurance premium rate at the time of shipment shall be for BUYERS' risks and account.

(3) SELLERS reserve the right to adjust the prices mentioned herein, if prior to delivery there is any substantial increase in the cost of raw material or component parts.

12. TAXES AND DUTIES, ETC.: Any duties, taxes or levies imposed upon the goods, or any packages, material or activities involved in the performance of the contract shall be for account of origin, and for account of BUYERS if imposed by the country of destination.

13. CLAIMS: In the event of any claim arising in respect of any shipment, notice of intention to claim should be given in writing to SELLERS promptly after arrival of the goods at the port of discharge and opportunity must be given to SELLERS for investigation. Failing to give such prior written notification and opportunity of investigation within twenty-one (21) days after the arrival of the carrying vessel at the port of discharge, no claim shall be entertained. In any event, SELLERS shall not be responsible for damages that may result from the use of

goods or for consequential or special damages, or for any amount in excess of the invoice value of the defective goods.

14. FORCE MAJEURE: Non-delivery of all or any part of the merchandise caused by war, blockage, revolution, insurrection, civil commotions, riots, mobilization, strikes, lockouts, act of God, severe weather, plague or other epidemic, destruction of goods by fire or flood, obstruction of loading by storm or typhoon at the port of delivery, or any other cause beyond SELLERS' control before shipment shall operate as a cancellation of the sale to the extent of such non-delivery. However, in case the merchandise has been prepared and ready for shipment before shipment deadline but the shipment could not be effected due to any of the abovementioned causes, BUYERS shall extend the shipping deadline by means of amending relevant L/C or otherwise, upon the request of SELLERS.

15. ARBITRATION: Any disputes, controversies or differences which may arise between the parties, out of, or in relation to or in connection with this contract may be referred to arbitration. Such arbitration shall take place in Taipei, Taiwan, Republic of China, and shall be held and shall proceed in accordance with the rule of the Arbitration Association of Republic of China.

16. PROPER LAW: The formation, validity, construction and the performance or this contract are governed by the laws of Republic of China.

IN WITNESS WHEREOF, the parties have executed this contract in duplicate by their duly authorized representative as on the date first above written.

BUYERS	SELLERS
XYZ CORPORATION	ABC CORPORATION
Manager	Manager

【註】

1. This contract is made...as below：這一段文字即為 preamble clause。"is made" 為締結之意，也可以 "is enter into" 代替。"this 15th day of July, 20–" 指締約日期，注意前面並無 "on" 一詞，假如要用 "on" 則應改為 "on July 15, 20–"。"by ABC Corporation..., and XYZ Corporation" 也可以 "between ABC Corporation..., and XYZ Corporation" 代替。
   hereinafter：在下文中。cf. hereinafter called "Buyer"（以下稱買方）。
   principal office：主要辦公室，即指總公司。

2. 本契約條款 1 至 9 即為本契約本文 (body) 的基本條款 (basic terms and conditions)，其內容與確認書的內容並無兩樣。

3. inspection：檢驗。
   independent inspector：獨立公證行。

4. IMF Parity = International Monetary Fund Parity：國際貨幣基金平價。

5. current rate：目前匯率。

6. for buyers' risks and account：風險及費用由買方負擔。

7. reserve the right：保留（調整的）權利。

8. component parts：組件；配件。

9. levies (pl.), levy (sing.)：課徵；賦課。

10. impose upon：課徵。

　　cf. impose upon 又解作「占⋯⋯的便宜」；「欺騙」。

　　　Don't let the children impose upon you.〔別讓小孩占你的便宜（別縱容小孩）。〕

11. in respect of：關於。

12. given in writing：以書面通知。

13. at the port of discharge：在卸貨港。

14. no claim shall be entertained：索賠不予受理。

15. consequential loss：間接損失；"consequential damages" 為間接損害。

16. defective goods：瑕疵貨物。

17. force majeure：不可抗力。

契約書

本契約由 ABC 公司——總公司設於中華民國臺灣省臺北市武昌街 1 段 19 號（以下簡稱賣方）與 XYZ 公司——總公司設於美國紐約州紐約市華爾街 30 號（以下簡稱買方）於 20—年 7 月 15 日訂定，雙方同意按下述條件買賣下面貨物：

1. 貨物：女用雙折洋傘。

2. 品質：雙節鐵質鍍鉻傘柄，單面鍍鉻傘骨架，塗矽銅防水素色尼龍傘布，同質尼龍布護套。

　　尺寸：$18\frac{1}{2}$ 英寸，10 支骨架。

　　依 20—年 6 月 30 日提供予買方的樣品為準。

3. 數量：10,000 打。

4. 單價及總金額：打 US$14 CIF 紐約，總金額 US$140,000 CIF 紐約。

5. 包裝：1 打裝 1 紙盒，10 紙盒裝 1 紙箱。

6. 裝船嘜頭：

NEW YORK
No. 1 & Up

7. 裝運：20—年 12 月 31 日前裝運，但以可接受的信用狀於 20—年 10 月底前開到賣方為條件，容許分批裝運及轉運。

8. 付款：憑一流銀行的不可撤銷的信用狀付款，信用狀以賣方為受益人，並照貨物金額 100% 開發。

9. 保險：賣方應洽保水險，投保 B 款險並加保遺失竊盜險及兵險，保險金額按發票金額的 110% 投保，並須規定如有索賠應在紐約以美金支付。

10.檢驗: 貨物須經一家獨立公證行檢驗,其出具品質及數量檢驗證明書應為最後認定標準。

11.運費、保險費、幣值等的變動:

⑴茲同意本契約內所列價格全是以目前國際貨幣基金平價匯率臺幣 35 元兌換美金 1 元為準。倘若這項匯率在押匯時有任何變動,則價格應根據這項變動比照調整及清償,俾賣方的臺幣收入不因而減少。

⑵契約中所列價格全是以目前運費率及(或)兵險和水險保險費率為準。裝運時運費率及(或)保險費率如有增加,應歸由買方負擔。

⑶交貨前如原料及組成配件的成本增加甚鉅,賣方保留調整契約中所列價格的權利。

12.稅捐等: 對於貨物或包件、原料,或履行契約有關活動所課徵的稅捐或規費,如由產地國課徵,歸由賣方負擔;如由目的國課徵,則歸由買方負擔。

13.索賠: 對所裝貨物如有索賠情事發生,則請求索賠的通知必須於貨物抵達卸貨港後即刻以書面提示賣方,並且必須給賣方有調查的機會。倘若運送船隻到達卸貨港後 21 天內沒有提示這項預先的書面通知以及提供調查機會,則索賠應不予受理。在任何情況下,賣方對於使用貨物引起的損害,或對於間接或特別的損害,或對於超出瑕疵貨物發票金額的款項均不負責。

14.不可抗力: 因戰爭、封鎖、革命、暴動、民變、民眾騷擾、動員、罷工、工廠封鎖、天災、惡劣氣候、疫病或其他傳染病、貨物因火災或水災而受毀壞、在交貨港因暴風雨或颱風而阻礙裝船,或在裝船前任何其他賣方所無法控制的事故發生,而致貨物的全部或一部分未能交貨,這未交貨部分的契約應予取消。然而,在裝運期限截止前,如貨物業經備妥待運,但因前述事故之一發生而致未能裝運,則買方於接到賣方請求時,應以修改信用狀方式或其他方式延長裝運期限。

15.仲裁: 有關本契約買賣雙方間所引起的任何糾紛、爭議或歧見,可付諸仲裁。這項仲裁應於中華民國臺灣臺北舉行,並應依中華民國仲裁協會規則處理及進行。

16.適用法: 本契約的成立、效力、解釋,以及履行均受中華民國法律管轄。

本契約書兩份業經雙方法定代理人訂定,於前文日期簽署。

買方	賣方
XYZ 公司	ABC 公司
經理	經理

第二十一章
信用狀交易 (Letter of Credit Transaction)

第一節　信用狀的意義及其關係人

一、信用狀的意義

信用狀是銀行循進口商的要求及指示，向出口商發出的文書，銀行在此文書中，與出口商約定：只要出口商提出合乎該文書中所規定的單據或匯票，即將妥予兌付。信用狀，英文稱為 "Letter of credit"，或簡稱為 "Credit" 或 "L/C"。

二、信用狀交易的關係人 (parties concerned under L/C transaction)

信用狀交易的關係人，至少有三方，即開狀申請人，開狀銀行及受益人。除此以外，往往尚有其他關係人介入。

1. Applicant for credit（開狀申請人），即向銀行申請開發 L/C 的人，通常即為 Buyer 或 Importer。開狀申請人在 L/C 中尚有下列幾種稱呼：Accountee（被記帳人），Grantee（授與人），Accreditor（授信人），Account party（被記帳的一方），Consignee（收貨人），Opener（開狀人）。

2. Opening bank（開狀銀行）又稱 Issuing bank, Establishing bank, Issuer, Grantor。

3. Beneficiary（受益人）通常即為賣方或出口商，又稱 Shipper（裝貨人），Accreditee（受信人），Addresee（擡頭人），Accredited party（受信的一方），Drawer, User。

4. Advising bank（轉知銀行）即轉知 L/C 的銀行，又稱為 Notifying bank, Transmitting bank。

5. Negotiating bank（押匯銀行）即自出口商購入信用狀項下匯票或單據的銀行，又稱 Discount bank。

6. Confirming bank（保兌銀行）即應開狀銀行的要求，就所開 L/C 承擔兌付的銀行，保兌銀行一經保兌 L/C，即須承擔與開狀銀行相同的責任。

7. Paying bank（付款銀行）又稱為 Drawee bank，即信用狀中規定匯票的被發票人 (Drawee)，付款銀行可能為開狀銀行本身，也可能為其他銀行。

8. Reimbursing bank（歸償銀行）又稱為 Clearing bank，即應開狀銀行的囑託，對押匯銀行償還其押匯票款的銀行。

茲將信用狀項下交易的關係人圖示於下。

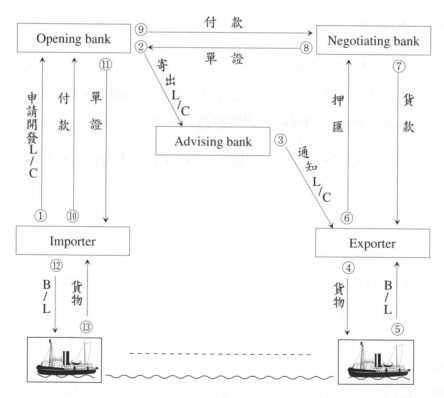

第二節　信用狀的種類

1. 依可否片面撤銷分：

⑴ Revocable L/C（可撤銷信用狀）：信用狀開出後，在受益人未押匯前，可隨時

片面地予以修改或撤銷，而不需有關當事人同意的信用狀。

⑵ Irrevocable L/C (不可撤銷信用狀)：信用狀受益人收到信用狀後，非經受益人、開狀銀行及保兌銀行（如經保兌）同意，不得隨意修改或撤銷的信用狀。

2.依有無保兌分：

⑴ Confirmed L/C（保兌信用狀）：經開狀銀行以外的另一銀行保兌的信用狀。

⑵ Unconfirmed L/C（未保兌信用狀）：未經開狀銀行以外的銀行保兌的信用狀。

3.依匯票期限分：

⑴ Sight L/C (即期信用狀)：規定受益人須憑即期匯票 (Sight draft) 取款的信用狀。

⑵ Usance L/C (遠期信用狀)：規定受益人須憑遠期匯票 (Usance draft) 取款的信用狀。

4.依可否押匯 (Negotiation) 分：

⑴ Negotiation L/C（可押匯信用狀）：允許受益人向付款銀行以外的銀行請求押匯的信用狀，又可分為：

　① General L/C（一般信用狀）：不限定押匯銀行。

　② Restricted L/C 或 Special L/C（限押信用狀）：限定押匯銀行的信用狀。

⑵ Straight L/C（直接信用狀）：規定受益人只能向 L/C 所指定銀行提示匯票或單證請求付款的信用狀。

5.依是否附有跟單分：

⑴ Clean L/C（光禿信用狀）：押匯或請求付款時，不需附上單證的信用狀。

⑵ Documentary L/C（跟單信用狀）：押匯或請求付款時，須附上一定單證的信用狀。

6.依可否轉讓分：

⑴ Transferable L/C（可轉讓信用狀）：受益人可將 L/C 轉給他人使用的信用狀。

⑵ Non-transferable L/C（不可轉讓信用狀）：不可將 L/C 轉給他人使用的信用狀。

7.依用途分：

⑴ Traveler's L/C（旅行信用狀）：專供作旅行用的信用狀。

⑵ Commercial L/C（商業信用狀）：專供貿易用的信用狀。

8.依主從分：

⑴ Master L/C（主信用狀）：由進口商開給出口商的信用狀。

(2) Back-to-back L/C（背對背信用狀）：又稱 Local L/C，出口商憑 Master L/C 另開給工廠或供應商的信用狀。

9. 依可否循環使用分：

(1) Revolving L/C（循環信用狀）：可循環使用的信用狀。

(2) Non-revolving L/C（不循環信用狀）：不可以循環使用的 L/C。

10. Stand-by L/C（擔保信用狀）：專供保證用的信用狀。

11. Red clause L/C（紅條款信用狀）：受益人在未提示 L/C 所規定單證之前，即可預支款項的信用狀。

12. Teletransmitted L/C（電傳信用狀）：以電傳開出的信用狀。如以郵遞方式開出者稱為 Mail L/C（郵遞信用狀）。

第三節　有關信用狀交易的信函

No. 145　催促進口商速開信用狀

Dear Sirs,

　　We wrote you a letter dated June 25, 20– confirming the receipt of your order for 1,000,000 sq. ft. of Lauan Plywood. In that letter, we enclosed a copy of your Purchase Order duly signed by us.

　　The Purchase Order stipulates shipment to be effected during August and L/C should reach us by the end of July. However, as of this date, we don't appear to have received your L/C. In order to book the shipping space at an earlier date, you are requested to have the L/C opened immediately. For this matter, we have despatched a telex to you today as follows:

　　"ELCEE FOR LAUAN PLYWOOD UNRECEIVED YET PLS OPEN IMMDLY"

　　We appreciate your immediate attention to this matter.

　　　　　　　　　　　　　　　　　　　　　　　　　　　　Yours very truly,

【註】

　1. as of this date：到今天為止。

　2. we don't appear to have received：我們似乎還沒有收到，是一種客氣的措詞，比 "we don't have received" 要客氣。

　3. ELCEE：即 L/C 的譯音。

4. UNRECEIVED：即 not received。

5. PLS：即 "Please" 的電傳交換文體 (telexese)。

6. IMMDLY：即 "immediately" 的 telexese。

No. 146　再催促進口商立即開發信用狀，否則取消訂單

Dear Sirs,

Your Order No. 123

We invite your attention to the fact that the L/C covering your order No. 123 has not reached us in spite of our repeated requests.

As we have specifically induced our maker to accept this order for an earlier shipment (than originally offered) as requested by you, on condition that the covering L/C be arrived here by the end of July, please do not fail to arrange an irrevocable L/C to be established immediately by cable (or teletransmission).

If your L/C fails to reach us by the end of July, we may be forced to cancel your order. However, we would prefer you to issue your L/C as stipulated so that we can continue our usual friendly business relations. For your information, there is every indication here that the prices show an advancing tendency.

Yours faithfully,

【註】

1. invite your attention to：請你注意，一種客氣的說法，也可以 "draw your attention to" 代替。

2. in spite of：不管；不顧 (notwithstanding)。

3. repeated requests：一再請求。

4. be forced = be compelled：被迫。

5. every indication：充分的跡象。

6. advancing tendency：上漲趨勢。

No. 147　說明廠家未收到信用狀之前不肯備貨

Dear Sirs,

We thank you for your letter of September 20.

Concerning your remark about the terms in your letter, we understand that customers are reluctant to open L/C until the last minute before shipment, but we ask your co-operation in getting them established within 14 days after the confirmation of sales.

In the first place, makers here would not proceed with any order until the arrival of the L/C for fear of possible cancellation. Their bitter experiences taught them to take this precaution.

The L/C will also help them financially, for it enables us to pay them in advance about 80% of the contracted amount. This favorably affects the prices they offer.

We ask you, therefore, to convince your customers that the sooner the L/C is opened, the earlier the shipment can be made and the better prices can be obtained.

Sincerely yours,

【註】

1. reluctant：不情願。

2. proceed with：著手；開始；處理。

3. bitter experience：痛苦的經驗。

4. take this precaution：預防；警戒。

cf. take precautions against：預防……；小心……。

No. 148　貨已備妥，催促進口商速開信用狀

Dear Sirs,

Your Order For Canned Asparagus

We refer to our letter to you dated June 1 confirming the captioned Order and trust that you would have already received it. In that letter, we have assured you that we can effect shipment of the total quantity of 900 cartons under the subject order during August provided your L/C reaches us 30 days prior to shipment.

Now the goods are ready for shipment and your L/C has not yet been received. In this regard, we despatched a telex to you on August 4, reading:

RYL MAY 25 OUR SALES NO. 89/345 CANNED ASPARAGUS READY SHIPMENT PLS OPEN LC BY CABLE IMMDLY

In order to enable us to ship the goods in time, we shall appreciate your complying with our request to have the L/C opened by cable immediately. For your information, market tone indicates prices are expected to advance before long.

Yours sincerely,

【註】

1. refer to：關於。

2. provided = if：*以……為前提*；*假使*。

3. in this regard：*關於此*。

4. RYL = refer to your letter。

5. market tone：*市況*；*市場景況*。*又寫成* "the tone of a market"。

6. before long = soon：*不久*。

No. 149　通知進口商速開信用狀以期能如約交貨

Dear Sirs,

　　This is to remind you that L/Cs to cover the following orders from you, calling for June shipment, will have to be opened soon:

Sale No.	Article	Quantity	Amount
A–122	Boys Shirts	624 doz.	US$7,488.00
A–123	Girls Shirts	185 doz.	US$1,921.50

　　According to our contracts with you, these L/Cs are supposed to be opened by the end of April. We shall be glad if you will arrange with your bank for opening them without delay so that we may ship the goods in June as contracted.

Yours faithfully,

【註】

1. to remind：*提醒*。

2. L/Cs 為 L/C 的多數形。letters of credit 的縮寫應寫成 Ls/C，但很多人寫成 L/Cs，其他類似情形如：B/Ls (Bills of Lading)，正確的寫法為 Bs/L。

3. call for：*要求*；*規定*。

4. be supposed to：*要*；*應該*。

No. 150　通知進口商迄未收到信用狀

Dear Sirs,

　　We don't appear to have received your L/C to cover 1,000 doz. of Ladies' Tetron/Cotton Blouses under your order No. A–543 dated January 10, 20–.

　　If you have not yet open the relative L/C, will you please do so immediately to enable us to effect shipment in April as arranged.

　　We are looking forward to your early advice by cable on the above matter.

Yours faithfully,

【註】

1. Tetron/Cotton Blouses: 帝特龍與棉花混紡的罩衫。

2. relative L/C: 相關的信用狀; 相應的信用狀。

No. 151　進口商向銀行申請開發信用狀

First Commercial Bank　　　　　　　　　　　　　　　　March 25, 20–
Foreign Dept.
Taipei, Taiwan

　　We enclose one copy of Application for Issuing Letter of Credit together with Exchange Settlement Certificate for Letter of Credit and shall appreciate your arranging to issue an L/C in favor of XYZ Motor Company at your early convenience.

　　Please note that all the details relating thereto are given in the Application. Meanwhile, you are authorized to charge our account with you any drawing under the L/C and your charges under advice to us.

Encl.: a/s　　　　　　　　　　　　　　　　　　　　　　　　Yours faithfully,

【註】

1. Application for Issuing Letter of Credit: 開發信用狀申請書。

2. Exchange Settlement Certificate for Letter of Credit: 信用狀項下結匯證實書。

3. 就我國目前情形而言,進口商請銀行開發信用狀時,並不另寫信給銀行。在美國則很多情形是將開狀申請書附上一封信,逕寄(或由信差 Messenger)銀行,即由銀行代開信用狀,公司的經辦人員不必到銀行洽辦。

No. 152 開發信用狀申請書

國外營業部或外區指定單位					
經理	副理	科長 襄理	副科長	經辦	

開發信用狀申請書
APPLICATION FOR ISSUING AN
IRREVOCABLE DOCUMENTARY CREDIT

受 理 單 位			
經理／主任	副襄理	經辦	

申請日期　　年　　月　　日

第一商業銀行 台照
TO: FIRST COMMERCIAL BANK

受理單位：　　　　　　辦、分行
辦事處

茲請貴行依下列條款開發不可撤銷信用狀一份
I/WE HEREBY REQUEST YOU TO ISSUE AN IRREVOCABLE
DOCUMENTARY CREDIT UPON THE FOLLOWING TERMS
AND CONDITIONS

[20] 信用卡號碼 Credit No.（由銀行填寫）, [31C] 日期 Date

[50] 申請人 Applicant（英文名稱及地址）

通知銀行 Advising Bank（倘未指定，則由銀行填寫）

To be advised by
□航郵 Airmail □簡電 Brief Cable □全電 Full Cable

[31D] 信用狀有效日期及地點 Expiry date and place

[59] 受益人 Beneficiary

[32B] 信用狀金額（小寫）：
Amount Say（大寫）：

左側直行說明：
一、請注意填列各點，並詳如銀行社址如因國保之外的信用狀M發證件將發生困難時，本行恕不負責。
二、航郵開發及電報開發，如本社如未填一項，概以「航郵」開發。
三、Partial shipments（分批裝運）是准許和本社如未填一項，概以「准許」開發。
四、Transhipment（轉運）是准許和本社如未填一項，概以 Marine Bill of Lading 時概以「不准許」開發。

[41m] 以讓購／付款／承兌／延期付款方式在任一銀行／通知銀行使用依受益人依商業發票全額簽發以貴行／貴行國外通匯行為付款人之匯票，並於
Available with Any Bank/Advising Bank by Negotiation/Payment/Acceptance/Deferred Payment of beneficiary's draft at

[42] □見票／提單簽發 .... 日後付款，並須符合下列件 ⊠記號之條件和檢附下列 ⊠記號之各項單據
□ Sight □____days after sight/shipment date for full invoice value drawn on you/your correspondent against the
following conditions and documents required:(marked with ⊠)

[78] 對付款／承兌／讓購銀行之指示
Instructions to the Paying/Accepting/Negotiating Bank:
延期信用狀利息由申請人／受益人負擔
Interest are for □ Applicant's □ Beneficiary's account.

□對外開發即期信用狀，惟對內向　貴行融資　　　　天。
□對外開發受益人信認　　　天利息之遠期信用狀，另
自匯票到期日起向　貴行融資　　　天。

[43P] 分批裝運 Partial shipments: □准許 Allowed □不准許 Prohibited

[43T] 轉運 Transhipment: □准許 Allowed □不准許 Prohibited

[44] 裝載港、交貨地·卸貨港、目的地；最後裝運日
Shipment from ____ for transportation to ____ not later than ____

[45A] 貨物內容 Covering:（請概括摘列，勿太冗長，但仍須能可能細記物品之數量及單價）
□ FOB □ CFR □ CIF □ FAS □ CIP □ ____（價格條件）

所需單據 Documents Required:

□ 1. 商業發票六份標明本信用狀號碼
Signed commercial invoice in six copies indicating this credit number.

□ 2A. 全數減一份海運提單以貴行為抬頭人，以繳處為被通知人，註明運費待付／付訖，並標明本信用狀號碼
Full set less one of clean on board marine Bills of Lading made out to the order of FIRST COMMERCIAL BANK notify
applicant, marked "Freight Collect/Prepaid" and indicating this credit number.

□ 2B. 全套海運提單以貴行為抬頭人，以繳處為被通知人，註明運費待付／付訖，並標明本信用狀號碼。
Full set of clean on board marine Bills of Lading made out to the order of FIRST COMMERCIAL BANK notify applicant.
marked "Freight Collect/Prepaid" and indicating this credit number.

□ 3. 空運提單以貴行為抬頭人，以繳處為被通知人，註明運費待付／付訖，並標明本信用狀號碼。
Clean air waybills consigned to FIRST COMMERCIAL BANK notify applicant, marked "Freight Collect/Prepaid" and
indicating this credit number.

□ 4. 快遞／海運郵包／空運郵包收據，以繳處為被通知人，註明快遞費／郵費　待付／付訖，並標明本信用狀號碼。
Courier/Sea parcel post/air parcel post receipt showing applicant as addressee marked "Courier charge/Postage Collect/
Prepaid" and indicating this credit number.

□ 5. 照發票金額百分之壹百壹拾投保以全套正本保險單，註明以同種貨幣賠償在台灣給付並作空白背書，其保險範圍包括：
Insurance policy or certificate all the originals, endorsed in blank for 110% of invoice value stipulating that claims are
payable in Taiwan in the same currency and incluling:
□ 1982 Institute Cargo Clauses (A)　　　1982 協會貨物保險條款(A)　　　Institute Cargo clauses 水險
□ 1982 Institute Cargo Clauses (B)　　　1982 協會貨物保險條款(B)　　　□F.P.A. 平安險 □W.A. 水漬險
□ 1982 Institute Cargo Clauses (C)　　　1982 協會貨物保險條款(C)　　　□All Risks 全險
□ 1982 Institute Cargo Clauses (Air)　　1982 協會貨物保險條款(航空險)　□Air Risks 航空險
□ 1982 Institute War Clauses (Cargo)　　1982 協會貨物保險兵變條款　　　□War Risks 兵險
□ 1982 Institute Strikes Clauses (Cargo)　1982 協會貨物保險罷工條款

□ 6. 包裝單 Packing list ____ signed by beneficiary.

□ 7. 其他單據 Others

□ 8A. 受益人證明書，證明各單據副本和一份正本運送單據已由受益人直接以航郵寄交信用狀申請人。
A Certificate signed by beneficiary stating that one non-negotiable set of the stipulated documents and one original
transport documents have been airmailed directly to the applicant.

□ 8B. 受益人證明書，證明各單據副本已由受益人直接以航郵寄交信用狀申請人。
A Certificate signed by beneficiary stating that one non-negotiable set of the stipulated documents has been airmailed
directly to the applicant.

[47A] 附特別條款如下 Special Instructions：
□ 本信用狀可轉讓並限由通知銀行辦理轉讓　　　This Credits is transferable and to be transferred by Advising Bank only.

左側直行說明續：
五、裝運港包括卸貨何方列者，受益人負擔，概以三個月計算。
六、開發信用狀申請書條款中紅字部份，務請填列其某一成期除不必委者，倘憑以開發。

[71B] 費用：所有國外費用包含補償銀行費用均由申請人／受益人負擔。
Charges：All Banking charges including Reimbursing Bank's charges outside Taiwan, If any, Are for □ Applicant's
□ Beneficiary's account.

[48] 提示期間：單據須於貨物裝運日後____日內且於本信用狀有效日期前提示。
Presentation period：Documents to be presented within____days after the date of shipment but within the validity
of this credit.

[49] 保兌 Confirmed：保兌費用由申請人／受益人負擔 Confirming charges are for □ Applicant's □ Beneficiary's account.

[53s] 補償銀行 Reimbursement bank（由銀行填寫）

迅006（26×42公分）88.9. 1,000 本

右側直行：
主管核對打字經辦

No. 153　臺灣的進口商通知國外出口商已開出信用狀

XYZ Motor Company   March 28, 20–

Pony Pick Up

Dear Sirs,

Your letter of March 10 confirming the order we placed with you for the subject Pick Up has been received. We are pleased to inform you that application for import licence has been approved.

On March 25, an L/C was opened in your favor for an amount of US$55,840.00 to cover the CIF value of this order by First Commercial Bank through Korean Exchange Bank, Seoul. For your information, the numbers of the import licence and the L/C are 66DHI/–003690 and 7DHI/00123/01 respectively. It will be appreciated if you will arrange to ship the total quantity not later than May 30, 20–.

Enclosed for your reference is one copy of the relative L/C we have opened for this order.

Yours sincerely,

【註】

1. application for import licence：輸入許可證申請書。在我國，官方的稱法是 application for import permit。輸入許可證在貿易界多稱做 IL（即 Import Licence）或 CBC（即 Central Bank of China 的簡稱），輸出許可證也常稱為 CBC。

2. Enclosed for your reference is one copy...：茲附上 (L/C) 抄本一份供參考。

No. 154　通知出口商已開出信用狀

Dear Sirs,

Subj: Our Order for Lauan Plywood

We confirm having received your telex and letter both dated August 2.

We take pleasure to inform you that we arranged with our bank for establishment of an L/C in your favor to cover the total value of subject order on August 5 and that the bank had it opened on August 6. We trust that you have received it by now from the advising bank on your side.

For your reference, we are enclosing a copy of the L/C. Thank you for your co-operation and look forward to your timely shipment.

Yours faithfully,

【註】

　1. on your side = at your end：在你那一方。

　2. timely shipment：適時的裝運。

No. 155　國外進口商通知本地出口商已開出信用狀 (See No. 78)

Dear Sirs,

Our Order for Canned Asparagus A1 Grade

　We have received your telex and letter both dated August 5, the contents of both have received our immediate attention.

　As requested, we have instructed the Dresdner Bank to establish by telex an L/C in your favor to cover the total value of this order. We trust that you will receive advice from the advising bank at your end very soon.

　Enclosed, for your reference and perusal, please find one copy of the relative L/C opened by our bank.

Faithfully yours,

Encl.: a/s

【註】

　1. as requested：遵囑；依照請求。

　2. perusal：細閱。

　3. enclosed please find... 的表現法雖然有些人認為是舊式的、陳腐的，但是到現在為止，用的人還很多，而且用起來也很方便。

　　cf. Enclosed for your reference is one copy of the L/C which we have opened for this order.

No. 156　通知銀行通知本地出口商收到電傳信用狀 (See No. 155)

Bank of Taiwan

Head Office, Foreign Department

Our Ref. No. 77/4213

Taipei, Taiwan
August 10, 20–

To: Taiwan Trading Co., Ltd.
　P.O. Box No. 123
　Taipei, Taiwan

Original credit issued by
  (Their Ref. No. 123   Dated:)
Advised by: (Ref. No.   Dated:)

                                           Amount: EUR28,950.00

Dresdner Bank Hamburg, Germany

                                           Shipping date:
                                           Expiry date:
                                           Covering:

Dear Sirs,

    We wish to inform you that we have received a telex from our correspondent (IT 12145) dated August 6 which decoded reads as follows:

    ORDER XYZ A.G. HAMBURG AIRMAILING CREDIT 123 EUR28,950 FAVOR TAIWAN TRADING COLTD POBOX 123 TAIPEI COVERING 900 CARTONS CANNED ASPARA-GUS AI GRADE

    Please note that this is merely an advice on our part and does not constitute a confirmation of this credit. This bank is unable to accept any responsibility for errors in transmission or translation of this cable or for any amendments that may be necessary upon receipt of the mail advice from the bank issuing this credit.

    Kindly acknowledge receipt by returning the attached form duly signed, and oblige.

                                     Yours faithfully,
                                     For the Bank of Taiwan

                                 _____
                                   AUTHORIZED SIGNATURE

【註】

    1. decode：將電傳暗碼譯成明碼稱為 decode，將明碼譯成暗碼稱為 code。

    2. A.G.：為德語「股份有限公司」(Aktiengesellschaft) 的簡稱，A.G. 的發音為 [a:ge:]。

    3. does not constitute...credit：不構成對本 L/C 的保兌。

    4. Kindly acknowledge...and oblige.：這一段是舊式表現法，但在銀行界、保險界仍常用。

No. 157　通知銀行通知國外開來的正本信用狀 (See No. 78)

                                         BANK OF TAIWAN
                                         HEAD OFFICE
                                         FOREIGN DEPATMENT

Our Ref. No. 77/4213

To:

TAIWAN TRADING CO., LTD.   Taipei, Taiwan
P.O. BOX 123   August 18, 20–
TAIPEI

Original Credit issued by:

（Their Ref. No. 123   Dated: August 6, 20–）
Advised by:（Ref. No.   Dated: ）

     Amount: EUR28,950.00

Dresdner Bank   Shipping date:
Hamburg, Germany   Validity: September 15, 20–
     Covering:

Dear Sirs,

At our correspondent's request, we are sending you herewith the captioned original credit.

Kindly acknowledge receipt by signing and returning us the copy of this letter.

Yours faithfully,
For BANK OF TAIWAN

_____
AUTHORIZED SIGNATURE

No. 158 國外開來的信用狀

DRESDNER BANK
P.O. BOX 123
Hamburg, Germany

Taiwan Trading Co., Ltd.   August 6, 20–
P.O. Box 123   VIA TELEX THROUGH
Taipei, Taiwan   BANK OF TAIWAN
R.O.C.   TAIPEI, TAIWAN, R.O.C.

Dear Sirs,

IRREVOCABLE LETTER OF CREDIT NO. 123

We hereby establish our IRREVOCABLE LETTER OF CREDIT in your favor for account of XYZ A.G., Hamburg, up to an aggregate amount of EUR28,950 (EUR DOLLARS TWENTY EIGHT THOUSAND NINE HUNDRED AND FIFTY ONLY) available by your draft(s) at sight drawn on us for 100% of the invoice value accompanied by the following documents:

1. Signed Commercial Invoice in triplicate.

2. Packing list in triplicate.

3. Marine Insurance Policy or Certificate in duplicate endorsed in blank for 110% of the invoice value, covering ICC (B) plus TPND and War.

4. Certificate of Origin in triplicate.

5. Full set of Clean On-Board Ocean Bills of Lading made out to order of Dresdner Bank, Hamburg, Notify Accountee, marked "Freight Prepaid".

Evidencing shipment of:

   NINE HUNDRED CARTONS OF CANNED ASPARAGUS, AI GRADE CIFC$_2$ Hamburg

Shipment from Taiwan to Hamburg not later than August 31, 20–.

Partial shipments are not permitted. Transhipment is prohibited. All drafts so drawn must be marked "Drawn under Dresdner Bank, Hamburg L/C No. 123 dated August 6, 20–."

The amount of any draft drawn under this credit must concurrently with negotiation, be endorsed on the reverse side hereof.

Negotiating bank is to forward all documents in one cover direct to us by airmail.

This credit expires on September 15, 20– for negotiation in Taiwan.

This is a confirmation of the credit opened by telex under today's date through Bank of Taiwan, Taipei.

This credit is subject to Uniform Customs and
Practice for Documentary Credits (1993 Revision)
International Chamber of Commerce   Yours faithfully,
Publication No. 500.   For Dresdner Bank

【註】

1. Dresdner Bank：開狀銀行名稱。

2. Taiwan Trading Co., Ltd.：受益人。信用狀形式上是一封信，受信人一般而言是受益人也即出口商，但須注意有時以通知銀行為受信人。

3. August 6, 20–：開狀日期。

4. via telex through Bank of Taiwan：說明曾以電傳經由 Bank of Taiwan 通知信用狀，所以本 L/C 係電報證實書 (cable confirmation)。

5. irrevocable letter of credit：不可撤銷信用狀（信的標題）。

6. we hereby establish...：說明開發信用狀的事實，"establish" 可以 "issue" 或 "open" 代替。

7. in your favor 也可以 "in favor of yourselves" 代替，「以你為受益人」之意，如 L/C 的受信人為通知銀行時以 "in favor of..." 表示，"in favor of" 後面即為受益人名稱。

8. for account of：「記入（進口商）的帳」之意，"for account of" 後面必為 L/C 申請人，通常為進口商。

9. up to an aggregate amount of...：總金額以……為度，為本信用狀可用金額。

10. available by your draft(s)...on us：得簽發以本行為付款人的即期匯票。

11. for 100% of the invoice value：匯票金額按發票金額全額開發，"value" 解做「金額」(amount)。

12. accompanied by the following documents：附上以下單證，這裡的 "documents" 係指 "shipping documents"（貨運單證；裝運單證）而言。

13. signed commercial invoice in triplicate：簽了字的商業發票三份。

14. packing list in triplicate：裝箱單三份。注意下列各詞：

duplicate （二份）	sextuplicate（六份）
triplicate （三份）	septuplicate（七份）
quadruplicate（四份）	octuplicate （八份）
quintuplicate（五份）	decuplicate （十份）

15. marine insurance policy...：空白背書海上保險單或保險證明書二份，按發票金額的 110% 投保水漬險及竊盜遺失險、兵險。

16. certificate of origin：產地證明書。

17. full set of clean on-board...：全套無瑕疵裝船提單，以 Dresdner Bank 為收貨人，以 L/C 申請人為到貨通知人 (Notify Party)，註明「運費付訖」。

18. evidencing shipment of...：證明裝運……（指以上各種單證須載明裝運……）

19. CIFC₂ Hamburg：貿易條件 (Trade Terms) 的一種。

20. shipment from...：由（裝貨港）至（卸貨港）。

21. not later than August 31, 20–：最後裝船日期為 20– 年 8 月 31 日。

22. partial shipments are not permitted：不准許分批裝運。

23. transhipment is prohibited：不准轉運。

24. all drafts so drawn must be marked "drawn under..."：簽發的所有匯票均須註明「憑漢堡 Dresdner Bank，20– 年 8 月 31 日開發的 123 號 L/C 簽發」字樣。

25. the amount of any draft drawn...hereof：憑本信用狀簽發的任何匯票金額，於押匯同時，必須在本信用狀背面記載。

26. negotiating bank is to forward...by airmail.：押匯銀行應將所有單證一次以航郵寄到本行（開狀銀行）。

27. this credit expires...in Taiwan：本 L/C 押匯期限將於臺灣 20– 年 9 月 15 日屆滿，即本 L/C 的

有效期限。

28. this is a confirmation...Taipei：本信用狀係本日以電傳經由臺北臺灣銀行開出的信用狀的證實書。

29. this credit is subject to...No. 500：本信用狀適用國際商會第 500 號公告所載 1993 年修訂的信用狀統一慣例規定。

30. 關於信用狀的問題，讀者如有興趣作進一步的瞭解，可參閱本書作者的另著：《信用狀理論與實務》。

至於國際商會制定的「信用狀統一慣例」，臺灣金融研訓院曾予以譯註，從事貿易的人士宜購讀，以求進出口押匯業務能順利進行。

No. 159　裝船期限內無直航船，要求展期

Dear Sirs,

L/C No. 123

We have received your letter of August 7, 20– and thank you for the establishment of the subject L/C in our favor. Your prompt compliance with our request is much appreciated.

Upon receipt of the L/C, we have contacted the shipping companies who have direct vessels sailing regularly between Taiwan and Hamburg and regret to inform you that the first available boat to Hamburg is scheduled to leave here on or about September 10. Since the shipment deadline as specified in the subject L/C is August 31 and since transhipment is not allowed, we are unable to ship the goods by a direct vessel before the deadline. This is why we despatched the following telex to you today:

"NO DIRECT VESSEL SAILING HAMBURG BEFORE AUGUST 31 PLS EXTEND SHIPT N EXPIRY LC 123 TO SEPT 20 N 31 REPCTVLY"

We are anxiously awaiting the amendment to the subject L/C.

Faithfully yours,
Taiwan Trading Co., Ltd.

【註】

1. direct vessel：直航船。

2. shipment deadline = latest shipment date：裝船最後日期。

3. 電文中的 telexese 說明：

PLS: please; SHIPT: shipment; N: and; REPCTVLY: respectively

No. 160　通知已裝出部分，並要求修改信用狀允許分批裝運

Dear Sirs,

Lauan Plywood under Your Order PO–123

Thank you very much for the L/C No. 1234 you have issued in our favor through Bangkok Bank for the subject order.

Apropos of delivery of this order, we confirm we have despatched the following telex to you today:

"RE YOUR PO123 DUETO SPACE LIMITATION ONLY 600,000 SQFT SHIPPED PLS AMEND LC ALLOWING PARTIAL SHIPT AIRMAILING"

Much to our regret, we could not deliver the total quantity by one shipment for lack of space of the carrying vessel. Since this is an accident unforeseen by us, we hereby request you to amend the L/C by deleting the clause reading "PARTIAL SHIPMENTS PROHIBITED". Your immediate compliance with our request will be greatly appreciated.

For the remaining 400,000 sq. ft., we will ship them by the first vessel available after receipt of your amendment to the L/C.

Faithfully yours,

【註】

1. apropos of = concerning = with regard to：關於。

2. 電文中的 telexese 說明：

   DUETO: due to

   AIRMAILING：詳情另航郵

3. for lack of space of the carrying vessel：由於載貨船的無艙位。這是由於裝船時才發現已無艙位，事出意外，所以下面一句用 "an accident unforeseen by us"（我們無法預料的意外事故）。

4. delete：刪除。也可以 "take out" 或 "cross out" 代替。

No. 161　要求修改准許轉運、分批裝運以及展延有效期限

Dear Sirs,

L/C 89/6053

We have received with thanks your contract No. 123 on 1,000 doz. of umbrellas amounting in total of US$12,000.00 on CIF basis.

The covering L/C No. 89/6053 has arrived in the amount of US$12,000.00. However, we are cabling you today asking for the following amendments:

1. Delete "Transhipment is prohibited"
   Transhipment will have to be made at Hongkong.

2. Amend to "Patial shipments allowed"
   We always like to have this proviso so as to eliminate the possibility of having one or two dozen shortshipped, which nullify the entire L/C.

3. "Extend expiry and shipment dates 15 days"
   The subject L/C requires shipment to be effected by Feb. 5. If we are lucky we hope to have this shipment go forward on a vessel leaving during the latter part of January, but if not, it will go on a vessel that will sail sometime between the 6th and 9th of February.

Your prompt attention to the foregoing would be much appreciated.

Yours faithfully,

【註】

1. amounting in total of 也可以 "amounting to" 或乾脆以 "for" 代替。

2. at Hongkong = at the port of Hongkong，照理一都市前面的介系詞應用 "in"，但實際上這裡是指港口，而港口前面則應用 "at"。例如 "at New York" 是指「在紐約港」之意，本來 "at" 為 "point of place" 的介系詞，所以在海上保險，表示 "claim" 的地點時，也以類如 "payable at London" 的方式表示。

3. amend = amend the L/C。

4. proviso：但書。

5. eliminate the possibility of...：消除……的可能。

6. shortship：短運。

7. nullify：使……無效。cf. null and void：無效。

8. go forward on...：裝……船。

No. 162　要求修改保險條款

Dear Sirs,

Thank you very much for your L/C No. 123 of the Bank of Montreal covering your order No. 245. We find it in order except the insurance clause, about which we have sent you a telex as per enclosed copy and are pleased to confirm as follows:

In our effort to meet the stipulation that the goods are to be covered "against all risks from any cause whatsoever irrespective of percentage of damage". We negotiated with the underwriters but learned that they would not insure the goods of this kind against such extensive risks even at a higher rate, which the bankers insist on the insurance policy strictly conforming to the L/C terms.

As a last resort we have requested you to amend the L/C by replacing the stipulation with "ICC (B) including war risks and TPNC". In the light of business practice also, this is the best possible coverage given to these goods. Moreover, all our shipments are so securely packed by our experienced hands that no inconvenience has ever been caused to any of our customers.

Your goods are ready for shipment and can be shipped within this month if the amendment to the L/C is telexed at once. If not, we would have to ask you to extend the validity and shipment time for one month.

We trust that you will see the reason of our request and amend (and, extend, if necessary) the L/C at once.

Yours faithfully,

【註】

1. against all risks...of damage：投保不論任何原因所致的一切危險，且免計損害所致的百分比。

關於 "Irrespective of percentage" 參閱本書第二十二章。

2. underwriters：保險商。cf. insurer：保險人。

3. extensive：廣泛的。

4. insist on：堅持。

5. in the light of：徵諸；按照；根據。

No. 163　信用狀郵遞耽誤，要求延展交貨期及有效期

Dear Sirs,

The L/C No. 10001 of the Bank of America to cover your order No. 104 reaches us today — 30 days after its establishment as is attested by the advising bank.

Apart from the question of who is to blame for this unusual delay, the immediate problem is that the manufacturers would not agree to May shipment since they require the relative L/C being received in ample time, usually two months before shipment.

We have been compelled, therefore, to ask you to kindly extend the shipment date and expiry of the L/C to June 30 and July 15 respectively.

We are sorry for the unfortunate outcome, but trust that you will understand our position and arrange the extension at once.

Yours faithfully,

【註】

1. apart from：撇開（對於這種不尋常的遲延應怪誰的問題）不說。

2. immediate problem：目前的問題。

3. ample time：充足的時間，"ample" 為 "quite enough"（足夠的；充分的）之意。

No. 164　因颱風無法如期交貨，要求展期

Dear Sirs,

L/C No. 123, Your Order No. 321

We are sorry to inform you that it has become impossible for us to complete shipment during September of the captioned order.

In fact, a terrible typhoon struck this part of the country on the 6th this month and our factory suffered serious damages, making it impossible to ship your order within the validity of the subject L/C which expires on September 15.

Under the circumstances, we hope you will agree to extend the credit till October 26 as we asked you by telex.

Though the delay is beyond our control, we are no less than sorry for it, and will do everything in our power to expedite manufacture. The expected date of shipment will be around October 15.

We trust that you will understand the situation and hope that you will comply with our request.

Yours truly,

【註】

1. complete shipment：完成裝運。

2. terrible typhoon：強烈颱風。

3. struck: pp. of strike：侵襲。

4. suffered serious damage：遭受嚴重損害。

5. no less than sorry for it：仍感到遺憾。

6. do everything in our power：盡我們所能。

7. expedite：加速；加緊。

No. 165　要求刪除提供領事發票

Dear Sirs,

Grey Cotton Twill

Further to our letter of August 10 informing the booking of s.s. "Nisho" to deliver the captioned goods, we are now requesting you to amend the L/C you have established in our favor by deleting the requirement to submit Consular Invoice along with other shipping documents.

As there is no Italian Consulate here in Taiwan, nor anyone representing it, we are unable to obtain this document required. Therefore, you are requested to amend the L/C accordingly.

Since the date of shipment is imminent, please have the L/C amended telegraphically. Your compliance with our request is hereby solicited.

Yours faithfully,

【註】

1. Consular Invoice：領事發票，又稱為領事發貨票。

2. imminent = very near：逼近。

3. solicit：請求。也可以 "request" 代替。

No. 166　短開信用狀金額，要求修改

Dear Sirs,

Thank you for your L/C No. 101 in our favor to cover your order of 1,000 sets transistor radios.

We find, however, that the L/C amount for US$2,000 is inconsistent with that listed in our Sales Confirmation No. 567. The correct amount should be US$2,500. Possibly, it was either a typographical error or an oversight on the part of your bank (or secretary). We wish to request you to ask your bank to amend the credit by cable to increase an amount by US$500 to it. In this regard, we have telexed you the following today:

"ELCEE NO 101 TOTAL AMOUNT INCORRECT PLS AMEND ELCEE BY CABLE TO INCREASE US$500 MAKING TOTAL US$2500"

The goods will be ready for shipment within 20 days as soon as we receive your cable amendment to the L/C.

We appreciate your immediate attention to this matter.

Yours truly,

【註】

1. is inconsistent with：與……不一致。

2. typographical error：打字錯誤。

3. oversight：疏忽；失察。

No. 167　由於工廠火災要求展延交貨期及有效期限

Dear Sirs,

"Jet Size" Sizing Stuff

We acknowledge with thanks the receipt of your L/C established through Bank of Tokyo for 2,000 kgs. each of Nos. 66 and 88–98 of the captioned stuff.

According to the stipulations of the L/C, the total quantity should be shipped not later than August 15, and no partial shipments are permitted. Much as we wish to comply with such conditions, we are, however, unable to do so because of a recent fire at our factory which has nearly destroyed all our stocks.

Under the circumstances, we hereby request you to have both the date of shipment and the expiry date of the L/C extended to September 10 and 25 respectively.

As fire is a force majeure beyond our control, we are sure you would comply with our request accordingly.

Yours faithfully,

【註】

1. much as we wish...：雖然我們很想……。

2. force majeure：不可抗力。

3. beyond our control：非我們所能控制。

No. 168　在交貨期內無直航船，要求展期

Dear Sirs,

Your Order for Grey Cotton Cloth

Your L/C No. 123 to cover subject order has been notified to us through Bank of America, Taipei. From the BOA's notification, we know that the deadline of shipment for the first batch of cloth was set on August 10, 20–.

Since receipt of the Bank's notification, we have been making every effort to book space to effect shipment within the specified time, but, much to our regret, we were told by the shipping company contacted that there would no vessels sailing to Marseilles direct before August 20. Since the L/C stipulates that transhipment is forbidden, we hereby request you to have the L/C amended by extending the shipment date and its validity to September 30 and October 15 respectively.

As this is an urgent matter, please amend the L/C telegraphically. Your compliance with our requests will be greatly appreciated.

Yours truly,

【註】

1. BOA：Bank of America 的簡稱。

2. first batch：第一批。

3. validity：信用狀的有效期。

4. telegraphically 也可以 "by telegram" 或 "by cable" 代替。

No. 169　無直航船要求修改允許轉運

Dear Sirs,

Subj.: Lathes

We have received your letter of credit No. 123 for US$20,000 opened in our favor through Bank of America to cover the CFR value of four sets of the subject lathes and thank you for it.

Concerning the shipment, we regret very much to inform you that, despite strenuous efforts having been made by us, we are still unable to book space of a ship sailing to Wellington direct. The shipping companies here inform us that, for the time being, there is no regular vessel sailing between Taiwan ports and Wellington. Therefore, it is very difficult, if not impossible, for us to ship these four sets lathes to Wellington direct.

In view of the difficult situation faced by us, you are requested to amend the L/C to allow transhipment of the goods at Hongkong where arrangements can be easily be made for transhipment. You can be sure that we will ship the goods to you with transhipment at Hongkong right upon receipt of the L/C amendment. Since this is beyond our control, your agreement to our request and your understanding of our position will be much appreciated.

We are anxiously awaiting the amendment to the L/C.

Yours truly,

【註】

1. strenuous efforts：很大的努力。

　cf. make strenuous efforts：盡最大努力。

2. regular vessel：定期航行的船隻。

3. right = at once：立即。

No. 170　向開狀銀行提出修改信用狀申請書

First Commercial Bank

Foreign Dept.

Taipei, Taiwan

April 14, 20–

Dear Sirs,

Re: L/C No. 7DHI/00123/01

In favor of XYZ Motor Company

We refer to the captioned L/C which was issued on March 25, 20– through your goodselves in favor of the subject beneficiary.

We have just received a letter from the beneficiary advising that a recent fire had nearly destroyed their stocks and requesting that the shipment date and L/C expiry date be extended to June 30 and July 15, 20– respectively.

Enclosed please find one copy of Application for Amendment of Credit for your processing. It will be appreciated if you will amend the credit and advise your correspondent accordingly.

Encl. a/s

Faithfully yours,

【註】

1. the subject beneficiary：該受益人。

2. application for amendment of credit：修改信用狀申請書。

3. amendment of credit 宜改為 "amendment to credit"。

4. correspondent：這裡指往來銀行。通匯銀行即 correspondent bank。

5. 在臺灣，進口商向銀行申請修改信用狀，只要提出修改信用狀申請書即可，並不另外寫信給銀行。

No. 171 信用狀修改申請書

國外營業部或外區指定單位					
經 理	副 理	科 長襄 理	副科長	經	辦

受 理 單 位			
經 理(主任)	副襄理	經理	辦

信 用 狀 修 改 申 請 書
APPLICATION FOR AMENDMENT OF DOCUMENTARY CREDIT

申請日期　年　月　日

第一商業銀行 台照
To: FIRST COMMERCIAL BANK

受理單位：

部、分行
辦事處

通知銀行 ADVISING BANK

【 20 】信用狀號碼 CREDIT NO.

【 21 】通知銀行通知號碼 ADVISING BANK REF. NO.

【 31C 】開發日期 ISSUING DATE

【 59 】原受益人名稱和地址 BENEFICIARY'S NAME AND ADDRESS

茲請貴行將原開發之上述信用狀內容以 □ AIRMAIL 郵航 □ CABLE 電報 方式修改下列註明 ⊠ 記號之項目：
WE HEREBY REQUEST YOU TO AMEND THE ABOVE MENTIONED CREDIT BY AIRMAIL/CABLE AS MARKED ⊠ BELOW:

□【 31E 】有效日期 EXPIRY DATE

□【 32B 】信用狀金額增加 CREDIT AMOUNT INCREASED BY

□【 33B 】信用狀金額減少 CREDIT AMOUNT DECREASED BY

□【 34B 】信用狀修改後金額 MAKING NEW TOTAL AMOUNT

□【 39 】信用狀金額增減比率 CREDIT AMOUNT TOLERANCE □ ABOUT □ ± _____ %

□【 44 】裝載港、收貨地、卸貨港、目的地
SHIPMENT FROM _____ TO _____

□【 44C 】最後裝運日期 LATEST DATE FOR SHIPMENT

□【 79 】1. 裝運方式以空運或海運 SHIPMENT TO BE EFFECTED BY □ AIRWAY □ SEAWAY
2. 價格條件 PRICE TERM
3. 分批裝運允許／不允許 PARTIAL SHIPMENT □ ALLOWED □ PROHIBITED
4. 轉船裝運允許／不允許 TRANSHIPMENT □ ALLOWED □ PROHIBITED
5. 新增貨品名稱、數量 SHIPMENT OF ADDITIONAL GOODS AND QUANTITY

6. 刪除須提示單據 DELETED REQUIRED DOCUMENTS:

7. 增加須提示單據 REQUIRED ADDITIONAL DOCUMENTS:

8. 其他 OTHERS

□【 　 】銀行間備註 BANK TO BANK INFORMATION

主管核對打字經辦

其餘條款維持不變
ALL OTHER TERMS WILL REMAIN UNCHANGED.
申請人茲同意絕不使貴行因本項修改或變更而發生任何損失糾葛，並確切保證承擔原「開發信用狀申請書」所擔保之全部責任。
WE HEREBY AGREE THAT WE SHALL NOT CAUSE YOU ANY LOSS OR TROUBLE WHATSOEVER IN CONSEQUENCE OF THE ALTERATION(S) AND UNDERTAKE TO ASSUME ALL OUR RESPONSIBILITIES AS PLEDGED IN THE ORIGINAL APPLICATION OF THIS DOCUMENTARY CREDIT.

摘　　要	金　　額	收 訖 費
手 續 費		
匯費及郵電費		
合　　計		

申請人 APPLICANT:

簽章 SIGNATURE OF APPLICANT(原結匯印鑑)
地址 ADDRESS:
電話 TELEPHONE:
日期 DATE:

核對印鑑

迪 011 (26.5×38cm) 89. 3. 500 本

No. 172 通知出口商已修改信用狀

Dear Sirs,

Re: L/C No. 7DH 1/00123/01

Your letter of April 10, 20– requesting us to extend the shipment date and L/C expiry date to June 30 and July 15, 20– has been received.

To comply with your request, we submitted our Application for Amendment of Credit to the opening bank, i.e., Central Trust of China, on April 14, and trust you will receive from the advising bank the amendment before long.

As we have informed you in our previous letters, we are in urgent need of these pickups. Therefore, please ship them as soon as possible. Enclosed please find a copy of amendment to the credit.

Yours faithfully,

【註】

amendment to the credit：信用狀修改書。注意用 "amendment to..." 而不用 "amendment of..."。

第四節　有關信用狀交易的有用例句

一、關於信用狀的動詞

1. To $\begin{Bmatrix} \text{issue} \\ \text{open} \\ \text{establish} \\ \text{arrange} \end{Bmatrix}$ a credit $\begin{Bmatrix} \text{through} \\ \text{with} \end{Bmatrix}$ a bank.

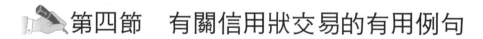

2. To cable a credit　　　　電開信用狀

3. To cancel a credit　　　　取消信用狀

4. To amend a credit　　　　修改信用狀

5. To extend a credit　　　　展延信用狀期限

6. To $\begin{Bmatrix} \text{increase} \\ \text{decrease} \end{Bmatrix}$ a credit　　$\begin{Bmatrix} 增加 \\ 減少 \end{Bmatrix}$ 信用狀金額

7. To send a credit 寄出信用狀

8. To take out a credit 請⋯⋯開發信用狀

二、催促開發信用狀

1. $\left\{\begin{array}{l}\text{As of this date}\\ \text{Up to now}\end{array}\right\}$, we have not yet received your L/C. In order to book the shipping space at an earlier date, you are requested to have the L/C opened immediately.

2. We will book the shipping space as soon as we receive your telex L/C.

3. The porforma invoice is enclosed. Please open L/C in our favor within 15 days after receipt of it.

4. Your order was confirmed by fax of August 5, subject to arrival of L/C within 20 days from date. The 20 days period having expired on August 25 without receipt of the L/C nor hearing any further advice from you. We faxed you today asking when the required L/C had been opened.

5. From your letter of June 12, we learn that the L/C has been arranged, but we regret to say that we have received no banker's advice yet. Please push your bankers for their soonest action.

6. With reference to your order No. 123, the manufacture of the goods has been completed and we are now arranging to ship from Keelung on the first available boat. In the meantime, we await your credit to cover.

三、通知已開發信用狀

1. We wish to inform you that we have $\left\{\begin{array}{l}\text{opened}\\ \text{issued}\\ \text{established}\\ \text{arranged}\end{array}\right\}$ a credit $\left\{\begin{array}{l}\text{with}\\ \text{through}\end{array}\right\}$ ABC Bank,

→ in $\left\{\begin{array}{l}\text{your favor}\\ \text{favor of yourselves}\end{array}\right\}$ for US\$... $\left\{\begin{array}{l}\text{covering}\\ \text{against}\end{array}\right\}$ our order No. 123, $\left\{\begin{array}{l}\text{available}\\ \text{in force}\end{array}\right\}$ until Dec. 26, 20–.

2. In order to cover this $\begin{Bmatrix} \text{order} \\ \text{business} \\ \text{transaction} \end{Bmatrix}$ we have arranged with ××× bank a credit in

your favor for US$...

3. We take pleasure in informing you that we have $\begin{Bmatrix} \text{submitted} \\ \text{sent} \end{Bmatrix}$ application to our

→ bankers for $\begin{Bmatrix} \text{establishing} \\ \text{issuance} \\ \text{opening} \end{Bmatrix}$ of an L/C in your favor to cover the total value of

this order.

4. An L/C is being established for a total of US$...covering the FOB price, stipulating shipment by November 20. The credit expires on December 5.

四、要求修改信用狀

1. We find that the L/C amount is inconsistent with that listed in our proforma invoice. The correct amount should be US$.... Possibly, it was a typographical error or an oversight on the part of your bank. Therefore, please request your bank to amend the credit by cable to increase US$...

2. The expiry date of the credit being November 30, 20–, we request you to arrange with your bank to extend it till December 20, 20–.

3. Please amend (adjust) L/C No. 123 as follows:

a. Amount to be increased $\begin{Bmatrix} \text{up to} \\ \text{by} \end{Bmatrix}$ US$... $\begin{pmatrix} 增加到 \\ 增\ \ 加 \end{pmatrix}$

b. The words "..." are to be $\begin{Bmatrix} \text{deleted.} \\ \text{replaced by "..."} \end{Bmatrix}$

c. Validity to be extended to April 30.

4. The L/C has been received this date, but without necessary amendment. Again, we must ask you to refer to our letter of July 3 in which our request was made to you regarding the following clause to be amended:

5. To our regret, however, this credit was found not properly amended on the following points despite our request:

6. As advised you by telex, the sailing of s.s. "..." was cancelled because of serious damage she sustained while at the berth of Keelung. This has made us impossible to ship the goods within the life of L/C which expires on June 2. Under the circumstances, we earnestly hope that you will extend the credit to the end of this month.

7. It is a matter of regret to in form you that the L/C valid until October 10 for order No. 123 became unusable owing to the fact that the careless mistake in the dyeing process necessitated our restart for manufacture. Please give us one more month for the completion of the goods and instruct your bankers for the extension of validity with ample time for enabling us to negotiate the draft under this credit.

8. Since this happening has been caused by force majeure over which we have had no control, we hope you will agree with us that we cannot be held responsible for this delay in shipment. We trust we may depend upon your generosity for the extension to shipping date and also of expiry date of the L/C as we requested by todays cable.

五、通知已修改信用狀，或憑 L/G 押匯

1. We will increase the L/C amount by 200 tons at the rate of US$150 per metric ton.

2. We have $\begin{Bmatrix} \text{instructed} \\ \text{requested} \end{Bmatrix}$ our bankers to amend the clause to read "partial shipments are permitted".

3. We are pleased to inform you that immediately we received your telex we arranged with our bankers to modify the L/C as requested, and we trust you have been able to dispatch 200 gross yard on July boat and 300 gross by direct vessel in mid-August.

4. Instead of our arranging for the increase of credit amount at costly cable charges, you may invoice to us through your bankers the said small difference of amount under your L/G. Meanwhile, we are instructing the issuing bank to accept such L/G at this end.

六、其 他

1. We regret that we could not ship the goods by a March vessel only because of the delay of your L/C. Please attend to this matter with all speed.

2. We would request you to take up this matter with the issuing bank at once and let us know what has become of the letter of credit, otherwise, contrary to our intention, the shipment may be delayed a great deal.

3. We presume the advice of L/C has gone astray in tranist from your bankers. If it has not been sent, repeat your instructions to your bankers.

4. Instead of asking you to arrange for the increase of the credit amount, invoicing was made to you based on accurate price calculation through the bankers here against our L/G and no balance account will be debited to you. Please appreciate our adjustment in regard to the above shipment.

   註：no balance account will be debited to you：差額不借入貴方帳（即放棄差額）。

5. To hasten the shipment, documentation was made out to you under our L/G invoicing at the unit price as originally agreed upon.

   註：a. make out：製作（單據）；b. under L/G：憑保結書；c. invoicing at...：按⋯⋯開發票。

6. Among the goods by s.s. "...", we are afraid that some pieces are of irregular widths, in the vicinity of 35″, despite of 36″ required in the credit. We will, therefore, submit an L/G to the negotiating bank here upon drawing the draft.

第二十二章
貨物水險(Marine Cargo Insurance)

國際買賣是隔地買賣，貨物從出口地運到國外進口商手中這一段運輸期間，難免會遭遇到天災人禍等等意外危險，以致受到損害。因此進出口商為防萬一貨品遭受損害時，可獲得補償，乃有將運輸中的貨物加以保險的必要。

第一節　損害及費用的類型

一、Total loss（全損）

即所承保的貨物全部滅失 (Loss) 的情形，可分為：

1. Actual total loss（實際全損）：即投保的貨物已經全部毀滅或受損程度到達已失去原有型態或其所有權被剝奪而不能恢復者。

2. Constructive total loss（推定全損）：即投保的貨物全損已無可避免或雖未及全損，但欲由絕對全損保全，其費用將超過其保全後的價值者。

二、Partial loss（分損）

即所承保貨物部分損失的情形，可分為：

1. General average（共同海損）：即基於船舶及裝載貨物所有人的共同利益，為避免共同危險，而船長故意及合理對於船舶或貨物加以適當緊急處分而生的犧牲 (General average sacrifice) 及費用 (General average expenditure)。此項犧牲及費用係保全所有利益而生，故應由全體利害關係人分擔，然後由其向保險人要求賠償。

2. Particular average（單獨海損）：即指承保貨物因不可預料的危險所造成的滅失或損害，這種損害並非由共同航海的財產共同負擔，而是由遭受損害的各財產所有人單獨負擔者，簡言之，分損無共同海損性質者，即為單獨海損。

三、Charges（費用）

1. Sue & Labor charges（損害防止費用）：貨物在遇險時，如被保險人或其代理人（如船長）或讓受人為之努力營救，以減輕損失程度，則保險人對這種費用支出應予賠償，這種費用即為 Sue & Labor charges。
2. Salvage charges（施救費用）：無契約關係或正式職責關係的第三者對於遭受危險的船貨，自願所使施救工作後，應由被保險人支付的報酬，保險人對此項費用負有賠償之責。

第二節　保險險類

海上貨物保險，根據保險人就海上損害發生的程度，所應承擔的責任加以區分，可分為下列三種基本險及附加險：

一、基本險

1. 協會貨物保險 A 款險 (Institute Cargo Clauses (A))，簡稱 ICC (A)。
2. 協會貨物保險 B 款險 (Institute Cargo Clauses (B))，簡稱 ICC (B)。
3. 協會貨物保險 C 款險 (Institute Cargo Clauses (C))，簡稱 ICC (C)。

二、附加險

常見的附加險有：1. War（兵險），2. Strike, riot and civil commotions（罷工、暴動、民眾騷擾），3. Theft, pilferage and non-delivery（偷竊、拔貨、遺失），4. Fresh water and rain damage（淡水、雨水損），5. Breakage（破損），6. Leakage（漏損），7. Hook hole（鉤損），8. Oil damage（油污），9. Contam ination with other cargoes（污染），10. Sweat and heat（汗濕、發熱），11. Washing overboard（浪沖），12. Mildew and mould（霉濕及發黴），13. Rat and vermin damage（鼠蟲害）等。

第三節　有關保險的信函

No. 173　向保險公司查詢費率

Fubon Insurance Co., Ltd.

Dear Sirs,

Please quote us your lowest rate for Marine Insurance, ICC (B) plus TPND and war risk, on a shipment of 900 cartons of Canned Asparagus, valued at EUR31,845, by the s. s. "Euryphates" from Keelung to Hamburg.

The ship is scheduled to leave Keelung on or about August 20 for Hamburg, and we hope to have your reply at your earliest convenience.

Yours faithfully,

【註】

1. value at：價值為……（元）。

2. is scheduled：預定。

3. 在我國，進出口商向保險公司查詢費率時多用電話辦理，很少用信函查詢。有時，也可透過保險經紀人 (insurance broker) 查詢費率。

No. 174　保險公司覆函

Taiwan Trading Co., Ltd.

Dear Sirs,

We acknowledge with thanks the receipt of your letter of August 10 inquiring about marine insurance rate to cover shipment of Canned Asparagus from Keelung to Hamburg.

In compliance with your request, we hereby quote you our lowest rate at 1% to cover ICC (B) plus TPND and War risks. Please note that this is the lowest rate we are able to offer and there will be no rebate whatsoever.

As the shipment is drawing nearer and nearer, we suggest that you contact us at the earliest in order that we may issue the policy in time.

We hope that you will pass us your business.

Yours faithfully,

【註】

1. rebate：回扣。按保險界慣例，保險公司往往按其所開費率，予投保人若干回扣。

2. pass us your business：將生意交給我們，意指由我們承保。

No. 175　向保險公司投保

Fubon Insurance Co., Ltd.

Dear Sirs,

　We thank you for your letter of August 15 quoting us marine insurance to cover ICC (B) plus TPND and War risks for shipment of 900 cartons Canned Asparagus from Keelung to Hamburg.

　In view of the fact that the rate you quoted is quite competitive, we have decided to entrust your company with the insurance of this shipment. Enclosed please find a copy of our Marine Insurance Application duly filled and signed by us. It will be appreciated if you will issue the Marine Insurance Policy as soon as you receive this letter and its enclosure.

　As for premium, we will arrange with the negotiating bank to pay you at the time of negotiation of draft.

Yours faithfully,

【註】

1. entrust your company with the insurance of this shipment.：將本批貨交給貴公司承保。

2. marine insurance application：水險投保單。

3. as for premium：至於保險費，用 "as to premium" 較普遍。

No. 176　水險投保單

富邦產物保險股份有限公司(原國泰産物保險公司)
Fubon Insurance Co., Ltd.

237, CHIEN KUO
S. ROAD SEC. 1
TAIPEI TAIWAN
TEL：(02)706-7890
FAX：(02)704-2915

貨 物 水 險 投 保 單
Marine Insurance Application

1.投保日期
Applying Date _____ 19. ____

※
紅框部份由本公司核保單位填寫

2.姓名或公司名稱 Name of Assured	
3.受 益 人 Beneficiary	
4.船名及航次 Name of Vessel	5.船 期 Sailing Date on/about
6.裝貨港／起運地 From	7.目的地 To
轉 船 地 點 Transhipment（if any）at	轉入船名 Into
8.內陸或最後目的地／經由 Thence To/Via	9.貨物代號 Cargo
10.發票號碼 Invoice No.	11.包裝代號 Package
12.信用狀號碼 L/C Number	13.輸入許可證號 I/L Number
發票金額 Invoice Value _____ + ____ %＝	14.保 額 Insured Amount

15.保險標的物 Subject-Matter Insured

16.數 量
Quantity

17.包 裝
Package

投保條件 18.Terms	F.P.A.	Clauses(C)	19.附加條款 Additional Clause
	W.A.	Clauses(B)	
	A.R.	Clauses(A)	
	War	War	
	S.R.C.C.	Strikes	

加 Supple 費 費	轉 船 T	小 船 L	舊 船 A	內 陸 Y

20.賠 款 地 點 及 幣 別
Claim（if any）Payable at _____ in _____ Currency

21.費率(%) M. _____ 總保費
Rate W. _____ Total
S. _____ Premium

22.保單號碼
POLICY No. _____

23.卸 貨 港
Discharging Port

該保證上列貨物特別載明外均裝艙內
Warranted Shipped under deck unless otherwise specified.

運 輸 方 式			進 出 口 別			外幣兌換率		押滙單		簽 單 日 期	保單份數	
海運 陸運 空運 郵包			進口 出口 國內			A.保額	P.保費	台銀	一銀	Issuing Date	正本	副本
S L A P	I E D							T	F	年 月 日	2	4

24.經 副 理 Manager or Deputy manager	科 長 Superint	經手人 Agents	投保人簽章 Signature

25.備 註 Remarks	聯絡電話： Tel. 收費地址： Address：

MUU-001(81 8.400本)廣興

No. 177　要求保險公司加保險類

Fubon Insurance Co., Ltd.

Dear Sirs,

Further to our letter to you of August 14, we have just received a telex from our customers asking us that, besides ICC (B) plus TPND and War risks, they also want added coverage against RFWD. Since this is added risk to be covered, we will pay extra premium that you may charge us therefor.

When issuing the Policy, you are requested to endorse this specific clause thereon. We will pay you the additional premium on behalf of our customers.

Your immediate attention to the above will be appreciated.

Yours faithfully,

【註】

　　1. RFWD: rain, fresh water damage：雨水、淡水損害險。

　　2. endorse：加批。意指在保險單上加批。

No. 178　保險公司通知予承保

Taiwan Trading Co., Ltd.

Dear Sirs,

As instructed by your letter of August 14 and 16, we have covered EUR31,845 on the shipment of 900 cartons Canned Asparagus from Keelung to Hamburg, ICC (B) plus TPND, RFWD and War risks.

We send you enclosed the policy together with a debit note and shall be obliged if you will request your negotiating bank to credit the amount to our account with Bank of Taiwan, Head Office, at an early date.

Yours faithfully,

【註】

　　debit note：借項通知單，即帳單。

　　貿易商有大批貨物分次裝運時，如於每次裝運一批貨物投保一次不僅麻煩，而且可能於裝運時疏忽忘記保險，於是，有預約保險 (Open policy) 的產生。換言之，貨主可預先購買一定金額的保險，這種保險只訂有概括的條件，至於船名保險金金額、啟航日期、貨物數量、包裝嘜頭則於每次裝出貨物時由投保人向保險公司申報 (Declare)。投保人每次申報保險金額，從預約保險單 (Open policy) 內扣減，直至 Open policy 所列保險金額用完為止。這種 Open policy 又稱為流動保單 (Floating policy)。

No. 179　向保險公司查詢預約保險條件

> Dear Sirs,
>
> 　Please quote your rate for an ICC (A) open policy for US$100,000 to cover shipments of general merchandise by APL's vessel from Taiwan ports to Atlantic ports in Canada and the United States.
>
> 　As shipments are due to begin on June 30, please let us have your quotation by return.
>
> 　　　　　　　　　　　　　　　　　　　　　　　　　　　　　　Yours faithfully,

【註】

　　1. APL's vessel: American President Line's Vessel：美國總統輪船公司的船隻。

　　2. Atlantic Ports：大西洋岸港埠。

　　3. due to+verb：預定……；即將……。

No. 180　保險公司的覆函（報價）

> Dear Sirs,
>
> 　We are replying your enquiry of June 10. Our rate for a US$100,000 ICC (A) open policy on general merchandise by APL's vessel from Taiwan ports to Atlantic ports in Canada and the United States is 2% of declared value.
>
> 　This is an exceptionally low rate and we trust you will give us the opportunity to handle your insurance business.
>
> 　　　　　　　　　　　　　　　　　　　　　　　　　　　　　　Yours faithfully,

【註】

　　declared value：申報金額。

No. 181　投保人接受購買預約保險

> Dear Sirs,
>
> 　Thank you for your letter of June 12 quoting your rate for an open policy of US$100,000 cover-

ing consignments on the routes named.

The rate of 2% is satisfactory and we shall be glad if you will now prepare and send us the policy and meanwhile let us have your certificate of insurance and statement of charges for the following, our first shipment under the policy:

10 c/s General Merchandise (Textiles), marked G
   Value US$5,000.

Yours faithfully,

【註】

   1. certificate of insurance：保險證明書。

   2. statement of charges：費用明細單，即指保險費帳單。

No. 182　投保人向保險公司申報 (Declaration)

Dear Sirs,

Open Policy No. 1234

Please note that under the above open policy, dated June 22, 20–, we have today shipped a third consignment, valued at US$3,000.00 by s.s. "Durham Castle," due to sail from Keelung tomorrow to Boston. We enclose the necessary declaration form. This leaves an undeclared value on the policy of US$ 90,000.00 and perhaps you will be good enough to confirm this figure.

Yours faithfully,

【註】

   1. declaration form：申報表格。

   2. undeclared value：剩餘保險金額，指預約保險單餘額而言。

按 FOB 或 CFR 交易，而進口商已購買 Open policy 時，往往要求出口商於起運時替進口商向其保險公司通知 (Declare)，其例函如下：

No. 183　起保通知書 (Insurance declaration)

To: ×××Marine Insurance Company

Dear Sirs,

Re: Your Open Policy No. 123

We, on behalf of...(name of assured or importer)...of (place) hereby declare below the particulars of a shipment to be made by us which is covered under the above Policy:

Conveyance: S. S. "President"

Voyage: from Keelung to New York

Sailing date: on or about August 5, 20–

Interest: Cold Rolled Steel Sheet

Marking: ◇ T

Value: US$12,000.00

Insurance coverage: ICC (A) plus war

Packing: in bundle

Please acknowledge receipt by signing and returning to us the duplicate copy at your earliest convenience.

Yours faithfully,

【註】

1. conveyance：運輸工具。

2. voyage：航程。

3. sailing date：啟航日。

4. interest：保險標的。

第四節　有關保險的有用例句

一、關於保險的動詞

1. To $\begin{Bmatrix} \text{insure} \\ \text{cover} \end{Bmatrix}$ a thing for US$...against $\begin{Bmatrix} \text{ICC (A)} \\ \text{ICC (B)} \\ \text{ICC (C)} \end{Bmatrix} \begin{Bmatrix} \text{with} \\ \text{in} \end{Bmatrix}$ China Insurance Company Limited.

〔就某東西向中國產物保險公司投保……元的 A 款險（等）。〕

2. To $\begin{Bmatrix} \text{effect} \\ \text{cover} \\ \text{place} \\ \text{provide} \end{Bmatrix}$ insurance on a thing for US$...against $\begin{Bmatrix} \text{ICC (A)} \\ \text{ICC (B)} \\ \text{ICC (C)} \end{Bmatrix} \begin{Bmatrix} \text{in} \\ \text{with} \end{Bmatrix}$ Cathay Insurance Co., Ltd.

3. To take out an insurance policy on a thing against $\begin{cases} \text{ICC (A)} \\ \text{ICC (B)} \\ \text{ICC (C)} \end{cases}$ for US\$...

4. To insure a thing at a low $\begin{cases} \text{premium} \\ \text{rate} \end{cases} \begin{cases} \text{in} \\ \text{with} \end{cases}$ the $\begin{cases} \text{insurance company.} \\ \text{underwriters.} \end{cases}$

5. To have a cargo insured. （將貨物付保。）

6. To carry insurance on a cargo. （將貨物付保。）

7. To $\begin{cases} \text{make out} \\ \text{draw up} \\ \text{issue} \end{cases}$ a policy of US\$... $\left(\begin{cases} 掣發 \\ 簽發 \\ 發行 \end{cases} ……元的保單。 \right)$

二、關於保險費率

1. Please $\begin{cases} \text{quote} \\ \text{let us know} \end{cases}$ your lowest $\begin{cases} \text{ICC (A)} \\ \text{ICC (B)} \\ \text{ICC (C)} \end{cases}$ rate for US\$... on notebook computers

from Keelung to New York.

2. Please quote us your best rate for a coverage of ICC (A) including War for the CIF value of US\$...plus 10% on electrical apparatus from Keelung to Melbourne.

三、關於投保

1. Please effect insurance for US\$...on 100 cases of plastic shoes against ICC (B) per s.s....due to leave Keelung for New York on or about October 15.

2. We wish to cover the following shipment against ICC (A) and War for the sum of US\$...Please send us your policy together with a debit note.

3. Attached are details of the shipment to be insured against ICC (B) for 110% of CIF value. The goods leave Keelung per M.S....for Singapore on or about May 20.

4. We are shipping 20 cases Rayon Silk to Messrs. Ishin Yoko, Hong Kong, by the M. S. "Tamba Maru" leaving Keelung May 3, and ask you to effect insurance for US\$2,000 ICC (B) on the goods.

四、關於預約保險

1. We shall shortly be making regular shipments of fancy leather goods to South America by approved ships and shall be glad if you will issue an ICC (A) policy for, say, US$100,000 to cover these shipments from our warehouse at the above address to port of destination.

2. We ask you to issue a covering note for the insurance.

   註：covering note：暫保單，美國稱為 "insurance binder"。

五、其　他

1. We regret, however, that we are unable to place any further line at this rate, our next best quotation being US$0.60 per cent.

2. We intend the insurance to be effected on our side, and as we shall not know the actual value of each cargo till after the vessel has sailed, we wish you to effect Open Cover ICC (C) on Cotton Yarn for about US$100,000.

3. We may mention that premiums are very much higher this year, as underwriters are asking increased rates all round.

   註：all round：全面地。

4. We shall be pleased to know whether you can undertake insurance of wines against ICC (B), including breakage and pilferage.

第二十三章
海上貨物運輸 (Marine Cargo Transportation)

國際貿易貨物的運輸，可依海上運輸、陸上運輸及航空運輸方式完成，其中大多數的貨物又以海上運輸方式完成。以下就海上運輸加以說明。

第一節　定期船運輸

海上運輸以輪船為運輸工具，但因貨物性質、數量的不同，其洽船方式也不同。如買賣貨物為一般雜貨 (General cargo) 或零星貨物，通常多委託定期船 (Liner, Liner vessel) 以搭載方式裝運。如係大宗散裝貨 (Bulk cargo)，則選擇不定期輪 (Tramper, Tramp vessel) 以傭船方式裝運。

在搭載運輸，出口商備妥貨物後，即可根據船公司 (Shipping company) 或船務代理行 (Shipping agent) 所印發的船期表 (Shipping schedule) 或報紙船期欄廣告，選擇適當的船隻，向船公司或其代理行洽訂艙位 (Booking space)。

一、以信函方式洽訂艙位

以信函方式向船公司或其代理行洽訂艙位的方式在臺灣並不流行，但在英美等國則常採用此方式，茲舉一例於下：

No. 184　向船公司查詢運價及洽訂艙位

Dear Sirs,

　　We shall be pleased if you will quote us the lowest rate of freight for 100 bales of Cotton Yarn weighing 400 1bs. each, to be shipped to Hamburg direct before September 25 from Keelung.

　　As the L/C stipulates that transhipment is not allowed, therefore, you must quote us for a vessel sailing from Keelung to Hamburg direct before the deadline as mentioned above.

Your early quotation will be appreciated.

Yours faithfully,

二、以 Booking note 方式洽訂艙位

在我國，洽訂艙位通常多採用 Booking note 方式，即由出口商（或委託報關行）向船公司索取空白的 Booking note、Shipping order 及 Mate's receipt 成套格式，依式套打填就後送請船公司就 Shipping Order (S/O) 予以簽署，日後憑此 S/O 將貨物交給船長，貨物裝上船後，大副 (Mate) 在 Mate's Receipt (M/R) 上簽字，並退還出口商，以便由其向船公司換領 B/L。

上述 S/O 及 M/R，在美國輪船公司，常以裝船許可證 (Shipping permit) 代替 S/O，而以碼頭收據 (Dock receipt) 代替 M/R，名稱雖不同，作用則同。

No. 185　託運單 (Booking note)

XX航業股份有限公司
XX NAVIGATION CO., LTD.
託運單 BOOKING NOTE

裝單號碼	日期	起運地點
S/O No. _____	Date _____	From _____
（本公司註）		
船名	航次	裝往地點
Ship _____	Voy. No. _____	Destination _____
	（本公司註）	

託運人
Shipper _____
受貨人
Consignee _____
Notify _____
運費率
Freight _____

下列貨物，記載均屬翔實，即請　配予艙位　為荷
You are requested to reserve the space for the undermentioned cargo, for which guarantee for the accuracy of all the declaration hereunder and agree to bind ourselves to the regulation and practice of your company.

標　誌 Marks &	包裝及件數 Packing &	貨　物 Contents	重量及呎碼 Weight &

Numbers	Packages		Measurement

報關行
TEL _____

Per _____

（託運人簽章）

Ref. No. _____

No. 186　裝貨單 (Shipping order)

××航業股份有限公司
×× NAVIGATION CO., LTD.
裝貨單 SHIPPING ORDER

裝單號碼
S/O No. _____
（本公司註）

日期
Date _____

起運地點
From _____

船名
Ship _____

航次
Voy. No. _____
（本公司註）

裝往地點
Destination _____

託運人
Shipper _____

受貨人
Consignee _____

Notify _____

運費率
Freight _____

茲有下列貨物，即請　簽收為荷

Please receive on board the undermentioned goods, in good order and condition, and sign the accompanying Receipt for same.

標　　誌 Marks & Numbers	包裝及件數 Packing & Packages	貨　　物 Contents	重量及呎碼 Weight & Measurement

×× 航業股份有限公司
×× NAVIGATION CO., LTD.

Hatch No. _____

Per _____

No. 187　收貨單 (Mate's receipt)

×× 航業股份有限公司
×× NAVIGATION CO., LTD.
收貨單 MATE'S RECEIPT

裝單號碼 S/O No. _____ 　　(本公司註)	日期 Date _____	起運地點 From _____
船名 Ship _____	航次 Voy. No. _____ 　　(本公司註)	裝往地點 Destination _____

託運人
Shipper _____
受貨人
Consignee _____
Notify _____
運費率
Freight _____

下開貨物，業經照數妥收裝船無訛，即請　憑發提單為荷
Received on board the undermentioned goods, in good order and condition, for which the ×× Navigation Co., Ltd.'s Bill of Lading is to be issued.

標　誌 Marks & Numbers	包裝及件數 Packing & Packages	貨　物 Contents	重量及呎碼 Weight & Measurement

Hatch No. _____

請在開船之前換發提單否則如有
延誤概由託運人或貨主自行負責

_____
大副 Chief Officer

No. 188　船公司對於查詢運價及洽船的覆函

Dear Sirs,

　　In response to your inquiry of...(date)...concerning the ocean freight to Hamburg for 100 bales of cotton yarn, we are pleased to inform you that the rate of freight is US$70 per cubic meter of 1,000 kilos, which is the agreed minimum rate of freight of the European Route Conference.

　　As to the steamers which will sail from Keelung to Hamburg, there are two regular liners, viz., s.s. "A" due to leave Keelung on September 15, to be followed by s.s. "B" on the 24th of the month.

　　We hope to have the pleasure of dealing with your shipment.

Yours truly,

【註】

1. US$70 per cubic meter of 1,000 kilos：意指一立方公尺或 1,000 公斤運費美金 70 元，至於按一立方公尺 (CBM) 或按 1,000 公斤計收，將視那一種對船方有利而定。Kilos 為 Kilograms 之意。CBM 為 Cubic Meter 的縮寫。

2. European Route Conference：歐洲航線運費同盟。

No. 189　申請加入運費同盟

Far Eastern Freight Conference
Taipei, Taiwan

Dear Sirs,

In order to avail ourselves of the good and swift services of vessels of the conference lines, we are writing you to apply for membership to the Conference and shall appreciate your passing our application on to your Hongkong Office for consideration.

For your reference, we are makers of umbrellas and, during the past three years, have exported large quantity of our products to North and South America, South and West Africa, as well as to Europe. The total tonnage of our umbrellas exported amounts approximately 3,000–3,500 measurement tons per annum, which is substantial enough to warrant us to join the Conference as a member.

Your prompt screening of our application and approval will be appreciated.

Yours faithfully,

【註】

1. freight conference = shipping conference：運費同盟是在一特定航線上有定期輪行駛的船公司，為限制或消除彼此間的競爭，而以協定方式結合而成的一種卡特爾 (Cartel) 組織。如果貨主加入運費同盟，則可享受運價上的優待。

2. Conference Lines：參加運費同盟的船公司。

3. tonnage：噸數；噸位。

4. warrant：證明……為正當，即 "give a right to" 或 "give a good reason for" 之意。

No. 190　覆申請加入運費同盟已獲准

Dear Sirs,

We have received your letter of September 12, 20– applying for membership to the Far Eastern Freight Conference and are pleased to inform you that your application has been approved by our Hongkong Office.

We are enclosing the blank forms of application and other relevant documents all in triplicate, which please fill in and return them to us at your earliest convenience.

Your observance of the required formalities is hereby requested.

Yours faithfully,

【註】

1. fill in：填寫，也可以 "put in 或 fill out" 代替。

2. formalities：手續。

3. 最後一段可譯為：「茲請查照辦理該手續為荷」，這是一相當公式化的句子。

參加運費同盟的會員貨主，原則上必須將其貨物交給同盟輪運輸，否則將受到處分。然而，有時因特殊原因無法交給同盟輪運輸，在此場合，應向運費同盟備案，取得諒解。

No. 191 通知運費同盟無法將貨交同盟輪裝運

Far Eastern Freight Conference
Taipei, Taiwan

Dear Sirs,

We are very much pleased to be admitted as a member of the Conference and assure you that we will, from now on, abide by all its regulations and rules.

Now, we are facing a quandary: our customers in Lagos insisted that the umbrellas we are to deliver to them be shipped by vessels of ABC Line, which is a non-conference line because they have contractual obligations with that particular line to have all the goods shipped to them per their vessels. Meanwhile, the L/C our customers have opened to us also calls for the goods to be shipped by vessels of the ABC Line.

Since this case is beyond our control, we can do nothing but to comply with our customers' request. As this is an exceptional case, we report it to you just for your information and reference.

Yours faithfully,

【註】

1. abide by: 遵守。

2. quandary: 窘境；困惑。cf. in a quandary: 進退兩難（維谷）。

3. call for: 規定；要求。

No. 192 出口商要求延展交貨期

Dear Sirs,

We have to inform you with the keenest regret that we find ourselves compelled to ask you for an extension of time in the execution of your order of the 20th September for 10 cases Cotton Pieces Goods.

Owing to the strike of operatives in our manufactory, our business has been completely disorganized, and though we have made strenuous efforts to fulfil our engagements, the nature of the trouble has rendered our attempts fruitless. We now find ourselves impossible to ship your order during the present month, as arranged, and have just sent you a cable, asking for your understanding for an extension of shipment.

Negotiations for reconciliation are proceeding apace, and we have every reason to believe that we shall be able to ship the goods in full by the s.s. "FORTUNE" due to leave here on the 10th November.

We tender you our apologies for the inconvenience you have been caused and hope that you will not allow these circumstances over which we have no control to influence you in your judgment of our business methods.

Yours faithfully,

Enc. 1 copy of telex

【註】

1. operatives：職工，主要指操作運轉機器的職工。

2. disorganize：（組織體制等）被毀；被擾亂。

3. make strenuous efforts：盡最大努力。

4. fulfil our engagements：履行我們的契約。

5. render our attempts fruitless：使我們的嘗試無結果，即「失敗」之意。

6. as arranged：依約。

7. understanding：諒解。

8. negotiations...apace：談判快速進行中。

9. you will not...business methods：（希望）你們不致因為這些我們無法控制的情事，影響到你們對我們辦事的看法。

第二節　傭船運輸

大宗散裝物資的運輸多利用不定期輪。不定期輪的洽船與定期輪不同。一般而言，貨主僱傭不定期輪時，須先向船公司或透過傭船經紀人 (Chartering broker) 詢價，然後由船公司報價，經過討價還價之後，傭船契約才告成立。契約一旦成立後，在雙方或經其代理人在未簽訂正式傭船契約書（Charter Party，簡稱 C/P）之前，常先簽立成交書或訂載書 (Fixture note)，並由船方、貨方、經紀人共同簽名後分別保存。隨後即根據這成交書作成正式的傭船契約。但如船方與貨方均在同一地方時，往往免簽成交書，而逕行協商簽立正式的傭船契約書。

不定期輪的傭船可分為三種：

1. Time charter（計時傭船）：即租傭一定期間，按每噸每日若干計費。

2. Voyage charter（計程傭船）：又稱為 Trip charter，即按航次計費的傭船。

3. Bareboat charter（空船租傭）：又稱為 Demise charter，即將空船交給租船人自行管理營運者。

No. 193　貨主向傭船經紀人詢價

Dear Sirs,

　　We have about 12,000 metric tons of Ammonia Sulfate packed in bags for shipment from Keelung to Bangkok and wish to charter a ship of about 6,000 metric tons. Would you please arrange a vessel and quote us the best charter rate for the shipment.

　　We should add that the vessel must be lying on the berth at Keelung on or before August 10, 20– ready to ship the cargo.

　　Please let us know whether you can arrange this for us and, if so on what terms.

Yours faithfully,

【註】

　　1. ammonia sulfate：硫酸亞。

　　2. charter rate：傭船價。

　　3. lying on the berth：停泊在碼頭船席。lying 為 lie 的現在分詞。

No. 194　傭船經紀人的報價

OFFER SHEET

Taiwan Fertilizer Co., Ltd.　　　　　　　　　　　　　　　　　　　　　June 15, 20–

Dear Sirs,

　　In reply to your enquiry of June 10, we are pleased to offer you subject to the following terms and conditions:

1. Vessel: to be nominated later.

2. Cargo & Quantity: 12,000 metric tons (±10%) of Ammonium Sulfate packed in bags.

3. Loading Port(s): one safe berth of one port Keelung, Taiwan.

4. Discharging Port(s): one safe berth of one port, Bangkok, Thailand.

5. Loading Date:

   a) Laydays: not to commence before July 10, 20–.

   b) Cancelling date: July 31, 20–.

6. Freight & Payment: At the rate of @ US$7.50 per metric ton based on FIO terms, @ US$9.00 per metric ton based on FO terms, and @ US$10.50 per metric ton based on Berth Terms.

   Freight payable at Taipei upon completion of loading.

7. Lighterage & Stevedorage: If any, for charterer's account and risk at both ends.

8. Loading & Discharging rate: Loading rate: 200 metric tons per gang per WWDSHEXUU.

   Discharging rate: CQD

9. Demurrage & Despatch money: Demurrage at US$2,000 per day or pro rata for any part of a day and half of despatch.

10. Commission & Brokerage: Owner's agents at both ends.

11. Agents:

12. Other terms & Conditions:

   a) A/P, if any, for charterer's account.

   b) Other terms and conditions as per "GENCON" C/P revised in 1922.

   This offer is subject to your reply reaching us before noon, June 20, 20–.

Yours truly,

Confirmed and accepted   Tien Kuang Shipping Co., Ltd.

By: By:

【註】

   1. to be nominated later: 船名另定。

   2. cargo & quantity: 貨物名稱及其數量。

   3. safe berth: 指可以安全停靠的碼頭、船席。

   4. loading date: 裝貨日期。

   5. laydays: 到達裝貨期限。

   6. not to commence before...: 某月某日以後開始……。

   7. cancelling date: 指備船人有權解約的日期。

   8. freight & payment: 運費率及其付款方法。

   9. FIO: 為 "free in and out" 的縮寫，係裝卸條件的一種。關於裝卸條件有下列幾種：

     liner terms: 又稱為 "berth term" 定期輪或埠頭條件。即裝卸費用均由船方負擔。

FI (free in)：裝船費用船方免責（即由貨主負擔）。

FO (free out)：卸貨費用船方免責（即由貨主負擔）。

FIO (free in and out)：裝卸費用船方免責（均由貨主負擔）。

FIOS (free in and out and stowed)：裝船卸貨以及艙內堆積費用船方免責（均由貨主負擔）。

FIOT (free in and out and trimmed)：裝船、卸貨以及平艙費用船方免責（均由貨主負擔）。

10. lighterage & stevedorage：駁船費及船上裝卸費。

11. at both ends：在兩端，即指在裝貨港及卸貨港。

12. loading & discharging rate：裝貨率及卸貨率（速度）。

13. WWDSHEXUU: Weather Working Days Sundays & Holidays Excepted Unless Used：為裝卸期限 (laytime or layday) 計算方式的一種，意指「星期例假除外，除非照常裝卸的天氣適宜工作日」，此外尚有：

WWD (Weather Working Day)：天氣適宜工作日。

WWDSHEX (Weather Working Days Sundays and Holidays Excepted)：星期例假除外之天氣適宜工作日。

CQD (Customary Quick Despatch)：照港口習慣速度。

14. demurrage & despatch money：延滯費及快速獎金。

15. commission & brokerage：手續費及經紀人佣金。

16. agents：指船務代理。

17. A/P 為 "Age Premium" 或 "Additional Premium" 的縮寫，即逾齡船保險費或額外保險費之意。

18. GENCON C/P：一般備船契約。

No. 195 訂載書

Taiwan Fertilizer Co., Ltd.   June 20, 20–
Taipei

Dear Sirs,

FIXTURE NOTE

This confirms engagement of freight spaces with you for a trip charter of M.V. "LICHON" or substitute as per following terms and conditions:

1. Cargo & Quantity: Ammonium Sulfate in bags for total 12,000 metric tons net (plus or minus 10%).

2. Type & Name of Vessel and Shipment: Tramp M.V. "LICHON" with Singapore flag building 1995 or substitute with two shipments to be carried into execution during August through October of 20– with each 6,000 metric tons 10% more or less at carrier's option.

3. Loading Port: one safe berth of one safe port, Keelung.

4. Discharging Port: one safe berth of one safe port, Bangkok.

5. Loading Rate: 200 metric tons per gang per WWDSHEXUU.

6. Discharging Rate: CQD

7. Demurrage/Despatch at loading port: Demurrage rate at the rate of US$2,000 per day or pro rata of a day. Despatch at the rate of US$1,000.00.

8. Laytime: Laytime to commence at 8 a.m. next regular working days after vessel's arrival at loading berth or anchorage within the commercial limits of the port and tender of Notice of Readiness during regular working hours.

9. Lighterage & Stevedorage: If any, at loading port to be for charterer's account, and at discharging port to be for consignee's account.

10. Freight Rate: US$7.50 per metric ton NET FIOS. Taiwan Transportation Tax, if any, for shipper's account. A/P, if any, for account of cargo.

11. Payment of Freight: Freight to be fully prepaid in NT Dollar or in U.S. Dollar at Taipei upon completion of loading on the Bill of Lading quantity, discountless and non-returnable whether ship and/or cargo lost or not. Exchange rate shall be official rate at the time of freight payment.

12. Brokerage: No brokerage.

13. Other Terms: As per Gencon Charter Party with modifications as revised 1922. New Jason Clause, Both to Blame Collision Clause, Institute War Risks Clause 1 & 2, and P&I Clause to be deemed fully incorporated in this Fixture Note.

Please sign the duplicate herewith enclosed and return to us at your earliest convenience.

Accepted by:

Yours very truly,

Tien Kuang Shipping Co., Ltd.

【註】

1. M.V. "LICHON" or substitute：裝載船名為 "LICHON" 輪或其代替船。

2. with Singapore flag：掛新加坡旗的，即船籍為新加坡。

3. 10% more or less at carrier's option：承運船有多運或少運 10% 的選擇權。

4. one safe berth of one safe port：一個可以安全停靠碼頭船席的安全港口。

5. laytime：裝卸時間。

6. regular working days：正常工作日。

7. notice of readiness：完成準備通知書。

8. for account of cargo：歸貨主負擔。

9. discountless：無折扣。

10. non-returnable：不退還運費。

11. official rate：官定匯率；官價。

12. New Jason Clause：新傑遜條款。

13. Both to Blame Collision Clause：雙方過失碰撞條款。

14. deemed：視為。

15. incorporated：併入。

 # 第三節　裝運通知與到貨通知

一、裝運通知

　　出口商於辦理裝船手續時，應將辦理裝船的事宜向進口商發出通知，這種通知稱為裝運通知（Shipping advice, Notice of shipment，英國則稱為 Advice of notice），裝運通知的內容包括：Order No.（或 L/C No.）、貨名、數量、船名、裝船日期（或預定開航日期，Estimated Time of Departure, ETD）、裝貨港、預定抵埠日期 (Estimated Time of Arrival, ETA)、以及嘜頭等。這種通知在 FOB, CFR 交易時尤為重要，通常多在裝船前以電傳向進口商發出通知，以便進口商可及時購買保險。

　　出口商於辦妥裝運手續取得 B/L 等裝運單證後，一方面可向銀行辦理押匯，他方面將裝運單證副本，寄給進口商，這也是屬於裝運通知。

No. 196　裝運通知確認書並附上副本貨運單證之一

Dear Sirs,

Re: Your order for canned asparagus

　　We are pleased to inform you that we have shipped today, as per copy of our fax enclosed, by the M.S. "Suez Maru" 900 cartons canned asparagus ordered by you on May 25. The boat is scheduled to arrive at Hamburg on September 30.

　　In order to cover ourselves for this shipment, we have drawn on Dresdner Bank, Hamburg, by sight draft for EUR28,950 under the L/C No. 123 and negotiated it with Bank of Taipei, Taiwan.

To facilitate your taking delivery, we are enclosing the following non-negotiable shipping documents, each in duplicate:

Commercial Invoice No. 123
Bill of Lading No. 231
Insurance Policy No. 567
Packing List No. 123
Certificate of Origin

We trust that you will find all these documents in order and the goods will arrive at Hamburg in perfect condition.

Yours faithfully,

【註】

1. the M.S. "Suez Maru"："M.S." 為 "motor ship" 的縮寫。船名前可以有 "the"，也可以沒有。但在一般文字裡通常有 "the"。用了 "the" 往往可以避免誤解。如 "Queen Mary has just started." 裡的 "Queen Mary" 也許指人，也許指船，但 "The Queen Mary has just started." 裡的 "The Queen Mary" 一定指船。準此。M.S. "Suez Maru" 或 S.S. "Taiwan" 前面可以有 "the"，也可以沒有 "the"。

2. in order to cover ourselves for this shipment：為了收回這批貨款。

3. to facilitate your taking delivery：為便利你的提貨。

4. non-negotiable shipping documents = copy shipping documents：副本貨運單證，因不能做為押匯之用，所以稱為 "non-negotiable"。

茲舉一 Fax shipping advice 於下：

No. 197 裝運通知

YOUR ORDER 123 CANNED ASPARAGUS 900 CARTONS SHIPPED PER MS SUEZ MARU ETD KEELUNG JULY TEN ETA HAMBURG SEPTEMBER THIRTY COPY DOCUMENTS AIRMAILING

No. 198 裝船通知確認書並附上副本貨運單證之二

Dear Sirs,

Re: Your Order No. 123 for canned mushroom

We are pleased to inform you that we have effected shipment of the first two items under the subject order per s.s. "Nisho Maru" which sailed from Keelung on March 15.

To facilitate your taking delivery, we are enclosing two copies each of the following non-negotiable shipping documents:

Commercial Invoice
Bill of Lading
Insurance Policy
Packing List
Certificate of Origin
Special Customs Invoice

For this shipment, we despatched a cable to you on March 15, reading:

"YOUR ORDER 123 SHIPPED FIRST TWO ITEMS TOTALING 1000 CARTONS SS NISHO MARU 15TH ETA NEW YORK APRIL 5"

We trust the goods will reach you in perfect condition and open up to your satisfaction.

Faithfully yours,

【註】

1. Special Customs Invoice：特種海關發票，簡稱 SCI。

2. open up to your satisfaction：開箱時會使你滿意，因為如以完好情況 (perfect condition) 抵達，而無破箱情形，則開箱時自將令人滿意。

二、到貨通知

當載貨船隻開抵進口港時，進口港的船務代理根據提單上 Notify party 的記載，向被通知人（一般而言，不是進口商就是其代理人），發出 Arrival notice（到貨通知），請其及時提貨。

No. 199　船公司對受貨人發出到貨通知

Dear Sirs,

Subject: 500 BALES AMERICAN RAW COTTON
B/L No. 123 EX M.S. "President"

We wish to inform you that the subject cargo, of which you are the Notify Party, arrived at Keelung on August 5, 20–.

Please produce the Bill of Lading to our agents, Messrs. Mo Lan Co., Ltd. in Keelung, in exchange for the necessary Delivery Order.

We take this opportunity to draw your attention that, as stated on the relative Bill of Lading, the carrier's responsibility for the cargo ceases immediately the cargo leaves the ship's tackle and thereafter all risks and expenses involved are the responsibility of the cargo. The cargo must also pay for coolie hire, landing, storing and delivery charges, and, if necessary, lighterage. According to Keelung Harbour Bureau regulations, the storage is payable commencing sixth day after the storing of the cargo in the Bureau's warehouse.

Customs regulations require that consignees complete customs clearance procedures within 15 days of the arrival of the vessel, failing which, overtime charge will be levied up to a limit of 30 days, if the delivery of the cargo is not taken at the end of the 30 days it may be auctioned.

Yours faithfully,

【註】

1. produce：提出（提單）。

2. in exchange for：交換（小提單）。

3. delivery order：小提單，簡稱為 D/O。在海運實務上，B/L 並不直接用於提貨，受貨人欲提貨，須憑 B/L 換取 D/O，作為實際報關提貨的憑證。

4. tackle：吊桿。

5. the responsibility of the cargo：即貨主負責之意。

6. coolie hire：搬運工人工資。

7. customs clearance procedures：通關手續。

8. overtime charge：滯報費。

9. auction：拍賣。

三、擔保提貨

載運貨物的船已到進口港，而經由銀行的正本提單還沒有寄到時，進口商為了及時提貨，可請銀行作保向船公司發出保證書，以便先行提貨，這種保證書稱為擔保提貨書 (Indemnity and guarantee for delivery of cargo without surrender of B/L)。俟正本提單寄達之後，由作保的銀行將其向船公司換回擔保提貨書，以便解除保證責任。由於保證責任重大，一般銀行並不隨便作保。當然，假如是憑信用狀交易時，進口商可請求開狀銀行作保，開狀銀行通常也不致拒絕。但以託收的 D/P 或 D/A 交易時，如進口商未事先與銀行洽妥，銀行不輕易作保，進口商不可不知。

　　進口商向銀行申請擔保提貨時，須提出擔保提貨申請書。

No. 200　擔保提貨書

INDEMNITY AND GUARANTEE
DELIVERY WITHOUT BILL OF LADING

From: _____
Address: _____
Date: _____

To:
MESSRS BUTTERFIELD & SWIRE
　　AGENTS:
　　　　KEELUNG

S.S. _____

Arrived_____ From_____

　　In consideration of your releasing for delivery to us or to our order the undermentioned goods of which we claim to be the rightful owner without production of the relevant bill(s) of lading (not as yet in our possession).

　　We hereby undertake and agree to indemnify you fully against all consequences and/or liabilities of any kind whatsoever directly or indirectly arising from or relating to the said delivery and immediately on demand against all payments made by you in respect of such consequences and/or liabilities, including costs as between solicitor and client and all or any sums demanded by you for the defence of any proceedings brought against you by reason of the delivery aforesaid.

　　And we further undertake and agree upon demand to pay any freight and/or General Average and/or charges due on the goods aforesaid (it being expressly agreed and understood that all Liens shall subsist and unaffected by the terms hereof).

　　And we further undertake and agree that immediately the Bill(s) of Lading is/are received by us we will deliver the same to you duly endorsed.

B/L No.	MARKS	QUANTITY & CONTENTS	SHIPPED BY	PORT OF SHIPMENT	PORT OF DISCHARGE

_____
Consignees

　　In consideration of you having accepted the above indemnity and undertaking at our request we hereby guarantee the due performance thereof by above consignees and agree that no time or other indulgence granted by you to the consignees shall discharge us from our liability hereunder.

Dated　　　　　　　20　　　　_____
Banker's Signature.

Indemnities with limited guarantees or bearing any
qualifying remarks whatsoever cannot be accepted.

No. 201　擔保提貨申請書

致: 中央信託局外匯業務公鑒　　　　　　　日期:
Central Trust of China　　　　　　　　　Date:
　　Foreign Department
　　　　Taipei

逕啟者:
Dear Sirs:

　　茲附奉擔保提貨書請　貴處惠予簽署以便向
　　We enclose herewith for your countersigning our Letter of Guarantee

　　　　　　請求提取下列貨物該貨物係由
addressed to _____calling for following cargoes, shipped from_____
裝　　　　　　輸運臺
per S.S. _____

信用狀號碼	提單號碼	嘜　頭	貨　品	價　值
L/C No.	B/L No.	Marks	Commodity	Value

　　因上列貨物之提單尚未寄到
　　The Bills of Lading of these cargoes have not yet arrived.

　　　貴處簽署上項擔保提貨書之一切後果均由本
　　In consideration of your countersigning this Letter of Guarantee, we

號 （　　　　）負責決不使　　貴處因此蒙受任
hereby agree to hold you harmless for all consequences that may arise from

何損失茲並同意一俟提單寄到即將上項擔
your so doing. We further agree that on receipt of the Bill of Lading for

保提貨書換回送還貴處註銷或由貴處
the above shipment we will deliver the said Letter of Guarantee to you

代勞將該項提單逕交船公司換回上項擔保提
for cancellation or you may deliver the Bills of Lading direct to the

貨書以便解除貴處之保證責任
steamship company on our behalf to release your Letter of Guarantee.

附批: 本件貨物之正式單據寄到時請煩掛號郵寄敝處，　　　申　請　人
　　　中途如有遺失，概由敝處負責，與貴處無涉。　　　　Yours truly,

簽　署　日　期　　　　餘　餘
Date of Countersigning　　　Balance

　　　　　　　　　　　　　　　　　　_____

簽章人　　　　　　　　　　　核對人
Signed by:　　　　　　　　　Checked by:

四、副提單背書

　　信用狀項下的貨物如從航程較近的地區進口，為配合實際需要，常於信用狀中規定出口商可將正本提單中的一份（稱為副提單）逕寄進口商，進口商收到此正本提單後，即可簽具「提單背書申請書」向開狀銀行申請背書，以便向船公司提貨。

No. 202　提單背書申請書

APPLICATON FOR BILL OF LADING ENDORSEMENT

CENTRAL TRUST OF CHINA　　　　　Date: _____
FOREIGN DEPARTMENT
TAIPEI　　　　　　　　　　　　　　Serial No. _____

Dear Sirs:

　　We enclose herewith for your endorsement the undernoted duplicate

Bill of Lading issued by _____

for delivery of the following cargoes, shipped from _____

per S.S. _____

L/C No.	B/L No.	Marks	Commodities	Quantity	Amount

　　In consideration of your endorsing this duplicate Bill of Lading, We hereby agree to hold you harmless for all consequences that may arise from your so doing.

Date Endorsed　　　　　　　Balance　　　　　　　Yours truly,

　　　　　　　　　　　　　　　　　　　　　　　_____

經副襄理　　　　主　任　　　覆　核　　　承　辦
　　　　　　　　副主任

第四節 有關貨物運輸的有用例句

一、洽 船

1. Please quote us your lowest rate of freight for 10,000 pcs. of plastic toys from Keelung to Seattle.

2. In confirmation of our conversation over telephone this morning, we ask you to reserve for us the space of 65 measurement tons on the s.s. "Oriental Despatcher" which is scheduled to sail from Keelung on May 25 for New York for shipment of plastic shoes.

3. We are glad to confirm our reservation of space for 65 measurement tons plastic shoes to New York per the s.s. "Oriental Despatcher" leaving Keelung May 25, at the rate of US$60 per ton.

4. We fear there would be no space available in our steamers to London before the end of December. We shall, however, make telex inquiry of our Kobe office to see if we can arrange to accept your cargo.

二、裝運通知

㈠開頭句

1. We are pleased to $\begin{Bmatrix} \text{inform} \\ \text{advise} \end{Bmatrix}$ you that we have shipped your order No....

→ for...(goods)... $\begin{Bmatrix} \text{by} \\ \text{per} \\ \text{on board} \end{Bmatrix}$ the s.s. "..." which sailed here on...(date).

2. We have today $\begin{Bmatrix} \text{forwarded} \\ \text{shipped} \end{Bmatrix}$ per s.s. "London" which $\begin{Bmatrix} \text{sailed} \\ \text{left} \\ \text{cleared} \end{Bmatrix}$ here today.

　　註：clear：出港；啟碇；開出，即 "leave port" 之意。

3. We {confirm / have the pleasure of confirming / are pleased to confirm} our {e-mail / fax / cable} just despatched informing you that we have shipped...

4. Your order No....has been forwarded, as per copy of our {cable / fax / telex} enclosed, by the s.s. "Taiwan" which left this port today and which is to arrive at your port on...(date).

5. We have sent you via airmail a yard of the shipment sample together with copies of shipping documents from which you will find that the article is of the required grade.

(二)條　件

1. Enclosed please find our invoice for US$...

2. We are {sending you herewith / enclosing} invoice and {air waybill / B/L} in duplicate with insurance policy.

3. We have {drawn / valued} on {Bank of America / you} {at sight / at 30 d/s} for $... under L/C No. 123 → issued by Bank of America, and negotiated it thru {our bankers. / Bank of Taiwan.}

4. In order to cover {this shipment / ourselves for this shipment}, we have {valued / drawn} on...Bank at sight for US$...under their L/C No. 123 thru our bankers, Bank of Taiwan.

(三)結　尾

1. We {trust / hope} the goods will reach you {in perfect condition / safe / safely / in good condition} and open up to

$$\text{your} \begin{cases} \text{complete} \\ \text{entire} \end{cases} \text{satisfaction.}$$

2. We trust the shipment will reach you $\begin{cases} \text{safely} \\ \text{in due time} \end{cases}$, and $\begin{cases} \text{hope} \\ \text{await} \end{cases}$ your

$\rightarrow$ further $\begin{cases} \text{order.} \\ \text{patronage.} \end{cases}$

3. We thank you for this order and trust the goods will reach you promptly and in good order.

4. We trust the high quality of our...(name of article)...will prove an inducement of further business.

5. We trust this purchase will bring you a good profit and result in your further orders.

6. If you have further need of this item, we shall be very much pleased to accept a reorder.

7. We are looking to $\begin{cases} \text{the continuation of our pleasant business relations.} \\ \text{a continued and increasing business with you.} \end{cases}$

8. We shall be pleased to receive your further orders, which will always receive our utmost careful attention.

三、延展船期

1. We regret to inform you that the shipment against your order No. 123 has been

$\rightarrow$ delayed $\begin{cases} \text{because of} \\ \text{as a result of...} \\ \text{due to} \end{cases} \begin{cases} \text{strike of the port-workers.} \\ \text{recent floods.} \\ \text{a rush of orders for Xmas.} \\ \text{mistake in the process of manufacture.} \end{cases}$

2. As a matter of fact, s.s. "Taiwan" was the boat booked first for your order, but we have been barred from using it owing to the seamen's strike at this end. We hope you will understand our position that the replacement of boat by s.s. "EASTERN ARGOSY" was the soonest possible action we could take.

3. Strikes of steel workers and coal miners are going on in this country with no prospect of immediate settlement. Therefore, we regret to tell you that there is little possibility of our being able to comply with your request for August shipment.

4. Owing to a strike in one of the factories with whom your order was placed, we regret having to confirm our cable just sent informing you that we were unable to ship order #123 by s.s. "Hupeh," which set sail from Keelung on June 2nd. We apologize for this unfortunate delay, which was unavoidable under the circumstances.

5. We had difficulty in obtaining necessary materials for manufacturing the machines you ordered. The shipment, therefore, will be delayed for two weeks.

6. We apologize for the delay in executing your order 112 for 500 bales of saxony cloth. The delay was caused by shortage of raw material and also by a backlog of orders.

7. Due to a serious shortage of shipping space we cannot deliver these machines until September 25.

8. Since there is no steamer for Lagos available until May 3, we cannot effect shipment before end of April.

9. We would appreciate it if you could allow us a one-month extension for delivery. In order to ensure high quality of the machine, we need at least two weeks for careful testing.

第二十四章
貨運單證 (Shipping Documents)

🖖 第一節　國際貿易貨運單證的種類

國際貿易的貨運單證 (Shipping documents) 繁多，但可分為基本單證與附屬單證兩大類。

1. Fundamental shipping documents：

(1)買賣方面：Commercial invoice （商業發票）

(2)運輸方面
- Ocean bill of lading （海洋提單）
- Air waybill （航空提單）
- Forwarder's cargo receipt （運輸承攬商貨物收據）
- Receipt for parcel post （郵政包裹收據）

(3)保險方面
- Insurance policy （保險單）
- Certificate of insurance （保險證明書）

2. Subsidiary shipping documents：
- Packing list （裝箱單）
- Weight/Measurment list （重量尺碼單）
- Consular invoice （領事發票）
- Customs invoice （海關發票）
- Certificate of origin （產地證明書）
- Inspection certificate （檢驗證明書）
- Phytosanitary certifcate （檢疫證明書）
- Fumigation certificate （薰蒸證明書）
- Health certificate （衛生證明書）

第二節　各種貨運單證實例

一、商業發票 (Commercial invoice)

　　商業發票（簡稱發票）為貨物運出時，賣方向買方所開出說明所出售貨物名稱、價格及有關費用的文件，國際貿易上使用的商業發票雖無一定的格式，但就其內容而言，通常都由首文、本文及結尾文構成。

No. 203　商業發票

```
          ╭─────╮        BURDA ENTERPRISES INC.
          │ ☀   │        Exporters-Importers-Manufacturers      REFERENCE BANK:
          │Burda│        E-MAIL: burda@ms1.hinet.net            FUBON COMMERCIAL BANK
          ╰─────╯        E-MAIL: burda1@ms22.hinet.net          HUA NAN COMMERCIAL BANK
                         5TH FL., 26 SEC. 3, JEN-AI ROAD
  PHONE : (02) 2705-9286 (10 LINES)   TAIPEI, TAIWAN, R.O.C.
  FAX   : (02) 2701-5235
          2701-5236
  MODEM : (02) 2705-9289
```

ABC TRADING CO., LTD.
INVOICE

No. 0107　　　　　　　　　　　　　　　　　　　Date: Oct.30,20-

INVOICE of 500 dozens of Sport Shirts

For account and risk of Messrs. XYZ TRADING CO., LTD.

Shipped by ABC TRADING CO., LTD.　　　Per S.S. "CAPE HENRY" Voy.No.13078

sailing on or Before Oct.30,20-　From Keelung　to New York

Issued Bank

Marks & Nos.	Description of Goods	Quantity	Unit Price	Amount
XYZ NEW YORK C/No.1-50 MADE IN TAIWAN R.O.C.	SPORT SHIRTS STYLE A Silk fabric for men's	500 dozs.	CIF NEW YORK US$25.72	US$12,860.00

SAY TOATL US DOLLARS TWELVE THOUSAND EIGHT HUNDRED SIXTY ONLY.

ABC TRADING CO., LTD.
Manager
Christen Lin

E. & O. E.

關於商業發票的用語

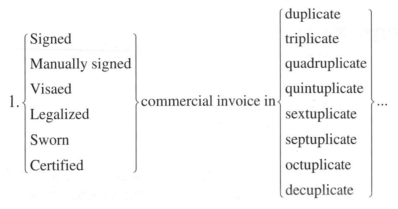

2. Signed commercial invoice indicating IL No....

3. Signed commercial invoice duly countersigned by Mr. A.

二、海洋提單 (Ocean bill of lading; Marine bill of lading)

海洋提單為船公司或其代理人所簽發證明收到貨物，並約定將貨物自一地運至另一地，交給提單持有人的一種物權證書 (Document of title)。海洋提單本質上為載運貨物的化身，係各種貨運單證中最重要者。

1.提單例示及說明：

　說明：

⑴ Bill of lading：表明本文件為「提單」。

⑵ Shipper：即託運人，通常即為出口商。

⑶ Consignee：受貨人，通常在此欄填上 "To order"（待指定）；"To order of shipper"（待託運人指定）；但也有以 "To order of ×××× Bank" 表示者。

⑷ Notify party：被通知人，通常為進口商或其代理人。

⑸ Ocean vessel：填入船名。

⑹ Port of loading：裝貨港。

⑺ Port of discharge：卸貨港。

⑻ Marks & No.：嘜頭及件數。

⑼ Description of goods：貨物描述。

No. 204　海洋提單

EVERGREEN
EVERGREEN MARINE CORPORATION

(1) **BILL OF LADING**

(2) Shipper/Exporter 　　XXX CORP	S/O No/Ref No　0051
	Export References
Shipper Code	
(3) Consignee 　TO ORDER OF 　SUEZ CANAL BANK	Forwarding Agent References
(4) Notify Party(Complete name and address) MAHMOUD COMPANY	Port and Country of Origin(for the merchant's reference only)
	Onward Inland Routing /Export instructions(for the Merchant's reference only)

Pre-carriage by　(6)	Notify code	Freight & Charges Indicated in Columns with appropriate Code P Prepaid　C Collect　N/A Not Applicable
Place of Receipt/Date	Port of Loading 　KAOHSIUNG	

		Iod Lond	Iod Handling	Oc Freight	Dis Handling	Dis Lond
		N/A	P	P	N/A	N/A

(5) Ocean Vessel　EVER FORWARD　Voy.No.　0002-032W

In Witness Whereof the undersigned, on behalf of Evergreen Marine Corporation, the Master and the Owner of the Vessel, has signed the number of Bill(s) of Lading stated below, all of this tenor and date, one of which being accomplished, the others to stand void.

(7) Port of Discharge　PORT SAID　Place of Delivery

Particulars furnished by the Merchant

(8) Container Numbers Marks & No	Number and Kind of Packages	(9) Description of Goods	measurement (m³) Gross Weight(KGS)
CONTAINER NO /SEAL NO EMCU2728800/20'/750844 　　.·.·.· ·.·.　HALAWA　.·.· 　　.·.·.· PORT SAID GEP-1387 C/NO.1-173 MADE IN TAIWAN R.O.C.	1 X 20'	SAID TO CONTAIN : 173 CARTONS MOTORCYCLE SPARE PARTS AND ACCESSORIES L/C NO.SC.35881 (6) "FREIGHT PREPAID" FULL LINER TERMS SHIPPER'S LOAD & COUNT	29.222 CBM 4,239.00 KGS

TOTAL NUMBER OF
CONTAINERS OR PACKAGES
IN WORDS　　ONE(1) CONTAINER ONLY

FREIGHT & CHARGES	Revenue Tons	Rate		Per	Prepaid	Collect
	AS ARRANGED					

(10) PARTICULARS OF SALE CONTRACT AND/OR ORDER AND/OR BANK LETTER OF CREDIT SHOWN HEREIN WERE INSERTED AT SHIPPER'S REQUEST AND FOR THEIR OWN PERSONAL CONVENIENCE IN ORDER TO FACILITATE THE NEGOTIATION OF THE B/L. SUCH PARTICULARS WERE NOT CHECKED BY CARRIERS AND/OR SHIP AGENTS NOR WERE ANY DOCUMENTS RELATING THERETO PRESENTED TO THEM. IT IS, THEREFORE, AGREED THAT THE INSERTION OF SUCH PARTICULARS OR THE LIKE IN THIS B/L MUST NOT BE REGARDED AS A DECLARATION OF VALUE OF THE GOODS SHIPPED.

B/L NO KSGPSD2051	(11) Number of Original B(s)/L 　THREE(3)	Prepaid at 　　TAIPEI	Payable at
	(12) Place of B(s)/L Issue/Date 　TAIPEI,TAIWAN NOV.04,2001	Exchange Rate US$1=NT$32.00	Exchange Rate
Service .Type 　FCL/FCL	Laden on Board the Vessel	EVERGREEN MARINE CORPORATION	
	(13) By		

TERMS OF BILL OF LADING CONTINUED FROM BACK HEREOF.　　　　　AS AGENTS FOR THE MASTER

COPY NON-NEGOTIABLE

(10) Freight & Charges：運費。

(11) Number of original B(s)/L：正本提單份數。

(12) Place of B(s)/L issue/date：提單簽發地、日。

(13) Signature：提單簽署人。

2.提單種類：

(1)按貨物是否已裝船分 { ① Shipped B/L　　　　（裝運提單）
② Received B/L　　　　（備運提單）

(2)按是否轉運分 { ① Direct B/L　　　　（直達提單）
② Through B/L　　　　（聯運提單）

(3)按受貨人表示方法分 { ① Straight B/L　　　　（直接提單）
② Order B/L　　　　（指示提單）

(4)按是否有瑕疵批註分 { ① Clean B/L　　　　（清潔提單）
② Unclean B/L　　　　（不潔提單）

(5)按是否詳載運輸條款分 { ① Short form B/L　　（簡式提單）
② Long form B/L　　（長式提單）

(6)其他提單 { ① Container B/L　　　　（貨櫃提單）
② Combined transport B/L　　（聯合貨運提單）

3.提單上的各種批註：

Top (Cap, Head) off	蓋子鬆落	Nails off	鐵釘鬆落
Seal broken	封印破裂	Seals off	封印鬆落
Bundle off	捆紮鬆弛	Hoop off	箍帶鬆落
Rattling	裡面商品動搖發出聲音	Leaking	漏出
Bented	彎曲	Hoop rusty	鐵及帶生銹
Bag torn	袋子破裂	Bale torn	袋子破裂
Renailed	補釘	Staves off	標蓋鬆落
Split	裂開	Mark mixed	嘜頭混淆
Mark indistinct	嘜頭不清楚	Dented	凹下

Cover chafed	擦傷	Old (Second-hand) case	舊箱
Used drum	舊鐵桶	Loose packing	包裝鬆弛
Crushed	壓壞	Label missing	標籤脫落
Two cartons in wet condition	兩箱潮濕	Two cases broken	兩箱破裂
Contents crushed	壓碎		

N/R for
- breakage（破損）
- mortality（死亡）
- wither（枯萎）
- melting（溶解）
- perishing（腐敗）
- damage（損害）

不負責

註：N/R 為 Not Responsible 的縮寫。

4.損害賠償約定書 (Letter of indemnity)：

　　一般的信用狀都規定出口商所提示的 B/L 必須為 Clean B/L，因此，B/L 上如有前述任何一種批註，就變成 Unclean B/L 而無法憑以押匯。然而，船公司往往循出口商的要求，憑其所出具的 Letter of indemnity 改發 Clean B/L，以便其能順利押匯，這是相當普遍的國際海運實務。然而這種做法是不合法的。所以船公司對善意的持單人 (Bona fide B/L holder) 仍需負全責。出口商一經出具 Letter of indemnity，將來如船公司被索賠，自應由出口商連帶負責。（有些國家的法院判決認為出口商不必負責。）

No. 205　損害賠償約定書

LETTER OF INDEMNITY

NAME OF COMPANY

ADDRESS:

Date:

EVERGREEN MARINE CORPORATION

Gentlemen:

　　We herewith acknowledge the following exception,

_____

which was taken by the receiving clerk and appears on the cargo receipt,

　　We request that this exception not be shown on the _____ v/____ shipment laden at _____, on _____, B/L No. _____ covering this shipment, and in consideration there of we and/or our principals, and/or others concerned hereby undertake and agree to hold the _____ her Master, Owners and Agent or any of them harmless and full indemnified against any and all consequences, loss, damage and expenses due to omitting the aforesaid exception from the bill of Lading.

　　If any claims are presented we will adjust them directly with the claimants and we will fully advise the ultimate consignee of this cargo as to the terms of this agreement.

Shipper: _____

cc. Port of Discharge　　　　　　　　　　　　By: _____
　　General Claims Agents,
　　Shipper
　　File　　　　　　　　　　　　　　　　　　　Title: _____

三、保險單 (Insurance policy)

　　保險單是證明保險契約成立的正式憑證，載明雙方當事人約定的權利及義務，如與保險公司訂有流動保單 (Floating policy)，則對各批貨物的保險，保險公司另簽發保險證明書 (Certificate of insurance)。

　　保險單的種類：

　⑴依船名是否已確定分

　　① Named policy（船名確定保單），又稱為 "Definite policy"。

　　② Unamed policy（船名未確定保單），又可分為：

No. 206 保險單

富邦產物保險股份有限公司
Fubon Insurance Co.,Ltd.
HEAD OFFICE: 237, CHIEN KUO SOUTH ROAD, SEC. 1, TAIPEI, TAIWAN
TELEPHONE: 7067890 CABLE: "SAFETY" TAIPEI TELEX:11143 SAFETY FAX: (02) 7042915

MARINE CARGO POLICY

POLICY NO. 008M-045590

Claim, if any, payable at/in
NEW YORK IN U.S.DOLLARS

ASSURED
ABC TRADING CO.,LTD.

Invoice No. 405067

Amount insured US$14,223.00

U.S.DOLLARS FOURTEEN THOUSAND TWO HUNDRED
TWENTY THREE ONLY.

Ship or Vessel	From	To/Transhipped at
S.S. "CAPE HENRY" Voy.No.13078	KEELUNG	NEW YORK

Sailing on or ~~about~~ before Oct.30,2001

Thence to

SUBJECT-MATTER INSURED

SPORT SHIRTS
TOTAL:500 DOZ.
PACKED IN 50 CTNS.
L/C NO.: SA-547/3796

SPECIMEN

Conditions
Subject to the following clauses as per back hereof
Institute Cargo Clauses (Ⓑ)
INSTITUTE WAR CLAUSES (CARGO)

Marks and Numbers as per Invoice No. specified above.   Valued at the same as Amount insured.

Place and Date signed in Taipei on Oct.29,2001   Number of Policies issued in Duplicate

☞ *The Assured is requested to read this policy and if it is incorrect return it immediately for alternation.*

INSTITUTE REPLACEMENT CLAUSE (applying to machinery)

In the event of loss of or damage to any part or parts of an insured machine caused by a peril covered by the Policy the sum recoverable shall not exceed the cost of replacement or repair of such part or parts plus charges for forwarding and refitting, if incurred, but excluding duty unless the full duty is included in the amount insured, in which case loss, if any, sustained by payment of additional duty shall also be recoverable.
Provided always that in no case shall the liability of Underwriters exceed the insured value of the complete machine.

LABEL CLAUSE (applying to labelled goods)

In case of damage from perils insured against affecting labels only, loss to be limited to an amount sufficient to pay the cost of reconditioning, cost of new labels and relabelling the goods.

CO-INSURANCE CLAUSE (applicable in case of Co-insurance)

It is hereby understood and agreed that this Policy is issued by CATHAY INSURANCE COMPANY, LIMITED, on behalf of the co-insurers who, each for itself and not one for the others, are severally and independently liable for their respective subscriptions specified in this policy.

Notwithstanding anything contained herein or attached hereto to the contrary, this insurance is understood and agreed to be subject to English law and practice only as to liability for and settlement of any and all claims.
This insurance does not cover any loss or damage to the property which at the time of the happening of such loss or damage is insured by or would but for the existence of this Policy be insured by any fire or other insurance policy or policies except in respect of any excess beyond the amount which would have been payable under the fire or other insurance policy or policies had this insurance not been effected.
We, CATHAY INSURANCE COMPANY, LIMITED, hereby agree, in consideration of the payment to us by or on behalf of the Assured of the premium as arranged, to insure against loss damage liability or expense to the extent and in the manner herein provided.
In witness whereof, *I the Undersigned* of CATHAY INSURANCE COMPANY, LIMITED, on behalf of the said *Company* have subscribed *My Name* in the place specified as above to the policies, the issued numbers thereof being specified as above, of the same tenor and date, one of which being accomplished, the others to be void, as of the date specified as above.

For FUBON INSURANCE COMPANY, LIMITED.

Examined ------   President

a.Floating policy（流動保單）：一種僅列明概括的條件，而船名和其他條件留待事後通知 (Declare) 的保險單。

b.Open policy（預約保單）：以預約方式一次承保未來多批貨物的保險單。

⑵依保險金額是否確定分

① Valued policy（定值保單）：即保險單上載明保險標的物價值（即保險價額 Insured value）者。

② Unvalued policy（不定值保單）：即保險單上僅訂有保險金額 (Insured amount) 的最高限度，而將保險價額留待日後保險事故發生時再予補充確定者。

四、裝箱單 (Packing list)

又稱為包裝單、花色碼單，英文又稱為 Packing specification, Specification of packages 或 Specification of contents，係說明一批包裝貨物，逐件內容的文件，為商業發票的補充文件。

包裝貨物如每件內容不同時，必須有裝箱單，否則受貨人除非拆包，將無法知悉各件包裝的內容，海關、公證行抽驗時也將因而遭遇困難。

No. 207　裝箱單

PACKING LIST

20_____
Place and Date of Shipment

To
Gentlemen:
　　　Under your Order No. _____ the material listed below
Was Shipped on _____ Via _____
To _____

SHIPMENT CONSISTS OF:		EXPORT SHIPPING MARKS
____ Cases ____ Packages		
____ Crates ____ Cartons		
____ Bbls ____ Drums		
____ Reels ____		

LEGAL WEIGHT IN WEIGHT OF ARTICLE PLUS PAPER, BOX, BOTTLE, ETC., CONTAINING THE ARTICLE AS USUALLY CARRIED IN STOCK.					
PACK-AGE	CONTENTS OF EACH PACKAGE	WEIGHT IN LBS. OR KILOS	DIMENSION IN	INCHES	QUANTITY

NUMBER		GROSS WEIGHT EACH	LEGAL WEIGHT EACH	NET WEIGHT EACH	HEIGHT	WIDTH	LENGTH	

Total Invoice price $ _____
Boxing charge, if extra, $ _____

(Shipper)_____

Remarks: Please return this Sheet Promptly if Contents of Case Do Not Agree with Items Given Above All Claims for errors or shortage must be made in writing within ten days after receipt of goods

五、重量尺碼證明書（單）(Weight/Measurement certificate/list)

係記載裝運貨物每一件重量及體積的文件，這種文件或由賣方出具或由公共丈量人 (Public weigher) 出具。

六、領事發票 (Consular invoice)

又稱為領事簽證貨單，是由出口商向駐在輸出國的輸入國領事申請簽發的一種單證，其作用為：

1. 供作進口國海關和貿易管理當局統計之用。
2. 代替產地證明書之用。
3. 限制或禁止某些非必需品或未經批准的貨品隨便進口。
4. 藉簽證課征規費，做為領事辦公費。

七、海關發票 (Customs invoice)

目前貨物銷往美國、加拿大、紐西蘭、澳洲、南非聯邦、西非各國時，出口商須提出海關發票，其作用：

1. 供作進口國海關統計之用。
2. 供作產地證明之用。
3. 供作進口國查核有無傾銷之用。

No. 208　　重量尺碼證明書（單）

Shipper	NIPPON KAIJI KENTEI KYOKAI
KAISEI SANGYO CO., LTD.	(JAPAN MARINE SURVEYORS & SWORN MEASURERS ASSOCIATION)
	FOUNDED IN 1913 & LICENSED BY THE JAPANESE GOVERNMENT

OKAMOTO FREIGHTERS LTD.

Certificate No.
1300-10420-0005386 (03)

Sheet　1

Certificate issued
YOKOHAMA　NOV.04,2001

Ref. No. (For our reference)
M2090-1538　004　(0339)　0239

CERTIFICATE AND LIST
OF
MEASUREMENT AND/OR WEIGHT

HEAD OFFICE:
Kaiji Bldg., No. 3-7, 1-chome, Hatchobori
Chuo-ku, Tokyo-104
JAPAN

Ocean Vessel	Port of Loading
VIRGINIA	YOKOHAMA

Port of Discharge	Date & Place of Measuring and/or Weighting
Kaohsiung	NOV.03,2001　　YOKOHAMA

Marks & Nos.	No. of P'kgs.	Kind of P'kgs.	Description	G. W.	Meast.
T F M C KAOHSIUNG KSCL-F116 MADE IN JAPAN C/NO. 1-2					
			"GOLD" VENEER ROTARY KNIFE	KG	CU.METER
	2 CASES ＊＊＊＊＊＊			680	0.265

DETAILS:	CASE	M CM L	M CM W	M CM H	KG	CU.METER
	1	2 89	0 27	0 18	360	0.140
	2	2 58	0 27	0 18	320	0.125

We hereby certify that the above measurements
and/or weights of the goods were taken by our
measurers solely for reasonable ocean freight
in accordance with the provisions of recognized
rules concerned.

(seal: NIPPON KAIJI KENTEI KYOKAI · JAPAN MARINE SURVEYORS & SWORN MEASURERS ASSOCIATION · N.K.K.K.)

No. 209 巴拿馬領事發票（正面）

巴拿馬領事發票

REPUBLICA DE PANAMA
MINISTERIO DE HACIENDA Y TESORO
DIRECCION GENERAL DE CONSULAR Y DE NAVES

Nº 00338 E

VALOR B/.3.00 EL JUEGO
LEY 52, ENERO 4, DE 1963
F C. 77 (30.000)

POR EFECTOS EMBARCADOS EN:

FACTURA CONSULAR
CONSULAR INVOICE

DISTRIBUCION
ORIGINAL : DEBERA SER PRESENTADO A LAS AU-
TORIDADES ADUANERAS
DUPLICADO : DIRECCION GENERAL DE CONSULAR
Y DE NAVES
TRIPLICADO : CONTRALORIA GENERAL
CUADRUPLICADO: ARCHIVO CONSULADO

PAIS DE ORIGEN: Taiwan, R.O.C. COUNTRY OF ORIGIN	VENDEDOR (ES): MAIN ... SELLERS OR SHIPPERS ENTERPRISE CO.,LTD.
PUERTO DE EMBARQUE: KEELUNG PORT OF SHIPMENT	CONSIGNADO: RODIVAN S.A. CONSIGNEE
PUERTO DE LLEGADA: CRISTOBAL PORT OF ARRIVAL	FECHA DE ZARPE: Jan. 5. 2001 DATE OF SHIPMENT
VAPOR: N.D.LLOYD ELBE NAME OF VESSEL Jan. 5. 2001	LUGAR DE DESTINO: FINAL DESTINATION

NO. CONOCIMIENTO DE EMBARQUE Y FECHA:
BILL OF LADING & DATE

Jan. 5. 2001

NO. DE MANIFIES.O Y FECHA
NO. OF MANIFIEST & DATE

NUMERO NUMBER	BULTOS NUMBER OF PACKAGES	CLASE DE BULTOS KIND OF PACKAGES	CAPACID. EN LITROS CAPACITY IN LITRES	PESO EN KILOS		DESCRIPCION DE LAS MERCADERIAS DESCRIPTION OF THE MERCHANDISE	VALOR	
				NETO NET WEIGHT	BRUTO GROSS WEIGHT		PARCIAL	TOTAL
C/No.		W/case	doz.	Kgs.	Kgs.	Cosmetics: "EXOTICA" BRAND	US$ (per doz)	US$
1-16	16		64	82	96	NP-900/F Make-up Kits	9 00	564 00
17-50	34		136	177	224	NP-555/B Make-up Kits	10 00	1,360 00
51-80	30		120	148	180	MSA-124 Mascara	7 64	916 80
81-92	12		48	73	102	M-108 Mascara	5 20	145 60
93-100	8		40	60	79	NP-252/B Make-up Kits	4 50	180 00
101-108	8		48	84	127	M-173/C Mascara	6 70	321 60
	108	W/cases	456	624	808	TOTAL		3,488 00

SAY: US DOLLARS THREE THOUSAND FOUR HUNDRED EIGHTY EIGHT ONLY.

SAY: TOTAL ONE HUNDRED AND EIGHT WOODEN CASES ONLY.

SAY: TOTAL FOUR HUNDRED FIFTY SIX DOZEN ONLY.

TOTAL 108 W/cases	VALOR TOTAL DE LA MERCANCIA	US$3,488.0

MARCAS MARKS			
RODIVAN CRISTOBAL ... PANAMA ...0707 3. 1-108 ... TAIWAN ... CHINA	FLETE INTERNO INLAND FREIGHT		
	MUELLAJE Y MANEJO HANDLING CHARGES		
	DEMORAS ZARPE DELAY IN SHIPMENT		
	OTROS GASTOS OTHER CHARGES		
	DESCUENTO DISCOUNT		
	COMISION COMMISSION		

TOTAL F.O.B.	US$3,488.00
FLETE FREIGHT	536.00
SEGURO INSURANCE	Nil
TOTAL CIF	(C&F) US$4,024.00

USO DE ADUANAS -
FOR CUSTOM ONLY

No DE LIQUIDACION

OFICIAL DE ADUANAS

AUDITOR:

NOTA : LAS FACTURAS CONSULARES Y DEMAS DOCUMENTOS QUE SE PRESENTEN PARA LA CERTIFICACION CONSULAR 8 DIAS
HABILES DESPUES DE LA FECHA DE EXPEDICION DEL CONOCIMIENTO DE EMBARQUE, PAGARAN UN RECARGO DEL 1%
ANTE EL CONSUL RESPECTIVO (ART 447 DEL CODIGO FISCAL)

IMP. CERVANTES 92431

八、產地證明書 (Certificate of origin)

產地證明書的作用有四:

1.供作享受優惠稅率的憑證。

2.防止貨物來自敵對國家。

3.防止傾銷。

4.供作海關統計。

九、檢驗證明書 (Inspection certificate)

為防止出口商裝出的貨物品質不合契約規定或短裝,進口商常要求出口商須提出檢驗證明書。此外,為符合輸入國海關規定,進口商也要求出口商提供檢驗證明書。

檢驗證明書除為保障出口品質,維護國家信譽或依輸入國海關的規定須由輸出國政府機構（例如我國標檢局或檢疫局）簽發外,大多由下列機構簽發:

1.製造廠或同業公會。

2.公證行。

3.進口商指定的代理人。

No. 210 美國特種海關發票

美國特種海關發票

DEPARTMENT OF THE TREASURY UNITED STATES CUSTOMS SERVICE 19 U.S.C. 1481, 1482, 1484	SPECIAL CUSTOMS INVOICE (Use separate invoice for purchased and non-purchased goods.)	Form Approved. O.M.B. No. 48-RO342

1. SELLER Burda Enterprises Inc. 1 Shingshen S. Road, Sec. 1 Taipei, Taiwan, Rep. of China	2. DOCUMENT NR.	3. INVOICE NR. AND DATE No.123, May 10, 20-
	4. REFERENCES	
5. CONSIGNEE The L. P. Henryson Company, Inc. 18W, 33rd St., New York , N.Y. 10001	6. BUYER (if other than consignee) same as consignee	
	7. ORIGIN OF GOODS Taiwan, Rep. of China	
8. NOTIFY PARTY Same as consignee	9. TERMS OF SALE, PAYMENT, AND DISCOUNT CIF, By Sight L/C	
10. ADDITIONAL TRANSPORTATION INFORMATION.		

11. CURRENCY USED US Dollar	12. EXCH. RATE (if fixed or agreed) US$1=NT$38	13. DATE ORDER ACCEPTED April 10.20-

14. MARKS AND NUMBERS ON SHIPPING PACKAGES	15. NUMBER OF PACKAGES	16. FULL DESCRIPTION OF GOODS	17. QUANTITY	UNIT PRICE 18. HOME MARKET	19. INVOICE	20. INVOICE TOTALS
L.P. NEW YORK C/1 - 100 MADE IN TAIWAN REP. OF CHINA	100	STRETCH NYLON YARN DYED 100/24/2 100 ctns. each contains 60 lbs.	6,000 lbs	C.I.F. New York US$1.50 oer lb.	US$1.50 er lb.	US$9,000.-

21 ☐ If the production of these goods involved furnishing goods or services to the seller (e.g. assists such as dies, molds, tools, engineering work) and the value is not included in the invoice price, check box (21) and explain below.	22. PACKING COSTS	US$100.

27. DECLARATION OF SELLER/SHIPPER (OR AGENT)	23. OCEAN OR INTERNATIONAL FREIGHT	US$400.-

I declare:
(A) ☐ If there are any rebates, drawbacks or bounties allowed upon the exportation of goods. I have checked box (A) and itemized separately below.

(B) ☐ If the goods were not sold or agreed to be sold, I have checked box (B) and have indicated in column 19 the price I would be willing to receive.

I-further declare that there is no other invoice differing from this one (unless otherwise described below) and that all statements contained in this invoice and declaration are true and correct.

(C) SIGNATURE OF SELLER/SHIPPER (OR AGENT): Burda Enterprises Inc.

24. DOMESTIC FREIGHT CHARGES	US$10.-
25. INSURANCE COSTS	US$90
26. OTHER COSTS (Specify below)	

28. THIS SPACE FOR CONTINUING ANSWERS

THIS FORM OF INVOICE REQUIRED GENERALLY IF RATE OF DUTY BASED UPON OR REGULATED BY VALUE OF GOODS AND PURCHASE PRICE OR VALUE OF SHIPMENT EXCEEDS $500. OTHERWISE USE COMMERCIAL INVOICE

Not necessary for U.S. Customs purposes

SHENG FA Printing Taipei Tel.781 1349 711-3055

Customs Form 5515 (8-80)

No. 211 產地證明書

臺灣省商業會

Taiwan Chamber of Commerce

P.O. BOX 1762, TAIPEI, TAIWAN   CABLE ADDRESS

THE REPUBLIC OF CHINA   "TCOC" TAIPEI

       TEL. 2313144. 2314152. 2319668

臺北市公園路三號三泰大樓七樓

CERTIFICATE OF ORIGIN   Sept. 9, 20–.

To whom it may concern:

We hereby certify the under-mentioned commodity is of Taiwan Origin.

Commodity: Tee Shirts for Men's Size Assortment from 32–42 White

Quantity: 1,500 doz. (38 Cartons)

Shipment: Shipped per s.s. "PONAPE MARU"/"TAHITI MARU" from Keelung Taiwan to Papeete with transhipment at Yokohama

Credit: L/C No. 942/74 issued by Banque De L'indochine, Papeete, Tahiti

Shipper: TAIWAN Textile Manufacturing Co., Ltd., Taipei, Taiwan

Shipping Marks:

<C.T.C.>

PAPEETE

VIA

YOKOHAMA

C/No. 1–38

MADE IN TAIWAN

REPUBLIC OF CHINA

No. 212 檢驗報告

經濟部

中華民國經濟部標準檢驗局

BUREAU OF STANDARDS, METROLOGY AND INSPECTION, MINISTRY OF ECONOMIC AFFAIRS REPUBLIC OF CHINA

1

優 甲	
甲 等	
乙 等	
一 般	

輸出檢驗合格證書

CERTIFICATE OF EXPORT INSPECTION

發證日期 Date of Issue: Oct.22,2001   證書號碼 Certificate No.: TT-77-405060

1. 申請人（商號）Applicant: Data Taxtile & Ready-Made Garment Manufacturer Co.,Ltd.
2. 生產者 Producer: Dada Taxtile & Ready-Made Garment Manufacturer Co.,Ltd.
3. 輸出者 Exporter: ABC Trading Co.,Ltd.
4. 品名 Commodity: SPORT SHIRTS   商品標準分類號列 C. C. C. Code: 61109010004
5. 規格 Specification: STYLE A, MEN'S OF SILK FABRIC
6. 數量 Quantity: 500DOZ.
7. 總淨重 Total net weight: 1,000KGS
8. 到達地 Destination: NEW YORK
9. 檢驗標識號碼 Inspection Label Nos.: T786543-987654321
10. 檢驗日期 Date of Inspection: Oct.22,2001

XYZ
NEW YORK
C/No.1-50
MADE IN TAIWAN
R.O.C.

ORIGINAL

11. 檢驗記錄 Inspection results:

12. 備註 Remarks:

13. 本證所載商品經檢驗合格限於　年　月　日以前出口，逾期本證無效
It is hereby certified that the commodity listed above has been inspected and passed. This certificate is valid only when the commodity is exported before Oct.31,2001

由經濟部標準檢驗局或所屬發證機關發證
This certificate shall be issued by B S M I or its branches.

本證書必須加蓋發證機關鋼印後生效
This certificate will become effective only when stamped with this Bureau's seal.

89. 10. 35,000

第二十五章
匯票 (Bill of Exchange)

第一節　匯票的意義

匯票 (Bill of exchange, Draft，簡稱為 Bill, Exchange，縮寫為 B/E)，依英國票據法，其意義為:

"A bill of exchange is an unconditional order in writing addressed by one person to another, signed by the person giving it, requiring the person to whom it is addressed to pay on demand or at a fixed or determinable future time a sum certain in money to the order of a specified person, or to bearer."

（所謂匯票，乃經一人簽署，給他人的無條件書面命令，要求受命人於見票時或在未來一定日期或在將來可確定的日期，支付一定的金額給一特定人或其指定人或執票人。）

我國票據法則謂「稱匯票者，謂發票人簽發一定之金額，委託付款人於指定之到期日，無條件支付與受款人或執票人之票據。」

第二節　匯票格式

國際貿易使用的匯票一般都用英文，其格式如下:

No. 213　匯票

Draft No. (1)   **BILL OF EXCHANGE**
                           (2)

For   (4)   Taipei, Taiwan   (3)

At   (5)   sight of this **FIRST** of Exchange (Second the same tenor and date being
                   (6)

unpaid) Pay to the order of **HUA NAN COMMERCIAL BANK. LTD.**

The sum of   (7)

                                       value received
                                       (8)

Drawn under   (9)

Irrevocable L/C NO.   dated

TO   (10)

                                 (11)

Draft No.   **BILL OF EXCHANGE**

For   Taipei, Taiwan

At   sight of this **SECOND** of Exchange (First the same tenor and date being

unpaid) Pay to the order of **HUA NAN COMMERCIAL BANK. LTD.**

The sum of

                                       value received

Drawn under

Irrevocable L/C NO.   dated

TO

【註】

(1)匯票號碼。

(2)Bill of Exchange：表明其為匯票。

(3)表明發票地點及日期。

⑷Exchange for：用阿拉伯字表示匯票金額。

⑸填上匯票期限，例如，即期時，"at sight"，90 天到期則 "at 90 days' sight"。

⑹其大意為：「憑本匯票第一聯（以旨趣及發票日期相同之第二聯匯票尚未付訖為限）見票（或見票後……天）支付華南商業銀行或其指定人……」。

⑺填上金額（以文字表示）。

⑻Valued received：價款收訖，意指發票人對付款人承認收到本匯票所載金額。

⑼"Drawn Under" 後面填上開狀銀行名稱，"Irrevocable L/C No." 後面填上信用狀號碼。"dated" 後面填上開狀日期。本條款稱為 "Drawn Clause"（發票條款），在憑 L/C 開發匯票時，都須填上，如非憑 L/C 者，免填。

⑽"To" 後面填上付款人名稱。

⑾填上發票人名稱及發票人簽字。

從法律觀點而言，匯票應記載下列各項：

1.表明匯票的文字 (Words express it to be a bill of exchange)。

2.一定的金額 (A sum certain in money)。

3.發票地點 (Place draft)。

4.發票日期 (Date of draft)。

5.匯票期限 (Tenor of draft)。

6.受款人名稱 (Payee's name)。

7.付款人名稱 (Drawee's name; Payer's name)。

8.無條件支付的命令 (Unconditional order of payment)。

9.發票人簽字 (Drawer and his signature)。

如係憑 L/C 簽發的匯票，尚須將 "Drawn clause" 載明。

第三節　匯票的種類

匯票從不同的角度來看，有種種的分類。

一、依匯票的發票人及付款人身分分

1.銀行匯票 (Banker's draft or bill)：銀行向銀行發出的匯票。通常用於順匯。

2.商業匯票 (Commercial draft, Trade bill)：商人向商人或銀行發出的匯票。

二、依匯票是否附有貨運單證分

1. 跟單匯票 (Documentary draft or bill)：附有貨運單證的匯票。這種匯票可向銀行申請押匯，所以又稱押匯匯票。
2. 光票 (Clean draft or bill)：未附有貨運單證的匯票。

三、依匯票期限分

1. 即期匯票 (Sight draft or bill, Demand draft or bill)：見票 (Sight) 即付或要求 (On demand) 即付的匯票。
2. 定期匯票 (Time draft or bill) 或遠期匯票 (Usance draft or bill)：即將來某一時日付款的匯票，可分為：

⑴發票後定期付款匯票：即發票日後一定日期付款的匯票，例如 "Ninety days after date" 即以發票日後 90 天付款。

⑵見票後定期付款匯票：即見票日後一定期間付款的匯票，如 "Ninety days after sight"，即以見票日後 90 天付款。

⑶定日付款匯票：即以某特定日付款的匯票。

四、依 L/C 的有無分

1. 憑信匯票 (Draft or bill with L/C)：即憑 L/C 開發的匯票，付款人通常為銀行。
2. 不憑信匯票 (Draft or bill without L/C)：即不憑 L/C 開發的匯票，付款人通常為進口商。

五、依交付單證方式分

1. 付款交單匯票 (Documents against payment draft or bill, D/P bill)：付款人付清票款後才交付貨運單證的匯票，又稱為付款押匯匯票 (Documentary payment bill)。
2. 承兌交單匯票 (Documents against acceptance draft or bill, D/A bill)：即匯票經付款人承兌 (Accept) 後即交付貨運單證的匯票，又稱承兌押匯匯票 (Documentary acceptance bill)。

六、依匯票張數分

1. 單一匯票 (Sola draft or bill)：即發票人只簽發一張匯票者，票面上往往印有 "Sola" 字樣。
2. 複數匯票 (Set draft or bill) 或成套匯票：即發票人簽發二張或二張以上成套的匯票者，稱為複數匯票，這種匯票每張均有效，但其中任何一張兌付，其餘各張即作廢。

 # 第四節　匯票的背書

背書 (Endorsement) 就是執票人以轉讓票據權利為目的，於票據背面簽名表示同意轉讓的行為。可分為：

一、空白背書 (Blank endorsement; General endorsement)

僅在匯票背面簽上背書人 (Endorser) 簽字，而不記載被背書人 (Endorsee) 名稱者。

C.y. Chang.

二、特別背書 (Special endorsement; Full endorsement)

即背書人除在匯票背面簽章外，並在簽章上面記載被背書人名稱者。

Pay to the order of John Wang

C.y. Chang.

三、限制背書 (Restrictive endorsement)

1.禁止再轉讓的背書：

Pay to George Ling only

C.y. chㅁ.

2.委任代收背書：

Pay to the order of ×× Bank for Collection

C.y. chㅁ.

四、保留背書 (Qualified endorsement)

背書人表明不負被追索責任的背書。

Pay to the order of George Wang without recourse

C.y. chㅁ.

第五節　匯票關係人及票據行為

一、匯票關係人

Drawer　發票人　　　　Payment　　　　　擔當付款人

Drawee〕	被發票人	Referee in case-of-need	預備付款人
Payer 〕	（付款人）	Acceptor for honor	參加承兌人
Payee	受款人	Payer for honor	參加付款人
Bearer	執票人	The party for whose honor acceptance or payment is made.	被參加人
Endorser	背書人		
Endorsee	被背書人	Prior party	前手
Holder	持票人	Preceding endorser	
Acceptor	承兌人	Subsequent party	後手
Surety	保證人		

【註】

1. Holder：為持有票據，有權要求支付票據金額的任何人，票據的持有人可能為⑴票據上所載 Payee，⑵經前手的特別背書而取得票據的 Endorsee，⑶執票人票據 (Bearer bill) 的持有人，⑷經前手的空白背書而取得票據的受讓人。依英美票據法，前述第⑶及第⑷的 Holder，特稱為 Bearer，我國票據法統稱為執票人，而並無 Holder 與 Bearer 之分。

2. Acceptance：遠期匯票的付款人（被發票人）同意發票人的支付命令而在匯票正面簽名的行為，稱為承兌 (Accept)，被發票人在匯票上簽名承兌時，此被發票人即為承兌人 (Acceptor)。

3. Payment agent：即代替付款人擔當支付票據的人。

4. Reference in case-of-need：指除付款人外，發票人或背書人另行指定的付款地的人，使其在付款人拒絕承兌或付款時，承擔承兌或付款責任。

5. Acceptor for honor：當遠期匯票被 "Drawee" 拒絕承兌時，由另一個人代付款人承兌的行為，叫做參加承兌 (Acceptance for honor)，這另一個人稱為 "Acceptor for honor"，原來的 "Drawee" 稱為被參加承兌人。

6. Payer for honor：包括預備付款人和其他任何人，在匯票遭拒付時，代發票人。背書人等對執票人付款的行為叫做 "Payment for honor"（參加付款），原來的 "Drawee" 就叫做被參加付款人。

7. Surety：票據債務人以外的第三者，在匯票上簽字保證票款的支付的人。

8. Presentment：又稱為 "Presentation"，即執票人將匯票送請 "Drawee" 承兌或付款的行為。

9. Recourse：票據不獲承兌或付款時，執票人對前手要求償還票款及費用的行為。

10. Protest：執票人可獲承兌或付款時，要求製作拒絕證書機構出具拒絕證書的行為稱為 "Protest"。所作成的文書，叫做拒絕證書 (Protest)。

二、票據行為

Draw, Drawing	發票
Endorse, Endorsement	背書
Negotiate, Negotiation	轉讓；讓購
Present, Presentment	提示
Accept, Acceptance	承兌
Pay, Payment	付款
Payment domiciled	擔當付款
Non-acceptance	拒絕承兌
Non-payment	拒絕付款
Acceptance for honor	參加承兌
Payment for honor	參加付款
Recourse	追索
Protest	做成拒絕證書

No. 214　承兌匯票

Draft No.   BILL OF EXCHANGE

For _____   Taipei, Taiwan _____

At _____ sight of this **FIRST** of Exchange (Second the same tenor and date being unpaid) Pay to the order of **HUA NAN COMMERCIAL BANK. LTD.**

The sum of _____

_____   value received

Drawn under _____

Irrevocable L/C NO. _____ dated _____

TO _____

Accepted on Oct. 12, 2004 due Dec. 12, 2004 Bank of Taiwan C.Y. Chen

第六節 有關匯票的有用例句

一、動 詞

1. To draw a $\begin{cases} \text{bill} \\ \text{draft} \end{cases}$ 簽發匯票

  → $\begin{cases} \text{on} \\ \text{upon} \end{cases} \begin{cases} \text{City Bank} \\ \text{a person} \end{cases}$ "on" 後面為被發票人

  → $\begin{cases} \text{in favor} \\ \text{to the order} \end{cases}$ of some person $\begin{cases} \text{"in favor"} \\ \text{"to the order"} \end{cases}$ of 後面 為收款人

  → at $\begin{cases} \text{sight} \\ \text{90 days' sight} \end{cases}$ "at" 後面為匯票期限

  → for US$12,345.00   "for" 後面為金額

  → against $\begin{cases} \text{documents} \\ \text{shipment} \end{cases}$ "against" 為「憑……交換」之意，即 in exchange for 之意

  → covering $\begin{cases} \text{order No. 567} \\ \text{the CIF invoice amount} \end{cases}$ "covering" 為抵付之意

  → under L/C No. 12 $\begin{cases} \text{opened} \\ \text{issued} \\ \text{established} \end{cases}$ by City Bank   "under" 為「依據，憑」之意

  → $\begin{cases} \text{through} \\ \text{with} \end{cases}$ Bank of Taiwan   "through" 為「經由，透過」之意

2. To $\begin{cases} \text{accept (sign)} \\ \text{bank} \end{cases}$ a bill   保證匯票的兌付

3. To $\begin{cases} \text{honor} \\ \text{protect} \\ \text{take up} \end{cases}$ a bill   兌付匯票

To $\begin{Bmatrix} \text{give} \\ \text{afford} \end{Bmatrix}$ one's protection ... 兌付匯票

4. To meet a bill upon presentation at maturity ... 到期提示時付款

   To pay a bill ... 支付票款

5. To $\begin{Bmatrix} \text{provide for a bill} \\ \text{prepare due honor for a draft} \end{Bmatrix}$ ... 準備兌付匯票

6. To $\begin{Bmatrix} \text{recommend} \\ \text{commend} \end{Bmatrix}$ a bill to one's protection ... 請某人兌付匯票

7. To ask one's protection for a bill ... 請某人兌付匯票

   To solicit the favor of one's acceptance of a draft ... 請某人承兌匯票

   To request a person to protect a bill ... 請某人兌付匯票

8. To endorse a bill ... 將匯票背書

9. To $\begin{Bmatrix} \text{discount} \\ \text{negotiate} \end{Bmatrix}$ a bill ... 將匯票 $\begin{Bmatrix} \text{貼現} \\ \text{讓購} \end{Bmatrix}$

10. To $\begin{Bmatrix} \text{withdraw} \\ \text{retire} \end{Bmatrix}$ a bill ... 贖票

11. To collect a bill ... 將匯票託收

12. To encash a bill ... 將匯票兌現

13. To present a bill for $\begin{Bmatrix} \text{acceptance} \\ \text{payment} \end{Bmatrix}$ ... 提示匯票 $\begin{Bmatrix} \text{承兌} \\ \text{付款} \end{Bmatrix}$

14. To bishonor a bill ... 拒絕兌付匯票

   To dishonor a bill by $\begin{Bmatrix} \text{non-acceptance} \\ \text{non-payment} \end{Bmatrix}$ ... 拒絕匯票的 $\begin{Bmatrix} \text{承兌} \\ \text{付款} \end{Bmatrix}$

15. To return a bill $\begin{Bmatrix} \text{accepted} \\ \text{unaccepted} \\ \text{unpaid} \end{Bmatrix}$ ... $\begin{Bmatrix} \text{承兌後} \\ \text{拒兌後} \\ \text{拒付後} \end{Bmatrix}$ 退還匯票

16. To $\begin{Bmatrix} \text{protest} \\ \text{note} \end{Bmatrix}$ a bill ... 作成拒絕證書

   To get a bill protested ... 使……作成拒絕證書

To cause a protest to be made   使⋯⋯作成拒絕證書

17. To fall due   到期

18. To be overdue   逾期

二、有關匯票的例句

1. We would $\left\{\begin{array}{l}\text{ask you to give}\\\text{commend}\end{array}\right\}$ our draft your kind protection.

2. We would $\left\{\begin{array}{l}\text{solicit the favor of}\\\text{claim}\end{array}\right\}$ your kind protection to our draft.

（惠請予以兌付〔承兌，支付〕我們所開的匯票。）

3. We have the pleasure to inform you that we have $\left\{\begin{array}{l}\text{drawn}\\\text{valued}\end{array}\right\}$ on you this day for US\$..., at 3 m/s, to the order of × × × Bank, which we commend to your kind protection.

   註：a. 3 m/s: 3 months after sight；b. 3 m/d: 3 months after date。

4. Against this shipment, we have drawn on you at sight for the invoice amount of US\$..., in favor of Bank of Taiwan, which please protect on presentation.

5. We have valued on your goodselves by 60 days' sight draft for US\$..., under L/C No. 123 issued by Bank of America in favor of Bank of Taiwan, to whom we handed full set of shipping documents.

6. Your draft, under date of June 10, at 30 days' sight to your own order, will be duly honored.

7. We are enclosing a draft for US\$...and request you to return it to us after being accepted.

8. We have drawn a sight draft on you for the payment of the invoice value US\$...under L/C 234 issued by City Bank and negotiated through Central Trust of China, Taipei.

9. According to the terms agreed upon, we have drawn on you at sight against the shipping documents through Bank of Taiwan. We ask you to protect the draft upon presentation.

第二十六章
收款與付款 (Collection and Payment)

由於國際貿易的雙方當事人遠隔兩地，國際貿易的付款方式與國內買賣的「一手交貨，一手付款」方式迥然不同。於是因國際貿易而生的帳目清理 (Settlement of account) 也與國內買賣大不相同。

一般而言，國際貿易貨款的清償通常多以押匯收款的方式完成。換言之，賣方於貨物運出後，開具匯票連同提單、商業發票、保險單等向外匯銀行押借款項，銀行則將匯票及貨運單證寄往進口地委託代理銀行收取貨款，這種押匯收款的方法，又可分為「憑 L/C 收款」及「憑 D/P, D/A 匯票收款」兩種。

買賣雙方如往來密切，信用可靠，或雙方互有進出口的買賣關係時，也有採用「記帳」(Open account) 方式或分期付款 (Instalment) 等方式，清理帳款者。

憑 L/C 收款時，出口商於發出裝運通知時，順便通知將憑 L/C 前往銀行辦理押匯外，通常多不另向進口商發出收款函。但是如採用 L/C 以外的方式收款，或結算因貿易而生的附帶費用時，通常多須發出收款函甚至催款函。而欠人的一方為清付帳款，也多須撰寫付款函。

第一節　憑 L/C 收款的函件

出口商在貨物裝出後，即可開具匯票備妥 L/C 規定的單證，連同 L/C 向當地外匯銀行申請押匯 (Negotiation)。向外匯銀行申請押匯時，出口商須提出押匯申請書，茲舉一例於下：

No. 215　向銀行提出押匯申請書

Huan Nan Commercial Bank

Foreign Department

Taipei
September 15, 20–

Dear Sirs,

We are presenting to you herewith for negotiation our draft No. Y123 for EUR28,950.– drawn under L/C No. 123 issued by Dresdner Bank accompanied by the following documents:

Commercial Invoice in triplicate
Packing List in triplicate
Marine insurance policy in duplicate
Certificate of Origin in triplicate
Bill of Lading in duplicate

For the proceeds, please have it settled in accordance with the regulations governing foreign exchange transactions.

In consideration of your negotiating the above documentary draft, we guarantee that you can receive the proceeds within two months, and further undertake to hold you harmless and indemnified against any discrepancy which may cause non-payment and/or non-acceptance of the said draft, and we shall refund you in original currency the whole and/or part of the draft amount with interest and/or expenses that may be accrued and/or incurred in connection with the above on receipt of your notice to that effect, and also verify that all advices relative to credit instruments including amendment advice(s), if any, have been submitted to you without failure.

Faithfully yours,

【註】這是一份出口押匯申請書，每家外匯銀行都備有這種申請書供出口商索取。

1. proceeds：票款。

2. in consideration of：因為。

3. undertake to hold you harmless：保證不使你受到損害。

4. indemnify against：使……免受損失、損害等。

5. non-payment：拒付。

6. non-acceptance：拒絕承兌。

7. to that effect：意旨；大意。notice to that effect：這種意旨的通知。

　　例：I have received a cable to the effect that：我收到一封電傳，大意是說……。He wrote to that effect：他寫的大意如此。

　　事實上，出口商初次向外匯銀行申請押匯時尚須簽具質押權利總設定書 (General letter of hypothecation)。如出口商所提出的貨運單證與 L/C 不符，換言之，單證有瑕疵

（Discrepancy），則外匯銀行可能不願接受押匯，但如出口商信用良好時，可著其提供 Letter of indemnity 之後，斟酌情形接受押匯，這種 Letter of indemnity 的內容如下：

No. 216　損害賠償約定書

Huan Nan Commercial Bank

Foreign Department　　　　　　　　　　　　　　　　　　　　　Sept. 15, 20–

Dear Sirs,

LETTER OF INDEMNITY

　　In consideration of your negotiating our Draft No....drawn on...for US$...under letter of credit No....issued by...Bank, which stipulates:

shipment before August 30, 20–

whereas the relative documents indicate:

shipped on September 3, 20–

　　We hereby undertake to indemnify you for whatever loss and/or damage that you may sustain due to the above-mentioned discrepancy.

　　　　　　　　　　　　　　　　　　　　　　　　　　　　　　Yours faithfully,

　　信用狀的受益人如非供應商，則須向 Maker 或 Supplier 購貨，在此情形下，如信用狀係可轉讓，則往往將信用狀轉讓給 Maker 或 Supplier。轉讓可由通知銀行辦理，也可由受益人自行出具轉讓書，連同 L/C 轉給 Maker 或 Supplier。

No. 217　信用狀轉讓書

Letter of Transfer

　　　　　　　　　　　　　　　　　　　　　　　　　　　　　December 12, 20–

To Whom It May Concern:

　　　　　　　　　　Re: Transfer of L/C No. 123
　　　　　　　　　　　　issued by Deutsche Bank, Frankfurt
　　　　　　　　　　　　dated November 3, 20–
　　　　　　　　　　　　amounting to EUR10,000

This is to certify that we have transferred the captioned L/C in full amount to:

ABC Co., Ltd.

1 Wuchang St., Sec. 1

Taipei

To Negotiating Bank: A signed certificate mentioning a set of copy shipping documents has been duly received by Taiwan Trading Co., Ltd. is required at the time of negotiation of drafts against this credit.

Taiwan Trading Co., Ltd.

【註】

letter of transfer：信用狀轉讓書。貿易界仍有不少人喜歡用 "letter of assignment" 這個舊式術語。

第二節　憑 D/P、D/A 匯票收款的函件

No. 218　出口商函告進口商已開出匯票並請其承兌——D/A

Dear Sirs,

Your order No. 123

We take pleasure in informing you that, in accordance with your order No. 123 of April 20, we have shipped you today by s.s. "President" from Keelung to New York, 500 cases of Fancy Goods, which we trust will reach you in good order and condition.

Herewith we hand you the following non-negotiable shipping documents for the shipment:

2 copies of commercial invoice
1 copy of bill of lading
2 copies of packing list
1 copy of certificate of weight and measurement
1 copy of special customs invoice

To cover this shipment we have drawn on you a 90 d/s draft for US$20,000 through our bankers. The original shipping documents are attached to the draft, and will be handed to you by our bankers through Bank of New York on your acceptance of the draft.

We shall be glad if you will duly protect the draft on presentation, and look forward to being favored with a continuance of your order.

Yours truly,

【註】

1. fancy goods: 精美貨品; 新奇貨品。

2. duly protect the draft: 妥予兌付匯票。"protect" 可以 "honor" 或 "take up" 代替。

3. "and look forward...order" 可以下列句子代替:

   "and assure you that any further orders you may place with us will always be carefully attended to"。

4. to be favored with a continuance of your order: 繼續惠賜訂單。

No. 219　出口商函告進口商已開出匯票並請其承兌——D/A

Dear Sirs,

Your order No. 234

We are pleased to inform you of our shipment, as the copy of our telex of the 19th enclosed, of your order No. 234 today by s.s. "Skyline".

> U.S.
>
> NEW YORK　　100 cases Ladies Pullover
>
> No. 1–100

For reimbursement, we have drawn on you for the invoice amount at 60 days after B/L date and passed the draft and shipping documents to our bankers. The documents—will be presented to you by the Chemical Bank, New York, against your acceptance of the draft. You are requested to protect our draft upon presentation.

We hope the goods will reach you sound and give you complete satisfaction.

Yours faithfully,

【註】

1. for reimbursement: 為了求償。

2. at 60 days after B/L date: 提單日期後 60 天 (付款的匯票), 即裝船後 60 天 (付款的匯票)。

3. against acceptance of the draft: 憑承兌該匯票 (交付單證), 所以本交易是 D/A 交易。

4. sound: 完好。

No. 220　出口商通知進口商已向其開出匯票並請其付款——D/P

Gentlemen:

We thank you for your letter of December 4, giving us shipping instructions together with marks and numbers, for 10 cases Cotton Goods on your order of November 5. We now have the pleasure of apprising you that the goods are being shipped per N.Y.K. steamer "KONGO MARU" sailing on the 14th December, as per our Invoice enclosed.

As arranged, we are drawing a draft upon your goodselves at sight, D/P, for the amount, viz., US$5,000, and ask you to protect it. The shipping documents will be surrendered by the Oriental Bank against your payment of our draft.

The present market is a trifle easier, but this is owing more to the time of the year, which is between seasons, than to any other cause, and no doubt prices will be firm again in about a month's time.

We trust our shipment will give you every satisfaction and induce you to place with us your renewed orders, to which we shall at all times be most pleased to give our every attention.

Yours very truly,

Inc. 1 Invoice

【註】

1. apprising：報告；通知。cf. apprising one of something：告知某人某事。

2. to surrender = to give up：交付。

3. a trifle easier = somewhat easier：有一點疲弱（軟），"easier" 為 "easy" 的比較級，easy= weak：意指行情趨下。

4. firm：意指堅穩。

   cf. fluctuating：恍惚，意指漲跌不定。

     strong：堅挺或放長。

     steady：穩定，意指價格一時不致有巨幅漲落。

No. 221　匯票到期未獲兌付發出警告

Dear Sirs,

We confirm our letter of 15th July informing you that according to advice from the Bank of America, our draft No. 123 for US$10,000, due on May 30, against Invoice No. 123 and order 77/123, had not yet been paid.

We had asked to attended to this matter at once, but to our astonishment we were informed by the Bank that, in spite of repeated requests, you still had not paid.

Through such delay considerable expenses are incurred, such as rent, moratory interest, additional insurance, etc., for which you are responsible.

We would remind you that our agreement with you is worded as follows: "We hereby agree to accept your Draft promptly and to meet same when due without regard to objections respecting this parcel, which shall be referred to arbitration. Should we not honor the Draft when presented or due, we hereby authorize you or your representatives to resell the goods by auction, private contract, or in such manner as may appear best to you, after giving us ten days' notice. We also agree to make good, on demand and without dispute, any deficiency arising from such sale."

In view of the above agreement recognized by you, we must ask you to honor our draft for US$10,000 within 10 days after receiving this letter, failing which we shall be compelled to resell the goods and hold you responsible for any loss.

We trust that you will now not neglect this last warning.

Yours faithfully,

【註】

1. attend to：注意。

2. rent：倉租，即因匯票拒付未能提貨而存儲於碼頭倉庫所生的費用。

3. moratory interest：延付所生利息。

4. remind：提醒。

5. is worded as follows：以如下的文字表達，即 "is expressed as follows" 之意。

6. when due：到期時。

7. without regard to objections：無異議。

8. parcel：一批貨，即 "a portion of a shipment of goods"，例如："We have been able to obtain the excellent prices of this parcel."。

9. shall be referred to arbitration：交付仲裁，即使有異議也須交由仲裁判斷。

10. in such manner as may appear best to you：依你們認為最佳的方法。此句子形容 "to sell the goods"。

11. to make good：補償，接 "any deficiency"（任何不足之數）。

12. on demand：一經要求。

13. failing which：否則。

14. hold responsible：負責。

15. warning：警告。

第三節　帳目的清理函件

一、帳目的清理 (Settlement of account)

在現代國際貿易，貨款的收付固然大多以 L/C 或 D/P、D/A 方式了結，但如雙方交易淵源深厚，或雙方互有進出口的買賣關係時，也間有採用 Open account 的結帳方式。即賣方將貨物運出後，將貨款記入買方帳，於一定時間後，雙方結算清理。除此之外，買賣雙方間或與代理商之間也有佣金以及其他種種債權債務暫予記帳，於一定時間後，再清理。這種帳款的發生原因約有下列數端：

　1.進口商欠出口商或代理商：

⑴短計貨款須補付。

⑵代理佣金。

⑶信用狀短開貨款金額。

⑷出口商代墊費用。

⑸樣品費。

⑹其他。

　2.出口商欠進口商或代理商：

⑴超收貨款須退還。

⑵代理佣金。

⑶進口商代墊費用。

⑷缺量或損失應由出口商負責者。

⑸索賠款項。

⑹其他。

上述欠人、人欠的結算時間或在每筆交易之後逐筆結算或規定每季、半年結算一次，結算或由付款人主動匯付，或由收款人主動收取。在國際貿易習慣上，一切人欠帳項可開製借項清單（通知單）(Debit note) 寄交對方，一方面為通知該項欠項已借入其帳戶，他方面則尚含有索欠之意。欠人者也宜開製貸項清單 (Credit note) 表示誠意。如買賣雙方互有人欠、欠人帳款，則可繕製往來帳單 (Statement of account) 以供核對及清付帳款之用。

No. 222　賣方寄出借項清單

Dear Sirs,

We regret to have to inform you that an unfortunate error in our invoice No. 832 of August 18 has just come to light. The correct charge for nylon shirts, medium, is US$15 per dozen and not US$14.50 as stated.

We are therefore enclosing a debit note for the amount undercharged, namely US$500.–.

The mistake is due to a typing error and we are sorry it was not noticed before we sent the invoice.

Yours very truly,

【註】

1. unfortunate error: 令人遺憾的錯誤。

2. has just come to light: 剛剛發覺。

3. correct charge: 正確的價款。

4. medium: 中號。

5. undercharge: 少計價款。

6. typing error = typographical error: 打錯。

No. 223　借項清單

DEBIT NOTE

Messrs. John Huges & Co.

112 Kingsway

Liverpool

August 22, 20–

We hereby advise you that we have placed the undermentioned amount to the debit of your account with us.

ORDER NO.	DESCRIPTION	AMOUNT
123	To 1,000 doz. nylon shirts, medium charged on Invoice No. 832 at US$14.50 per doz. should be 15.00 per doz. difference	US$500.–
	The amount debited to your account	US$500.–

For Taiwan Trading Co., Ltd.

【註】

在 Debit Note 中的 Description 欄以 "To" 開始，意指 "Dr. to"（即 debit to）。

No. 224　貸項清單

CREDIT NOTE

Messrs. John Huges & Co.　　　　　　　　　　　　　　September 20, 20–

112 Kingsway

Liverpool

　　We hereby advise you that we have placed the undermentioned amount to the credit of your account with us.

ORDER NO.	DESCRIPTION	AMOUNT
123	By ten doz. returned. Charged to you on Invoice No. 832	US$150.–
	The amount credited to your account	US$150.–

For Taiwan Trading Co., Ltd.

【註】

　　Credit Note 中的 Description 欄則以 "By" 開始，意指 "Cr. to"（即 credit to）。

No. 225　賣方寄出往來帳單

Dear Sirs,

　　We are enclosing a Statement of your account up to and including May 31, 20–, showing a balance in our favor of US$125.40.

　　We suppose you will prefer to send us a check, as this is too small to draw a bill for.

　　　　　　　　　　　　　　　　　　　　　　　　　　　　　Yours very truly,

【註】

　　in our favor：有利於我方，即欠我方。balance in our favor：我方順差。

二、催款函 (Collection letter)

　　如果發出 "Statement of account" 函或請求付款函，未獲覆或逾期未付款時，應發出催款函。寫催款函（文稿催收函）是一種藝術，寫得不夠藝術就傷感情 (Hurt feeling)。欠人者，不付款大約有三種原因，①忘記，②裝聾作啞，③財務困難。不論是那一種，都可採下列三步驟：

1.催促付款 (Urge payment)：這種信函為催付函 (Reminder) 語氣略帶堅定，但仍應
有禮，其內容包括：

⑴提及前次發出的 Statement of account，或 Letter of asking for payment。

⑵假定發生了特殊事故，才使付款延誤。

⑶附上前寄信函副本。

⑷請匯下帳款。

No. 226　催促付款信之一

Dear Sirs,

As you are usually very prompt in settling your account, we wonder whether there is any special reason why we have not received any information with regard to the statement of account submitted on May 31.

We think you may not have received the statement of account and we enclose a copy and hope it may have your early attention.

Yours faithfully,

【註】

1.in settling your account：可以 "in paying your bills" 代替。

2.第二段 "We think...."（以為你沒有收到我們的帳單。）是一種有禮貌的表現法。

No. 227　催促付款信之二

Dear Sirs,

In checking our accounts, we find there is a balance of US$100 due us for purchases made in August, 20–.

As you have no doubt overlooked this bill, we are bringing it to your attention.

Will you please send your check at once so we can clear this indebtedness from our books and bring your account up to date.

Yours very truly,

【註】

1. check account：查帳。

2. due us：欠我們。cf. due to us。

3. overlook this bill：沒注意到這筆帳款。

4. clear this...books：將這筆債款（從我們帳簿中）清理。

2.堅持要求付款 (Insist on payment)：如果發出催付函，仍不發生作用，只好再寫第二次措詞較強硬的信，信中應以堅強的語氣要求對方付款，信中內容包括：

(1)重述以前所寄的收款函。

(2)提出一個付款的最後期限，堅持要求在此期限內付訖。

No. 228　堅持要求付款信之一

Dear Sirs,

　　Owing to the fact that you have not replied to our communications of May 31 and July 5 respectively on the subject of your outstanding balance of US$125.40, we can only write again to remind you.

　　As you have considerably exceeded the terms of credit usually allowed we must insist on receiving payment by the 20th July, 20–.

Yours very truly,

【註】

1. communications = letters。

2. outstanding balance：未清理（償）的餘款，也可用 "outstanding account" 代替。

3. remind：提醒。

4. terms of credit：賒帳期限。

5. insist on：堅持。

No. 229　堅持要求付款信之二

Dear Sirs,

　　We are disappointed not to have received any word from you in answer to our letters concerning the bill of US$100.– which you owe us.

　　As you know, the terms of our agreement extend credit for one month only. This bill is now two months overdue. Surely you don't want to lose your credit standing with us and with others, nor do we want to lose you as a customer.

We therefore insist on receiving a check at once. We urge you to keep your account on the same friendly and pleasant basis it has always been in the past.

Of course if there is some reason why you cannot pay this bill, or can pay only part of it now, we would be very happy to talk it over with you...and perhaps we could be helpful.

Yours very truly,

【註】

1. owe us：欠我們。

2. as you know：如你所知。

3. the terms...only：欠帳的期限只有 1 個月。

4. overdue：逾期。

5. lose credit standing：失去信用。

6. lose you as a customer：失去你這麼一個客戶。

7. urge：催促。

8. keep your account...basis：保持你的帳戶建於友善而愉快的基礎。

9. perhaps...helpful：或許我們可以幫忙。

No. 230　堅持要求付款信之三

Dear Sirs,

We regret to note that you appear to have ignored our letter of December 10, calling your attention to an outstanding balance of US$...on your account.

You will recognize that, as this sum is two months overdue, we are at least entitled to some expression of your intentions in the matter. Your failure to pay causes us particular inconvenience, as the prices of the goods were cut so low as to leave only the narrowest margin of profit; the delay in payment threatens to turn this small profit into a loss.

We accordingly enclose a duplicate statement of account and must ask you to meet this engagement without delay.

Yours faithfully,

【註】

1. outstanding...on your account：貴方欠帳。

2. are entitled to some expression of your intentions in this matter：有權利知悉你對此事的企圖。

3. margin of profit：利潤。

4. threaten...into a loss: 使此一微利大有成為損失之虞。

5. without delay: 不得遲誤、毋延遲。

3.要求立即付款 (Demand payment): 三催四請，對方仍然裝聾作啞，相應不理時，祇好發出最後通牒，但不要向對方恐嚇、威脅、漫罵，以免犯法。這種信的內容如下：

⑴表示幾次催款皆未獲覆甚遺憾。

⑵要求立即付款。

⑶如果仍不付款，將採法律行動（交給律師處理）。

No. 231　要求立即付款信之一

Dear Sirs,

　　We are much disappointed in your ignoring our repeated requests for payment of your outstanding account of US$124.50.

　　It is with the utmost regret that we have now reached the stage when we must demand immediate payment. We have no wish to be unreasonable, but failing payment by September 15, we are afraid you will leave us no choice but to place the matter in the hands of our attorney, but we sincerely hope this will not become necessary.

　　　　　　　　　　　　　　　　　　　　　　　　　　　　　　　Yours faithfully,

【註】

1. ignore: 忽視。

2. we have no wish...unreasonable: 我們無意不講理。

3. leave us no choice = leave us no alternative: 使我們別無其他辦法。

4. but to place...our attorney: 除將本案交給我們律師處理外（別無他法）。

No. 232　要求立即付款信之二

Dear Sirs,

　　We cannot tell you how sorry we are, you have not answered any of our letters about the US$ 100 you owed us. We must now regretfully assume you do not want to pay, and we have no other choice than to turn the matter to our collection attorneys.

　　This is most distasteful to us, especially in your case, and we are therefore making one last request for your check, or for a letter of explanation.

We are holding up proceedings for five days in the hope of hearing from you. Please do not disappoint us!

Very truly yours,

【註】

1. we cannot...sorry we are: 我們感到有說不出的遺憾。

2. assume: 認為。

3. have no other choice than to 也可以 "have no other choice but to" 代替。

4. turn the matter...attorneys: 交給律師辦理。

5. this is...to us: 這是對我們最不愉快的事。

6. making...your check: 最後一次要求你寄上支票。

7. hold up proceedings: 止住訴訟；不提起訴訟。

No. 233　幽默的催款信

Dear Mr. Wang:

NO NEWS IS GOOD NEWS...

Doesn't always hold true. Especially, so, when we have sent you a number of reminders about your past-due account, and have received no response.

And certainly not, when the absence of any word from you could adversely affect your CREDIT STANDING—ONE OF YOUR MOST VALUABLE ASSETS.

It is our earnest wish to cooperate with you in every possible way, but you make it difficult to do by not answering any of our reminders.

May we expect some good news...by return mail?

Yours Very truly,

【註】

1. no news is good news: 沒有消息就是好消息。

2. doesn't always hold true: （這句話）有時候並不正確。

3. past-due account = overdue account: 逾期欠款。

4. adversely affect: 不利地影響。

5. earnest wish: 熱忱的希望。

No. 234　債權人的律師向債務人發出要求付款

Gentlemen:

My client, William Taylor & Co., has informed me that letters regarding your indebtedness to them have not received your attention.

Therefore, I must inform you that unless full payment of your arrears is made within 10 days, I shall find it necessary to follow the instructions of my client and institute collection proceedings without further delay.

Very truly yours,
Charles C. Y. Chang
Attorney-at-law

【註】

1. arrears：拖欠款項。

2. follow the instructions：依照……的指示。

3. institute...proceedings：將……訴之於法。cf. institute legal proceedings：訴之於法。也可以 "to have recourse to legal proceedings" 代替。

三、催付樣品費、佣金等

在國際貿易，昂貴的樣品，往往向進口商索取樣品費。然而，進口商往往裝糊塗。再者，出口商應付佣金者，也往往拖延不付。於是，不得不寫信催付，以下數例，可供參考。

No. 235　催付樣品費之一

Dear Sirs,

Subject: Sample Charges

On October 12, 20– we submitted to you a Debit Note No. 123 for the sample charges. However, up to this date, we don't appear to have received your remittance.

In the cirsumstances we are forced to conclude that our Debit Note has not reached you, and we now enclose a duplicate of the Debit Note for reminding you of this outstanding account.

May we draw your immediate attention to this matter? Your prompt remittance to us will be appreciated.

Yours faithfully,

【註】

    1. sample charges：樣品費。

    2. are forced to conclude：不得不推斷；不得不作……的結論。

No. 236　催付樣品費之二

Dear Sirs,

    You seem to have ignored our previous letters of October 12, November 12 and December 12 requesting for settling of the long overdue account.

    You should have not forget that you asked us for samples at cost. Enclosed is a photostatic copy of your letter for your attention.

    As you are considered as one of our good customers, you don't want to lose for sure your credit with us.

    Please keep your promise and send us your remittance at once. Don't disappoint us this time.

Yours very truly,

【註】

    1. long overdue account：過期很久的帳款。

    2. for sure = certainly。

    3. keep your promise：守信。

No. 237　催付佣金

Dear Sirs,

    We wish to draw your attention to the fact that our commission of $2,100 on the two sales effected through us and shipment was made during May as per your Invoices 101 and 102, totaling $42,000 in value, has not been remitted into our account with our bank, Bank of Taiwan, Taipei, though it was due in March. We believe that the relative drafts were negotiated at your bank in due course.

    In the past you were always punctual in remittance, and therefore the delay in this instance is presumed to have been caused by some clerical error. At any rate, we are inconvenienced by the existence of the outstanding account.

    If you have not yet arranged for remittance, please do so at once. If you have already done, you may disregard this letter.

    Your prompt reply will be appreciate.

Yours faithfully,

【註】

1. in this instance：在這種情況下。

2. is presumed：假定。

3. clerical error：小錯誤，另有「筆誤」之意。cf. clerical mistakes：筆誤。

4. at any rate＝in any case：總之；無論如何。

四、請求延付 (Customer requests time to pay)

No. 238　請求延緩付款

Dear Sirs,

　　Everyone has his individual problems, and today, on account of changing conditions, each of us is seemingly confronted with more financial problems than ever.

　　We have been having one bad break after another over the past few months. On account of the shortage of raw material, we have not been working in high gear. As a result, we have no alternative but to ask you forget our account for thirty days. Things are looking up and, barring unforeseeable difficulties, we should be able to pay you before....

　　The world has lucky and unlucky people. It is too bad that misfortunes came to us in a row, but it is our good fortune to have friends like yourselves, whose unfailing assistance in time of need has helped us to tide over our difficulties.

　　　　With very good wishes,

　　　　　　　　　　　　　　　　　　　　　　　　　　　Very sincerely yours,

【註】

1. everyone has his individual problems：每個人都有他自己的問題，即家家有難念的經。

2. on account of changing conditions：由於情況的改變。

3. confront with：遭遇到。

4. than ever：比以前。

5. one bad break after another：一連串的不幸。cf. one after another。

6. working in high gear：全速生產。

7. forget our account for thirty days：忘記我們的帳款 30 天。意指延遲 30 天付款。

8. things are looking up：情形好轉中。

9. barring：除非。

10. unforeseeable difficulty：不可預料的困難。

11. the world has lucky and unlucky people：世上有幸運的人和不幸運的人。

12. it is too bad that misfortunes came to us in a row：很不幸我們遭遇到一連串的不幸。in a row：接連，一連串之意。

13. unfailing assistance：無止境（不斷）的協助。cf. unfailing friendship：永久的友誼。an unfailing friend：一個可靠（忠實）的朋友。

14. in time of need：在困難之時。cf. A friend in need is a friend indeed.（患難之交始為真朋友。）

15. to tide over difficulties：度過困難。

五、請求先付部分款

No. 239　請求先付部分款

Dear Sirs,

We have your letter of December 20, and note with regret your views on the subject of our slow payment.

To some extent we feel that these are justified, but we would point out that we are not entirely to blame. Our inability to settle our account is due to the dificulty of marketing the goods bought from you, a direct result of the high figures at which you priced them and the poor condition in which arrived. It is not our wish to emphasize the justification for our action, but we must remind you that it does not seem quite fair that the entire onus of responsibility in what is, after all, a joint transaction, should be thrown on us.

In the circumstances we trust you will accept our banker's draft for US$...on account, subject to payment of the balance next month.

Yours faithfully,

【註】

1. the subject of slow payment：遲付的問題。

2. to some extent：在某種程度內。

3. we are not entirely to blame：不能完全歸咎於我們。

4. marketing the goods：銷售貨物。

5. the high figures at which you priced them：你們所訂價格過高。

6. justification：理由；口實；辯解。

7. it does not seem quite fair：不似很公平。

8. the entire onus of responsibility in what is：關於……的全部責任。

9. after all：究竟。

10. a joint transaction：共同的交易。

11. throw on us：推諉於我們。

六、對於先償還部分款要求的答覆

No. 240　同意先償還部分款的要求

Dear Sirs,

　　We have your letter of December 27, and thank you for the draft for US$...enclosed. In view of the fact that hitherto your payments have been made promptly, we agree to your proposal that payment of the balance of US$...should be postpones.

　　It must be understood, however, that our agreement does not amount to an acceptance of your opinion of the goods supplied. While regretting your failure to effect a sale, we take no responsibility for it. Further, we do not regard our prices as excessive, and would point out that adequate allowance was made for damage in transit. It must be clearly understood also that we dissociate ourselves from the loss on resale, which definitely forms a new transaction.

　　A receipt for your payment is enclosed, and we must insist on receiving the balance strictly in accordance with your undertaking.

　　　　　　　　　　　　　　　　　　　　　　　　　　　　　　　　　Yours truly,

【註】

1. in view of the fact that：鑒於（你在此以前→ hereto）。

2. amount to：等於。

3. effect a sale：銷售。

4. allowance：補償。

5. dissociate...from：不相涉；無關。

6. undertaking：諾言。

No. 241　不同意先償還部分款的要求

Dear Sirs,

　　We thank you for your letter of November 5 enclosing a draft in part-payment of your account, but would point out that the sum still outstanding is considerable.

As we work, to a very great extent, on a small-profit basis, extended credit with the consequent loss of interest tends to absorb the small figure which accrues to us.

In the circumstances we think you will agree that long-term credit is impracticable and that in asking for an immediate settlement we are not making an exceptional request.

Yours faithfully,

【註】

1. in part-payment of your account：支付你的部分帳款。

2. sum still outstanding：尚欠付的帳款。

3. considerable：相當可觀。

4. to a very great extent：極大部分（的交易）。

5. on a small-profit basis：以薄利的基礎（交易）。

6. consequent loss of interest：利息的損失。

7. absorb the small figure which accrues to us：奪去我們所能獲得的微薄利潤。

8. long-term credit：長期賒欠。

9. impracticable：不切實際。

10. not making an exceptional request：非不情之請。

七、函送帳款

No. 242　以銀行匯票償付帳款

Dear Sirs,

We are pleased in enclosing a bankers' draft No. 123 for US$100 dated October 15 drawn by Bank of Taiwan on the City Bank, New York in settlement of your commission for the order No. 321.

We shall be pleased if you will send us your official receipt.

Yours faithfully,

【註】

1. 為了萬一遺失時易於掛失，宜將匯（支）票號碼、金額、日期、發票人、付款人等均予以寫明。

2. bankers' draft = bank draft：銀行匯票。

3. in settlement of：用以清償。

4. official receipt = formal receipt：正式收據。

八、函送收據

No. 243　寄送收據

Dear Sirs,

　　Thank you very much for your letter of October 15 enclosing a bankers' draft No. 123 for US$100.00 on the City Bank, New York in settlement of our commission for order No. 321, and we are pleased to send herewith an official receipt.

　　We hope you have been satisfied with our service and that we may solicit more orders for you.

Yours faithfully,

OFFICIAL RECEIPT

No. 123

US$100.00　　　　　　　　　　　　　　　　　　　　　　　　October 20, 20–

RECEIVED from Messrs. Taylor & Co., New York, NY the sum of U.S. Dollars One Hundred Only in settlement of commission for order No. 321.

[Stamp]

Taiwan Trading Co., Ltd,

(Signature)

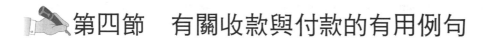

第四節　有關收款與付款的有用例句

一、關於 L/C、D/P、D/A

1. We are surprised to receive a notice from our bankers here to the effect that the draft for US$...drawn on ABC Bank under L/C No. 321 covering your order No. 123 has been dishonored (unpaid) for the following reason which we cannot admit. The said

draft has already negotiated and we are in a very embarrassing situation.

2. We are surprised to receive a notice from our bankers that draft for US$...drawn on you have been dishonored by you without any reason.

二、關於帳目清理

1. $\begin{cases} \text{We have pleasure in enclosing} \\ \text{We are enclosing} \\ \text{We enclose} \end{cases}$ a statement of account, showing a balance in our

favor of US$...

2. We $\begin{cases} \text{are enclosing} \\ \text{enclose} \end{cases}$ a statement of your account up to and including June 30,

→ showing a balance in our favor of US$..., which we hope you will find $\begin{cases} \text{correct.} \\ \text{in order.} \end{cases}$

三、關於催收帳款

第一次催收

㈠開頭句

1. We notice that your account, which was due for payment on..., $\begin{cases} \text{is still outstanding.} \\ \text{still remains unpaid.} \end{cases}$

2. May call your attention to our outstanding account for US$..., settlement of which is now considerably overdue.

3. We shall be glad if you will give attention to our account dated...,

→ which still remains $\begin{cases} \text{unsettled.} \\ \text{unpaid.} \end{cases}$

4. We wish to draw your attention to our invoice No....,

→ for the sum of $...which $\begin{cases} \text{we have not yet received.} \\ \text{is still unpaid.} \end{cases}$

5. We are writing to remind you that we have not yet received the balance of our

September statement, amounting to US$..., payment of which is now more than a month overdue.

6. We $\begin{Bmatrix} \text{call} \\ \text{draw} \end{Bmatrix}$ your attention to our account of $..., due May 30, as per statement tendered on June 15. Knowing the promptness with which you generally meet your obligation, we presume the above has escaped your notice.

7. Your prompt attention to the above account which is now overdue will be deemed a favor.

8. It is doubtless that the rush of business at the period of the year has caused you to overlook the payment of our account $..., which was due on March 2.

9. $\begin{Bmatrix} \text{Permit} \\ \text{Allow} \end{Bmatrix}$ us to remind you that our account rendered to April 30, amounting to $...is still outstanding, and immediate settlement of the same will be appreciated.

(二)條　件

1. According to the stipulation of the agency contract, you are requested to remit us US$..., by the end of December in settlement of the commission.

2. You must understand that profit on this transaction is very small, and if your remittance is not forthcoming we shall fall into an difficult situation.

(三)結尾句

1. We hope to receive your check $\begin{Bmatrix} \text{by return.} \\ \text{within the next few days.} \end{Bmatrix}$

2. We look forward to your remittance $\begin{Bmatrix} \text{by return.} \\ \text{within the next few days.} \end{Bmatrix}$

3. As the amount is now more than a month overdue, we trust you will settle it within the next few days.

4. As our statement may have gone astray we enclose a copy and shall be glad if you will deal with it promptly.

5. $\begin{Bmatrix} \text{Kindly} \\ \text{Please} \end{Bmatrix} \begin{Bmatrix} \text{inform us} \\ \text{let us know} \end{Bmatrix} \begin{Bmatrix} \text{immediately} \\ \text{promptly} \end{Bmatrix}$ when we may expect

$$\to \text{the} \begin{cases} \text{settlement of your outstanding account.} \\ \text{your remittance for the commission.} \end{cases}$$

6. We shall thank you to kindly $\begin{cases} \text{send us your check for} \\ \text{remit us} \end{cases} \to \text{US\$}... \begin{cases} \text{in settlement of} \\ \text{in payment of} \end{cases}$

$\begin{cases} \text{the agent commission} \\ \text{our invoice} \end{cases}$ at your earliest convenience.

　　註: payment of our invoice: 支付發票上所載貨款。

第二次催收

(一)開頭句

1. We do not appear to have had any reply to our request of...for settlement of the amount due on our invoice No....of...

2. We regret not having received a reply to our letter of...reminding you that your account, already more than a month overdue, had not been settled.

3. We are at a loss to understand why we have had no reply to our letter of asking you to settle the amount outstanding on our November statement.

(二)結尾句

1. We trust you will now attend to this matter without further delay.

2. We must now ask you to settle this account $\begin{cases} \text{by return.} \\ \text{within the next few days.} \end{cases}$

3. We regret that we must now press for immediate payment of the amount still owing.

4. We hope it will not be necessary to send you any further reminders.

5. As the amount owing is considerably overdue, we shall be grateful to receive your check by return.

第三次催收

(一)開頭句

1. We note with surprise and disappointment that we have had no replies to our two previous applications for payment of your account.

2. We wrote to you on...and again on...concerning the amount owing to us on our invoice No....

3. As we have had no reply to our previous requests for payment of our invoice dated..., we must now ask you to remit the amount due by the end of this end.

4. Owing to the fact that you have ignored our repeated application for settlement of your outstanding balance US$...we can only assume that you are not prepared to effect a payment.

5. Repeated application for payment have been made for the enclosed account for our agent commission under the wheat transaction, but no reply has been received from you.

(二)結尾句

1. Unless we receive your $\begin{Bmatrix} \text{check} \\ \text{payment} \end{Bmatrix}$ in full settlement by the end of this month,

$\rightarrow \begin{cases} \text{we shall instruct our solicitors to recover the amount due.} \\ \text{we shall take legal proceedings.} \\ \text{we shall have no choice but to take other steps.} \\ \text{we shall take legal action for the recovery of the amount due.} \\ \text{we shall have no alternative but to recourse to legal proceedings.} \end{cases}$

2. Unless we receive your check in full payment by...we shall be compelled to take steps to enforce payment.

3. We still hope you will discharge this account without further delay and thus save yourself the inconvenience and considerable costs of a legal action.

4. This is our third application for payment of the enclosed account, and unless a remittance is received within...days, we shall be compelled to place

$\rightarrow \begin{cases} \text{the account into other hands for collection.} \\ \text{the matter in the hands of our attorneys for collection.} \end{cases}$

5. Previous applications for payment of the enclosed account having been ignored, we must definitely inform you that unless our claims are satisfied by October 30,

$\rightarrow$ we shall immediately $\begin{Bmatrix} \text{take legal action} \\ \text{have no alternative but recourse to legal proceedings} \end{Bmatrix}$

for the recovery of the amount.

四、對於催款的覆函

1. In settlement of your account of commission, we enclose a check No....for US$... issued by Bank of Taiwan, on Bank of America, for which please send receipt.

2. We enclose a cheque for US$...to settle the accounts to the end of May,

→ and $\begin{cases} \text{shall be glad if you will acknowledge its receipt.} \\ \text{your acknowledgement of receipt will be appreciated.} \end{cases}$

3. We are enclosing a $\begin{cases} \text{sola draft} \\ \text{check} \end{cases}$ No. 123 for US$540.00 issued by Irving Trust

→ Company, payable to your order, on Bank of Taiwan, which we $\begin{cases} \text{hope} \\ \text{trust} \end{cases}$ you will

find in order.

4. Covering the amount of commission, we have today remitted US$1,520 by

→ $\begin{cases} \text{mail payment order} \\ \text{cable} \end{cases}$ on Bankers' Trust Company in New York in favor of your-

selves through our Bankers, Taipei Bank.

5. We have pleasure in enclosing our check in $\begin{cases} \text{full payment} \\ \text{part payment} \end{cases}$ of your

→ $\begin{cases} \text{shipment} \\ \text{statement} \\ \text{invoice} \\ \text{commission} \\ \text{sample charges} \end{cases}$ of October 20. Please sign the enclosed form and return it to us.

6. The financial difficulties from which we are suffering at present are cause of our inability of meeting your draft at maturity. Would you kindly allow us a further extension of the payment, say another one month?

註：at maturity：（支票，匯票）到期，不是 "on"maturity。

7. We shall be much obliged if you will give us a little more time in settlement of your account.

8. We trust that you can see your way to accommodate us in this instance.

9. We regret that we are unable to settle the account as originally promised, because money is very tight with us now.

10. Owing to the stagnation of trade our customers have been slow in meeting their obligations, with the result that our reserves have been considerably depleted.

第二十七章
索賠與調處 (Claims and Adjust-ments)

第一節　索賠的概念及其種類

一、索賠的概念

從事貿易，自尋找交易對象，洽談交易以至交貨、付款完成一筆交易，通常多需要經過一段漫長的時間。而且買賣雙方遠隔兩地，其交易的過程大多有賴於函電的往返接洽。而往返的函電有不明、錯誤、誤解之時；貨物自出口地運到進口地曠日費時，貨物輾轉搬運，途中難免遭遇到意內或意外之事；因語言殊異，法律風俗習慣及傳統心理的不同，買賣雙方之間，難免發生齟齬。

由於上述的種種情事，而受到委屈或損失的一方，自將向對方提出抱怨 (Complaint) 或要求賠償等等。所謂索賠 (Claim) 廣義地說，就是指這些抱怨及要求賠償等而言。

二、索賠的種類

造成委屈或發生損失的原因，或由於故意，或由於疏忽或由於意外事故，或由於不可抗力。所以其責任，並不以進出口商為限，大體說來，貿易上的索賠可大別為兩類。

1.貨物損害索賠 (Claim for loss and damage of goods)，又可分為：

⑴運輸索賠 (Transportation claim)。

⑵保險索賠 (Insurance claim)。

2.買賣索賠 (Trade claim or business claim)。

以下分別介紹有關買賣索賠信、運輸索賠信及保險索賠信的寫法。

第二節　買賣索賠信的寫法

一、概　說

　　買賣索賠是買賣雙方當事人間的索賠，以相對人為索賠對象。在商業往來上，因不滿意而引起爭議乃是家常便飯而不可避免的。在進口商方面，發現貨物品質低劣 (Inferior quality)、規格不符 (Different specifications)、數量短少 (Shortage)、包裝不良 (Bad packing)、貨物破損 (Breakage)、延誤裝運 (Delay shipment) 等延不交貨等情事時，自必去函（電）責問或要求賠償損失，出口商對於這些事件，則有予以補正或解釋的必要。

　　反之，假如進口商簽發訂單之後，遲遲不開發信用狀或開出不正確的信用狀，或任意取消訂單或拒不付款等等，則出口商也必將去函（電）責問進口商或要求損害賠償，進口商對於這種索賠，自有理楚的必要。

二、撰寫索賠信的要領

　　索賠人 (Claimant) 的態度應力求平心靜氣，措詞應委婉適當，切忌肆意譏訕，或詞鋒激厲，否則不但轉圜無望，而且可能使雙方決裂，惡感愈深，結果無補實益，徒遭損失。撰寫索賠信時，應注意下列七點：

　　1.迅速提出索賠，尤應在規定索賠期限內提出。延誤索賠，將削弱索賠的立場，並且將使對方難於找出發生事故的原因。

　　2.索賠理由務求明確。具體指出索賠的原因，如有證明文件尤佳。

　　3.引用案號、日期及有關資料，以便其查辦。

　　4.提出希望解決的方法。

　　5.不要貿然認為對方有錯，也許對方有理由。

　　6.措詞誠摯有禮，意誠詞婉，避免粗魯以免償事。

　　7.要求早日解決。

三、答覆索賠信的撰寫要領

　　處理索賠有兩大目標：一是提出使對方滿意的解決辦法；二是排除不愉快，維持良好的關係。當接到令人不快的索賠信時，應以冷靜的態度，就事論事，不感情用事。須知得罪客戶容易，爭取客戶卻不簡單。茲將撰寫答覆索賠信的要領，簡述於下：

1.假如錯在我方時：

⑴儘速答覆：可使對方覺得受到重視，由而減少不悅的情緒。

⑵表示謝意：收到客戶的 Claim letter，仍宜表示高興，並致謝意。

　①因為由於客戶的來函，可以知道對方有什麼不滿之處。

　②使我方有解釋的機會，或糾正錯誤，由而保持良好關係。

　③使我方獲得改善的機會。

⑶對於所發生的問題表示遺憾，並表示願意解決。

⑷解釋所以發生問題的原因。

⑸敘述有條有理，避免含糊其詞。

⑹顧客永遠是對的 (Customer is always right)，因此即使對方出言不遜，也應謙恭地作答。

⑺最後，保證今後再也不發生此類情事，並表示仍願意繼續合作。

2.假如錯不在我方時：

⑴儘速答覆。

⑵對於提醒我方注意其不滿表示謝意及遺憾。

⑶表示同情對方的處境。

⑷將事實作充分的解釋，並提示有關資料，促其了解責任的歸屬。

⑸表示我方對此事雖無責任，但仍很關心，並表示願意協助解決。

3.雖然我方無錯，但仍同意酌給撫慰金 (Consolation money)：有時錯誤雖不在我方，但基於同情，願酌給付撫慰金，以安慰對方，由而維持良好的關係。在此情形，Adjustment letter 的寫法，應注意下列各點：

⑴感謝對方的提醒我方注意其不滿，及表示對這件事感到遺憾。

⑵表示同情及體諒。

⑶把事實真相有條有理地說明，使對方了解責任不在我方，不必告訴對方本應怎麼做，或本不應怎麼做，或未曾怎麼做。

⑷表示將願意酌給撫慰金。

⑸建議對方將來應如何防範此類事情的再發生。

四、索賠的原因

1.關於品質方面：

(1) Inferior quality (Bad quality, Poor quality)　　品質不佳

(2) Inferior grade　　劣等品質

(3) Defective quality, Defects in quality　　瑕疵品質

(4) Inferior quality mixed in　　混雜不良品

(5) Different specifications　　規格不符

(6) Different quality　　品質不符

(7) Different type　　式樣不同

(8) Deterioration in quality　　變質

(9) Excess of moisture　　水分過多

(10) Misweaving　　誤織

　　Different printing　　印花不同

　　Wrong color　　顏色錯誤

　　Different shade　　色緻不同

(11) Different size　　尺寸不符

(12) Short size　　尺寸不足

2.關於數量方面：

(1) Shortage　　短失

(2) Shortlanding　　短卸

(3) Shortshipment　　短裝

(4) Non-shipment　　漏裝

(5) Over-shipment　　溢裝

3.關於包裝方面：

(1) Bad packing　　包裝不善

　　Incomplete packing　　包裝不全

(2) Wrong packing　　包裝錯誤

(3) Loose packing　　包裝鬆弛

Nails off	釘鬆落
Top off	蓋子鬆落

⑷ Breakage　　　　　　　　　　　　　破損

⑸ Leakage　　　　　　　　　　　　　漏損

⑹ Hoop rusty　　　　　　　　　　　　鐵皮生銹

⑺ No shipping mark　　　　　　　　　漏刷嘜頭

⑻ Insufficient mark　　　　　　　　　嘜頭不全

⑼ Different mark, Wrong mark　　　　　嘜頭不符

⑽ Marked mixed　　　　　　　　　　　嘜頭混亂

4.關於交貨方面：

⑴ Delayed delivery, Delayed shipment　　遲延交貨

⑵ Transhipment　　　　　　　　　　　轉運

⑶ Demurrage　　　　　　　　　　　　延滯費

⑷ Surcharge　　　　　　　　　　　　附加費

5.關於保險方面：

Negligence in effecting insurance　　　保險的疏忽

6.關於貨款方面：

⑴ Non-payment　　　　　　　　　　　不付款

⑵ Over-payment　　　　　　　　　　　溢付款

⑶ Non-acceptance　　　　　　　　　　不承兌

⑷ Non-opening of L/C　　　　　　　　不開 L/C

⑸ Delay in opening L/C　　　　　　　遲開 L/C

⑹ Opening of wrong L/C　　　　　　　開出不當的 L/C

五、有關買賣索賠的函件

　　進口商訂了貨之後，除非因不可抗力或出口商有過錯，不能隨意取消訂貨。然而，由於種種原因，不得不取消訂貨時，其取消訂貨信中應說明：

　　⑴取消的原因，以求其諒解，並說明如因而有損失時，願意將來多訂貨以補償對方。

　　⑵但取消原因，可歸責於對方者，不但不予補償，且還可要求對方補償損失。

No. 244　存貨尚多，擬取消訂貨

Dear Sirs,

　　This refers to your Sales Note #321, our Order #123 for 1,000 tennis rackets, to be delivered at the end of May. Owing to persistent bad weather, the sales have been affected and we now find that our present stock will probably be ample for the current season. We therefore very much regret having had to send you this morning the following cable to ask you to cancel our order:

　　"CANCEL ORDER#123 YOUR SALES NOTE#321 FOR TENNIS RACKETS DETAILS FOLLOWS"

　　We are sorry to make this request so late, but hope that, because of our long-standing connection, you will see your way to agree. Should sales improve we would get in touch with you again and shall not fail to indemnify you by placing larger order with you.

　　We look forward to a favorable reply.

Yours faithfully,

【註】

1. tennis rackets：網球拍。

2. persistent bad weather：持續的壞天氣。

3. sales have been affected：銷售受到影響。

4. be ample = be quite enough：足夠。

5. see your way to agree：能同意。to see one's way 為「知道做某事的可能性」之意。cf. I don't see my way (clear) to help you.（我不知道怎樣幫助你。）

6. get in touch with：與……聯繫。

7. shall not fail：一定會……。

8. indemnify：賠償；補償。

假如出口商不同意進口商的要求取消訂貨，則其拒絕取消訂貨的信應：

(1)表示收到來函。

(2)提及契約上的條款。

(3)解釋為何不能同意取消的原因。

(4)表示歉意。

No. 245　不同意取消訂單

Dear Sirs,

Your order No. 123

Our sales note No. 321

In reply to your cable of May 1, we wired you this morning as follows:

"SORRY BUT CANNOT ACCEPT CANCELLATION OF ORDER #123 LETTER FOL-LOWS"

We are sorry that you find it necessary to make the request of cancelling your order No. 123, especially at this late stage. To be able to meet our customers' needs promptly we have to place our order with manufacturers well advance of the season and in estimating quantities rely very largely upon the orders we ourselves have received.

We always dislike refusing requests of any kind from regular customers, but regret that on this occasion we have no choice but to do so. Your cable reached us when your order had been almost half finished. Part of it had been virtually completed, and the rest was in various stage of manufacture. We hope therefore you will understand why we must hold you to your contract.

Had we received your request earlier we should have been able, and would have been glad, to help you. For your information, we are now arranging shipment, and we hope we may effect shipment on or about May 25.

Yours faithfully,

【註】

    1. wire = cable：拍出電報。

    2. at this late stage：在這麼遲的階段。

    3. well advance：很早以前。

    4. rely upon：依賴。

    5. on this occasion：在這次。

    6. we hope you will understand...hold to your contract：希望你能體諒我們必須請你能堅守契約。

        "hold to" 為堅守之意，「堅守諾言」為 "hold to promise"。

No. 246　責問賣方逾期交貨

Dear Sirs,

We confirm that we have sent you today a telex which reads as follows:

"TELEX REPLY WHEN U SHIPD ODR 123 IF UNSHIPD YET SEND GOODS BY AIR BEFORE MAY 15 N 70% AIRFREIGHT SHALL BE UR A/C"

Your attention is invited to our order which stipulates shipment must be effected within 60 days after receipt of our L/C, and please also pay your careful attention to our previous communications of April 2 and 12 stressing that punctual shipment is essential. In spite of this ten days have passed from the latest shipment date as stipulated in the L/C and we have not yet received your shipment advice.

If the umbrellas have not yet been shipped, we must request you to ship them to us by air before May 15, instead of by ship and 70 percent of the airfreight shall be for your account.

Thank you for your cooperation and look forward to your immediate reply by telex.

Yours faithfully,

【註】

1. which reads as follows：內容如下。注意這種句子的用法，reads 的 "s" 不能省，"as follows" 不能改為 "as following"。

2. 關於 telex 的簡體字說明：U: you; SHIPD = shipped; UNSHIPD = unshipped; N = and; UR = your; A/C = account。

3. your attention is invited：敬請注意。

4. in spite of：儘管；雖然。

5. for your account：算你的帳；歸你負擔。

No. 247　覆責問逾期交貨函

Dear Sirs,

This is to confirm that we have sent you today a telex which reads as follows:

"YX 2 UR ODR 123 FOR 1000DZ UMBRELLAS WILL BE SHIPD BEFORE MAY 15 PER AIR DETAILS AIRMG"

We are very sorry for the delay which has occurred in the shipment of your order #123 occasioned by a serious breakdown in our machinery which brought all our work to a standstill for nearly two weeks. The damage, however, has been repaired, and to recoup for the loss in time we will ship the goods to you by airfreight before May 15 as requested. You may expect to receive the goods by May 20 or so. We are also agree to absorb 70% of the airfreight.

Meanwhile, please extend the shipment date and expiry date of the L/C to May 15 and May 25 respectively. Of course, amendment charges, if any, shall be for our account.

We regret the inconvenience you have sustained, but trust this unavoidable accident will not influence you unfavorably in the matter of future orders.

Yours faithfully,

【註】

1. 關於 telex 的內容：

   YX = your telex；2 = 2 日，DZ = dozen; AIRMG = airmailing。

2. occasioned by：肇因於……。

3. breakdown：損壞。

4. bring our work to standstill：使我們的工作停頓。

5. to recoup for the loss in time = to atone for the delay in delivery：為了彌補時間上的損失。

6. or so = about。cf. one hundred or so：100 左右。

7. absorb：負擔。

8. the inconvenience you have sustained：你遭受的不便。

9. unavoidable accident：不可避免的意外事件。

10. influence... order：對日後的訂貨引起不良影響。

No. 248　關於短裝的索賠

Gentlemen:

Re: 1,000 cartons Canned Mushroom
<u>shipped per s.s. "NISHO MARU"</u>

Please be informed that the subject goods shipped by the captioned vessel arrived at New York on April 20.

Upon taking delivery of the cargo, we have found that there were only 920 cartons against 1,000 cartons shipped by you. When checking with the shipping company, we were told that only 920 cartons had been loaded on board the carrying vessel. Since the loss is not negligible you are hereby requested to make up the 80 cartons shortshipped when you deliver the last two items to us. In the meantime, it will be appreciated if you will check at your end and to make sure if all these 1,000 cartons had been loaded.

We are looking forward to receiving your findings soon.

Yours faithfully,

【註】

1. at New York：照文法來說，大都市的前面應用 "in"，這裡用 "at"，乃係指港口之故，其意為 "at the port of New York"。

2. take delivery of cargo：提貨。

3. loaded on board the carrying vessel：裝上載貨船。

4. not negligible：非同小可。

5. make up：（將不足的予以）補償。"make up" 的意義很多，例如：

(1) The committee is made up of seven members.（組成）

(2) He made up an excuse.（捏造藉口）

(3) You have to make up your 2nd year English.（重修二年級英文）

6. shortshipped：短裝。

No. 249　對短裝索賠的覆函

Dear Sirs,

Re: 80 cartons of Canned Mushroom
shortshipped per s.s., Nisho Maru

Your letter of April 25, 20– concerning the subject shipment has received our immediate attention. We regret very much that this incident has caused you much inconvenience. Much as we are eager to help you straighten out this matter, however, we regret to inform you that we are not the party to blame. According to the B/L issued to us by the shipping company, there is noted clearly that 1,000 cartons have been loaded on the carrying vessel. It is quite a regrettable matter that the New York Office of the shipping company has failed to tell you the fact, thus causing you to file the claim against us.

According to the clauses on the back of the B/L which forms the base of freight contract entered into between the carrier and us, the shipping company is responsible for any cargo shortlanded at the port of discharge. In view of this, you are requested to lodge your claim with the shipping company immediately and ask them to compensate you therefor. We trust that the carrier certainly will make up the losses incurred by you under this shipment.

Please let us know as soon as the case has been satisfactorily settled and if there is anything we can do to help settle this case, please just write us. We will comply with your instructions wholeheartedly.

Yours faithfully,

【註】

1. eager：渴望。接 "for"，"after"，"about" 或不定詞 "to"。

2. straighten out = settle：解決。

3. we are not the party to blame：我們不是要負責的一方；不能歸咎於我們。

4. file the claim against us = lodge the claim with us = lay claim to us = render the claim against us = make the claim on us = set up the claim to us。

5. shortland: 短卸。

6. the losses incurred by you: 你所遭到的損失。

7. wholeheartedly: 全心盡力地。

No. 250　向賣方提出貨物損壞的索賠

Dear Sirs,

Subj.: Damage on Canned Asparagus
Shipped per M.S. "Suez Maru"

We acknowledge with thanks the receipt of your telex and your letter dated September 1 and 2 respectively together with the shipping documents for the subject shipment.

Upon the arrival of the goods at Hamburg we have taken delivery of these 900 cartons and have asked Far East Superintendence Co., Ltd. to have them inspected. It is very regrettable that many of the cartons have been found broken, thus causing a large number of cans inside the broken cartons hollowed. We are sure that the asparagus contained therein would have been damaged. We enclose a copy of a survey report, No. 999 issued by Far East Superintendence Co., Ltd. from which you will note the details of the damage inflicted on the goods under this shipment.

We shall appreciate it if you will compensate us the $CIFC_2$ value of all the hollowed cans as shown in the report.

Since this is an urgent matter, your early attention thereto is hereby requested.

Faithfully yours,

【註】

1. Far East Superintendence Co., Ltd.: 遠東公證公司。

   茲列出若干國際性著名公證行於下，供讀者參考。

   Beckmann & Jorgensen, Ltd.

   Bristish Inspecting Engineers Ltd.

   Bureau Veritas

   Cargo Superintendents Pty., Ltd.

   China Corporation Register of Shipping

   General Superintendence Co., Ltd.

   Japan Inspection Co., Ltd.

   Lloyd's Register of Shipping

   Netherlands Superintending & Sampling Co., Ltd.

Robert W. Hunt Co., Ltd.

Superintendence Co., Ltd.

U.S. Consultants, Inc.

Engineering Consulting Office

United States Testing Co., Ltd.

Int'l Inspection Co., Ltd.

Intertechnic GmbH

Daiichi Kensa Kabushki Kaisha

Nichnan Engineering Corp.

Underwriters Adjustment Co., Inc.

Germanischer Lloyd

2. hollowed：凹陷。

3. the damage inflicted on the goods：貨物遭受的損害。

4. under this shipment：本批貨載。

No. 251　覆貨物損壞索賠函

Dear Sirs,

Re: Damage on Canned Asparagus
shipped per M.S. "Suez Maru"

　　Your letter of October 5, 20– along with a copy of survey report issued by Far East Superintendence Co., Ltd. has been received and we have given it our immediate attention.

　　Although we are very regretful to be advised that damage has been inflicted on the shipment we made to you, we, however, are not responsible for the losses and, consequently, are not in the position to compensate you therefore as requested. As arranged with the insurance company, the marine insurance coverage for this shipment includes all risks and the damage as mentioned in the survey report has already been covered by the insurance policy. Therefore, instead of claiming against us, you are requested to submit your claim together with all necessary supporting documents to the insurance company—China Insurance Co., Ltd., whose address is as follows:

China Insurance Co., Ltd.
50 Wu Chang St., Sec. 1
Taipei, Taiwan, R.O.C.

　　In case you need our further assistance regarding this case, please let us know. We will do everything we can to help you.

Faithfully yours,

【註】

1. consequently = therefore; as a result。

2. supporting documents：供佐證的單證。

3. in case = if。

No. 252　責問賣方運交錯誤貨物

Gentlemen:　　　　　　　　　　　　　Indent No. 15

　　With reference to the 5 cases of assorted goods ex s.s. "Arabia" we are greatly surprised at the unbusiness-like way in which you are handling our order. You have sent us, instead of the "Commonwealth" Rubber Boots with black-fleeced lining which we distinctly ordered, 1,000 doz. pairs "Ideal" Rubber Boots with cloth lining, which are quite unfit for the market they are intended for.

　　By referring to the above Indent dated June 21, you will find that we have impressed upon you the especial importance of the boots being of warm lining, and that any delay in delivery will have serious consequence upon us. You made a similar mistake in the last shipment putting us to considerable inconvenience and annoyance. As you are well aware, competition of the home-made goods is so keen that should we miss the best season for the sale of these goods, there is a fear that they will hang upon our hands as dead stock until next year. What is worse still, we shall lose our ground by continually disappointing our customers, and indeed we fear that our reputation with our clients for prompt execution of orders is now at stake, and this entirely through your neglect.

　　If you will value our further orders, you will please oblige us by sending, with all speed, the "Commonwealth" Rubber Boots with warm lining on receipt of this letter, at the same time cabling us the approximate date of their arrival in Kobe. When writing please also give us your instructions as to the disposal of the goods sent in error.

　　We trust you will pay more attention to our commands in the future, otherwise we shall have to go elsewhere for our future supplies.

　　　　　　　　　　　　　　　　　　　　　　　Faithfully yours,

【註】

1. to be surprised 不宜以 "to be astonished" 或 "to be astounded" 代替。

2. unbusiness-like：無效率的；無條理的。

3. black-fleeced lining：黑色羊毛質襯裡；黑絨襯裡。

4. cloth lining：布質襯裡。

5. unfit for = unsuitable for。

6. impress upon：使銘記；使記住。

7. warm lining：暖質襯裡。

8. home-made goods：手工製品；本國製品。這裡指後者而言。

9. hang upon our hands：留存手頭中。

10. dead stock：dead 為 "unproductive" 之意。"dead stock" 未售出的存貨。

11. to lose one's ground：失去信用。

12. value：重視。

13. with all speed：全速。

14. commands = orders。現在很少用 "commands" 此字。

No. 253　對運交錯誤貨物索賠的覆函

Gentlemen:

Your Indent 15

　　We sincerely regret the mistake we have made in the execution of your order. There is no doubt that, according to your Indent, the "Common-Wealth" Rubber Boots with black-fleeced lining were ordered. We have therefore hastened to make the exchange and the right goods have just been sent forward by express to Keelung for shipment per s.s. "Fushimi Maru" which is scheduled to sail on October 3, arriving at Kobe on the 7th of the same month; at the same time we have cabled you accordingly.

　　As we do not wish to let you bear the consequences of an error on our part, we are sending you herewith the Invoice corrected and would ask you to dispose the goods sent to you by mistake at the best possible prices on a consignment basis. Should you have any stock remaining on your hands at the end of December, you may ship back to us the balance at our expenses.

　　As you know, our shipping department fall into somewhat disorganized condition in consequence of the sudden and untimely death of our head clerk at the busiest season of the year, but fortunately the vacancy left by him was filled a few days ago to our satisfaction.

　　Your cooperation in this instance is very much appreciated and we trust the replacement will reach you in due course.

Yours faithfully,

【註】

1. hasten to make exchange：趕快掉換。

2. sent forward：運出。

3. express：快捷。

4. bear the consequences of：承擔……的後果。

5. dispose：處分。

6. on a consignment basis：以寄售方式。

7. at our expenses：費用由我們負擔。

8. shipping department：貨運部門。

9. fall into：陷入……狀態；變成……。

10. somewhat：有一點。

11. disorganized condition：紊亂狀態。

12. in consequence of：由於；因為……的緣故。

13. untimely death：死得非其時。

14. vacancy：遺缺。

15. filled：補實。

16. replacement：掉換的貨物。

No. 254　抱怨賣方未寄發裝運通知

Dear Sirs,

In spite of our last letter of the 20th of this month, we have today been informed by the Everett Shipping Lines, Keelung, that the cargo in question, i.e., 25 tons Crude Rubber as shipped by you against our Contract #123, has arrived on board the M/V "Hugheverett".

Usually even if you did not inform us of your shipment, you sent us copies of shipping documents by air, and we knew by which steamer you shipped a contract. As written to you some time ago, we need your advice or copy of invoice for effecting insurance on this side, beside of course we want to know the whereabouts of the contract.

Please therefore do not forget to give us a shipping advice by any means convenient for you as soon as a contract is shipped. Airmail is very quick these days, and you never need to cable us if you airmail the necessaries. We might just as well add that buyers sometimes would make trouble as though the contract had been delayed when your advice did not come in within reasonable time after the promised shipment time.

Yours faithfully,

【註】

1. ship a contract：運出某契約項下貨物。

2. whereabouts of the contract：該契約項下貨物的下落。

3. necessaries = necessary copy shipping documents。

4. might just as well add：再說……或許得當。

　　as well = also, too, in addition：add = 再言；附言。

5. make trouble：找麻煩。

6. as though = as if：好像；恰如。

No. 255　貨品不符的索賠

Gentlemen:

30 pcs. Silk Crepe "Taiyo Maru" Shipment

Your letter of September 10 brought us the invoice of your last shipment; and, with the exception of the item forming the subject of this letter, we have found all the goods quite satisfactory.

With regard to the 30 pcs. No. 100 Silk Crepe, however, we much regret to find ourselves under the necessity of rejecting them, as they are wrong in both quality and width. The quality is far below the pattern on which we placed the order with you; the width is narrower by 2 inches than that ordered. These errors on your part cause us to disappoint our important customers, to whom the goods must be delivered, and we are now placed in a very awkward position.

We hope you will put the matter right at once, letting us know by cable what you are going to help us out. Carelessness of this kind makes us inclined to give our orders to your competitors.

Yours faithfully,

【註】

1. with the exception...this letter：除了本信主題所指貨品外。

2. under the necessity of rejecting：必須拒收。

3. wrong：不符。

4. to put (or set) the matter right：改正；矯正。在這裡也可以 "to adjust the matter" 代替。

5. what you...out：你要怎樣辦，即如何 "to help us out of this awkward position" 之意。

6. to be inclined to：想；有意。

No. 256　對貨品不符索賠的覆函

Gentlemen:

We thank you for calling our attention in your letter of September 28 to the fact that the wrong goods were shipped on your order of August 10.

Upon tracing we find that owing to the pressure of business our shipping clerk shipped the goods of Pattern No. 105 instead of those of Pattern No. 100 on which your order was placed. We sincerely regret that we have much troubled and inconvenienced you through our oversight, and we assure you that every effort will be made in future to prevent any repetition of such mistakes.

In order to adjust the matter we have just cabled, asking you to accept the goods at an allowance of 20%, though we sustain a great loss, the price of No. 105 being lower by 10% than that of No. 100.

We must ask you to accept our apologies for the inconvenience you have been caused and grant us further opportunities to regain your confidence.

Yours faithfully,

【註】

1. ship on your order = ship against your order：根據你的訂單運出。

2. trace：追查。

3. pressure = rush（繁忙）。

4. through oversight：由於疏忽。

5. every effort...mistakes = every possible step will be taken to prevent any similar occurrence in future = severe measures will be adopted to secure the most careful attention to your orders in future。

6. we must ask you to accept our apologies：務請包涵；敬請原諒。

7. regain your confidence：恢復你（對我們的）信任。

No. 257　品質低劣的索賠

Dear Sirs,

We have received a very serious complaint from our clients respecting the goods you shipped for us in execution of our Indent No. 123. They write us that the goods are not an exact match in shade to the sample; and further the quality is decidedly inferior. Naturally, we rely on you to supply the exact shade and quality when we send you samples to match, and for any deviation from such samples we hold you responsible. Our clients claim an allowance of at least 50 cents per yard, and inform us that failing this, they will re-ship the goods back to us.

We should be glad to have your full explanation as to why such goods were shipped, as you will readily understand that we must be satisfied that our orders are faithfully carried out, otherwise we shall run the risk of losing the business from this particular quarter, as it is not very likely that our friends will send us repeat orders for these cloths after once being bitten by receiving inferior articles.

Please let us have your reply by return, so that we may explain to our clients.

Yours faithfully,

【註】

1. exact match in shade to the sample：色度與樣品完全相符。

2. deviation from：與……不同；有偏差。

3. claim an allowance：要求折價。

4. failing this = in default of this：如果不這樣；否則。

5. we must be satisfied：必須使我們滿足；我們必須確信。

6. bitten: p.p. of "bite"：受騙。

No. 258　對品質低劣索賠的覆函

Dear Sirs,

We are indeed surprised to receive your complaint respecting the goods shipped against your Indent No. 123. We keep a reference sample of such special orders, and upon referring to the sample we find that it is a very good match indeed to your sample both in quality and in shade, and we cannot understand the complaint.

We should be inclined to think that your clients wish to return the goods for some reason best known to themselves, as they must know very well that no manufacturers would make such an allowance as 50 cents per yard on the cloth delivered. We enclose cuttings from your own sample and from the reference sample which we kept from the pieces supplied, and we ask you to submit them to any judge of such goods for their impartial opinion.

We have been in the trade for more than ten years, and never in the whole of our experience have we had a claim so unfair made upon us. We do not for a moment doubt your own good faith in this matter, but we are convinced that, when you have compared the samples we are sending you, you will come to the same conclusion as ourselves—viz., that the claim is preposterously unfair, and that the goods have been delivered up to sample in every respect.

Yours faithfully,

【註】

1. reference sample：交運時，由交運的貨物中抽出一部分寄給買方的樣品。買方收到此樣品而未提出異議時，即視為承認（接受）該貨物，日後不得以與 original order sample 不符為由拒收，或提出 claim。如貨物為 soft goods 時稱為 reference pattern；如貨物為 raw material 或 semi-manufactured goods 時稱為 shipping sample。

2. to be inclined to think：以為。cf. to be inclined to believe：相信。

3. for some reason best known to themselves：為了他們自己的理由。

4. a claim so unfair：比 "so unfair claim" 強而有力。

5. viz.：即；就是（拉丁語 videlicet 之略，通常讀為 namely）。

6. preposterously：荒謬地；反常地。

7. up to = equal to。

No. 259　與樣品不符的索賠

Dear Sirs,

Re: Your contract No. 122 for

100 B/S Wool Yarns

Under the captioned contract, we have taken delivery of 100 bales of wool yarns shipped per s.s. "Eugene Lykes".

Upon unpacking the bales, we have found that 20 bales of the lot are much inferior to your sample. This error has been apparently made by the carelessness of your shipping clerk, and can be easily found out by examining the remaining stock on your side. But, for your reference, we have asked FESCO to draw out samples from the 20 bales in question and have airmailed them to you today by parcel post.

For the inferiority in quality, you are requested to make a compensation allowance in price, the amount of which you will please telex to us, after examination of the sample sent you.

Yours faithfully,

【註】

1. wool yarns：（羊）毛紗。

2. of the lot：lot 為 "shipment"（貨載）或 "consignment"（貨物）之意。out of = from。

3. inferior to：注意用 "to"，解做 "than"。

4. apparently = no doubt。

5. shipping clerk：貨運承辦人。

6. remaining stock：（手邊）剩下的商品。

7. FESCO = Far East Superintendence Co., Ltd.。

8. draw out：抽取。

9. inferiority in quality：也可用 "inferior quality" 代替。

No. 260　對貨樣不符索賠的覆函

Dear Sirs,

We confirm we have telexed you today the following message:

"YOUR CLAIM SAMPLE UNDER CONTRACT NO. 123 RECEIVED UPON EXAM THE GOODS ARE WELL UP TO STANDARD GRADE IN QUALITY EXCEPT A LITTLE

MORE DUST THAN ORDINARY CASES WE OFFER ONE PERCENT ALLOWANCE"

We have duly received your samples of the wool yarns shipped by s.s. "Eugene Lykes", which you sent to us protesting that they are different from our samples previously sent.

On careful examination of the samples sent by you, the wool yarns in question have been found well up to the standard grade in quality, except the dust existing in them is a little more than in ordinary cases. We should like, therefore, to make an allowance of 1% for the excess presence of dust. Please note that this is the best allowance we can offer.

If, however, you are not satisfied with this offer, we wish to submit the case to an arbitration and to abide by its decision.

Yours faithfully,

【註】

1. sample previously sent：也可以 "previous sample" 代替。

2. in question：繫爭的；爭議中的。

3. well up to the standard grade：很夠標準級。

4. in ordinary cases：普通的場合。

5. not satisfied = not content with，也可用 "not agreeable to"。

No. 261　貨物受損的索賠

Dear Sirs,

Re: Shipment of our order No. 123
per M.V. "Oriental Despatcher"

The goods you shipped against our order No. 123 per M.V. "Oriental Despatcher" arrived at Keelung on May 15.

Upon examination immediately after taking delivery, we found that many of the goods were severely damaged, though the cases themselves showed no trace of damage.

Considering this damage was due to the rough handling by the shipping company, we claimed on them for recovery of the loss, but investigation made by the surveyor has revealed the fact that the damage is attributable to the improper packing. For further particulars, we refer you to the surveyor's report enclosed.

We are, therefore, compelled to claim on you to compensate us for the loss, US$250, which we have sustained by the damage to the goods. We trust you will be kind enough to accept this claim and deduct the sum claimed from the amount of your next invoice to us.

Yours faithfully,

【註】

　　1. no trace：無跡象。

　　2. considering = thinking：以為。

　　3. rough handling：粗魯的處理。

　　4. to claim $\left\{\begin{array}{l} \text{on} \\ \text{upon} \\ \text{against} \end{array}\right\}$ a person：向某人索賠。

　　5. has revealed the fact：揭開了某事實，即發現。

　　6. attributable：可歸因於……的。

　　7. improper packing：包裝不當。

　　8. for further particulars：進一步的詳情。

　　9. to be compelled to = to be forced to：不得不。

　　10. which 指 loss 而言。

　　11. sustained by = suffered from：蒙受。

　　12. to accept this claim = to admit this claim：承認此索賠，即同意賠償。

　　　　cf. advance a claim
　　　　　　put forward a claim $\Big\}$ 提出索賠。

　　　　　　entertain a claim：受理索賠。

　　　　　　to dismiss the claim of...on the ground that...：基於……原因，駁回索賠。

　　　　　　relinquish (withdraw) a claim：撤回索賠。

　　13. next invoice to us：下次開給我們的發票。

No. 262　對貨物受損索賠的覆函

Dear Sirs,

　　We have received your letter of May 17 informing us that the goods shipped to you against your order No. 123 arrived damaged on account of the imperfectness of our packing.

　　This is the first time that we have received such a complaint from our customers, although we have been shipping the goods for five years in the past, packing them in the similar manners as we shipped the goods to you.

　　Furthermore, we would point out that we hold a copy of clean B/L from shipping company, which relieves us of all responsibilities. We are, therefore, convinced to think that the present damage was due to extraordinary circumstances under which they were transported to you. We are, therefore, not responsible for the damage, but as you must have insured the shipment at your end, we would suggest that take up the matter with the shipping company or lodge your claim with the insurance company if you have insured the goods against All Risks. We shall, of course, place at your disposal any documents necessary to substantiate your claim.

While we are sorry for the inconvenience you have suffered, we believe the above explanation will prove satisfactory to you.

Yours very truly,

【註】

1. arrived damaged = reached you damaged，即 "arrived in damaged condition"。

2. imperfectness：不完善。

3. in the similar manners = in the same manners。

4. to relieve...responsibilities：免除……的責任。

5. are convinced to think：堅信。

6. due to = caused by。

7. extraordinary circumstances：特殊情事。

8. under which：在此狀態下，which 指上述「特殊情事」。

9. take up the matter with：將此事向……提出。

10. to lodge a claim：提出索賠。

11. all risks：全險。

12. to place at your disposal：聽你使喚；聽你使用；聽你支配。

13. to substantiate：作證；供作證明。

　　cf. substantiated claim：正當的索賠（要求）。

14. prove：成為……；使（你滿意）。

No. 263　貨物包裝不良而索賠

Dear Sirs,

Re: 2,500 M/T's Bonlai Rice
shipped per s.s. "Nagasaki Maru"

With reference to the subject shipment, we confirm that we have received your cable of August 10 which reads:

"LC 123 BOOKED SS NAGASAKI MARU FOR 2500 M/TS SWC SUGAR ETD KAO-HSIUNG 12TH ETA BANGKOK AUGUST 20TH"

We have just taken delivery of the goods and found that about 10% or, to be exact, 2,500 bags have been found broken with at total loss of approximately 2,000 kgs of rice. According to the surveyor's landing report, a copy of which is enclosed for your reference, the breakage of the bags is due mainly to the fragility of the gunny bags. The surveyor has further found that the bags are woven with jute of very inferior quality. Therefore, your Corporation, instead of either the shipping company or the insurance company should be held responsible for the loss of the goods.

Considering the preceding, you are requested to compensate us for the total loss of rice at the FOB Stowed value of US$250 per metric ton.

Your early settlement of this case will be appreciated.

Yours very truly,

【註】

1. landing report：卸貨報告。
2. fragility：脆弱。adj.：fragile。
3. woven with jute：用黃麻編織的。

貨物包裝如不適當，以致在運輸中受損，則承運人 (carrier) 固然不負責，保險公司也不予理賠。所以出口商對於包裝是否適當應予相當的注意。

No. 264　對貨物包裝不良索賠的覆函

Dear Sirs,

Re: Shipment of 2,500 M/T's Bonlai Rice
per s.s. "Nagasaki Maru"

Your letter of August 24, 20– together with a copy of surveyor's landing report has reached us. Much to our surprise considerable loss has been inflicted on the subject shipment owing to the fragile gunny bags used for packing the rice.

Upon receipt of your letter, we have immediately instructed our laboratory to give a strict test of the durability of the gunny bags. We are quite surprised that the findings of our laboratory happened to be the same as stated in the landing report, namely, the gunny bags have been found not strong enough to bear weight of 100 kgs.

Based on the findings, we agree to compensate you for the total loss and enclose a Sola Draft No. 123 for US$500 issued by Bank of Taipei to pay therefor. Please let us have your formal receipt in due course.

We trust that you would be satisfied with our prompt settlement of this case.

Yours faithfully,

【註】

1. durability：堅固。

2. not strong enough to bear weight of 100 kgs：不足以負荷 100 公斤的重量。

3. sola draft = sola exchange：單張匯票。國際貿易上，匯款用的匯票多係單張，而押匯用者，多係一式二張。市場上往往將 "sola draft" 誤稱為 "sola check"。

4. in due course：在適當期間。

No. 265　向賣方提出箱件遺失的索賠

Dear Sirs,

Re: Our order No. 123

We have just taken delivery of the umbrellas of our order No. 123, but we regret to have to tell you that ten cartons of the articles are missing because we received only one hundred and forty cartons which were inconsistent with the entries of one hundred and fifty cartons on the packing list and on the invoice.

We have informed the shipping company at this end to trace the whereabouts of the missing goods. Enclosed is a copy of the enquiry to the shipping company for your reference. You are requested to take proper measures immediately to check up with your factory and the shipping company on your side and let us know your findings about it before we make a claim.

Thank you for your cooperation and look forward to your speedy reply.

Yours faithfully,

【註】

1. missing：失蹤。

2. to be inconsistent with：與……不符。

3. entry：記載。

4. take proper measures：採取適當措施。

5. check up with...：與……核對。

6. findings：結論；結果。

7. speedy reply：迅速的答覆。

No. 266　對箱件遺失索賠的覆函

Gentlemen:

Subj.: Your order No. 123

We greatly regret to be informed that you have received only 140 cartons of umbrellas under captioned order instead of 150 cartons.

Upon receipt of your letter, we have given this matter our immediate attention. We are quite willing to get the case cleared up if we should be responsible for. However, you will understand that we are not the party to blame as you will see that according to the B/L issued by the shipping company, 150 cartons have been loaded on board the vessel. It is apparent that 10 cartons were missing due to the carrier's improper handling. Therefore, you are requested to file your claim against the shipping company if they fail to find out the whereabouts of the missing articles. Based on the clauses stipulated on the back of the B/L, the shipping company is responsible for any goods shortlanded at the port of discharge. In view of this fact, we trust that the shipping company will give you satisfactory settlement of the claim.

Meanwhile, we have passed a photostatic copy of your letter together with a copy of the B/L on to the shipping company for their attention. Should you need our further assistance to clear up this matter, please do not hesitate to let us know. We promise to give you a hand.

Very truly yours,

【註】

1. clear up = solve = settle。

2. photostatic copy：影印本。又稱 "xeroxed copy"。

3. do not hesitate to let us know：請通知我們，勿庸猶豫。

4. give you a hand：幫助你一下。

No. 267　將索賠案件提交仲裁

Dear Sirs,

We hasten to inform you that we have instructed our Hongkong Branch to adjust your claim for the defects in our goods shipped in execution of your order No. 50 of the 3rd May. We are now surprised to note from your letter of the 7th June that you are not prepared to consider the offer of a 15% allowance made by our Branch to compensate you.

Though we consider our offer adequate, and even generous, we extremely regret that our offer has been refused. As it is not likely to come to amicable settlement between us, we suggest that we have to submit the matter to arbitration, according to the stipulations in Business Agreement.

We would recommend on the ground of economy a joint arbitrator, but should you prefer to have one appointed by each of us, and a third called in with a casting vote in the event of disagreement, we would be prepared to fall in with your wishes.

Yours faithfully,

【註】

1. amicable settlement：友善的解決；和解。

2. arbitration：仲裁。有關仲裁的用語，列舉若干於下：

   arbitrator：仲裁人。umpire：判斷人；公斷人。當仲裁人有兩人而其意見不一致時，由他們再選一人，作最後的決定，此人稱為 "umpire"。arbitration award：仲裁判斷書。to settle the matter by arbitration：以仲裁解決事件。to submit (refer) the matter to arbitration：將事件提交仲裁。

3. Business Agreement：「交易條件協議書」，請參閱第十一章。

4. joint arbitrator：（由爭執的買賣雙方協議選定的）共同仲裁人。

5. a third = a third person：第三人，即上述的 umpire。

6. to call in：聘請。

7. casting vote：裁決權。指二位仲裁人意見不一致時，判斷人 (umpire) 的最後決定權。

8. disagreement：（二位仲裁人的意見的）不一致。

9. to fall in with = to meet; to agree to：同意。

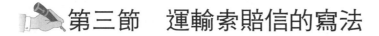

第三節　運輸索賠信的寫法

一、貨物短損型式

貨物於運輸過程中、裝卸作業時均有發生破損及短少的可能。對於這些短損，船方是否應予賠償，胥視損毀或短少發生的原因，是否為船方依法應負責而定。

貨物發生短損的情形，大致可歸納為下列幾類：

1. Shortland（短卸）。

2. Shortage（短失）。

3. Damage：

(1) Breakage　　　　　　　　(2) Sweat（汗濕）

(3) Rain and fresh water damage　(4) Wet by sea water

(5) Cover torn　　　　　　　(6) Scratch

(7) Rust　　　　　　　　　(8) Leakage

(9) Scorch　　　　　　　　(10) Stain

(11) Bending & Denting　　　　(12) Collapse of cargo

二、索賠的要領

1.一發現貨損、滅失應即以書面通知。提貨時：

⑴如貨物的損害、滅失顯著者，受領人應於受領貨物時即刻以書面通知。

⑵如貨物的損害、滅失不顯著者，受領人應於提貨後三日內，以書面通知。

⑶受領人也得不以書面通知，而在收貨證件上註明損害或滅失。

⑷將通知副本抄送出口商、保險公司等有關方面。

2.索賠時應備齊有關文件：

⑴ Claim letter ⑵ B/L

⑶ Invoice ⑷ Packing list

⑸ Damage & Shortage report（短損報告） ⑹ Debit note

⑺ Survey report

3.不斷催請處理結案。

4.提出訴訟：應於卸貨後一年內為之。

No. 268　初步索賠通知 (Preliminary notice of claim)

Dear Sirs,

> Preliminary notice of claim for
> stain damage to 200 bales cotton
> ex s.s. "President"

　　We regret to inform you that stain damage is found in connection with the shipment of the below-mentioned cargoes:

B/L	MARKS	NO. OF PACKAGE	DESCRIPTION OF GOODS
123	CTC LOT 1 1/200	200 bales	AMERICAN RAW COTTON

shipped by Hohenberg Bros. Co., from Galveston on board s.s. "President" consigned to us under B/L No. 123, and arrived at Keelung on August 5, 20–.

　　We hereby declare that we reserve the right to file a claim with you for this damage when the details and amount of the damage are ascertained. We will apply to Lloyd's Agents for survey on August 8, 20–.

Please acknowledge the receipt of this letter.

Very truly yours,

【註】

1. stain damage: 油損。

2. arrive at Keelung: 船到達某港埠時用 "at"，不用 "in"。

3. reserve the right to file a claim $\left\{ \begin{array}{l} \text{with} \\ \text{against} \end{array} \right\}$ you: 保留向你索賠之權。

4. ascertained: 確定。

No. 269　向船公司提出短卸索賠

Dear Sirs,

Re: Notice of shortlanding of Dye-stuff
shipped per s.s. "Loide Equador"
under B/L No. 6 arrived at Keelung
Dec. 10, 20–

We are the consignees of the captioned Dye-stuff totaling 100 drums of which ten drums are shortlanded.

Enclosed pleased find a photostatic copy of a Shortlanded Report issued by the Keelung Harbour Bureau and certified by your Keelung Office. It will be appreciated if you will trace the whereabouts of these ten missing drums and let us know the findings at your earliest convenience.

We are awaiting your speedy reply.

Faithfully yours,

【註】

1. shortlanding: 短卸。

2. dye-stuff: 染料。

3. shortlanded report: 短卸報告。

4. Keelung Harbour Bureau: 基隆港務局。

No. 270　請求公證行做公證

Dear Sirs,

Application for survey

You are hereby requested to conduct a survey of the following goods consigned to us which have arrived damaged:

Shipper: Hohenberg Bros. Co., Memphis, Tenn., USA

Description of goods: 200 B/S American Raw Cotton

Marks & No.:   CTC
                   Lot 1
                   1/200

Date of landing: August 5, 20–
Date of delivery to us: August 10, 20–
Nature of damage: stain damage by oil
Numbers of packages/units to be examined: 200 bales
Location stored: No. 18 warehouse, Keelung Harbour

Your attention to this matter and issue to us your Survey Report in triplicate will be appreciated.

Yours faithfully,

【註】

1. conduct a survey：查勘；鑑定。也可以 "make a survey" 代替。

2. location stored：存儲地點。

3. 貨物如有受損，進口商應請求保險公司及船公司同意的公證行做公證，並取得公證報告 (survey report) 以便索賠。

No. 271　正式索賠──污損

Dear Sirs,

Re: Claim for stain damage to 200 B/S cotton ex s.s. "President"

With reference to our letter of August 11 we now submit our claim for the captioned amounting to US$340.25 as per Debit Note No. C–123 attached, and shall be glad to receive settlement at your earliest convenience.

In support of this claim, we also enclose one copy each of the following documents:

1. Shipper's signed invoice.
2. Shipper's packing list/weight list.
3. Damage certificate issued by Keelung Harbour Bureau.
4. Survey Report issued by Robert W. Hunt Co.
5. Receipt for inspection fee.
6. B/L.

Yours faithfully,

【註】

1. settlement：此處解釋做「賠款」。

2. damage certificate：短損證明書。

3. receipt for inspection fee：檢驗費收據。

4. 當進口商提貨時，如發現貨物有受損應及時向船公司提出索賠。但由於公證費時，所以，通常先提出初步索賠通知 (preliminary notice of claim)。俟取得公證報告後，再提出正式索賠 (final claim)。

No. 272　正式索賠──破損

Dear Sirs,

Re: 100 bags ABS Resin
shipped per s.s. "May Flower"

We have taken delivery of the subject ABS Resin and regret to inform you that ten bags thereof have been found broken with a total weight of 400 kgs. only.

To recover for the loss in weight, we enclose one photostatic copy each of the following documents for your processing:

1. Invoice issued by American Chemical Co.
2. Weight list issued by the same shipper.
3. Survey Report issued by General Superintendence Co., Ltd.
4. B/L No. 123 issued by APL.

Enclosed pleased also find our Debit Note, in quadruplicate, for US$240 which sum is arrived at with the following computation:

10,000 (net wt. of whole shipment in kg.) ÷ 100 (No. of bales) = 100 kgs. (average net wt. per bag)

100 kgs. x 10 (No. of bags broken) − 400 kgs. (total wt. of the ten broken bags) = 600 kgs. (total loss in weight)

600×US$0.40 (CIF value net per kgs) = US$240.00

> We shall appreciate your compensating us this amount at your earliest convenience.
>
> Faithfully yours,

【註】

1. ABS Resin：一種樹脂。

2. to recover for = to recoup for = to make up for：補償。

3. APL = American President Lines。

4. arrive at = obtain = get：獲得。

5. computation：計算。

No. 273　正式索賠──破損

> Gentlemen:
>
> Subject: Claim for loss of soybean oil
> per s.s. "Pioneer" arrived
> at Keelung July 25, 20–
>
> We regret to inform you that upon taking delivery of the 50 drums of the captioned goods which were discharged at Keelung from s.s. "Pioneer", we have found drum No. 8, 11, 12, 26, 44 were partly broken, resulting in a loss of 180 lbs. or 81 kgs. of its contents. We have, therefore, to claim on you for compensation for this loss.
>
> In support of our claim, we enclose the relevant documents as follows:
>
> 4 copies of debit note.
> 1 original survey report made by the China Survey Co., Ltd.
> 1 copy of invoice for inspection fee.
> 1 copy of shipper's invoice.
> 1 B/L No. 15 of s.s. "Pioneer".
> 1 damage certificate issued by Keelung Harbour Bureau.
>
> We shall appreciate it if you will kindly let us have your check for US$304.20 in settlement of the above claim at your earliest convenience.
>
> Yours very truly,

【註】

1. in support of our claim：為了支持我們的索賠。

2. relevant：相關的，即 relative。

No. 274　正式索賠──短損

Messrs. American President Lines

Dear Sirs,

Re: Claim for shortage of W/S Kip Skin

per s.s. "President"

Claim is hereby filed with you for shortage of W/S Kip Skin shipped by Kanfamnn Trading Corp. from New York on board the s.s. "President Grant," consigned to us under Bill of Lading No. 31, and arrived at Keelung on June 27, 20–

The details and the amount of the claim are as follows:

Marks & No.	B/L No.	Description	Invoice value	Quantity
PARCHMENT TAG AX	No. 31	W/S Kip Skin	$568.08	15 B'dles

Claim Amount:

Shortage... 1 B'dle

$568.08×1/15 = $37.87

Total Claim Amount: $37.87

You are requested to investigate the matter immediately, and your earliest compensation will be highly appreciated.

Attachment:

B/L copy

Invoice copy

Survey Report

Debit Note

Yours truly,

TAIWAN TRADING CO., LTD.

(Signed)

Manager

【註】

1. Kip Skin：小獸皮（牛犢、羊羔等的生皮）。

2. B'dles 為 Bundles（捆）的略字。

3. Debit Note 的格式並無一定的型式，下面 No. 275 的格式只是一例而已。

No. 275　借項通知單

DEBIT NOTE

To American President Lines
   Keelung

                                    6th September, 20–

CLAIM: Shortage of W/S Kip Skin
      Per s.s. "President Grant"

Marks & No.	B/L No.	Description	Invoice value	Quantity
PARCHMENT TAG AX	No. 31	W/S Kip Skin	$568.08	15 B'dles

Claim Amount: Shortage... 1 B'dle
              $568.08×1/15 = $37.87

              Total Claim Amount: $37.87

                    TAIWAN TRADING CO., LTD.

                        (Signed)

                        Manager

No. 276　船公司的理賠函

Gentlemen:

                S.S. HAWAII BEAR VOY. 16–W
                B/L No. 19 & B/L No. 34;
                Claim a/c Alleged Shortage of 8 p'cs Hides

    Reference is made to your claim letter of March 3 and our reply of the 6th, captioned as above.

    Please be advised that our cargo tracers on eight pieces Hides have been returned stating that same were not overlanded at any of the vessel's ports of call.

    We therefore acknowledge our liability and responsibility for shortage of (8) pieces Hides in amount of $74.36 and are enclosing our check in that amount in full settlement thereof.

                        Very truly yours,

                    United States Lines Company

             Agents: PACIFIC FAR EAST LINE, INC.

【註】

    1. cargo tracers：貨物追查信。

2. overland：誤卸。即 "misland" 之意。

3. port of call：停靠港。

No. 277 覆收到船公司賠款函

American President Lines, Ltd.

Dear Sirs,   Your ref. No. K−616−1

We have pleasure in acknowledging receipt of your cheque for US$37.87 in settlement of our claim for shortage of one bundle Kip Skin ex s.s. "President Grant" V/20, New York/Kee B/L No. 31.

Please find your vouchers duly signed as requested by your goodselves.

Thank you for your cooperation.

Yours truly,

TAIWAN TRADING CO., LTD.

【註】

1. ex：係介系詞，出自拉丁文，有「由」，「自」之意，cargo ex s. s. "President Grant" 為由格蘭總統輪運來（即由該輪取出）的貨物。

2. vouchers：收據。

No. 278 催請船公司理賠函

Dear Sirs,

Re: Claim for stain damage to 200 B/S
American Raw Cotton ex. s.s. "President"

With reference to our letter of August 20, we would remind you that the captioned claim is still outstanding, and we shall be glad to receive settlement at your earliest convenience.

Yours faithfully,

No. 279 船公司拒絕賠償函之一

Dear Sirs,

S.S. "TJIPONDOK"/47

BELAWAN DELI/KEELUNG B/L No. 2
332.2 M/T PALMOIL IN BULK, 2.7838 m/t shortage

We acknowledge receipt of your letter dated 10th August, 20– dealing with above-mentioned shortage.

We may invite your kind attention to clause 9 (a) printed on the reverse side of the covering Bill of Lading which we quote below:

"As the carrier has no reasonable means of checking the weight of bulk cargo, any reference to such weight in this Bill of Lading shall be deemed to be for the convenience of the shipper only, but shall constitute in no way evidence against the carrier."

Furthermore we may refer to the clause printed on the face of the Bill of Lading reading in part:

"Contents and...weight...unknown, any reference in this Bill of Lading to these particulars is for the purpose of calculating freight only."

We regret that in view of the above we are not in a position to assume liability for the shortage.

Please note that in passing the above information on to you, we do so without prejudice to our defences under the terms and conditions of the contract of affreightment.

Yours faithfully,

【註】

1. bulk cargo：散裝貨。

2. deemed：視為。

3. constitute：構成。

4. in no way：絕不。

5. against carrier：對抗運送人。

6. contents and weight unknown：內容及重量不詳。

7. without prejudice = without detriment to existing right or claim：對於現存權利或要求無影響或無損；不侵害權利；不使權利受到損害。

8. defences：抗辯。

9. contract of affreightment：運輸契約。

No. 280　船公司拒絕賠償函之二

Gentlemen:

Re: Claim for Damage to Tallow in bulk
ex. m.s. "Akagi Maru" Voy. No. 16–Home
Los Angeles/Kaohsiung B/L No. 13

We have received your notice of claim regarding the above.

In the same connection, our records show that the cargo in question was discharged at this port on February 20, 20– in apparent good order and condition with no exception noted on Cargo Boat Note.

We, therefore, regret to have to advise you that we are unable to entertain your claim for damage to the above cargo.

Yours faithfully,

【註】

1. Cargo Boat Note：我國又稱為 Delivery & Receiving Cert. 貨物授受證。

2. entertain your claim：受理你的索賠。

No. 281　船公司拒絕賠償函之三

Dear Sirs,

B/L No. 16 4094 990 Bundles Black Steel Sheets
B/L No. 11 4090 500 Bundles Black Steel Sheets
Ex S.S. "KNOWSLEY HALL" May 10, 20–

We have before us yours of the May 21, requesting us to examine the above cargo alleged to have been landed damaged at Kaohsiung.

Although you state in the letter that many sheets are extraordinarily torn on the edges and that some sheets have been cut through by the wire sling during discharging, we have to refuse your request in view of the fact that the steamer holds a clean receipt for the goods, and that the same is unprotected cargo.

We regret that we are unable to accede to your request.

Yours faithfully,

【註】

1. we have before us yours：這是一種舊式的表現法，現在已經很少用，可以 "we have received your letter" 代替。

2. sling：吊索。

3. clean receipt：無瑕疵收據（受貨人受領貨物時簽發給船方的收據，如在其上面未註明所收到貨物有任何瑕疵時此收據即為 clean receipt）。

第四節　保險索賠信的寫法

　　進出口商所以將貨物投保保險，目的在於貨物受到損害時可由保險公司獲得補償，所以懂得如何投保而不懂得如何索賠，仍不實用。

一、保險索賠一般注意事項

　　1.取得公證報告 (Surveyor's report)。

　　2.迅速通知保險人或其代理人 (Notice of claim)。

　　3.掌握索賠權時限及時向事故責任人索賠。

　　4.索賠文件必須齊全。

二、全損索賠應提出的文件

　　1. Claim letter。

　　2. Insurance policy。

　　3. Commercial invoice。

　　4. Packing list/Weight certificate。

　　5. Sea protest copy（海難證明書副本）。

　　6. Certificate of total loss from the carrier（運送人全損證明書）。

　　7. Survey report。

　　8. B/L。

三、單獨海損索賠應提出的文件

　　1. Statement of claim（索賠計算書）。

　　2. Insurance policy。

　　3. Claim letter。

　　4. Survey report。

　　5. B/L。

　　6. Commercial invoice。

7. Packing list/Weight certificate。

8. Damage and shortage report。

9. Others。

四、共同海損索賠應提出的文件

申請保險公司繳納保證金或簽發保證函應提出的文件：

1. Copy of notice from carrier。

2. Insurance policy。

3. Commercial invoice。

4. B/L。

5. Packing list/Weight certificate。

6. Average bond（共同海損保證書）。

7. Others。

No. 282　向保險公司提出油污索賠通知 (Notice of claim)

Dear Sirs,

> Re: Preliminary Notice of damage of
> American Raw Cotton per s.s. "President"
> Your marine insurance policy No. 123

We regret to inform you that the American Raw Cotton consigned to us and covered by the subject policy arrived damaged.

> Shipped from: Galveston, U.S.A.
> On board: s.s. "President"
> Arrived at : Keelung on October 5, 20–
> Nature of loss: stain damage by oil

If you have no objections, we will apply to Lloyd's Agents for survey and the documents in respect of the formal claim will be forwarded to you in due course.

A copy of our preliminary notice of claim against the carrier is enclosed and your attention to this matter will be appreciated.

Yours faithfully,

【註】

    1. stain damage by oil：油污。

    2. apply to = ask：要求。

    3. Lloyd's Agents：Lloyd's 為 "The Corporation of Lloyd's"（勞依茲公司）的簡稱，是英國保險市場中一個特殊組織。Lloyd's Agents 為勞依茲公司分布全球各處重要商埠的代理處，協助處理報導海上運輸或保險有關的業務（查勘公證等）或消息。

    4. preliminary notice of claim：初步索賠通知，與 "formal claim" 相對而言。

No. 283 　向保險公司提出油污正式索賠 (Formal claim)

Dear Sirs,

Re: Claim for stain damage to American
Raw Cotton under your policy No. 123

    With reference to the captioned claim, of which we sent you a preliminary notice of claim on October 6, we hereby file the claim amounting to US$512.10 as per the enclosed Debit Note No. 345.

    In support of our claim, we enclose also the following documents:

    1. one original policy No. 123 duly endorsed.
    2. one original survey report No. 456 with a receipt for survey fee.
    3. one copy of shipper's invoice.
    4. one copy of B/L No. 789 of s.s. "President".
    5. one copy of our letter to carriers and their reply.
    6. one copy of packing/weight list.
    7. statement of claim.

    We shall appreciate it if you will investigate the matter promptly and let us have your cheque in settlement of the above claim at your earliest convenience.

Yours faithfully,

【註】

    就索賠程序而言，應先向船公司索賠，如船公司不予理賠，則憑其覆函，向保險公司提出索賠。萬一船公司相應不理，不覆函也不理賠，則貨主可免附船公司覆函，而以其致船公司的索賠函副本代替之。

No. 284 　貨物遺失，向保險公司提出正式索賠

Gentlemen:

Re: Claim for non-delivery of goods

under your Marine Insurance Policy No. 12
Our ref. No. 45

Reference is made to our letter of June 16, 20– addressed to Messrs. Jardine Matheson & Co., Taipei in connection with the subject claim, with a copy to your office.

The United States Lines have already compensated us with a sum of US$522.80 representing the CIF value of the goods. We wish now to claim on you for (1) the difference between the insured value and the CIF value and (2) the surveyor's fee, viz., US$120.56 as per Debit Note enclosed.

In support of our claim, we enclose the relevant documents as follows:

3 copies of Debit Note.
1 survey report made by the Overseas Merchandise Inspection Co.
1 copy of invoice for inspection fee.
1 original and 1 duplicate of Insurance Policy.
1 copy of shippers' invoice.
1 B/L No. 55 of s.s. "Pioneer Wave".

We shall appreciate it if you will please let us have your cheque for US$120.56 in settlement of the above claim at your earliest convenience.

Very truly yours,

【註】

1. non-delivery：未送達，即遺失之意。

   船公司對於承運的貨物如無法按 B/L 所載件數交出的話，應負賠償責任，但對於每一包件 (package) 的賠償金額依法有一限度 (limitation)。依我海商法每件 SDR666.67 單位或每公斤 SDR2 單位計算所得的金額，兩者較高者為限（海商法第 70 條），因此，被保險人如自船公司所獲賠償金額不足以彌補其損失時，尚可就其差額要求保險公司補償。

2. claim on you：正確的說法應為 "claim against you" 或 "claim from you"。如 "claim" 為名詞時，"lodge a claim on (= upon = against) you for..."。

No. 285　包裝破損，請保險公司理賠

Dear Sirs,

Re: 100 bags ABS Resin
shipped per s.s. "May Flower"
Your Policy No. 123

Enclosed please find one copy of the following documents with regard to the above-mentioned shipment:

1. Survey Report issued by INTECO.
2. Invoice issued by American Chemical Co.
3. Weight list issued by the same supplier.
4. Insurance policy.
5. B/L issued by APL.
6. Damage report issued by Keelung Harbour Bureau.

From the survey report enclosed herein, you will note that there are ten bags of the 100 bags of this shipment with cover torn, and contents partly exposed and split.

In view of the damages, you are requested to compensate us the CIF Value plus ten percent of these 10 broken bags at your earliest convenience.

Yours faithfully,

【註】

1. cover torn：包裝破裂。

2. exposed：暴露。

3. split：散失。

No. 286　保險公司對於包裝破損索賠的覆函

Dear Sirs,

Re: Your claim for loss of ABS Resin
shipped per s.s. "May Flower"
covered by our Policy No. 123

We have received your letter of May 5, 20– submitting the subject claim together with the relative documents and have duly noted the contents thereof.

After our careful deliberation of this case, we regret to reply that you should file the claim first against the carrier. In case shipping company refuses to compensate you for the loss and their refusal found justified, we will have the case our further consideration. Returned herewith are the relative documents submitted by you.

Please be assured that, if the loss is found within the coverage of our policy, we will compensate you for the loss immediately.

Yours faithfully,

【註】

1. have duly noted the contents thereof：妥予注意到其內容。舊式用法。

2. deliberation = careful thought：熟慮。因此 "after careful deliberation" 中的 "careful" 毋寧是多餘的。cf. after long deliberation。

3. justified：證明為正當；有道理。

4. compensate you for 可以 "make up the loss"，"make up for the loss" 代之。

No. 287　經船公司拒賠後向保險公司索賠

Dear Sirs,

Re: Your policy No. 123

We acknowledge the receipt of your letter of May 8, 20– concerning our claim under the subject policy and have duly noted its contents.

Following your instructions, we have referred this case to the shipping company and filed our claim against them for the loss we have suffered. Now we have received their reply and enclose one photostatic copy thereof for your reference and perusal.

Considering that the coverage of the captioned policy includes risks against leakage and breakage, you are requested to give our claim further and favorable consideration. In order to facilitate you to settle this claim, we submit again all the relative documents required.

We are looking forward to your early settlement of this case.

Faithfully yours,

【註】

1. considering = in view of the fact：鑑於……。

2. coverage：擔保範圍；承保範圍。

3. risks against leakage and breakage：漏損及破損險。

4. early settlement 也可以 "speedy settlement" 代替。

No. 288　短損的索賠

Dear Sirs,

Re: Your marine insurance policy No. 123

We are the consignees of 300 B/S of American Raw Cotton shipped per s.s. "Kagoshima Maru" which arrived at Keelung on May 6.

Upon taking delivery of the cargo, we have asked Far East Superintendence Co., Ltd. to weigh all these 300 B/S and regret to inform you that there is a loss in weight of 1,050 lbs. Enclosed please find a copy of Survey Report No. 567 along with following documents:

one copy of shipper's invoice.

one copy of weight sheet issued by the same shipper.

one original copy of the subject insurance policy.

one copy of B/L issued by NYK.

one copy of letter from carrier.

In view of the shortage, we hereby file our claim against you for US$493.50 which sum is arrived at with the following computation:

1,050 lbs. (total loss in weight)×US$0.47 (CIF value per lb.)=US$493.50

We are also enclosing our Debit Note, in duplicate, for your processing and shall appreciate your speedy settlement of this case.

Yours very truly,

No. 289　保險公司對於短損索賠的覆函

Dear Sirs,

Re: Our policy No. 123

We have received your letter of May 8th, 20– together with a copy of Survey Report and relating documents filing claim against us under the subject policy for loss in weight totaling 1,050 lbs. of American Raw Cotton shipped per s.s. "Kagoshima Maru".

After examination of the documents you submitted, we agree to make up the loss as requested and are enclosing our check No. C234 for US$493.50 drawn on Bank of Taipei.

Enclosed please also find two copies each of our Subrogation Receipt, Loss Subrogation Receipt and Marine Cargo Loss Receipt. We shall appreciate it if you will fill out these receipts properly and return them to us duly signed at your early convenience.

Yours faithfully,

【註】

1. subrogation receipt：代位求償權收據。

2. loss subrogation receipt：賠款授權書收據。

3. marine cargo loss receipt：海上貨物保險賠款收據。

No. 290　船公司通知貨主宣布共同海損

Dear Sirs,

<u>Re: General Average of s.s. "Produce"</u>

We regret to inform you that the above ship collided with s.s. "Central City" outside Kobe on April 2, and general average has been declared.

Delivery of cargo to consignees must be withheld until receivers comply with the usual General Average security requirements. In addition to presentation of the covering original B/L, it is necessary that receivers prepare and submit:

1. Lloyd's Average Bond in suplicate.
2. Certificate copy of commercial invoice.
3. Deposit in US Dollars of an amount representing 15% of CIF destination value.

Underwriter's guarantee in lieu of cash deposits are acceptable.

As vessel is scheduled to arrive at Keelung on/about April 15, 20–, kindly give these matters your prompt attention in order to prevent unnecessary loss of time and extra storage and handling expenses. Further information may be obtained by contacting this agency in Taipei.

Very truly yours,
WATERMAN STEAMSHIP CORPORATION
C. F. SHARP & COMPANY, INC.

as Agent

【註】

1. collide with：與……碰撞。
2. general average has been declared：宣布共同海損。
3. Average Bond：指 General Average Bond（共同海損保證書）而言。
4. Deposit：指 General Average Deposit（共同海損保證金）而言。
5. Underwriter's Guarantee：指 Underwriter's General Average Guarantee（保險人出具的共同海損保證函）而言。
6. in lieu of：in place of; instead of（代替）。

No. 291　貨主請保險公司簽發共同海損保證函

Dear Sirs,

<u>Re: Your Marine insurance policy No. 123</u>

We have been informed by C. F. Sharp & Company, Inc. that s.s. "Produce" carrying 500 bales of Raw Cotton consigned to us and covered by the captioned policy has collided with s.s. "Central City" and general average has been declared.

We enclose one photostatic copy of C. F. Sharp & Company, Inc.'s letter together with one copy each of the invoice and B/L for this shipment for your reference and perusal. As stated therein, you are requested to issue a General Average Guarantee on our behalf in order that we, as consignee, need not pay the deposit in cash amounting to 15% of the CIF value of the cargo.

Your immediate attention and action will be appreciated.

Faithfully yours,

【註】

　　1. general average guarantee：共同海損保證函。

　　2. on our behalf：替我們；為我們。

 ## 第五節　有關買賣索賠的有用例句

一、提出買賣索賠的有用例句

(一)開頭句

1. The goods $\begin{cases} \text{we ordered} \\ \text{which you shipped} \end{cases}$ on May 7 have arrived $\begin{cases} \text{damaged.} \\ \text{in damaged condition.} \end{cases}$

2. We have duly received...(names of goods)...ordered from you on March 4,

   → but $\begin{cases} \text{we find to our regret} \\ \text{regret to say} \end{cases}$ that...

   註：a. find to our regret：歉然發現；b. regret to say：遺憾的是。

3. Upon our taking delivery of the goods on arrival of s.s. "Produce" at Keelung, we find that...

4. We have to inform you that...(goods)...ordered from you on August 7 has not arrived here. Nor have we heard anything from you concerning the shipment.

   （茲通知貴公司關於本公司 8 月 7 日訂購的……迄未抵達本地，關於本批貨的裝運情形也未收到貴公司任何通知。）

5. With reference to our order #123 dated June 9, for...(name of goods), we shall be

→ $\begin{Bmatrix} \text{glad} \\ \text{pleased} \end{Bmatrix}$ to know when we may $\begin{Bmatrix} \text{expect shipment} \\ \text{expect delivery} \end{Bmatrix}$, as the goods

→ are $\begin{Bmatrix} \text{most urgently.} \\ \text{in most urgent need.} \end{Bmatrix}$

6. We have just received the 150 C/S of chinaware shipped per M.S. "Amazon" on our order #315, but regret to inform you that cases Nos. 3 & 6 are broken and their contents badly damaged through faulty packing.

註：faulty packing：包裝不良。

㈡索賠理由

I. 裝運遲延

1. Our order No. 567 of fishmeal is now $\begin{Bmatrix} \text{considerably} \\ \text{long} \end{Bmatrix}$ overdue. This delay has caused us great inconvenience.

註：considerably overdue：逾期相當久。

2. As the goods are urgently $\begin{Bmatrix} \text{needed} \\ \text{required} \end{Bmatrix}$, we must ask you to $\begin{Bmatrix} \text{dispatch} \\ \text{ship} \end{Bmatrix}$

→ them $\begin{Bmatrix} \text{by air.} \\ \text{without further delay.} \\ \text{immediately.} \\ \text{on or before August 10.} \end{Bmatrix}$

註：without further delay：勿再延誤；勿再稽延。

3. When we placed our order with you, we pointed out that prompt $\begin{Bmatrix} \text{delivery} \\ \text{shipment} \end{Bmatrix}$

→ was $\begin{Bmatrix} \text{essential} \\ \text{absolutely necessary} \end{Bmatrix}$. However, we have not yet received the goods or

→ any advice when we may expect $\begin{Bmatrix} \text{delivery} \\ \text{shipment} \end{Bmatrix}$. Your delay will threaten the loss of

→ one of our $\begin{Bmatrix} \text{old} \\ \text{new} \end{Bmatrix}$ customers.

註：threaten the loss：使失去……。

4. You wrote us that our order No. 123 was almost ready for shipment and that your shipping advice would soon follow. Nearly a month has passed since then, yet we have heard nothing from you about the shipment.

　　註：your shipping advice...follow：將隨即通知裝船。

5. You will remember that we stressed the importance of punctual shipment and you will understand that your delay in the circumstances give us a right to sue for the damage caused.

　　註：a. stress：強調；b. punctual shipment：準時裝運；c. give us...caused：給我們對貴方的遲延有權訴求損害賠償。

II. 品質不符

1. $\left\{\begin{array}{l}\text{Upon} \\ \text{On} \\ \text{When}\end{array}\right\}$ unpacking the $\left\{\begin{array}{l}\text{consignment} \\ \text{cases} \\ \text{bales} \\ \text{packages} \\ \text{bags}\end{array}\right\}$, we found that the goods did not agree with

$\rightarrow$ the original $\left\{\begin{array}{l}\text{sample.} \\ \text{pattern.} \\ \text{swatch.}\end{array}\right\}$

　　註：a. unpacking：拆開；b. did not agree with：與……不符。

2. When the $\left\{\begin{array}{l}\text{case} \\ \text{package}\end{array}\right\}$ was unpacked, we found that the goods did not agree with the original sample.

3. $\left\{\begin{array}{l}\text{On} \\ \text{Upon}\end{array}\right\}$ $\left\{\begin{array}{l}\text{inspection} \\ \text{examination}\end{array}\right\}$ we have found that the $\left\{\begin{array}{l}\text{color} \\ \text{finish}\end{array}\right\}$ is not satisfactory.

　　註：finish：修飾。

4. The goods invoiced on...(date)...are so poor in quality that they cannot be delivered to our customers.

　　註：the goods invoiced on...：××日發票所列貨物。

5. You used a much inferior quality of the stuff, which makes the products look very clumsy and not at all as fine as the original sample.

註：a. stuff：原料；b. look very clumsy：看起來很差；c. not at all：毫不。

6. We stipulated in our order that they should be Nos. 7 & 8, but instead, the two boxes arrived were Nos. 5 & 6. It is impossible for us to make use of these, especially as we are well-stocked in No. 6.

7. You have evidently sent us the wrong goods, and, as we are in great hurry for the shirts that we ordered, this error is very inconvenient and annoying to us.

8. Believing this to be caused by a mistake on your part, we await your prompt reply, meantime warehousing them at your expenses.

9. From the survey report which was held on our premises, you will readily admit that the goods are much inferior in quality to FAQ. Our chemists report that the material contains at least 12% acid. Acid in this compound is known to reduce its value by about 15%.

註：a. held on our premises：在我們公司場地舉行；b. readily admit：容易承認；c. chemists：化學人員；d. acid：酸；e. compound：合成物。

10. Frequent complaints have been received from our customers to the effect that the pens leak and will not write without blotting. Quite frankly, they fall far below our standard.

註：a. to the effect that = purporting that：大意是說；b. will not write without blotting：一寫就沾污紙張；c. quite frankly：坦白地說。

11. A comparison of the cuttings enclosed will convinced you of the reasonableness of our proposition.

（請比較隨函附上的剪布，即可使你認為我們的提議是合理的。）

III. 數量短失

1. $\begin{Bmatrix} \text{In checking the goods against your invoice} \\ \text{Upon examination} \end{Bmatrix}$, we $\begin{Bmatrix} \text{discovered} \\ \text{found} \end{Bmatrix}$ a considerable shortage in the number of toy animals and toy pianos.

2. $\begin{Bmatrix} \text{Upon} \\ \text{On} \end{Bmatrix}$ examination, we found that all the cases weigh short by 5 to 12 lbs.

3. Case 21 was found to be 5 packages short. As the case was in good shape and does not appear to have been tampered with, we surmise that they must have been short-

shipped.

註： a. case was in good shape: 箱的形狀良好； b. does not appear to have been tampered with： 無動過手腳的跡象； c. surmise: 認為； d. shortshipped: 短裝。

4. Upon the boxes being opened, we found that there was a shortage of 12 doz. of the first item. While the second item was over-supplied by as many dozen. We believe you mixed up these two items when reading the order.

5. Since the loss is not negligible, we requested that you make up the shortage promptly.

註： not negligible: 非同小可。

IV. 貨物損壞

1. We are surprised to find that some of the goods have been $\begin{cases} \text{damaged.} \\ \text{broken.} \\ \text{cracked.} \end{cases}$

2. On unwrapping the cases, we found the goods were partly soaked by rain.

註： soaked by rain: 被雨水淋濕。

3. Upon examination, we found that many of the goods were $\begin{cases} \text{severely} \\ \text{seriously} \end{cases}$ damaged, though the cases themselves show no trace of damage.

4. The damage was apparently caused by $\begin{cases} \text{poor} \\ \text{improper} \\ \text{insufficient} \end{cases}$ packing. A machine of this size and weight should be blocked in position inside the export case.

（這種笨重的機器，應在出口用木箱內部加以固定起來。）

5. In our opinion, the export cases used were not sufficiently strong to protect these instruments. The wood should have been at least 7/8″ thick.

6. The goods had been packed loose in the case without sufficient padding in it, thus causing the breakage of some.

7. The casks were not apparently strong enough for the purpose they were used for; the result was that several casks sustained a leakage whilst in transit and the contents had all run out when delivered.

註： run out = leak out = be lost。

(三)要求處理事項或結尾語

I. 對於遲延裝運的要求

1. We must accordingly insist on $\begin{cases} \text{your fixing a definite date for shipment.} \\ \text{your informing us by telex of the earliest date you} \\ \text{can ship the goods.} \end{cases}$

2. Please $\begin{cases} \text{let us know} \\ \text{inform us} \end{cases}$ by $\begin{cases} \text{cable} \\ \text{telex} \end{cases}$ when we can expect delivery.

3. Please give this matter your $\begin{cases} \text{urgent} \\ \text{immediate} \end{cases}$ attention.

4. Please cable us at once whether you can deliver the goods by the end of July. If your answer is in the negative, we shall have to cancel the order as we cannot possibly wait any longer.

5. Unless the goods arrive by the end of this month, we will cancel the order.

6. The demand for these goods is seasonable, as you know. We shall, therefore be forced to cancel this order and buy from other source unless we can get immediate shipment.

II. 對於品質不符的要求

1. We are holding the cases and their contents just as we received them and ask you to instruct your agents to call on us to adjust our claim.

2. Please let us know if you will take the goods back or make us an allowance for the inferior quality.

3. We ask you, therefore, either to send us a credit note for the amount of these goods together with the duty paid on them, or to pass the duty to our credit and send us a replacement at your expenses.

4. Though the goods are of very inferior quality, we shall retain them, but only at a substantially reduced price.

5. Unless you can give us an assurance that you will in future provide us with first-class quality, we regret we shall reluctantly have to go elsewhere.

6. We regret to have to return these goods, and shall be glad if you will substitute the right goods for them as soon as possible.

7. Please let us know $\begin{cases} \text{how you wish us to return the goods.} \\ \text{when we may expect the correct goods.} \end{cases}$

III. 對於數量短失的要求

1. As this shortage is too heavy to overlook, we must ask for your credit note for $... representing the value of the lost oil.

2. Please investigate the matter, and ship the goods to make up the deficiency as soon as possible.

IV. 對貨物損壞的要求

1. Will you please arrange for an inspector to examine the damage.

2. In view of this, please replace the broken items as per enclosed list.

3. We are compelled, therefore, to request you to make up for the loss of $...which we have sustained by the damage to the goods.

㈣結尾句

1. Please look into this matter and let us know your $\begin{cases} \text{instructions} \\ \text{decision} \end{cases}$ immediately.

2. Please investigate this matter and adjust it $\begin{cases} \text{without delay.} \\ \text{as soon as possible.} \end{cases}$

3. We are most anxious to have this matter cleared up and request you, therefore, to go into it and write us promptly.

4. We reserve the right to claim compensation from you for any damage.

5. We trust that there will be no repetition of the trouble.

二、答覆買賣索賠的有用例句

㈠開頭句

1. Immediately upon receipt of your $\begin{cases} \text{letter} \\ \text{cable} \end{cases}$ of...inquiring about your order..., we consulted our files and records.

註：consult our...records：查卷。

2. We acknowledge receipt or your $\begin{cases} \text{cable} \\ \text{letter} \end{cases}$ of...

3. Thank you for $\begin{cases} \text{calling our attention to...} \\ \text{notifying us so promptly of...} \end{cases}$

    註：call our attention to：提醒我們注意到……。

4. As soon as we received your letter, we got in touch with the $\begin{cases} \text{packers} \\ \text{shipping agents} \\ \text{manufacturers} \end{cases}$ and

    asked them to look into the matter.

5. We are $\begin{cases} \text{sorry} \\ \text{very sorry} \\ \text{extremely sorry} \\ \text{very concerned} \end{cases}$ to $\begin{cases} \text{learn from your letter that...} \\ \text{have your complaint regarding...} \end{cases}$

6. $\begin{cases} \text{We are sorry to learn} \\ \text{We note with regret} \\ \text{We very much regret to learn} \end{cases}$ that you are not satisfied with the goods supplied to

    your order of...

7. We note with surprise from your $\begin{cases} \text{letter} \\ \text{cable} \end{cases}$ of...

8. We are very much surprised to learn from your $\begin{cases} \text{letter} \\ \text{fax} \\ \text{cable} \end{cases}$ of...

(二)賣方承認過失而接受對方索賠

I. 說明發生原因

1. We have looked into the matter and find that your claim is perfectly justified.

    註：perfectly justified：確有道理。

2. We are very sorry for the delay in the $\begin{cases} \text{shipment} \\ \text{execution} \end{cases}$ of your order.

3. $\begin{cases} \text{This} \\ \text{The} \end{cases}$ delay is $\begin{cases} \text{due to causes beyond our control.} \\ \text{entirely attributable to the recent} \rightarrow \end{cases}$

    $\rightarrow \begin{cases} \text{strike.} \\ \text{typhoon.} \\ \text{shortage of raw materials.} \\ \text{breakdown in factory machinery.} \end{cases}$

註：a. attributable to：可歸因於……的；b. breakdown in factory machinery：工廠機械發生故
障。

4. After a careful investigation, we have discovered that by some unaccountable carelessness your order was misplaced and was not attended to. We are wholly to blame
for the delay.

註：a. unaccountable carelessness：無法說明的疏忽；b. misplaced：誤置；c. not attended to：
未予處理；d. we...for the delay：對此遲延自應負全責。

5. We deeply regret to find that the wrong goods have been shipped thru a mistake on
the part of our shipping clerk. We are temporarily understaffed and had to hire new
hands, but all this is no excuse.

註：a. shipping clerk：發貨員；b. understaffed：人手不足；c. to hire new hands：雇用新手；
d. all this is no excuse：這些都不是藉口。

6. We frankly admit that delivery was delayed, but it was really beyond our control
since it was caused by a fire in our works. We asked for your understanding then
and you kindly made allowance by extending the relative L/C.

註：a. fire in works：工廠火災；b. asked for your understanding：徵求你的諒解；c. you kindly
made allowance：蒙你惠允；蒙你原諒。

7. On careful examination of your samples, the cotton in question has been found well
up to standard in quality, except that it is a little more spotted than in ordinary cases.

註：a little more spotted than...：污點比……稍多一點。

8. The shortage of 4 out of the 60 cases of No. 18, we understand, must have been
caused thru unskilled packing for which please accept our profound apologies for
the inconvenience you have been put to by this irregularity.

註：unskilled packing：不熟練的包裝。

9. As the articles were packed with the utmost care, we can only conclude that the
damaged cases has been stored or handled carelessly. We have reported your claim
to our insurance company.

註：stored or handled carelessly：保管或搬動不小心。

II. 解決方案的提示

1. We fully appreciate your position in the matter, and will do our best to dispatch the

goods before May 15.

2. We regret that such a delay has been caused when you need them most urgently, and in order to show you that we are anxious to avoid putting you to any further inconvenience, we have put aside other standing orders and have arranged to dispatch your goods by m.s. "Taiwan" leaving Keelung on May 5.

註：put aside other standing orders：擱置其他經常性訂單。

3. We regret these faulty sets were sent to you, and have today sent a replacement of 21 sets. We hope you will be pleased with the new lot.

註：a. faulty sets：不良的貨品（成套的貨品）；b. new lot：新品。

4. We offer our sincere apologies for the error in our $\begin{cases} \text{invoice} \\ \text{statement} \\ \text{shipment} \end{cases}$ and are sending you

→ our correct $\begin{cases} \text{invoice.} \\ \text{statement.} \\ \text{shipment.} \end{cases}$

5. We agree to compensate for the total loss and enclose one bankers' draft for US$... to pay therefor.

6. We have asked the shipping company to collect these goods from you and return them to us at our expenses.

7. We are, therefore, taking the matter up with the agents of the...Lines. In the meantime we are sending you a replacement today and hope it will arrive in good order.

註：a. take the matter up with：將本案向……提交；b. Lines：輪船公司或航空公司。

8. As requested, we agree to pay five cents per yard for the quantity specified above to compensate you for the loss.

9. If you will meet us and keep the goods, we shall offer you an allowance of 10% off the invoice value.

註：meet us = meet our request。

10. Please let us know what adjustment you think is satisfactory in the circumstances. We shall be glad to consider it.

註：a. adjustment：解決方案；b. in the circumstances = under the circumstances：在此情形之下。

III. 結尾句

1. We thank you for calling our attention to this error. We have taken necessary precautions to prevent a recurrence of similar mistakes in future.

   註：recurrence：再發生。

2. Please accept our sincere apologies for the $\begin{cases} \text{delay} \\ \text{error} \\ \text{mistake} \end{cases}$ and the inconvenience it has

   caused you.

3. We are very sorry this mistake occurred. You may be sure that we

   $\rightarrow \begin{cases} \text{will make every effort} \\ \text{shall do everything in our power} \end{cases}$ to see that such mistake does not happen

   again.

   註：a. you may be sure that：請相信；b. to see that：設法使……。

4. We believe that the matter is now settled to $\begin{cases} \text{your satisfaction.} \\ \text{our mutual satisfaction.} \end{cases}$

   註：mutual satisfaction：（使）雙方滿意。

5. We thank you for the opportunity given us to rectify our error.

   註：rectify our error：改正我們的錯誤。

6. We appreciate the leniency you have shown in keeping the wrong goods and trust that you will give us an opportunity to supply you with further goods.

   註：a. leniency：寬容；b. keep the wrong goods：收下不良貨品。

7. We apologize for the inconvenience this transaction has caused you and assure you of our better attention to your future orders.

三、過失在買方或不在賣方

I. 情況說明

1. We have closely examined the sample taken from our last consignment and find it is no way different in quality from the TP–123 that we have here in stock. We can only surmise that there be a mistake somewhere.

2. We have made a careful investigation but have failed to find that your letter was

ever received by us. It is possible, of course, that it went astray in the mail.

註：go astray in the mail：郵途中遺失。

3. You claim that the quality is inferior to the original sample, and request us to credit you with 5 cents per 1b. This does not appear to be very reasonable as we sent you an advance sample prior to shipment, and, not hearing from you to the contrary, presumed it to be in order.

4. We are enclosing the report to show you that your order has been despatched from this end as promised. The shipping agents are now tracing it; they will check at the port of transhipment also.

5. Your order No. 123 was received on May 15 and shipped by the first available boat on June 2. Under normal circumstances, you should have received the shipment by the end of June. Apparently the order has been delayed in transit.

6. The difference in quality is usual, and indeed unavoidable in this kind of goods, and the conditions of your order have been fully complied with.

7. In making our examination, we do not find any evidence of inferior quality or workmanship and the dye used was of the highest quality.

8. We are, however, unable to explain the damage, since we took the best possible care in packing the cases and the shipping company received the whole lot in perfect condition as can be seen from the clean B/L we obtained.

9. We can assure you that the goods were in good order when left here and, the damage must have occurred during transit.

10. The goods were in good condition when they were shipped on board the vessel as shown in the clean B/L issued by shipping company.

11. The shipping company is to be responsible for the shortage referred to in your letter. We suggested that you take up the matter with the shipping company.

12. We regret that we cannot see our way clear to make as exception in that case. The time limit for the claim expired a fortnight ago.

註：see our way clear：認為適當（可能）＝ regard as suitable or possible（主要用於否定句或疑問句）。

13. We are pleased to enclose a certificate of inspection to show that your order has been fully complied with.

14. When goods left Keelung, they were in perfect condition; the clean B/L issued by... shipping company proves it.

II. 解決方案的提示——拒賠或妥協

1. With this point in view, we regret to have to ask you to reconsider the claim.

2. We must, therefore, refuse to accept the goods if returned, and shall have no alternative but to insist upon an early settlement in full.

3. Much as we would like to be of service to you, we are unable, in this instance, to accept your claim.

4. As you consider our proposals $\begin{Bmatrix} \text{unacceptable} \\ \text{unsatisfactory} \end{Bmatrix}$, we suggest that the matter be submitted to arbitration.

5. We suggest that this claim should be settled in accordance with the arbitration rules of the International Chamber of Commerce.

6. We wish to settle this dispute in a friendly way, and we, therefore, suggest that we submit it to arbitration and agree to abide by the arbitrators' decision.

7. We, therefore, are unable to consider your demand for a $\begin{cases} \text{discount.} \\ \text{reduction in price.} \end{cases}$

8. If you will look at the matter impartially, you will no doubt understand that we cannot see our way to make any allowance.

9. Under these circumstances, we trust you will agree that we cannot be expected to grant a free replacement of these damaged machines. However, we are ready to compromise by replacing them at a reduced price of US$500 each.

10. Granting that you suffered a certain loss, we do not think it fair that we have to bear the whole responsibility. We fully realise, however, the awkward position in which you are placed, and in view of our friendly relations, we will offer you an allowance of 3%.

III. 結尾句

1. We hope the above explanation has convinced you that we have been abiding by our agreement with you.

2. We regret the difficulty you have had and wish to be of any possible assistance to you.

3. Under these circumstances you will readily understand that we cannot possibly see our way clear to assume any liability whatever in the matter.

4. Under the circumstances, we cannot see our way clear to $\begin{cases} \text{fall in with} \\ \text{agree to} \\ \text{yield to} \end{cases}$ your view in

respect of the present claim.

　　註：a. fall in with = agree to；b. yield to = agree to：順從。

四、待查明後再議

1. In order to make a detailed report, however, certain investigation will have to be made. These will require a week to conclude.

2. We are giving due consideration to the matter and will let you have a reply probably next week.

3. We propose to have the goods inspected immediately. If the inspection confirms the accuracy of your estimate, generous compensations will be allowed at once.

　　註：a. generous compensations：大幅補償；b. allowed：予以……。

4. In order to settle the matter, we should be pleased to send a representative to conduct an inspection.

5. We are immediately investigating the matter and just as soon as we have anything definite to report, we shall write you again.

6. We are having the consignment traced and as soon as we have located the trouble, we shall inform you by cable.

第二十八章
挽回客戶 (Regaining Lost Customers)

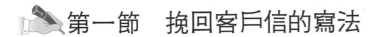 第一節　挽回客戶信的寫法

　　爭取新客戶固然重要，但是保住老客戶更重要。做生意如能一方面與老客戶保持繼續往來，他方面又能積極爭取新客戶，是最理想的事，然而往往事與願違。在交易往來過程中，或由於誤會、不滿、齟齬，或其他種種原因，往往交易了一兩次之後，音訊就不繼。精明的生意人，一方面開拓新客戶，他方面則絕不慢待老客戶，經常與老客戶保持連繫，對老客戶表示關心，並隨時詢問有什麼不周到或需要改進的地方。這樣才能保住客戶。

　　當發覺客戶中止訂購或訂單越來越少時，應即檢討為什麼會發生這種現象，並進而採取適當的措施，以挽回失去了的或將失去的客戶。挽回客戶的最經濟的方法就是寫一封令他感動的信。這種信的內容包括下列各項：

　　1.直率地或婉轉地提醒客戶很久沒有接到訂單。

　　2.誠懇地詢問客戶為什麼不再惠賜訂單，以便檢討、改進。

　　3.誠懇地表示願意恢復以往愉快、密切的往來關係。

　　4.假如有什麼使對方不滿的事，即告訴客戶，以後不致再發生類似情事，並要求客戶給予機會，以便能提供最佳服務。

　　5.最後，表示謝意並祈惠覆。

 第二節　挽回客戶信函

No. 292　久未收到訂單，查詢何故

Dear Sirs,

　　We have not had either your order or your letter for almost six months. It seems to us that you have gone away from us. Surely, there must be some reasons for your having done so, and it is our sensitive assumption that possibly we have failed to satisfy you in some way.

As the saying goes: "Customer is always right." Your frank comments will be a help for us to renew and promote the good relationship between our two firms. Would you please let us know what is in your mind?

We shall appreciate receiving your kind reply.

Yours faithfully,

【註】

1. go away from us：離開我們。

2. sensitive assumption：敏感的假定。

3. in some way：在某一點（方面）。cf. in some ways：在某些方面。

4. as the saying goes = as the saying is：俗語說得好。

No. 293　中止訂貨，查詢何故

Dear "Old customer":

An old saying goes: "Old friends are like the ticking of a clock." You get so used to hearing the tick that you rarely notice it until it stops.

We get used to doing business with old customers, too. So much so that now and then we assume that everything is running along smoothly and we sometimes fail to express our appreciation as often as we should. And then—suddenly the clock stops and we find an OLD CUSTOMER has stopped buying.

That is just the position in which we find ourselves with you. Your orders have stopped and we are wondering if you would tell us frankly just what the trouble has been—whether there is something we did not do that we should have done, and whether there is anything we can do NOW to get you back on our list of regular customers. If there is, we surely want to do it.

Now just give us one minute of your time to say: "We haven't sent our order, because: ....
....

Please fill in your answer and return this sheet to us.

Thank you for your co-operation.

Yours very truly,

【註】

1. ticking of a clock：鐘的滴答聲。

2. get used to = be accustomed to = be used to：習慣於……。

3. so much so that：……到要……；因為非常……以致……。cf. He was very weak, so much so that he could not walk.

4. now and then：有時候；時常 (= occasionally; from time to time)。

5. suddenly the clock...stopped buying：突然間，我們發覺老客戶中止訂貨了，正像我們發覺大鐘停止了擺動。

6. That is just the position...with you.：貴公司與我們目前的關係正是那樣。

7. What the trouble is?（毛病在那裡？）

No. 294　訂了一次貨後即音訊不繼時

Dear Mr. Ford,

　　Greetings.

　　It has been some time since we received your initial order. I think it's time I drop a get-acquainted note.

　　One thing that make this company different from most manufacturing concerns is our spirit of service from top to bottom. Our workers, office staffs and all of our executives—they all do their parts to make your selling our products more profitable.

　　I am keenly interested in helping you to promote your sales. In fact, my job is to help you. And if you can help me to do a better job, you will really be helping yourself. Let me hear from you!

　　　　　　　　　　　　　　　　　　　　　　　　Sincerely yours,

【註】

1. greetings：您好！

2. drop a get-acquainted note：寫一封增進認識的信。

3. manufacturing concerns：製造廠。

4. spirit of service：服務精神。

5. from top to bottom：從上到下。

6. executives：主管。

7. do their parts：盡他們的職責。

本封信私人化 (Personalized)，有時效果很好。

No. 295　　問客戶久未訂貨的原因

Dear Sirs,

　　We have not received your order quite a long time. We are eager to find out what the trouble has been. Did we do something wrong to your firm? Was our service satisfactory to you? Whatever it may be, could you kindly let us know?

　　It is our policy to render the best service to customers. You have always been considered as one of our regular customers, for you have given us remarkable patronage. We need you and we do not want to lose your business. We hope to have the pleasure of serving you again soon.

　　Your kind reply will be much appreciated.

Yours sincerely,

【註】

　　1. to render the best service：提供最佳服務。

　　2. remarkable patronage：惠顧很多。

No. 296　　問客戶久未訂貨的原因

Dear Sirs,

　　We have missed you very much and we have been disturbed greatly at your having not placed with us any order since our delivery of your order No. 123. We are just wondering why you have stopped buying from us.

　　To recall the pleasant relationship between our two companies in the past, we think it is a treasure which we want to cherish. And your comments on our products, service, etc. will be of great help to maintain the good friendship between our two companies.

　　Will you please let us know the reason why you have discontinued buying from us?

Sincerely yours,

【註】

　　1. miss you very much：很想念你。

　　2. disturbed = annoyed：困擾。

　　3. recall = recollect：回想。

　　4. a treasure which we want to cherish：我們欲珍視的實物。

　　　cf. a. I treasure your friendship. （我珍視你的友誼。）

b. He cherishes friendship. (他珍視友誼。)

No. 297　問老客戶為何很久沒有訂貨

Dear Mr. Johnson,

　　I wish it were possible for me to step into your place and have a heart-to-heart talk with you. But as this is not possible, I am using this letter to keep up the contact between us.

　　The most unpleasant thing I have discovered in a long time is that we have not had an order from you for more than 100 days. There is a reason, of course. Won't you tell me what it is? If anything is wrong—whatever it is—I will fix it.

　　As you know, there isn't more pleasant feeling than to know you have a pleased customer. A satisfied customer is the best asset any business can have. Errors cannot always be helped, but they can be corrected. That's what I am here for. If you're not satisfied with our service from any angle, I am not doing my job right.

　　Your account is always appreciated. We consider it so valuable that we will do whatever is necessary to keep it alive.

　　An addressed envelope is enclosed to make it easy for you to tell if anything is wrong—and if you will send an order along at the same time, I will personally see that you get special attention.

Yours sincerely,

【註】

1. step into your place: 到你那裡。

2. heart-to-heart talk: 坦率的洽談；促膝相談。

3. keep up the contact: 保持聯繫。

4. fix: 處理。

5. As you know, ...a pleased customer. (你知道，沒有比知道你有一個滿意的客戶更能使人感到愉快的了。)

6. business: 指「事業」。

7. errors cannot always be helped: 錯誤是無可避免的，"help" 當「避免」解釋時總與 "can't" 或 "can" 連用。

8. That's what I am here for. (那就是我的職責所在。)

9. from any angle: 從任何角度看。

10. I am not doing my job right: 我沒有盡到責任；我沒有做好工作。

11. Your account is always appreciated. (我們經常感激你的照顧。)

12. keep it alive：保持住（你的照顧）。

13. I will personally...attention：我將親自加以特別的留意。

No. 298　寫了一封信沒有反應，再接再厲

Dear Sirs,

　　We are still wondering why we seem to have lost an old friend in you. A short time ago, we wrote to ask why we're not receiving your orders recently, but as yet we have had no reply.

　　It is possible that something we may have done—or did not do—has disturbed you. This could be a misunderstanding.

　　The matter is of great importance to us. Frankly, we want to keep the business of old customers like you, and we are sure it can't be too late to restore the pleasant relations that we formerly enjoyed. In this old world, we can't get anywhere without old friends.

　　So once again we ask: won't you please take a minute to jot down on the back of this letter the reason we haven't been hearing from you.

Yours sincerely,

【註】

1. We are still...in you.（我們還在想：為什麼我們會失去了一個像閣下的老朋友。）

2. as yet = so far：到目前為止。

3. disturbed you：得罪了你。

4. restore：恢復。

5. In this old world, ...without old friends.（在這古老的世界中，我們如果沒有老朋友，一定不會有什麼成就。）

6. jot down = write briefly：摘記下來。

寫了一封信沒有反應，應該再接再厲，鍥而不舍。

No. 299　再接再厲，鍥而不舍

Dear Sirs,

　　We have written you repeatedly in the hope that you will let us know why you have discontinued buying our products for almost four months but as of this date we still have not had a word from you.

　　We are much concerned about this matter just because we don't want to lose the good business of our old customer like you. Moreover, you are one of our oldest friends, and we do need you as our success in business attributes to patronage, friendship, and co-operation of our customers.

You may be sure that we will do better than before if only you will tell us why you have stopped your business with us.

May we have your reply right away?

Yours sincerely,

【註】

1. as of = as it is at：到……為止。as of this date：到今天為止。

2. concerned：關心。

3. attribute to：歸因於。

4. right away：立即。

No. 300　老朋友才是好朋友

Dear Sirs,

You will agree that old friends are the best friends. If an old friend suddenly seems to forget you, you want to know the reason.

That's just the situation we are up against.

Several months have passed since that last order. Of course, we would like to know the reason.

A frank discussion of any problem is the quickest and best way to a satisfactory solution. It is possible that something we may have done—or did not do—has annoyed you. Often it is just a mis-understanding.

In this old world we can't get anywhere by ourselves. We've got to have someone else's help, so won't you tell us just why we haven't heard from you? Be absolutely frank.

We very much appreciate your business and it is our sincere hope that we may renew a relation-ship with you that has been most pleasant.

May we hear from you soon?

Yours sincerely,

【註】

1. That's just the situation...up against.（我們正面臨這種情形。）

2. In this old world...by ourselves.（在此古老的世界裡，我們不能獨善其身。）

如因我方錯誤致遭到客戶指責，以後，再也沒有收到他們的訂單時，我們如想拉回生意，則寫信的方式有二，一為向他們認錯，請他們原諒，二為裝迷糊，不知有何錯誤，上面一封信就是裝迷糊的寫法。

No. 301　拉回因糾紛而失去的客戶

Dear Sirs,

Your orders have discontinued since the settlement of a claim you approved for the damaged goods of your order No. 123.

We still feel sorry for the trouble that has caused you much inconvenience. We reiterate that we will make every effort to avoid similar mistake in our future transactions.

If one of your good friends once did something wrong in spite of his paying best attention when handling some case for you, but he apologized to you for that and expressed sincerely to do a better job for you in the future, would you forgive him? Now we are of the same situation. As the saying goes: "To err is human, to forgive divine." therefore, could you give us opportunities to provide you with our better service?

Welcome your challenge! We are sure that we are in a position to do outstanding service for you. We shall not fail to meet your satisfaction in executing any of your orders in the future.

We are expecting your kind reply.

Yours sincerely,

【註】

1. reiterate：重申。

2. To err is human, to forgive divine.：men are apt to make mistakes and to forgive mistakes is a divine character，意指「犯錯是人之常情，寬恕錯誤才是超凡的」。

3. challenge：挑戰。

4. shall not fail to：必定。

No. 302　挽回因不滿意而失去的客戶

Dear Sirs,

We have not had your patronage for a long time and there must be some specific reasons that made you go away from us.

Through a thorough looking over the communications exchanged between us in our file, we have found that you have enough reason to ignore us as we did not provide you with satisfactory service during the period when Mr. Sherman T. Wu, our former Directing Manager was in power. We are extremely sorry for the many mistakes our firm made to cause your dissatisfaction.

For your information, we are pleased to inform you that the situation of our firm is now completely different. Under the leadership of Mr. Charles C. Chang, we have won remarkable reputation from overseas customers owing to our fine service. Attached here are four photostatic copies of letters received from some of our customers, expressing their satisfaction to our service. We trust this is instrumental to bringing back your confidence in us.

It is our honor to let you know that we have developed some new products recently. We take pleasure in enclosing a catalog for your perusal. Should they interest you, please feel free to write us. We will be glad to send free samples to you for your evaluation upon hearing from you.

We hope to serve you soon again and look forward to receiving your favorable response.

Yours faithfully,

【註】

1. ignore = pay no attention to = disregard：忽視；不理睬。

2. be in power：當權。

3. under the leadership of：在……領導下。

4. be instrumental to：有助於。

5. bring back your confidence in us：恢復你對我們的信心。

6. feel free to：不必客氣。please feel free to write us：竭誠歡迎來函。

No. 303　詢問客戶有何需改進之處

Dear Mr. Charleston,

In looking over accounts, we are very glad to note the nice business you have given us, and want to assure you of our appreciation.

However, in view of the somewhat larger order received from you last year, we are just wondering if any difficulty has arisen which is within our power to remedy.

If you have not received full satisfaction from us in every way, either in product or service, we would deeply appreciate your letting us know about it; not only because of our earnest desire to serve you well individually, but to help us avoid disappointing other customers.

Please understand that we appreciate your continued patronage, but we want to be sure that we are rendering the kind of service you deserve, and will appreciate a word from you as to whether or not anything may have gone wrong.

Sincerely yours,

【註】

1. 當客戶訂單減少時，固然要寫信爭取訂單，但是訂單很多時，也要寫信表示謝意，這樣才能保住客戶。上面的一封信，就是一個例子。

2. look over accounts：看了帳目。

3. nice business you have given us：你給我們的好生意（意指很多生意）。

4. within our power to remedy：我們能力範圍內所能補救的……。

5. in every way：在各方面；在任何方面。

6. rendering the kind of service you deserve：給你應得的服務。

7. go wrong：差錯。

第二十九章
代理關係的建立　(Establishment of Agencyship)

　　產品的銷售需要靠人去推銷，尤其在國際行銷時為然。當廠商著手選擇產品銷售通路之前，首先要考慮的是應經那一類商人去推銷最佳。廠商將其產品推銷到國際市場的通路固然很多，但在很多場合，以透過代理商 (Agent) 優點較多。

　　反之，進口業者為確保貨源以及採購的順利，也往往在出口地指定採購代理 (Buying agent) 代辦採購業務。

　　以下就較常見的售貨代理 (Selling agency)、寄售代理 (Consignment agency) 及採購代理 (Buying agency) 有關的書信往來分別舉例說明。

第一節　售貨代理信的寫法

一、申請充任售貨代理函件的撰寫要領

　　要求一廠家指定本公司為售貨代理商的函件，其撰寫原則與買賣客戶的招攬函件撰寫原則類似。所以這種 Agency application 的內容大致如下：

　　1.如何知悉對方（如係初次認識）。

　　2.具體說明本公司的優點，例如組織、資本、經驗、能力以及商業關係等。

　　3.表示願充任其售貨代理。

　　4.請其提示代理條件（或主動提出代理條件）。

　　5.表示願衷誠合作，共同開拓業務。

售貨代理商 (Selling agent) 可分為：

1. Exclusive (Sole) agent（獨家代理）。

2. Non-exclusive agent（非獨家代理）。

一般而言，代理商以能取得獨家代理較佳。

No. 304　希望指定本公司為售貨代理商

Dear Sirs,

We owe your name to CETRA, through whom we have learned that you are seeking an agent here. Therefore, we write to ask if you are interested in extending your export business to our country by appointing us agents for the sale of your products.

We are sure that you know very well the advantages of representations. This is especially necessary when foreign exporters want to do large volume of business with our country, because most of our government, military and industrial purchases are done through open tenders. It is only through a regular representative that will have a chance to offer or bid in time and effectively.

We, the Taiwan Trading Company, are a well-established firm with 20 years of history and integrity in Taiwan foreign trade. We not only have very good connections with the government procurement agencies and government enterprises, but also have close business relations with the domestic private enterprises. We know how to meet tender requirements and could handle them effectively and successfully. We feel confident that if you give us any opportunity to deal in your products, the result will be entirely satisfactory to both of us.

So, we shall be very much pleased to act as your sole agents in Taiwan for your products. If our request is accepted by you, we shall thank you to let us know your terms and conditions under such agency agreement, and forward quotations, samples and other helpful literature, with a view to getting into business in the near future.

For any information you may desire regarding our standing, we are pleased to refer you to the following bank:

Bank of Taiwan
Chungking S. Road, Sec. 1
Taipei

We hope that this letter will be a forerunner to many years of profitable business to both of us, and look forward to the pleasure of hearing from you.

Yours faithfully,

【註】

1. we are sure that：我們相信。類似的用語有：
   we have the confidence that...
   we firmly trust that...
   we feel confident that...
   we are confident that...

2. representation：代表；代理。

3. open tender：公開招標。

4. bid：投標；報價。

5. government procurement agencies：政府採購代理機構，在我國為中央信託局 (Central Trust of China)。

6. government enterprises：政府經營的企業。

7. private enterprises：私人經營的企業。

8. act as：充任。

9. forerunner：先驅；預兆。

No. 305　同意指定對方為售貨代理商

Dear Sirs,

　　Thank you for your letter of October 11 in which you indicate your desire to act as our sole agents for our products in Taiwan.

　　After completion of our credit files and careful consideration of your proposal, we have now made our decision to accept your proposal and to appoint you our sole agents in Taiwan district for a period of one year.

　　For the guidance and to help towards pleasant relations between us in our transactions, we have prepared an Agency Agreement, which is enclosed for your approval. Any suggestions you may have to make on it would be welcome. If you have no objection to any of its articles please return us the duplicate duly signed.

　　It is our firm belief that this agency relationship would prove mutually beneficial, and we trust that you will be prepared to co-operate with us closely.

Yours faithfully,

【註】

1. sole agent：獨家代理，又稱為 exclusive agent。

2. completion of credit files：完成信用檔卷，即指已調查過代理商的信用情況。

3. Agency Agreement：代理契約。

4. objection to its articles：對（代理契約的）條款有異議。

5. firm belief：堅信。

6. mutually beneficial：互相有利。cf. mutual benefit：互相的利益。

當雙方同意建立代理以後，即可簽立代理契約（Agency contract, Agency agreement

或 Agency arrangement）規定雙方的權利義務。至於代理契約的內容，視情形而定。有的規定得很簡單，有的則規定得很詳細。但一般而言，代理契約的內容應包括下列各項。

1.契約的性質：言明係代理，而非買賣或經銷。

2.代理性質：言明係獨家代理抑係非獨家代理。

　　　　　　言明係媒介代理抑係締約代理。

3.契約生效日期及有效期限。

4.代理商的義務及權限。

5.代理區域。

6.代理商品項目。

7.佣金的計算及支付方式。

8.訂單的處理方式。

9.準據法。

10.其他事項。

No. 306　獨家售貨代理契約

Suitable for exclusive and sole agents
representing manufacturers overseas

An Agreement made this............day of........................ 20...........between......................whose registered office is situated at........................... (hereinafter called "the Principal") of the one part and...........................(herein after called "the Agent") of the other part.

Whereby it is agreed as follows:

1. The Principal appoints the Agent as and from the............to be its Sole Agent in......................(hereinafter called "the area") for the sale of............manufactured by the Principal and such other goods and merchandise (all of which are hereinafter referred to as "the goods") as may hereafter by mutually agreed between them.

2. The Agent will during the term of............years (and thereafter until determined by either party giving three months' previous notice in writing) diligently and faithfully serve the Principal as its Agent and will endeavour to extend the sale of the goods of the principal within the area and will not do anything that may prevent such sale or interfere with the development of the principal's trade in the area.

3. The Principal will from time to time furnish the Agent with a statement of the minimum prices at which the goods are respectively to be sold and the Agent shall not sell below such minimum price but shall endeavour in each case to obtain the best price obtainable.

4. The Agent shall not sell any of the goods to any person, company, or firm residing outside the area, nor shall he knowingly sell any of the goods to any person, company, or firm residing within the area with a view to their exportation to any other country or area without consent in writing of the Principal.

5. The Agent shall not during the continuance of the Agency constituted sell goods of a similar class or such as would or might compete or interfere with the sale of the Principal's goods either on his own account or on behalf of any other person.

6. Upon receipt by the Agent of any order for the goods the agent will immediately transmit such order to the Principal who (if such order is accepted by the Principal) will execute the same by supplying the goods direct to the customer.

7. Upon the execution of any such order the Principal shall forward to the Agent a duplicate copy of the invoice sent with the goods to the customer and in like manner shall from time to time inform the Agent when payment is made by the customer to the Principal.

8. The Agent shall duly keep an account of all orders obtained by him and shall every three months send in a copy of such account to the Principal.

9. The Principal shall allow the Agent the following commissions (based on FOB United Kingdom values)......in respect of all orders obtained direct by the Agent in the area which have been accepted and executed by the Principal. The said commission shall be payable every three months on the amounts actually received by the Principal from the customers.

10. The Agent shall be entitled to commission on the terms and conditions mentioned in the last preceding clause on all export orders for the goods received by the Principal through Export Merchants, Indent Houses, Branch Buying offices of customers, and Head Offices of customers situate in the United Kingdom of Great Britain and Northern Ireland and the Irish Free State for export into the area. Export orders in this clause mentioned shall not include orders for the goods received by the Principal from and sold delivered to customers' principal place of business outside the area although such goods may subsequently be exported by such customers into the area, excepting where there is conclusive evidence that such orders which may actually be transmitted via the Head Office in England are resultant from work done by the Agent with the customers.

11. Should any dispute arise as to the amount of commission payable by the Principal to the Agent the same shall be settled by the Auditors for the time being of the Principal whose certificate shall be final and binding on both the Principal and the Agent.

12. The Agent shall not in any way pledge the credit of the Principal.

13. The Agent shall not give any warranty in respect of the goods without the authority in writing of the Principal.

14. The Agent shall not without the authority of the Principal collect any moneys from customers.

15. The Agent shall not give credit to or deal with any person, company, or firm which the Principal shall from time to time direct him not to give credit to or deal with.

16. The Principal shall have the right to refuse to execute or accept any order obtained by the Agent or any part thereof and the Agent shall not be entitled to any commission in respect of any such refused order or part thereof so refused.

17. All questions of difference whatsoever which may at any time hereafter arise between the parties hereto or their respective representatives touching these presents or the subject matter thereof or arising out of or in relation thereto respectively and whether as to construction or otherwise shall be referred to arbitration in England in accordance with the provision of the Arbitration Act 1975 or any re-enactment or statutory modification thereof for the time being in force.

18. This Agreement shall in all respects be interpreted in accordance with the Law of England.

As Witness the hands of the Parties hereto the day and year first hereinbefore written.

(Signatures)

（中譯）

　　本契約係於 20＿＿＿ 年＿＿＿ 月＿＿＿ 日由當事人一方

＿＿＿＿＿＿＿＿＿＿＿＿＿＿＿＿＿＿＿＿＿＿＿＿＿＿＿＿＿＿ 其已登記的辦公地址在

＿＿＿＿＿＿＿＿＿＿＿＿＿＿＿＿＿＿＿＿＿＿＿＿ （以下簡稱本人）

與他方當事人＿＿＿＿＿＿＿＿＿＿＿＿＿＿＿＿＿＿＿ （以下簡稱代理商）

所簽訂。茲約定事項如下：

　　1.本人茲任命代理商自＿＿＿＿＿＿＿＿＿＿＿＿＿＿＿ 起為 ＿＿＿＿＿＿＿＿＿＿＿

＿＿＿＿＿＿＿＿＿＿＿＿＿＿＿＿＿＿＿＿＿ （以下簡稱代理地區）的獨家代理

　　以推銷本人所製造的＿＿＿＿＿＿＿＿＿＿＿＿＿＿＿＿＿＿ 以及以後雙方

　　互相同意的其他貨物及商品（以下簡稱商品）。

　　2.代理商在其＿＿＿＿＿ 年任期內（以及其後由當事人一方在 3 個月以前書面通知他

　　方所決定期間內）將竭誠為本人服務並努力在代理地區內推銷本人的商品，並

　　將不作任何有礙此一推銷的行為或干擾本人在代理地區內所作的貿易推廣。

　　3.本人將經常向代理商提供每種商品最低售價表，代理商不得以低於表定價格出

售，並應設法在每一場合爭取最好的價格。

4.代理商未經本人書面同意，不得將商品的任何部分售與代理地區以外的任何個人、公司或行號，亦不得故意將商品售給代理地區內的個人、公司或行號以便其再行輸往其他國家或地區。

5.代理商在本代理契約有效期間內，不得以自己名義或代表任何其他個人、公司或行號銷售與本人商品類似的商品，因而干擾本人商品的銷售或與之競爭。

6.代理商接到商品訂單後，應立即將訂單寄交本人，本人如果接受，則將商品直接運交顧客。

7.本人履行每一訂單後，即將開往顧客的發票副本寄交代理商，並當顧客向本人付款時，以同樣方式隨時通知代理商。

8.代理商所獲得的一切訂單均應記帳並每隔 3 個月向本人寄送一份帳單。

9.代理商在代理地區所直接獲得的一切訂單經本人接受並履行後，由本人給予代理商佣金＿＿＿＿＿＿＿＿＿（按 FOB 英國計算）。佣金每 3 個月付一次，按本人收自顧客實際金額計算。

10.本人經由出口商、購貨代理商、顧客的採購辦事處，以及顧客在英國、北愛爾蘭與愛爾蘭自由邦的總公司，輸往代理地區的一切出口訂單，代理商均得依照上條所述之條件請求支付佣金。惟顧客在代理地區以外的營業處所寄往本人的商品訂單縱然此等商品爾後再由顧客從該營業處所運往代理區，亦不包括在本條所稱的出口訂單以內。但如有確切證明，此等實際上可以經由英國總公司的訂單，係由於代理商向顧客爭取所致者，不在此限。

11.如因本人付給代理商的佣金數目發生爭執時，將以本人的稽核員所決定的數目為準，稽核員的證明對本人及代理商均有最後的拘束力。

12.代理商不得對本人的信用能力作任何保證。

13.代理商未獲本人的書面授權，不得對商品作任何保證。

14.代理商未獲本人授權不得向顧客收取任何金錢。

15.曾經本人指示代理商不得對某些個人、公司或行號授予信用或與之交易者，代理商不得對此客戶授予信用或與之交易。

16.本人有權拒絕履行或接受代理商所獲得的訂單或訂單的一部分，而代理商對被拒之訂單或其中的一部分無任何佣金請求權。

17.當事人間，或其各別代表人間，今後有關本契約所指的事項或有關解釋等所引起的一切異議，均應在英國依照 1975 年仲裁條例或任何新制定或法定修正本，以仲裁解決。

18.一切有關本契約的解釋，以英國法律為準。茲證明本契約係雙方當事人於前述的年月日簽訂。

（簽字）

No. 307　拒絕指定對方為售貨代理商

Dear Sirs,

Thank you very much for your proposal of October 20 offering your services to act as our agent for sale of our products in Nigeria. Although we greatly appreciate your offer, we regret to inform you that we have already made our decision to appoint Rammy Trading Company, Lagos, our agent in your area prior to receiving your proposal.

As it has been our policy of setting, one agent for one country up to the present time, we are obliged to decline your kind offer. We will, however, keep your name and address in our file so that we may communicate with you when our activity becomes more extensive in your country.

Thank you again for your proposal and hope a friendly business relationship between us will be possible at some future time.

Yours faithfully,

【註】

1. prior to = before。

2. keep in file：存卷；歸檔，同樣的用法尚有 "keep on file" 或 "keep in a file"。cf. put this letter in the file：將此信件放入檔卷中。

3. at some future time：將來有一天。

　　cf. in the future：在將來。

　　　　in future：下一次；嗣後。

No. 308　謀任獨家售貨代理

Gentlemen:

Your name having been supplied to us by The Taiwan Chamber of Commerce, we are writing to ask if you would be good enough to consider our proposal to appoint us Sole Agents for Taiwan.

In explanation we would like to let you know that this firm was formed early in 1985, and consists of two sleeping partners and W. Smith, which latter party is in charge of the affairs of the firm. We would add that both two sleeping partners are well-known and respectable Chinese gentlemen, while W. Smith is a British subject.

The firm having been established for more than twenty years, as we explain, we carry some weight and influence in Taiwan. Yours is a line we can handle well, and we are of the opinion that your products would sell extremely well in this market. We are willing to attend to the sale of your goods on commission basis and can give you a reasonable guarantee to sell US$100,000 a month.

We earnestly hope you will give the foregoing your best attention and shall be pleased to receive a full range of samples of your various products.

For reference we would cite the following:

    Bank of Taipei

    Park Road

    Taipei

Yours very truly,

【註】

1. be formed early in 1985：1985 年初創立。

2. sleeping partners：隱名合夥人。

3. latter party：後者，指 W. Smith 而言。

4. in charge of the affairs：經理業務。

5. respectable：有聲望的。

6. British subject：英國人。

7. carry some weight and influence：有相當分量與勢力（的人）。

8. yours is a line we can handle well：貴業是我們最善經營的事業。

9. attend to the sale：承銷。

10. best attention = best consideration：最審慎的考慮。

11. for reference we would cite the following：關於備詢人，謹列舉於下。

No. 309　提議充任保信代理

Dear Sirs,

The demand for toilet preparations in the United Arab Republic has shown a marked increase in recent years and we are convinced that there is here a considerable market waiting to be developed.

There is every sign that an advertising campaign, even on a modest scale, would produce very good results if it were backed by an efficient system of distribution.

We are well-known distributors of over ten years' standing, with branches in most of the principal towns, and have the knowledge of local conditions, the experience and the resources necessary to bring about a marked development of your trade in this country. Reference to the Embassy of the United Arab Republic and to Middle East Services Sales Ltd. would enable you to verify our statements.

If you were to appoint us as your agents, we should be prepared to discuss the rate of commission, but as the early work on development would be heavy, we feel that 10% on orders placed during the first twelve months would be a reasonable figure. As the market would be new to you and customers largely unknown, we would if you wished be quite willing to act on a del credere basis in return for an extra commission of 2% to cover the additional risk.

We hope you will see a worthwhile opportunity in our proposal and that we may look forward to your early decision.

Yours faithfully,

【註】

1. toilet preparations: 浴室用具。

2. marked increase: 顯著的增加。

3. there is here a considerable market: 這裡有很大的市場。

4. there is every sign: 有充分的跡象。 "every" 是加強語氣的字。

5. advertising campaign: 廣告活動。

6. on a modest scale: 規模不太大。

7. efficient system of distribution: 有效的經銷制度。

8. standing: 期間。 cf. a friend of many year's standing: 多年的朋友。

9. resources necessary to bring about...: 發展（貴公司業務的）必要的財力。"resources" 當做財力用時，用多數。

10. verify: 證實。

11. reasonable figure: 合理的數字。

12. Del Credere: 保信，即保證支付貨款，保證支付貨款的代理商稱為 "Del Credere Agent"。保信代理商責任較重，所以要求多付佣金，這種佣金稱為 "Del Credere Commission"。

13. in return: 以為報答。

14. see a worthwhile opportunity in our proposal: 從我們的提議中你能看出有價值的機會。

No. 310　對於謀任保信代理的覆函

Dear Sirs,

We are interested in your proposals of July 9 but, though favorably impressed by your views, feel that local conditions may not at present make worth while the expense of even a modest advertising campaign. We therefore suggest that we first test the market by sending you a representative selection of our products, for sale on our account.

In the absence of advertising we realize that you would not have an easy task, but the experience gained would provide a valuable guide to future prospects. If the arrangement proved to be a success, we would consider your suggestion for a continuing agency. Meanwhile, if you are willing to receive a trial consignment, we will allow commission at 10%, with an additional 2% del credere commission, making 14% in all, commission and expenses to be set against your monthly payments.

Yours faithfully,

【註】

1. worth while：值得。

2. modest：中等的。

3. representative selection of our products：具有代表性的本公司產品。

4. sale on our account：以我們名義出售（盈虧歸我們負責）。

5. in the absence of advertising：由於沒有廣告。

6. future prospects：未來賺錢的可能性。

7. continuing agency：繼續代理。

8. to be set against your monthly payments：從你每月付款中扣除（抵付）。

No. 311　謀任佣金代理

Dear Sirs,

We have a well-developed sales organization in Tanzania and are represented by a large staff in various parts of the country. From their reports it seems clear that there is a good demand for nylon textiles and as we believe you are not directly represented in Tanzania we are writing to offer our service as commission agent.

We have numerous connections throughout the country and there are good prospects of a very profitable market for your manufactures. Provided satisfactory terms could be arranged, we would be prepared to guarantee payment of all amounts due on orders placed through us.

You will naturally wish to have information about us. For this we refer you to the Barminster Bank, Dares Salaam and to Electrical Household Appliances Ltd., Taipei, with whom we have held the sole agency in Tanzania for the past three years.

　　We hope to hear favorably from you and feel sure that we could come to a satisfactory arrangements as to terms.

Yours faithfully,

【註】

1. well-developed sales organization：指「很好的推銷網」。

2. large staff：很多職員（指推銷員）。

3. there is a good demand for nylon textiles：這裡需要大量的尼龍紡織品。

4. not directly represented：無直接代表。

5. commission agent：佣金代理商。

6. good prospects：有很好的前途；前途樂觀。

7. manufactures：製品。

8. arrangements as to terms：有關代理的條件。

　　要求增加佣金不是容易的事，特別是契約中已有規定時為然。因此，任何要求增加佣金的信，應將其所以要求增加的理由具體地說明，而且要技巧地提出，以博得受信人的諒解與同情。

No. 312　代理商要求增加佣金

Dear Sirs,

　　We should be glad if you would consider some revision in our present rate of commission. The request may strike you as unusual since the increase in sales last year resulted in a corresponding increase in our total commission.

　　Marketing your goods has proved to be more difficult than could have been expected when we undertook to represent you. Since then, German and American competitors have entered the market and firmly established themselves. Consequently, we have been able to hold our own only by putting pressure on our salesmen and increasing our expenditure on advertising.

　　We are quite willing to incur the additional expense and even to increase it still further because we firmly believe there is business to be had if the required effort is made; but we do not think we should be expected to bear the whole of the additional cost without some form of compensation. You may feel that this could most conveniently take an increase in the rate of commission by, say, 2%. We suggest this figure after carefully calculating the increase in our selling costs.

> You have always been considerate in your dealings with us and we know we can rely on you to consider our present request with understanding and sympathy.
>
> Yours faithfully,

【註】

1. strike you as unusual：使你認為不尋常。

2. corresponding increase：相應的增加；相稱的增加。

3. more difficult than could have been expected：遠較我們預期為難。

4. undertake to represent you：承受代理你；接下代理你的任務。

5. hold our own = maintain our position in the market：維持我們的地位。

6. put pressure on：加以壓力。

7. incur = to be responsible for：負責。

8. there is business to be had：有生意可作。

9. you may feel：你可能認為。

10. say：譬如說。

11. considerate in your dealings with us：與我們交易時你們顧慮週到。

12. rely on：靠（你）；信賴（你）。

No. 313　對於代理商要求增加佣金的覆函

> Dear Sirs,
>
> Thank you for your letter of August 28.
>
> We note the unexpected problems presented by our competitors and wish to say at once that we appreciate the extra efforts you have made with such satisfactory results.
>
> In the long run we feel that the high quality of our products and the very competitive prices at which we offer them will ensure steadily increasing sales despite the competition from other manufacturers. At the same time we realize that in the short term this competition must be met by more active advertising and agree that it would not be reasonable to expect you to bear the full cost. To increase commission would be difficult as our prices leave us with only a very small profit. Instead, we propose to allow you an advertising credit of US$500 in the current year towards your additional costs, this amount to be reviewed in six months' time and adjusted according to circumstances.
>
> We hope you will be happy with this arrangement.
>
> Yours faithfully,

【註】

1. unexpected problems presented by：由……引起的預料之外的問題。

2. extra efforts：額外的努力。

3. in the long run = in the end = eventually：最後。

4. ensure：保證。

5. steadily：穩定地。

6. in the short term：短期內。

7. be met by：以……應付。

8. bear the full cost：負擔全部費用。

9. to allow：給予。

10. in the current year：今年內。

11. adjust according to circumstances：隨情況而調整。

供應商想提出降低佣金率的意見時必須慎審，否則將引起惡感而破壞感情。這種信應具體說出所以要降低的原因，藉以說服代理商。

No. 314　提議降低佣金

Dear Sirs,

We write this letter with the utmost regret. It is to ask you to accept a temporary reduction in the agreed rate of commission. We make the request because of an increase in manufacturing costs due to additional duties on our imported raw materials and to our inability either to absorb these higher costs or, in the present state of the market, to pass them on to consumers. In the event, our profits have been reduced to a level that no longer justifies continued production.

We hope that in this disturbing, but we trust purely temporary, situation you will accept a small reduction of, say, $1\frac{1}{2}\%$ in the agreed rate of commission, with our promise that as soon as trade improves sufficiently, we shall return to the rate originally agreed.

Yours faithfully,

【註】

1. We write...utmost regret.：以最大的歉意寫這封信。

2. to absorb these higher costs = to accept without raising prices：負擔這些增加了的成本而不漲價。

3. pass them on to：轉嫁給（用戶）。

4. In the event, our profits...continued production.（結果，我們的利潤已經低到無法繼續生產下去。）

5. disturbing situation：令人困擾的情況。

No. 315　建議以買方語言印行型錄

Dear Sirs,

　　We feel we must draw your attention to the harmful effects on your trade here of issuing your catalogs in English and your quotations in US dollars.

　　Most buyers here have little or no knowledge of English and your literature is of very little value as a result, except perhaps that the illustrations sometimes attract interest. But, as you well know, professional buyers look for something more. They want to know about the goods and this means that our staff have to spend much of their time in giving information that should be obtainable from the catalogs.

　　Quotations in US dollars present an even more serious problem since movements in the rate of exchange make it impossible to work out prices correctly and rather than face possible losses on exchange rates, buyers often choose to place their orders with those manufacturers who quote in the local currency. If you were to follow the practice of these manufacturers and issue your catalogs and price-lists in the language and currency of the area, we are confident that you would greatly strengthen your competitive position here.

Yours faithfully,

【註】

1. harmful effects：不良影響。

2. literature：印刷廣告物。

3. professional buyers look for something more：職業性買主所需的不止於此。

4. movements in the rate of exchange：匯率變動。

5. work out prices：算出價格。

6. and rather than...rates：不用說可能遭受兌換損失之事。

7. to follow the practice：遵循……的做法。

8. strengthen your competitive position：加強你的競爭地位。

No. 316　代理商建議低價政策

Dear Sirs,

　　Referring to your quotation No. 123 of January 20, we now enclose our customer's indent No. 456 for card-index and filing equipment.

To secure this order has not been easy because your quoted prices were higher than those which our customer had stated he was prepared to pay. We eventually persuaded him to accept the quotation on the ground of your reputation for quality, but we think we should warn you of the growing competition in the office-equipment market here. Agents of German and Japanese manufacturers are now active in the market and as their products are of good quality and in some cases cheaper than yours, we shall find it very difficult to maintain our past volume of sales unless you can reduce your prices. For your guidance we are sending you by airmail such copies of the price-lists of competing firms as we have been able to collect.

Concerning the present shipment, our customer will issue an L/C for the net value of the indent upon our receipt of your confirmation.

Yours sincerely,

【註】

1. indent：訂單。經由代理商，尤其佣金代理商發出的訂單往往稱為 "Indent"。

2. secure order：獲取訂單。

3. on the ground of：基於……。

4. warn you：警告你。

5. active：活躍。

No. 317　建議改變付款方式

Dear Sirs,

After studying the catalog and price-list received with your letter of March 31, we have no doubt we could obtain good orders for many of the items, but feel you are placing both yourselves and us at a disadvantage by adopting an L/C settlement basis. Nearly all business here is done on D/A basis, the period varying from 90 days to 180 days. Your prices on the other hand are reasonable and your manufactures sound in both design and quality. We therefore suggest that you could afford to raise your prices sufficiently to cover the cost and fall into line with your competitors in the matter of payment method. In our experience, to do so would be sound policy and greatly strengthen your hold on the market. With the best will in the world to serve you, we are afraid it would be neither worth your while, nor ours, to continue business on L/C basis.

If it would help you at all, we should be quite willing to assume full responsibility for unsettled accounts and to act as del credere agents for an additional commission of $2\frac{1}{2}$%.

Yours faithfully,

【註】

1. good orders：大量訂貨。

2. placing both yourselves and us at a disadvantage：使我們處於不利地位。

3. sound：好。

4. could afford to raise your prices：能提高價格。

5. to cover the cost：抵補（利息）成本。

6. fall into line with：與……（採）同一步調。

7. sound policy：穩健的政策。

8. strengthen your hold：加強你的據點。

9. with the best will in the world to serve you：極願為你服務。

10. neither worth your while, nor ours：對你我都不值得。

11. unsettled accounts：未清帳目。

No. 318　對改變付款方式建議的覆函

Dear Sirs,

We thank you for your letter of April 10 and are glad you think that a satisfactory market could be found for our goods, but are not altogether happy at the prospect of transacting all our business on D/A basis.

To some extent your offer to act in a del credere capacity meets our objections and for a trial period we are prepared to accept it on the terms stated, namely, an extra commission of $2\frac{1}{2}\%$, provided you are willing either to provide a guarantor acceptable to us, or to lodge adequate security with our bankers.

Yours sincerely,

【註】

1. are not altogether happy at：對……不盡愉快；對……不敢苟同。

2. prospect：期望。

3. to some extent = more or less; considerably。

4. meets our objections：不願意；我們有異議，即不同意。

5. provide a guarantor：提供一位保證人。

6. lodge adequate security：存入（我們的銀行）適當保證金。

No. 319　賣方抱怨代理商產品滯銷

Dear Sirs,

We are very concerned that your sales in recent months have fallen considerably. At first, we thought this might be due to the disturbed political situation in your country, but on looking into the matter more closely, we find that the general trend of trade during this period has been upwards.

It is of course possible that you are facing difficulties of which we are not aware. If so, we should like to know of them since it is always possible for us to take measures that would help. We therefore looking forward to receiving from you a detailed report on the situation and also any suggestions of ways in which you feel we may be of some help in restoring our sales to at least their former level.

Yours faithfully,

【註】

1. very concerned：很關心。

2. at first：起初。

3. disturbed political situation：混亂的政治情況。

4. general trend of trade：一般貿易趨勢；一般商業趨勢。

5. take measures that would help：採取有用的措施。

6. former level：以往水準。

第二節　採購代理信的寫法

進口商委託出口地的商號就地代辦採購的交易，稱為委託購買 (Indent)。在委託購買交易中，委託商稱為 "Indentor"，受委託商稱為 "Indentee"，一般又稱為採購代理商 (Buying agent)。

No. 320　申請擔任採購代理

Dear Sirs,

Your name has been recommended to us by Bank of America, San Francisco, as one of the leading importers of handicrafts. If you are not already represented in this market, we venture to inquire whether you are disposed to accept us as Buying Agents for Taiwan.

We have been established here in a fairly large way for the last ten years, and from our intimate knowledge of the business we believe that we can entirely meet your requirements. As our business has extended considerably of late, we have decided to devote our principal attention to electronics and consider that the reputation we enjoy here and the connections we have all over this island, justify us at this stage in enlarging our operations and in shipping to U.S. importers.

We, therefore, feel confident that if you entertain our proposal, the result will be perfectly satisfactory to both sides.

We shall be glad to learn your views in the matter, and, to assist you in coming to a decision, we subjoin the name of several firms in U.S. whom we are regularly supplying what they require. We are sure that they will be pleased to reply to any inquiries from you concerning us.

For the guidance and to help towards pleasant relation between us in our transactions we enclose a copy of Agency Agreement for your approval, and, as we are from the first putting you on a very advantageous footing, we doubt not that you decidedly accede to our proposal.

Yours faithfully,

【註】

1. venture：冒昧。

2. are disposed to：有意。

3. in a fairly large way：相當大規模地。

4. intimate knowledge of：熟悉。

5. of late = recently。

6. devote to：專心致志於……；專營。

7. enlarge our operations：擴大營業。

8. in shipping to U.S.：運銷美國。

9. entertain our proposal：接受我們的提議。

10. assist you in coming to a decision：協助你們裁決。

11. subjoin：附上。

12. putting...footing：把你放在一種極有利的地位。

No. 321　採購代理契約

Agreement for Buying Agency

THIS AGREEMENT, made and entered into this 30th day of June, 20–, by and between The American Trading Company, Inc., of New York, NY, U.S.A., party of the first, and Taiwan Trading Co., Ltd., Taipei, Taiwan, party of the second part, witnesseth:

1. The party of the first part (hereinafter called the Principal) here by grants to the party of the second part (hereinafter called the Agent) the exclusive right to buy and ship its requirements in Tai-

wanese electronics for the period of three years from the date hereof unless sooner terminated as hereinafter provided.

2. The Agent covenants that it will use every reasonable diligence in executing all orders placed with it by the Principal and will not, without the consent of the Principal, deviate from any instructions given by the Principal in regard to the purchase and shipment of its orders.

3. On all orders executed by the Agent during the continuance of this agreement, the Agent shall be paid the following commission:

on an order amounting to US$10,000 or less: 3%;

on an order amounting to more than US$10,000: 2.5%.

4. The amount of an invoice rendered by the Agent, including its commission and its disbursements except postages and petties, shall be paid by the Principal under Irrevocable L/C against the delivery of shipping documents.

5. This agreement may terminated by either party upon three months' written notice delivered or sent by registered mail to the other, and may be terminated at any time, without such notice, upon breach of any its terms and conditions.

IN WITNESS WHEREOF, the parties hereto have caused this agreement to be executed in quadruplicate, two for each party, as of the date first above written by their duly authorized representative.

The American Trading Company, Inc.

(signed)

Taiwan Trading Co., Ltd.

(signed)

【註】

1. made and entered into：締結。"made" 與 "entered into" 重覆，是法律用語。

2. party of the first：第一當事人。

3. witnesseth：證明，為 "witnesses" 的古語用法，但法律文件卻喜用此語，難怪 (no wonder) 法律文件都晦澀難懂。

4. exclusive right：獨占權利。

5. terminate：（契約）終止。

6. deviate from：逸出；違反。

7. disbursements：開支。

8. petties：零星支出。

9. written notice：書面通知。

10. in witness whereof：茲證明……。

No. 322　同意指定對方為採購代理

Gentlemen:

　　We have received your letter of August 12 together with its enclosures and have taken your proposal to act as our Buying Agents for Taiwan into serious consideration.

　　You will readily understand that we wish to proceed without undue haste. We, therefore, suggest that the proposed Agreement for Buying Agency shall hold good at first for a period of twelve months only and be terminated on 30 days' notice from either side.

　　If, as we have no reason to doubt, the results are such as may justly regarded as satisfactory, we shall, after the lapse of that period, be prepared to confirm the agreement for a longer time.

　　We trust we may both have reason to congratulate ourselves upon this departure.

Very truly yours,

【註】

　　1. have taken...into serious consideration：對……做了審慎的考慮。

　　2. readily = easily。

　　3. we wish to proceed without undue haste：我們願意不過於草率的進行。

　　4. hold good：有效。

　　5. justly：公正地；理由充分。

　　6. after the lapse of that period：滿期之後。

　　7. congratulate...upon：恭喜。

　　8. this departure：這個新的發軔；這個新行動的開始。

第三節　寄售代理信的寫法

一、寄售的意義

　　寄售 (Consignment) 又稱為寄銷。在國際貿易，寄售乃寄售商 (Consignor) 為開拓銷路，先將貨物運交國外商號委託其伺機代為銷售，貨物售出後才收回貨款的交易。受託代銷的國外商號稱為受託商，或代銷商或承銷商，英文稱為 "Consignee"。受託商就其代銷行為，向寄售商酌收佣金。上述寄售商又稱為委託商或貨主 (Principal)。

在寄售交易，受託商只不過是代理他人銷售商品，從中賺取佣金的商號，所以，本質上，為佣金商 (Commission merchant) 的一種。自法律上而言，寄售是一種委託行為，而寄售商與受託商之間的關係是一種委任及代理的關係。因此，寄售交易應受委任、代理以及行紀等有關法律的拘束。

按寄售並非普通的銷售，兩者的區別為：普通銷售，先有訂貨或先訂有買賣契約，而後將貨物運出，商品所有權乃由賣方移轉給買方；至於寄售，未有訂貨或未訂有買賣契約（換言之，尚未有買主），即將貨物運出，貨物在受託商未代售出之前，其所有權仍屬於寄售商。此一區別甚重要，因為：

1. 寄售並非出售，故商品轉移行為並不產生利潤，而且受託商未將商品售出前，無任何價值的獲得。

2. 商品所有權仍屬於寄售商所有，所以，寄售商品仍係寄售商的存貨。

3. 萬一受託商破產，寄售商仍可收回寄售商品，受託商的債權人，對此寄售商品不能主張權利。

No. 323　提議寄售交易

Dear Sirs,

　　From our bankers, Bank of Taiwan, Taipei, we have learned that you are one of the firms of the highest respectability in Nigeria and have been doing a very good business in our line. We are now on the lookout for a reliable and competent firm, who would like to take care of our business in your market. At their suggestion, therefore, we inquire whether you would be disposed to work with us on a consignment basis.

　　As manufacturers and exporters of transistor radios we stand in the front rank. For a good many years we have been supplying our African friends with our goods. We should like to introduce our goods, notably our new products, further in your market, and are convinced that, if you fall in with our ideas, a large turnover will result to our mutual advantage.

　　We are sending you, under separate cover, a representative range of samples together with a price-list. We ask you to select from them the items which you think to be best suited for your market and give us a report on them. If, as we hope, you will agree to our proposal, and on receipt of your report, we shall immediately proceed to ship you a sample consignment.

　　Meanwhile, we hope to ship our products to your market under either of the following two ways:

　1. We ship our products to you on consignment and commission basis, without your financing.

2. Under consignment and commission arrangement, we hope that you would finance us 70% of the invoiced cost when you are in receipt of shipping documents.

While each of the foregoing two suggestions differs in meaning and operation, we, as shippers, prefer the second method which, with your financial assistance, will enable us to make more shipments.

As to our standing, please refer to:

Bank of Taiwan
Chungking S. Road, Sec. 1
Taipei

We are quite prepared to listen to any suggestions you may have to make in the matter and trust that you will be able to work in concert with us.

Yours faithfully,

【註】

1. on the lookout for：尋找。

2. competent：能勝任的。

3. stand in the front rank：站在前排。意指第一流的公司。

4. turnover = volume of business。

5. representative range of samples：具有代表性的樣品。

6. a sample consignment：樣品性質的寄售品。為試探市場而運出的少量寄售品。

7. work in concert with：與……協力工作。

No. 324　同意寄售交易

Dear Sirs,

We acknowledge receipt of your letter of April 10, requesting us to handle transistor radios on consignment basis for you.

The samples which you sent to us were also received. We regret that our reply has been delayed. We have just got around to an inspection of the samples. The samples, in our opinion, are well manufactured. They are similar to those originated from Japan. We figure these are better than those originated from Hongkong. The question is what brand should you pack these products. In other words, are we dealing with a known brand?

As one of the leading and experienced Nigerian importers who has been in this field for more than five years, we are quite glad to handle your products on consignment basis, and we are confident in securing considerable orders for you.

We have also noticed the two ways under which you would operate. At present time, we prefer the first method, i.e., you ship the goods to us on consignment and commission basis, without our financing.

We have drafted a consignment agency agreement and enclose two copies for your approval. We should like to know if you would fall in with our terms and conditions. This will form the basis upon which we do business and apply to all your consignment to us, unless otherwise prearranged in each transaction. Any suggestions you may have to make on them would be welcome.

We thank you for your giving us the opportunity to handle your products.

Yours very truly,

【註】

1. get around to：找到時間做……；注意及……（美國口語）。

2. figure：認為（俗語，口語）。

3. known brand：大家知道的牌子。

4. draft：草擬。

5. consignment agency agreement：寄售代理契約。

6. apply to：適用於……。

7. prearrange：事先約定；事先安排。

No. 325　寄售契約書

Suitable for agents appointed overseas to whom stocks are shipped on consignment account

An Agreement made this........................ day of................................ 20..............................between
.... whose registered office is situated at.. (here-inafter called "the Principal") of the one part and.. (hereinafter called "the Agent") of the other part.

Whereby It Is Agreed As Follows:

1. The Principal hereby appoints the Agent as their del credere Sole Agent for the sale in (the territory).............. of.............. (the goods) and the Agent accepts the appointment on the terms hereinafter defined.

2. This Agreement shall come into operation as from.............. and shall continue in force for one year certain and thereafter until terminated by either party giving to the other.............................. calendar months' notice in writing expiring on any date provided always that either party may determine it summarily if the other party fails to carry out the terms of the Agreement, or goes into liquidation

other than voluntarily for the purpose of reconstruction, or is wound up, or makes a composition with his creditors.

3. The Agent shall at all times use his best endeavours to develop the sale of the Principal's goods the subject of the agency within the territory and undertakes not to sell, or offer for sale, any goods competing with those of the Principal.

4. The Principal shall be free to sell the aforesaid.............. without the territory of the Agent and shall not be obliged to pay commission to the Agent on.............. shipped into the said territory unless sold by the Agent.

5. The Principal shall ship on consignment reasonable quantities of the goods to the Agent against indents. All such goods on consignment together with the necessary containers shall remain the absolute property of the Principal until delivered to the Buyer under safe contract. The Agent shall pay landing charges, duty, cartage, warehousing costs and insurance on account of the Principal and shall recover these by debiting the Principal in current account.

6. The Agent undertakes that no charge, or lien, shall be created on the goods so consigned.

7. As consideration for the Agent's service hereunder the Principal shall pay the Agent commission at the rate of............................per cent which shall be calculated on the cif invoice value of all......................... sold in the said territory by the Agent. The Agent undertakes to indemnify the Principal against all default by purchasers in payment for the Principals' goods after sale and delivery.

8. The Agent undertakes not to pass on to the buyer any part of his commission, but may divide it with his appointed distributors where he can control their selling prices.

9. The Agent shall keep and maintain proper books of account and shall remit quarterly to the Principal statements and the sterling equivalent of the net proceeds of sales of the goods which have been made by him subject to retention by him of the commission provided for in Clause 7 hereof and those charges payable by him under Clauses 5, 10, and 11 and properly to be debited to the Principal.

10. Each of the parties hereto shall bear his own expenses other than those specifically payable by the other except that charges incurred by the Agent for cables and telegrams properly dispatched in the conduct of the business shall be refunded to him by the Principal.

11. Sales literature shall be supplied free to the Agent by the Principal and a reasonable allowance for advertisement and propaganda to be mutually agreed upon shall be paid for by the Principal.

12. The Agent shall be free by nameplate at his office, or on his letter-heading, or in other manner approved by the Principal, to inform the public that he is the Sole Agent of the Principal during the continuance of this agreement.

13. Determination of this agreement shall relieve the Principal of any right of claim against them by the Agent for any commission upon sales made by the Principal to customers introduced to them by the Agent during the existence of the agreement, or in respect of deliveries made against contracts effected before the determination of the agreement.

14. This agreement shall be interpreted according to the Laws of England and any dispute between the parties shall be settled by arbitration as provided for by the rules laid down by the Arbitration Act 1975 or any statutory modification thereof.

AS WITNESS the hands of the Parties, etc.

【註】

1. whereby it is agreed as follows：爰（茲）約定如下。

2. come into operation：生效。

3. as from：自……開始。

4. continue in force：繼續有效。

5. for one year certain：1年整。"certain"為「不多也不少」之意。

6. calendar month：曆月。

7. in writing：以書面。

8. summarily：迅速地。

9. carry out：履行。

10. go into liquidation：破產。

11. is wound up：結束。

12. make a composition with his creditors：和債權人說和；和解。

13. safe contract：可靠的契約。

14. cartage：搬運費用。

15. current account：往來帳；交互計算帳。

16. lien：留置權。

17. consideration：報酬。

18. allowance：補助。

19. nameplate：名牌。

20. determination：終止。

No. 326　通知運出寄售貨物

Dear Sirs,

　　We have received your letter of April 18 together with two copies of draft Agreement for Consignment Agency and thank you for your willingness to work with us on our transistor radio sets consignment to Nigeria.

We have carefully read through your proposed agency agreement and we find the terms and conditions in it are quite acceptable to us. We return herewith one copy of the agreement duly signed by us.

Being anxious to materialize an initial business we have shipped to you on April 30 per s.s. "Sky Lykes" 1,000 sets of transistor radio sets. We enclose the following documents for your processing:

1. Invoice in triplicate.
2. B/L in duplicate.
3. Marine insurance policy in duplicate.
4. Packing list in triplicate.
5. Customs invoice in triplicate.

We hope that the goods will arrive in good condition, and look forward to the relations now being established between us will last long and become mutually profitable.

Yours faithfully,

【註】

　　1. transistor radio sets：即指 "transistor radios"。

　　2. materialize：實現。

No. 327　覆收到貨運單證

Dear Sirs,

We have received your letter of May 2 together with duly signed Consignment Agency Agreement and the shipping documents covering the 1,000 sets transistor radios you consigned to us.

Upon arrival of the goods, we shall use our best endeavours in disposing of them to your satisfaction.

At the moment, we have several inquiries for the line of your consignment, and we feel certain that your goods will realize profitable business.

We thank you for your giving us a trial consignment, and we assure you of our best services at all times.

Yours very truly,

【註】

1. use our best endeavours: 盡我們最大努力。

2. line = item: 項目；貨品。

3. your goods will realize profitable business: 貴方貨品將促成有利的生意。"realize" 為 "gain" 之意。

No. 328　報告寄售貨物已售出

Dear Sirs,

We have the pleasure of informing you that the transistor radio sets you consigned to us have been profitably disposed of.

You will see in the Account Sales, which we enclose, that all the goods were cleared at better prices than we expected. We are pleased that in this first instance we have been able to give you such a satisfactory report.

The net proceeds amounting US$2,165.00 have been remitted to you today by mail transfer through our bankers, Bank of America, Nigeria, who will request Bank of Taiwan, Taipei to pay them to you.

We assume you will receive the proceeds in due course of time.

We are very glad to say that your products are quite welcome here. So, please consign to us further 2,000 sets as soon as practicable.

Yours very truly,

【註】

1. account sales: 簡稱 A/S，售貨報告。

2. clear: 售完；賣光。cf. wheat cleared rapidly: 小麥一下子就賣完了。

3. in this first instance: 首先；第一；一開始。cf. I thought so too in the first instance.

4. net proceeds: 淨售價款，即扣除代理商佣金及代墊費用後的貨款。

5. as soon as practicable = as soon as possible。

No. 329　售貨報告

Account Sales of 1,000 sets Transistor Radios received ex s.s. "Sky Lykes" from Keelung sold by Nigerian Trading Co., Ltd. for account & risk of Taiwan Trading Co., Ltd.

1,000 sets Transistor radios		US$2,600.00
@ US$2.60 per set		
Charges		
Customs clearing & storage	US$ 50	
Import duty	100	
Fire insurance, cable charge	25	
Commission: 10%	260	435.00
Net Proceeds		US$2,165.00
Mail Transfer through		
Bank of America, Nigeria		
E.&O.E.		
Lagos, July 5, 20–		
Nigerian Trading Co., Ltd.		

第四節　有關代理的有用例句

一、謀任代理

(一)開頭句

1. We should be glad if you would consider our application to act as agents for the sale of your...(products)...

2. We are acting as agents for a number of American publishers and are wondering whether your firm is represented in...(country)...

   註：your firm is represented in...：你的行號（是否）有代表在……。

3. We thank you for your letter of...(date)...enquiring whether our firm is represented in...(country)...

4. We have received your letter of...(date)...and shall be glad to offer you a sole agency for the sale of our products in...(country)...

(二)結尾句

1. We hope our handling of this first order will lead to a permanent agency.

2. I hope to hear favorably from you and feel sure we should have no difficulty in arranging terms.

註：arranging terms：指代理條件的安排。

3. If you give us the agency we should spare no effort to further your interests.

註：a. spare no effort：盡力；b. to further your interests：增進你的利益。

4. If you are interested we can give you first-class references.

註：first-class references：最佳的備詢人（或信用資料）。

二、任命代理

㈠開頭句

1. Thank you for your letter of...(date)...offering us the sole agency in...(country)... for your manufactures.

註：offering：授與。

2. We thank you for your letter of...(date)...and are favorably impressed by your proposal for a sole agency.

3. We thank you for offering us the agency in...(country)...for your products and appreciate the confidence you have placed in us.

㈡結尾句

1. We already represent several other manufacturers and trust you will allow us to give you similar service.

2. We accept your terms and conditions set out in the draft agency agreement and look forward to a happy and successful working relationship with you.

第三十章
國際投標 (International Tender)

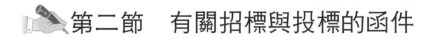 第一節　招標與投標的意義

　　國際貿易的交易方式買賣雙方除採協議 (Negotiation) 方式進行外，也可以公開招標 (Open tender) 的辦法。所謂公開招標乃買方或賣方利用公開競爭方式，邀請多數的商號出價 (Bid)，從而以最理想的條件買進或售出貨物的行為，上述多數商號的出價行為稱為投標 (Bidding)。

　　報紙上常有 "Invitation-to-bid" 的啟事，這個 "Invitation-to-bid" 詞，照字面解釋，是「邀請出價」之意，而實際上即為「招標」之意。「標」即是「標價」(Bid or price quotation)，也即「報價」(Offer or quotation) 之意。所以，"Invitation to bid" 又可寫成 "Invitation to offer" 或 "Invitation to tender"。因其公開邀請多數商號競標，所以稱為「公開招標」，但自參加競標商號的立場而言，這種交易方式，又稱為「公開投標」(Open bidding)。

　　如果買方以招標方式購入貨物，則就買方立場而言，可稱為公開標購 (To purchase through open tender)；如賣方以招標方式出售貨物，則就賣方而言，可稱為公開標售 (To sell through open tender)。

　　招標在法律上乃為一種向不特定人的買賣要約引誘行為。在招標之前，招標商首先要擬定招標單。所謂招標單乃規定標購或標售貨物名稱、規格、數量、交貨、付款、驗收、投標商資格、開標時地、押標金、決標方法者。擬定招標單之後，即可公告 (Announcement)。公告除張貼於招標機構布告欄外，通常都在報紙上刊登廣告。

第二節　有關招標與投標的函件

No. 330　招標單 (Form of invitation-to-bid)

CENTRAL TRUST OF CHINA
TRADING DEPARTMENT
INVITATION-BID-CONTRACT

Invitation No. CTCTD 123   Date: Sept. 20, 20–

Contract No. _____

Sealed bids in triplicate, which shall be valid for at least six(6) hours and subject to the terms and conditions of the attached sheets, will be received at CTC Trading Dept. 4th floor, 49, Wu-Chang Street, Taipei, not later than 3:00 p.m., Oct. 9 (Tuesday), 20– (Taipei Time), and then they will be publicly opened in the presence of bidders. The successful bidder shall be informed individually by CTC for negotiation, if necessary, within the validity of the offer, CTC will disclose the result of bidding to the unsuccessful bidders. CTC reserves the right either to award or to reject any or all bids, or to negotiate with the lowest acceptable bidder.

1. Commodity:

   U.S. Yellow Soybeans.

2. Quality & Specifications:

   U.S. Grade No. 2, of 20– crop

   Oil contents: 18.5% min.   Bushel weight: 54 lbs. min.

   Moisture: 14.0% max.   Foreign material: 2.0% max.

   Splits: 20.0% max.   Damaged total (including heat damaged): 3.0% max.

   Heat damaged: 0.5% max.   Brown & black kernels: 2.0% max.

3. Quantity:

   10,000 metric tons, 5% more or less in bulk, up to 10% in bags for stowage purpose if required, gross for net, at ship's option.

4. Price:

   Price to be offered either on FOB ex-spout basis (stowing & trimmings ship's account) or on CFRFO Kaohsiung basis at CTC's choice.

5. FOB Award:

   CTC will arrange the vessel(s) at the following terms and conditions: (All detailed shipping conditions shall be settled before awarding by bidder & shipowner directly, CTCTD serves only as a witness, without any responsibility on CTCTD's part)

   a) Loading rate: CQD

   b) Carrying charges: In the event that the vessel contracted could not arrive at the loading spouts within the shipping period as stipulated above, any carrying charges and interest thus incurred shall be for shipowner's account. However, in case that the vessel arrives at the loading port in accordance with the shipping period as stipulated, but could not load soybeans due to port congestion or other causes which are beyond the control of the vessel, no carrying charges and interest shall be paid by the shipowner. Carrying charges, if any, shall be borne by the seller, irrespective of force majeure clause under Article 25 of conditions.

c) The shipowner or his agent shall notify the seller of the ship's name, her ETA to loading port, at least seven days prior to the arrival of the vessel. In counterpoint, the seller shall make a penalty payment at the rate as stipulated in Article 18 of general conditions, should the seller fail to load the commodities within the stipulated shipping period after receipt of Notice of Readiness forwarded by the shipowner or his agent.

d) Other terms & conditions: According to Gencon Charter Party with overtime clause.

6. CFR Award:

(I) Terms of carrier(s):

a) Bidder must specify vessel(s) name, age, speed, tonnage, number of hatches, draft etc. otherwise the bid is not acceptable.

b) Vessel(s) over 20 years of age and below 12 knots per hour are not acceptable. Vessel(s) should be sufficiently and efficiently equipped with derricks/winches to handle the loading and discharge of cargo for each and every hatch.

c) Bidder/owner must guarantee that the carrier(s) used must be able to berth alongside the Kaohsiung wharf for discharge alongside the wharf, otherwise all losses incurred should be for Bidder's/owner's account.

(II) Discharging rate:

Cargo to be discharged at the rate of 200 MT per hatch per weather working day, Sundays and Holidays excepted, even it used, provided that each hatch is equipped with one set of efficient derrick with adequate power. However, CQD discharge is preferred.

(III) Discharging laytime:

Laytime to commence at 1 p.m. if Notice of Readiness to discharge is given to the consignee before Noon and at 8 a.m. next working day if Notice of Readiness is given during office hours after Noon, whether in berth or not. If work commenced earlier than the time stipulated above, laytime still to be counted from the time stipulated not counted earlier.

(IV) Demurrage & Despatch at discharging Port:

Discharging US$200/US$100 per hatch day or pro rata. Others according GENCON Charter party, with BFC Saturday clause & overtime clause.

(V) Vessel's overages insurance premium shall be for seller's account.

7. Shipment:

Loading port: US Gulf port, one safe berth port only.

Discharging port: Kaohsiung.

Shipping period: November, 20– or earlier.

8. Payment:

By Irrevocable Letter of Credit in full amount available against at sight draft and documents, to be established before shipment in favor of seller.

9. Bid Bond:

One percent Bid Bond shall be posted before 17:00 of the day previous to the bidding date and shall be valid until 60 days after the bidding date.

10. Performance Bond:

Five percent performance bond shall be posted with CTC within ten working days after awarding. The performance bond shall be valid until 90 days after the contracted shipment date and shall be extended accordingly in case the shipment is delayed or the contracted shipment date is extended.

11. Insurance:

To be covered by CTC, but seller shall notify CTC by fax of all necessary data before loading for insurance coverages.

12. Documentation:

a) Official grade certificate issued at time of loading certifying that all soybean loaded are strictly in conformity with the specifications as stipulated in article 2 of this Invitation, is required, and will be considered final as to quality of the soybean shipped.

b) Weight Certificate issued by a licensed weighmaster at the point of loading on vessel is required and will be considered final as to quantity of the soybean shipped.

c) Official Phytosanitary Export Certificate is required.

d) Official Certificate of Origin is required.

e) First & Second Originals with copies of clean-on-board Bill of Lading, ten copies of seller's detailed Invoice, and stowage plan are required.

f) As soon as available, the seller shall telex CTC Trading Dept.
(Fax: 886–2–2363–2964) the following information:
Invitation & Contract numbers, brief description of quantity shipped, ship's name, port of departure, ETD, ETA, Kaohsiung, Invoice value, a copy or this telex shall constitute a part of the required documents for payment.

g) At the time of loading, the seller shall forward by registered airmail two complete sets of duplicates of documents stipulated as Articles a) to f), to CTC Trading Dept. immediately.

h) All expenses of quality/weight inspection and analysis, together with the fee of all documents required in the L/C, etc. except that expenses paid to the independent surveyor will be for

CTC's account, shall be for the seller's account.

13. CTC reserves the rights to award the contract to any bid or reject all bids. In case the terms quoted by the bidders cannot fully meet with CTC's requirements, CTC shall request all bidders to submit their firm offer(s) according to the same terms and conditions of this Invitation-Bid-Contract to CTC, in sealed envelop marked "Firm Offer" day by day consecutively or any day at their choice, within the next ten days after the tendering date subject to CTC's prior acceptance, if acceptable, of such Firm Offer as stated above.

14. General Instructions & Conditions:

(I) "Attached General Instructions & Conditions" are incorporated hereto to form a part of this Invitation-Bid-Contract, if any contradiction in-between, the latter shall govern.

(II) CTC's award of this tender is valid subject to the approval of BOFT.

【註】

1. sealed bid: 密封標單。

2. bidders: 投標商。

3. successful bidder: 得標商，又稱為 awarded bidder。

4. negotiation: 議價。

5. disclose: 公開。

6. reject any bids: 否決任何「標」。

7. lowest acceptable bidder: 可接的最低標投標商。

8. foreign materials: 夾雜物；雜質，又稱 foreign matter。

9. in bulk: 散裝。

10. for stowage purpose: 做為積載之用。

11. gross for net: 以毛重代替淨重，意指按毛重計價。

12. at ship's option: 意指多裝或少裝 5% 由船方決定。

13. FOB ex-spout: 利用噴注管的船上交貨價，"spout" 為用以將穀類注入船艙的噴注管。

14. stowing & trimmings: 堆積及平艙費。

15. FOB award: 按 FOB 決標。

16. carrying charges: 加計費用或持有費。按 FOB 成交時由於買方未能在約定期間內裝船時，買方應負擔的各種費用：包括⑴儲存費用，⑵利息。

17. shipowner: 船東。

18. in counterpoint: 對應。

19. overtime clause: 逾期條款。

20. hatch：艙口。

21. draft：（船的）吃水 (=draught)。cf. a ship of 10 feet draft：吃水 10 呎的船。

22. knot：海里。

23. derrick/winch = derrick：起重機，winch：絞盤。

24. overage insurance premium：逾齡保險費。

25. bid bond：押標金。

26. posted：繳存。

27. performance bond：履約保證金。

28. documentation：單證的提供，本條規定賣方押匯時應提供的單證。

29. official grade certificate：官方出具的等級證明書。

30. licensed weighmaster：特許（指有執照的）重量檢定人。

31. official Phytosanitary Export Certificate：官方出具的植物檢疫證明書。

32. stowage plan：積載計畫。

No. 331 招標布告 (Announcement for open tender)

INVITATION-TO-BID

The Central Trust of China, Trading Department, intends to purchase the following listed commodity through open tender. Any registered and qualified firms or manufacturers, if interested in this tender may obtain from this office Invitation-to-bid and submit their bids to this office on or before the date specified hereunder:

Inv. No.	Description	Area of source	Invitation issuing date	Tender opening date
CTCTD 123	US Grade No. 2 Yellow Soybean 20– crop	U.S.A.	Sept. 20, 20–	10:00 a.m. Oct. 9, 20–

【註】

1. purchase...through open tender：以公開招標方式購買……。

2. registered and qualified：登記有案的合格（廠商）。

3. invitation-to-bid：招標單。

4. area of source：貨物來源地區，即指採購地區。

5. invitation issuing date：招標日期，即印發標單日期。

6. tender opening date：開標日期。

No. 332　　Agent 通知供應商招標消息

Gentlemen:

We confirm we have sent you the following Fax today:

CTC TRADING DEPT WILL OPEN TENDER 10:00 A.M. OCT 9 FOR 10,000 MT US GRADE NO. 2 YELLOW SOYBEANS 20– CROP NOV SHIPMENT PLS SEND US YOUR FIRM BID IN FOB EX SPOUT AND CNF KAOHSIUNG DETAILS AIRMAILING

We are pleased to inform you that CTC Trading Department intends to purchase 10,000 metric tons US grade No. 2 yellow soybean, 20– crop through open tender at 10:00 a.m. on October 9, 20–, Invitation-to-Bid for which is enclosing for your information and study. If you are in a position to make the supply, please send us your firm bid on FOB ex spout Gulf Port and CFRFO Kaohsiung. You are requested to get your offer here before October 8, our time.

In addition to the particulars carried in the Invitation-to-Bid, we want to call your attention to the following:

1. Bid Bond: one percent (1%) bid bond should be deposited by you either in cash, L/C or Bank guarantee before 17:00 of the day previous to the bidding date and shall valid until 60 days after the bidding date.

2. Performance Bond: Five percent (5%) performance bond should be deposited by you in cash, L/C or Bank guarantee within 10 working days after the award of the contract.

3. Penalty for delay shipment: the buyer has made it known that suppliers must make a penalty payment to buyer on the basis of 0.1% of the contract value of the delayed portion for each one day's delay.

Please note that your offer should include one percent (1%) commission for us on FOB ex spout value or 0.8% on CFRFO value.

We are hoping to receive your most competitive offer with details in due course.

Yours faithfully,

【註】

1. CNF=CFR。

2. U.S. Grade No. 2 yellow soybeans：美國二級黃豆。

3. 20– crop：20– 年生產的作物。

4. Gulf port：這裡是指美國的 Gulf port，包括 Mobile, New Orleans, Beaumont, Houston, Galveston 等港口。

5. get your offer here：將你的報價寄達此地。

6. our time：意指此地時間 (October 8)。

7. bidding date：指開標日而言。

No. 333　供應商通知 Agent 應標

Dear Sirs,

Re: Open Tender of Yellow Soybeans

Thank you very much for your fax and your letter of September 20 and Invitation-to-Bid with which you called our attention to the forthcoming bid of the CTC Trading Department for 10,000 M/T of yellow soybeans.

After due consideration, we have decided to authorize you to participate in this tender on behalf of ourselves in accordance with the terms and conditions stipulated for subject tender.

As you will note from the attached enclosures, the price is left open. We will fax you one day before the tender opening date.

In the meantime, we shall ask our bankers, Chemical Bank, New York, to issue a letter of Guarantee favoring CTC, Trading Department through CTC, Foreign Department as a bid bond for the subject tender.

We trust that our offer will enable you to secure this bid for us and thank you for your kind cooperation.

Yours faithfully,

【註】

1. participate in this tender：應標。

2. price is left open：價格未決定。

3. secure this bid：奪得此標。

No. 334　供應商以 Fax 通知 Agent 價格

REFER TO YOUR LETTER SEPT 20 OUR LETTER SEPT 30 CONCERNING CTCTD'S INVITATION NO 123 FOR YELLOW SOYBEAN WE OFFER FIRM VALID UNTIL 4:00 PM OCT 9 USD 220 PER MT FOB EX SPOUT GULF PORT AND USD 236 PER MT CFRFO KAOHSIUNG INCLUDING YOUR COMMISSION STOP PLS FAX US RESULT SOONEST

　　Bid bond 為投標人因投標而向招標人繳存的保證金。投標人於得標後如拒絕簽約時，招標人即可沒收該項保證金。參加國際投標時多以 Bank letter of guarantee 或 Stand-by L/C 代替現款或票據。

No. 335　中信局規定的押標保證標準格式

STANDARD BID BOND FORM (SPECIMEN)

Letter of Guarantee No. _____ Date of Issuance _____

for (Name of Currency & Amount in Figure) _____

as Bid bond

for Your Invitation No. _____

To:　　　　　　　　　　　　　　Advised Through

Central Trust of China　　　　　(Name of Issuing Bank's Correspondent Bank in Taipei)

Trading Department

49, Wu Chang Street, Sec. 1

Taipei, Republic of China

　　Upon request of Messrs. (name of Accountee), we hereby agree to issue this irrevocable letter of guarantee up to the aggregate amount of (name of currency & amount in words) only in your favor for account of (name of Accountee) as (name of Bidder)'s bid bond for your Invitation No._____ and engage ourselves to pay you immediately without recourse upon receipt of your written statement certifying that (name of Bidder) have withdrawn their bid in its entirety or part thereof before its expiration including its extension, if and and/or after being accepted by you, or their failure in entering into contract after being awarded, or their failure in posting the performance bond within fourteen (14) working days after the date of the notice of award, accompanied by your simple receipt(s) or sight draft(s) drawn on (Issuing Bank or Accountee to be indicated by Issuing Bank)

　　We bind ourselves to obtain necessary permit from relevant authorities to transfer the claimed amount to you in case it is confiscated.

　　This letter of guarantee will expire on (date of expiration—30 days after tender opening date) Taiwan time.

Authorized Signature

No. 336　Agent 向招標商報價

Central Trust of China

Trading Department

Dear Sirs,

Re: Invitation No. CTCTD 123

In response to the captioned Invitation and subject to all the terms and conditions thereof, we are pleased to offer firm 10,000 metric tons, 5% more or less, of US grade No. 2 yellow soybeans for account of Cargill Inc., New York valid until 4:00 p.m. October 9 at price quoted and on conditions as stipulated in our bid. Enclosed please find our bid in duplicate for your evaluation. Under separate cover, we are submitting samples of yellow soybeans which were received from our principals.

We shall appreciate it very much if you will give our bid your favorable consideration.

Faithfully yours,
Taiwan Trading Co., Ltd.

No. 337　招標商通知 Agent 已得標

Taiwan Trading Co., Ltd.

Dear Sirs,
Subj: Your Bid No. NY–123
Our Invitation No. CTCTD 123

We are pleased to inform you that contract has been awarded to you for the 10,000 M/T, 5% more or less, yellow soybeans under the subject invitation and shall appreciate your advising the award to your principals and asking them to post performance bond within 10 working days after the date of this notice.

We are awaiting your immediate confirmation.

Faithfully yours,
Central Trust of China
Trading Department

【註】

principals：與 agent 相對之詞，法律上稱為「本人」，這裡是指 "seller" 或 "supplier" 而言。得標之後，投標商一方面與招標商訂約，一方面須繳存履約保證金。

No. 338　Agent 告知招標商已將得標消息轉告 Principal

Central Trust of China
Trading Department

Subj: Invitation No. CTCTD 123

Dear Sirs,

We acknowledge with thanks the receipt of your Notice of Award for 10,000 metric tons, 5%

more or less, of yellow soybeans and wish to inform you that, right upon receipt thereof, we have despatched the following fax to our principals in New York:

RE CTCTD'S INVITATION NO 123 CONTRACT AWARDED 10,000 M/TS 5% MORE OR LESS TO US PLS ARRANGE POSTING P-BOND EQUAL 5% OF CONTRACTED VALUE WRITING

We will write you again when we are informed that the performance bond has been posted.

Yours faithfully,
Taiwan Trading Co., Ltd.

【註】

　　電文中的 Telexese：p-bond: performance bond。

No. 339　　Agent 通知 Principal 得標，並請繳履約保證金

Cargill Inc.
New York, NY

Dear Sirs,　　　　　　　　Re: Your Bid No. NY-123
　　　　　　　　　　　　CTC's Invitation No. CTCTD 123

This confirm our fax dispatched to you today regarding to the captioned bid.

We are pleased to inform you that we have, on your behalf, submitted your bid to CTC, Trading Department for 10,000 M/Ts (±5%) yellow soybeans with a total value of US$2,200,000, and the contract has been awarded to us.

According to CTC's regulations, the suppliers to whom a contract is awarded should post performance bond equivalent to 5% of the contract value with them by cash, L/C, or L/G. Therefore, you are requested to post the bond with them on or before October 19. Meanwhile, please let us be informed as soon as the performance bond has been posted.

One copy of the original contract has been airmailed to you today for your reference and perusal.

Faithfully yours,
Taiwan Trading Co., Ltd.

【註】

　　1. supplier：供應商。

　　2. post：繳存。

No. 340　　Principal 通知 Agent 已繳履約保證函

Taiwan Trading Co., Ltd.

Dear Sirs,　　　　　　　　Re: CTC's Invitation No. CTCTD 123

　　We are very much pleased to learn from your fax and your letter both dated October 9 that the contract for 10,000 M/Ts yellow soybeans has been awarded to us and appreciate your efforts to obtain it. We have also received the original copy of the contract to which we have given our careful attention.

　　As requested, we asked the Chemical Bank, New York, on October 16, to issue a letter of guarantee for US$110,000 through CTC, Foreign Department as our performance bond to guarantee our performance under the contract. Enclosed please find one copy of letter of guarantee for your reference. Please pass this information on to CTC, Trading Department for their reference.

　　We are now going to start preparing the commodity and trust that we can deliver them within the period as stipulated in the contract.

Yours faithfully,
Cargill Inc.

　　Performance bond 為賣方與買方簽約時，為防賣方無力履行契約，而要求賣方繳存的保證金。在國內買賣，多用現款或票據繳存，但在國際貿易則多以 Bank letter of guarantee 或 Stand-by L/C 代替，如賣方不履約，則買方可沒收 Performance bond。

No. 341　　中信局規定的履約保證標準格式

(SPECIMEN)
STANDARD PERFORMANCE BOND FORM

Date of Issuance _____

Letter of Guarantee No. _____

for (Name of Currency & Amount in Figure) as
Performance bond
for Your Contract No. _____
under Your Invitation No. _____

Advised Through
(Name of Issuing Bank's Correspondent Bank in Taipei)

To:
Central Trust of China
Trading Department
49, Wu Chang Street, Sec. 1

Taipei, Republic of China

Upon request of Messrs (name of Accountee), we hereby agree to issue this irrevocable letter of guarantee up to the aggregate amount of (name of currency & amount in words) only in your favor for account of (name of Accountee) as (Name of Contractor)'s Performance bond for your Contract No._____ & amendment(s), if any, under your Invitation No._____ and engage ourselves to pay immediately to you without recourse upon receipt of your written statement certifying that (name of Contractor) have failed to fulfil the contractual obligations, accompanied by your simple receipt(s) or sight draft(s) drawn on (Issuing Bank or Accountee to be indicated by Issuing Bank).

We bind ourselves to obtain necessary permit from relevant authorities to transfer the claimed amount to you in case it is confiscated.

This letter of guarantee will expire on (date of expiration—120 days after contract specified shipment date) Taiwan time.

Authorized Signature

　　由招標而投標以至簽約，文書作業 (Paper work) 甚繁雜，為簡化起見，可以將招標 (Invitation)、投票 (Bid) 以及契約 (Contract) 合併成一定的格式，稱為 "Invitation, Bid and contract"，將有關各種條件或規定，印妥備用，甚為方便。

　　茲將中信局購料處所用的格式列示於下頁：

No. 342   Invitation, Bid and contract

FORM B
Bid Bond is required

INVITATION, BID & CONTRACT

INVITATION NO. _____   CONTRACT NO. _____
CENTRAL TRUST OF CHINA. PROCUREMENT DEPARTMENT. 49, WU CHANG ST. SEC. 1. TAIPEI. 100, REP.
OF CHINA, CABLE ADDRESS: TRUSTPRO TAIPEI, TELEX: 11377 CENTRUST, ON BEHALF OF THE CLIENT
_____   ADDRESS: _____
                                                            DATE _____

INVITATION
         Sealed bids subject to the instructions and conditions on the attached sheets will be received at this office un-
til _____ o'clock, _____ and then publicly opened for furnishing the following supplies from _____
_____ for shipment to _____ on or before _____
                    CENTRAL TRUST OF CHINA, PROCUREMENT DEPARTMENT

Bids to be given on _____ basis valid for ____ days.   MANAGER

ITEM/ CODE NO. (1)	DESCRIPTION OF SUPPLIES (2)	QUAN- TITY/ UNIT (3)	UNIT COST FOB/ FAS (4)	TOTAL COST FOB/ FAS (5)	UNIT OCEAN FREIGHT (6)	TOTAL OCEAN FREIGHT (7)	INSUR- ANCE PREMIUM (8)	TOTAL CFR/ CIF (9)
		GRAND TOTAL:						

Insurance _____
Inspection by _____
L/C Opening Bank _____

BID   DATE _____
         In response to the above invitation and subject to the instructions and conditions thereof, the undersigned offers and
agrees, if this bid be accepted within days from the date of the opening, to furnish any or all of the items, upon which
prices are quoted, at the price set opposite each item and deliver at the point(s) specified in accordance with the delivery
schedule as shown below or in the attached continuation sheet.
Shipment on or before _____
To be shipped by Liner Vessel/Tramp Steamer/Parcel Post/Airlift _____
Port of Shipment _____ Source of Origin (indicate country) _____
Name of L/C Beneficiary _____
Address _____
Supplier _____   Address _____
Manufacturer _____   Address _____
Bidder _____
Address _____
Telephone No.   Telex or Cable address   By _____
                                    (Signature and title of person authorized to sign)

Opening of bids witnessed by		
CTC	Auditor	Client

CONTRACT
ACCEPTED BY CENTRAL TRUST OF CHINA   DATE _____
1. Accepted as to items numbered: _____   CENTRAL TRUST OF CHINA
        Total Price _____   PROCUREMENT DEPARTMENT
2. Shipment on or before _____
                to _____
3. Inspection by _____
4. Markings to include " _____ "
   in addition to those required in Article 13 of attached Conditions.   MANAGER
5. "On deck" B/L acceptable. Transhipment allowed.
6. Partial shipments allowed, but not more than shipments.

No. 343　Agent 通知招標商已繳履約保證金

Central Trust of China
Trading Department

Dear Sirs,　　　　　　　　　Re: Invitation No. CTCTD 123
　　　　　　　　　　　　　　Contract No. 321

　　With reference to captioned contract, we are pleased to inform you that we have just received a letter from our principals, Cargill Inc., informing us that they have requested Chemical Bank, New York, to issue a letter of guarantee in your favor through CTC, Foreign Department as the performance bond required under the subject contract. Enclosed please find one photostatic copy of said letter of guarantee. The amount posted is US$110,000, which is equivalent to five percent of the total contract value of US$2,200,000.

　　Please also be informed that our principals have already started processing this contract and promised to ship the soybean as stipulated in thereof.

　　Your attention to the above will be appreciated.

　　　　　　　　　　　　　　　　　　　　　　　　Faithfully yours,
　　　　　　　　　　　　　　　　　　　　　　　　Taiwan Trading Co., Ltd.

No. 344　Agent 通知招標商裝船事宜

Central Trust of China
Trading Department

Dear Sirs,　　　　　　　　　Re: Contract No. 321
　　　　　　　　　　　　　　10,000 M/Ts Yellow Soybeans

　　Since the captioned soybeans is sold on an FOB ex spout Gulf Port basis, we are giving hereunder all the necessary information in order to enable you to declare to your marine insurance company:

Invitation No. CTCTD 123
Contract No. 321
Commodity: US Grade No. 2 Yellow Soybean, 20– Crop
Quantity: Ten Thousand (10,000) Metric Tons
Name of Ship: s.s. "Union Faith"
Port of Departure: Houston, U.S.A.
ETD: November 2, 20–
ETA: December 10, 20–
Total Invoice Value: US Dollars Two Million Two Hundred Thousand (US$2,200,000) Only.

Please be informed accordingly.　　　　　　　　　Yours faithfully,
　　　　　　　　　　　　　　　　　　　　　　　　Taiwan Trading Co., Ltd.

No. 345　Agent 將貨運單證副本轉交招標商

CTC, Trading Dept.

Dear Sirs,　　　　　　　　　　　Re: Contract No. 321
　　　　　　　　　　　　　　　10,000 M/TS Yellow Soybeans

　　We are pleased to inform you that our principals have effected shipment of the captioned yellow soybeans per s.s. "Union Faith" sailed from Houston on November 3.

　　For this shipment, we are transmitting to you two complete sets of the following non-negotiable shipping documents, as required in the captioned contract, for your records:

　　Detailed Invoice.
　　Bill of Lading.
　　Official Grade Certificate.
　　Weight Certificate.
　　Official Phytosanitary Export Certificate.
　　Official Certificate of Origin.
　　Copy of Telex Despatched to You.
　　Suppliers' Certificate.

　　We trust that you will find all of the above in order.

　　　　　　　　　　　　　　　　　　　　　Faithfully yours,
　　　　　　　　　　　　　　　　　　　　　Taiwan Trading Co., Ltd.

No. 346　Agent 要求招標商發還履約保證函

CTC, Trading Dept.

Dear Sirs,　　　　　　　　　　　Re: Contract No. 321
　　　　　　　　　　　　　　　　Letter of Guarantee

　　We refer to our letter to you of October 21, 20– informing you that our principals, Cargill Inc., have posted the performance bond in letter of guarantee form as required under the subject contract.

　　Now, the yellow soybeans have arrived and been inspected to the fullest satisfaction of yourselves. In view of the fact that the supplier have performed satisfactorily the obligations under the contract, you are requested to release the performance bond as early as possible.

　　Your compliance with our request will be appreciated.

　　　　　　　　　　　　　　　　　　　　　Yours faithfully,
　　　　　　　　　　　　　　　　　　　　　Taiwan Trading Co., Ltd.

【註】

1. have performed satisfactorily the obligation under the contract = have fulfilled satisfactorily the contractual obligations。

2. release：解除；"release the performance bond" 為解除履約保證之意，也即發還履約保證函之意。

第三十一章
貿易電傳(Business Teletransmission)

做國際貿易，由於進出口雙方遠隔兩地，所以利用電傳 (Teletransmission)，即電報 (Cable)、電報交換 (Telex)、電話傳真 (Telefacsimile，簡稱 Fax) 或電子郵件 (E-mail) 交易的情形相當普遍。諸如詢價、報價、還價、簽約，乃至交涉索賠等等，都可以利用電報、電報交換、電話傳真、電子郵件進行。在此情形下，書信則往往不過是用來確認或補充說明而已。在國際貿易用 Cable 或 Telex 往往僅以寥寥數十語，即可做成數千百萬元的交易，其貢獻，不可謂不大。

然而國際電報或電報交換，費用較貴，假如對於電文的撰寫不得要領，則不但交易不能順利達成，而且還浪費金錢。因此，要成為一個適任的貿易從業員，必需對於電報及電報交換，尤其電報文體 (Cablese) 及電報交換文體 (Telexese) 應具有相當的知識。

第一節　國際電報

「電報」英文寫成 "Telegram" 此字原先用於陸上傳遞的電報，至於以海底電纜傳遞的電報則稱為 Cablegram 或簡稱為 Cable。現在無線電發達，所以無線電報則稱為 Radiogram。雖然，這些字各有不同意義，但時至今日，Telegram, Cablegram, Cable 及 Radiogram 等字，大家都不再作嚴格的區分，而彼此交替使用。

國際電報 (International telegram) 是指國與國之間的電報往來而言。從事貿易的人士，對於國際電報的有關知識，如能諳熟，對於業務的發展，必有很大的助益。

一、國際電報的用語

國際電報的用語可分為明語 (Plain words, Plain language) 和密語 (Code words, Secret language) 兩種。二者可單獨使用，也可混合使用。

1.明語：是以國際電報通訊所准用的各國文字繕成而且每字及每一辭句均保存其

所屬文字的原來意義者。凡電文及署名完全用明語書寫的，稱為明語電報。明語電報內有下列各種情形時，亦作明語論：

①用字母或阿拉伯數字書寫的數目，而無秘密意義的。

②電報掛號字樣。

③商務通訊上習用的名詞及編號而無秘密意義的。

④電文第一字母所用的對號（查對、押碼）字或數目。

2.密語：是由文字或數字加以組合而成，每個字都有特定的意義，但已失去文字的原來意義。

二、國際電報的種類

國際電報按處理緩急，可分為下列三種：

1.尋常電報 (Ordinary or Full-rate telegrams)：是一種快遞電報，電報局收到拍發電稿後，必須立即遞往國外目的地。這種電報優先拍發的權利，僅在加急電 (Urgent telegram) 之後。尋常電須照全費率 (Full-rate) 計算，所以無納費標識。我國規定報費最少以 7 個字起算。經營國際貿易的商人常在日間利用這種快電拍往同一時區 (Time belt) 之外國顧客，使對方在營業時間內收到電報，並能在當日營業時間內獲得覆電。尋常電文可用明語、密語或兩種混合者。

2.加急電報 (Urgent telegrams)：加急電為尋常電的加速，享有優先傳遞及投送之權，報費按尋常電加倍收，以 7 個字起算，在收報人姓名地址之前加 "Urgent" 納費標識，並作一字計算。但須注意有些地區及國家不適用加急電。

3.書信電報 (Letter telegrams)：凡電文冗長而不急於傳遞的電報，多用書信電。這種電報投遞較遲，但費用較廉，故在商業通訊上應用最廣。書信電報最為經濟，在我國按尋常電減半收費，並規定最少以 22 字起算。須在收報人姓名地址前加註 "LT" 納費標識，並作一字計算。

茲將三種電報收費比較如下：

電報種類	電報種類標識	費用比	最低收費字數	用語限制
加急電報	URGENT	2	7	明・密
尋常電報	–	1	7	明・密
書信電報	LT	1/2	22	明

三、電報字數的計算方法

1. 業務標識、收報人姓名地址、收報局名、電文及署名，概按 10 個字碼以內作 1 個計費字數計算，每逾 1 至 10 個字碼，加算一個字。
2. 明語電報的電文各字，應按照各種明語文字的標準字典所載單字書寫，其他各種不規則的縮寫或湊合字，應隔開分別計字。

四、電報實例說明

ZCZC JAF846 PTM672 LHB18 ①
CHIA HL JATKO23 ②
TOKYO ③ 27/26 ④ 12 ⑤ 1530 ⑥
LT ⑦
⑨
20– JUNE 12 PM 3 40
TAICO ⑧
TAIPEI ⑩
PLEASE OFFER ⑪ BEST CFR YOKOHAMA 10000DOZ LADIES UMBRELLAS AS PER YOUR SAMPLE MAY10 PAYMENT SIGHT LC JULY AUGUST SHIPMENT CABLE REPLY JATRA ⑫
COLL 10000 ⑬

茲依編號順序說明於下：

①電報經由路線
②呼號
③發報地
④電報字數：實際數字為26字，收費數字為27字，因為10,000 Doz 應算2個字
⑤發報日期：本月 12 日
⑥發報時間：下午 3 點 30 分
⑦電報種類：LT

⑧收報人電報掛號
⑨收報時間：收報局收到電報時間，通常用橡皮戳表示
⑩收報地
⑪電文
⑫發報人電報掛號
⑬電文複核：凡電文中涉及阿拉伯數字或發電可能錯誤的文字，再行核對一遍

五、撰寫電報文體 (Cablese) 的要領

1.第一、二人稱代名詞可省略:

書信體	電報體
We confirm having accepted your order No. 10 dated October 5.	YOURS 5 ORDER NO 10 ACCEPTED

2.省略標點符號: 非絕對必要時,標點符號不必使用,尤其地址中的符號為然。

書信體	電報體
454, Market Street, San Francisco, California, USA	454　MARKETSTREET　SANFRANCISCO CALIFORNIA

3.省略縮寫字及 Apostrophe 的符號: 縮寫字及 Apostrophe 的符號是單獨計算為一字,因此應省略。

書信體	電報體
Mr.	MR
Mrs.	MRS
Don't	DONT
Can't	CANT
F.O.B.	FOB
U.S.A.	USA

4.省略冠詞: 冠詞如 a, an, the 均可省略。

書信體	電報體
Send us a sample.	SEND US SAMPLE

5.省略介系詞:

書信體	電報體
In compliance with your cable of 12th, we have opened L/C today.	YOURS 12TH OPENED LC TODAY

6.省略助詞: 如 will, shall, can, have been 等可予省略。

書信體	電報體
We shall send you the letter of confirmation.	CONFIRMATION LETTER FOLLOWS

7. 以 "un-"，"in-"，"dis-" 等接頭詞 (Prefix) 簡化否定詞。

書信體	電報體
(1) We are not interested in it.	(1) UNINTERESTED
(2) We cannot obtain offers.	(2) OFFERS UNOBTAINABLE
(3) Goods have not been sold.	(3) GOODS UNSOLD

8. 以 "-able" 代替可能性，許多表示可能性的字，如 "can"，"possible" 等字，都可在適當動詞之後加 "-able"，而有同樣意義。

書信體	電報體
(1) We can accept your terms.	(1) TERMS ACCEPTABLE
(2) We can obtain orders.	(2) ORDERS OBTAINABLE

9. 用命令式：

書信體	電報體
Please inform us by cable whether you have opened L/C.	CABLE WHETHER LC OPENED

10. 以現在分詞表達未來的行動：

書信體	電報體
(1) We shall effect shipment on July 10.	(1) SHIPPING JULY 10
(2) We shall write for the result.	(2) RESULT WRITING

11. 以被動代替主動：

書信體	電報體
We have obtained import licence.	IL OBTAINED

12. 以單字代替數個字：

書信體	電報體
Do not agree	DISAGREE
Do your best	ENDEAVOUR
At your earliest convenience	SOONEST
Send by air mail	AIRMAIL
By cable	TELEGRAPHICALLY
Letter of Credit	LC
Per piece	APIECE
On condition that	PROVIDED
As soon as possible	SOONEST
Be put off	SUSPENDED
In the middle of May	MIDMAY
At the end of April	ENDAPRIL
Will send you a letter	WRITING
Speed up	EXPEDITE
In spite of	DESPITE

13.將過去分詞放在句末以省略動詞：

書信體	電報體
(1) We would like to ask you to effect shipment promptly.	(1) PROMPT SHIPMENT REQUESTED
(2) We look forward to hearing from you as early as possible.	(2) EARLY REPLY SOLICITED

14.地址使用電報掛號：如 "Central Trust of China, 49, Wuchang Street, Section 1, Taipei" 共有 10 個字，如使用電報掛號 "CENTRUST TAIPEI" 則只有兩個字。

15.數字 (figure) 宜予拼出 (spell out)：

書信體	電報體
We offer 100 dozen	OFFER ONEHUNDRED DOZ

六、貿易電報英文實例

1.詢價：

書信體	電報體
We are informed that you are the sole agent for	INFORMED YOU ARE AGENT FOR NING

書信體	電報體
Ningta Bicycles and we shall be pleased if you will send us your price list and state your best terms.	TA BICYCLES STOP SEND PRICELIST WITH BEST TERMS

2.寄型錄及價目表：

書信體	電報體
We have sent you today our illustrated catalog of the bicycles suitable for your area together with the price list.	SENT TODAY BICYCLES CATALOG SUITING YOU WITH PRICELIST

3.寄樣品：

書信體	電報體
In comliance with your request of May 15, we have sent you today our samples under separate airmail cover. Shipment will be effected within 30 days after receipt of your L/C. Other terms and conditions remain unchanged.	YOURS FIFTEENTH AIRMAILED TODAY SAMPLES SHIPMENT THIRTY DAYS AFTER RECEIPT L/C OTHERS UNCHANGED

4.請報穩固價：

書信體	電報體
With reference to your letter of March 10, please make your best possible firm offer for Taiwan Refined Camphor Tablets 1/8 oz and 1/4 oz for earliest shipment indicating the quantity available at present.	YOURS TENTH CABLE OFFER EARLIEST SHIPMENT CAMPHOR TABLETS ONEEIGHTH ONEQUATER

【註】①如雙方對於數量、金額的單位都已明瞭，則電文中可省略。

　　②"indicating the quantity available at present" 電文中所以未表明，乃因既要求 firm offer，則在 firm offer 中自然會表明，所以電文中不必寫出。

5.報穩固價：

書信體	電報體
With reference to your cable of March 19, we offer firm subject to immediate reply 100 cases each of Taiwan Refined Camphor Tablets 1/8 oz and 1/4 oz, each cases containing 100 1bs.,	OFFER ONEHUNDRED CASES EACH CAMPHOR ONEEIGHTH EIGHTYTHREE ONEQUARTER SEVENTY FIVE CENTS POUND CIF SANFRANCISCO EACH CASE

at $0.83 and $0.75 respectively CIF Sanfrancisco, shipment early April, packing and other particular as per letter. If you accept, please cable irrevocable L/C immediately.	ONEHUNDRED POUNDS EARLY APRIL PACKING AND PARTICULARS SAME OUR LETTER SUBJECT CABLING LC PROMPTLY

6.報穩固價：

書信體	電報體
With reference to your letter of June 5, we are pleased to offer firm for your reply reaches here by August 1, 1,000 dozen of pullovers at US$15 per dozen CFR New York, for December shipment.	YOURS FIFTH OFFER FIRM SUBJECT REPLY HERE AUGUST 1 ONETHOUSAND DOZ PULLOVERS FIFTEEN USDOLLARS DOZ CANDF NEWYORK DECEMBER SHIPMENT

7.報非穩固價：

書信體	電報體
In compliance with your cable of July 20, we are pleased to make you an offer subject to our final confirmation as follows: commodity: sport shirts, style A. quality: as per sample submitted to you on July 10. quantity: 500 dozen only. price: US$20.34 per doz CIF New York. packing: export standard packing. insurance: All risks plus war. shipment: during October. payment: by L/C which must reach us by the end of September. exchange risks: for your account.	YOURS TWENTIETH OFFER SUBJECT CONFIRMATION 500 DOZ SPORT SHIRTS STYLE A SAMPLED JULY TENTH USD 20.34 PER DOZ CIF NEWYORK COVERING AR AND WAR EXPORT STANDARD PACKING OCTOBER SHIPMENT LC MUST REACH US END SEPT EXCHANGE RISK YOUR ACCOUNT

8.發出訂單：

書信體	電報體
With reference to your offer of July 20 we are pleased to send you our order for 500 dozen of sport shirts style A, at US$20.34 per dozen CIF New York.	YOURS TWENTIETH ORDER 500 DOZ SPORT SHIRTS STYLE A USD 20.34 PER DOZ CIF NEWYORK

9.催開信用狀：

書信體	電報體
Referring to your order for camphor tablests, we have not yet received your letter of credit. Please open it urgently. Otherwise, we can not ship the goods in time, as the steamer is leaving within a week.	CAMPHOR LC UNRECEIVED STEAMER LEAVING WITHIN WEEK EXPEDITE OTHERWISE UNSHIPPABLE

10.請求延期裝運：

書信體	電報體
In view of longshoremen strike which is exactly a case of force majeure, we regret there is no alternative but to request you to kindly agree that our shipment of your order No. 123 to be delayed for a period of 30 days.	OWING LONGSHOREMEN STRIKE REQUEST AGREE ORDER NO 123 DELAYING SHIPMENT 30 DAYS

11.通知裝運：

書信體	電報體
We have shipped today 200 cases of camphor tablets per s.s. "Pacific Transport" which is scheduled to sail from Keelung July 10 and to arrive at Kobe July 15.	SHIPPED CAMPHOR TWOHUNDRED TODAY PACIFIC TRANSPORT ETD KEELUNG JULY 10 ETA KOBE JULY 15

12.通知已付保：

書信體	電報體
In compliance with your request, we have effected the insurance with China Insurance Company Ltd. for US$11,000 on 1,000 dozen of garment. The copy of which has been sent to you.	AS REQUESTED INSURED USDOLS ELEVENTHOUSAND ONETHOUSAND DOZ GARMENT COPY POLICY SENT

13.索賠：

書信體	電報體
We have taken delivery of the garment you shipped against our order No. 123. However, upon checking, we have found the quality is	REGRET GARMENTS UNDER ORDER 123 QUALITY MUCH INFERIOR TO SAMPLE STOP AIRMAILED YOU SAMPLE

much inferior to the sample you submitted to us on July 10. One piece of the sample is air-mailed to you under separate cover.	

14.覆索賠：

書信體	電報體
We are surprised to hear that the garments we shipped to you against your order No. 123 are much inferior to the sample we submitted to you on July 10.	SURPRISED GARMENTS UNDER ORDER 123 MUCH INFERIOR TO SAMPLE

15.覆已延展 L/C：

書信體	電報體
With reference to your letter of July 20, we have requested our bankers to extend L/C No. 123 for shipping and expiry dates to December 15 and 25 respectivly.	YOURS TWENTIETH EXTENDED LC 123 SHIPPING AND EXPIRY DECEMBER FIFTEENTH AND TWENTYFIFTH

16.查問何時裝運：

書信體	電報體
Please advise immediately when you will ship our order No. 123.	ADVISE WHEN SHIP ORDER 123

17.催裝運：

書信體	電報體
The marked here is so active that our stocks are almost exhausted. Please send the goods as soon as possible.	MARKET ACTIVE STOCKS EXHAUSTED SHIP IMMEDIATELY

第二節 國際電報交換

一、概 說

貿易廠商為了節省時間、費用，避免到電信局拍發電報的麻煩，可在自己辦公室裡利用預先裝就的電傳打字機 (Teletypewriter) 和國外裝有同樣設備的客戶，直接交換信息。這種國際間利用電傳打字機的交換電信，稱為電報交換，英文為 Teletypewriter exchange，簡稱為 Telex 或 TLX。

需要利用 Telex 的用戶，要事先向國際電臺申請裝設專線和 5 單位啟閉式電傳打字機全套，由國際電臺列入電報交換用戶，並編列號碼（猶如電話簿）。這編號稱為呼叫號碼 (Call number)，另由用戶選定 3 至 8 個英文字母當作回呼電碼 (Answer-back code)。呼叫號碼及回呼電碼一併列入國際電報交換號碼簿後，就可經掛號手續，與國外 Telex 用戶直接通報。Telex 對於電報數量較多的公司行號，既經濟及便利。

二、Telex 的優點

1.自動紀錄：憑電傳打字收發電信、電文內容可自動紀錄下來，並可產生副本。

2.直接筆談：可以打字方式和對方直接交談，與電話通話相似。

3.可進行不在現場通訊：如電傳打字機無人看管時，也可自動收錄電文，對於時差較大的國際通訊，尤屬便利。

4.費用低廉：Telex 按通訊時間計費，較電報按字計費者，低廉甚多。冗長電文使用 Telex 尤為經濟。

5.通信文體自由：明語、密語、簡體字均可使用。

6.可在辦公室內直接收發。

7.可與無 Telex 設備的客戶通訊：電報交換用戶需要拍發普通電報至國外尚未裝設電報交換機 (Telex set) 客戶時，可利用 Teletypewriter 先叫通國際電臺，再由其將電文轉發國外收報地點，這種電信叫做專線電報 (Private Tieline, PTL)。

三、Telex 呼叫接線的種類

Telex 呼叫的接線種類，視各被呼叫用戶所在國（或所在地）的交換設備情形，可分為下列二類：

1. 人工交換接線 (Manual operation)：即呼叫用戶先將被呼用戶的 Telex 呼叫號碼 (Call number) 及所在地名向國際電臺 Telex 席值機員掛號後，由電臺值機員請被呼叫用戶的交換值機員將被呼叫用戶接出後，相互通報。

2. 全自動交換接線 (Fully automatic connection)：呼叫用戶可直接選撥國外被呼叫用戶的 Telex，呼叫號碼後，即可直接將被呼叫用戶接出，直接通報。

四、Telex 費率的計算方式

Telex 的計費方式有二種。

1. 3 分 1 分制：與未開放全自動交換作業系統區域通報時採用。每次基本計費時間為 3 分鐘，未滿 3 分鐘也以 3 分鐘計算，如超過 3 分鐘，則超過部分以 1 分鐘為計費單位，不足 1 分鐘也以 1 分鐘計費。

2. 6 秒 6 秒制：與已開放全自動交換作業系統區域的通報時採用。每次基本計費時間為 6 秒鐘，未滿 6 秒鐘者以 6 秒鐘計算，超過 6 秒鐘者，超過部分也以 6 秒鐘為計費單位，不足 6 秒鐘，以 6 秒鐘計費。

五、撰寫電報交換文體 (Telexese) 的要領

Telex 係按時間計費，因此 Telex 在文體方面，應力求簡潔，譬如在 Cable, Accumulate（1 字）比 Heap up（2 字）有利，但在 Telex 則用 Heap up 反而划得來。因此，乃有 Telexese 的產生。茲將撰寫 Telexese 的要領，列舉於下：

1. 省略主詞、助詞、介系詞、冠詞。

2. 以現在分詞代替未來式。

3. 以形容詞 "-able" 代替可能動詞。

4. 將主動改為被動。

5. 以 "un-"，"in-"，"mis-"，"dis-" 表示否定。

　　以上可參閱 "Cablese" 部分。

6.使用 Telex 用略語，例如：

As Soon As Possible → ASAP

Refer to Your Letter → RYL

Refer to Our Letter → ROL

7.單字簡縮。

因 Telex 係按時計費，所以應力求「字要簡短」，「字數要少」，以下就單字簡縮要領加以說明。

⑴略去母音字母，保留重要子音字母，以及最後一個子音字母。但第一個字母一律保留

Accept	ACPT
Cable	CBL
Dollars	DLS
Freight	FRT
Limited	LTD
Manger	MGR

⑵保留第一音節

①保留第一音節及第二音節第一個子音字母

Answer	ANS
Captain	CAPT
Document	DOC
Exchange	EXCH

②保留第一音節及第二音節

Avenue	AVE
Memorandum	MEMO
Negotiation	NEGO

③保留第一音節及第二、三音節第一個子音字母

Approximate	APPROX
Immediately	IMMED

Manufacturer   MANUF

④保留第一音節及後面重要子音字母

Airfreight   AIRFRT

Consignment   CONSGT

Department   DEPT

Government   GOVT

⑶保留最後一個音節全部，其他保留重要子音字母或全部子音字母（重音在最後音節者，通常按此法簡寫）

Addressee   ADRSEE

Cancel   CCEL

Guarantee   GTEE

Transfer   TRSFER

⑷保留首尾兩字母

Bag   BG

Bank   BK

Debtor   DR

From   FM

⑸音譯

Are   R

Business   BIZ

Light   LITE

New   NU

You   U

⑹過去分詞字尾 "ed" 以 "d" 代替

Arrived   ARVD

Confirmed   CFMD

Included   INCLD

Received   RCVD

⑺以 "g" 代替字尾的 "ing"

Airmailing	AIRG
Establishing	ESTABG
Manufacturing	MFG
Packing	PKG

六、貿易電報交換英文實例

1. Message:

Please open L/C immediately at contract price, otherwise we cannot apply for export license.

Cablese:

PLEASE OPEN LC PROMPTLY AT CONTRACT PRICE OTHER WISE EL UN-AVAILABLE

Telexese:

PLS OPN LC IMMDLY AT CONT PRC OZWS EL UNAVLBL

註：OZWS = otherwise。

2. Message:

We have not yet received the shipping documents. Please investigate and reply.

Cablese:

DOCUMENTS UNRECEIVED CHECK AND REPLY

Telexese:

DOC UNRCVD PLS CHCK N RPL

註：CHCK = check; N = and。

3. Message:

Please inform us by cable whether import license is obtained or not.

Cablese:

CABLE WHETHER IL OBTAINED (OR CABLE IL SITUATION)

Telexese:

TLX WHZ IL OBTND

註：WHZ = whether; OBTND = obtained。

4. Message:

We are pleased to inform you that we have accepted your order of April 16th. Therefore, please open letter of credit immediately and airmail the design sample to us.

Cablese:

YOURS 16TH ACCEPT PLEASE OPEN LC PROMPTLY AND AIRMAIL DESIGN SAMPLE

Telexese:

YS 16TH ACPT PLS OPN LC IMMDLY Y AIR DSN SMPL

註：YS = yours; Y = and; DSN = design; SMPL = sample。

5. Message:

With reference to your order No. 105, we have not yet received the design sample. Since unless the design sample arrives here by the end of May, June shipment will be impossible, would you please approve July shipment and open your L/C accordingly.

Cablese:

ORDER 105 DESIGN UNRECEIVED UNLESS IT ARRIVES END MAY JUNE SHIPMENT IMPOSSIBLE PLEASE APPROVE JULY AND OPEN LC ACCORDINGLY

Telexese:

ODR 105 DSN UNRCVD UNLESS IT ARV ENDMAY JUNE IMPSBL PLS APPR JULY N OPN LC ACDGLY

6. Message:

Messrs. Jones & Co., the buyer of order No. 105, informed us that the printed shirting you shipped on this order was not quite up to your usual standard.

The finish was not bright enough and the color was slightly different from the design sample. Though they did not demand any compensation, it would be advisable for us to do something about it. Since they are very important customers to us both, your kind consideration would be appreciated.

Cablese:

ORDER 105 BUYERS JONES AND COMPANY COMPLAIN FINSH DULL AND COLOR SLIGHTLY DIFFERENT WHAT CAN YOU DO JONES BEING MUTU-

ALLY IMPORTANT CUSTOMERS

Telexese:

ODR 105 BUYERS JONES N CO COMPLAIN FINISH DULL N COLS SLIGHTLY DIFFERENT WHAT CAN U DO JONES BEING MUTUALLY IMPTNT CUS-TOMERS

7. Message:

Please refer to our letter of October 12, concerning difficulty of allowing discount as requested. However, we have obtained a five percent reduction from the maker although the market here is advancing. They emphasized such reduction would not apply to the shipments made after November.

Cablese:

OURLET OCT 12 CONCERNING DISCOUNT DIFFICULTY MAKER ALLOWED SIX PCT BUT UNAPPLICABLE FOR SHIPMENTS AFTER NOVEMBER

Telexese:

OURLET OCT 12 ABT DISCOUNT DIFFICULTY MAKER ALLOWED 6% BUT UNAPLYCBL SHIPTS AFTER NOV

七、常用 Telexese

A			
About	ABT	Account	A/C, ACCT
Accept(ed)	ACPT(D)	Acknowledge(d)	ACK(D)
Accordingly	ACDGLY	Application	APLCTN
Additional	ADDL	Applicable	APLICBL
Addition	ADDN	Advise	ADV
Agree	OK	Already	ALRDY
All right	OK	Approximate	APPROX
Amount	AMT	Approve	APPR
And	N	Airmail	AIR
April	APR	Airmailing	AIRG
Arrange	ARRNG	Airmailed	AIRD
Arrive(d)	ARV(D)	August	AUG
Attention	ATTN	Average	AVRG
Altered	ALTRD	Answer	ANS

		As soon as possible	ASAP
B			
Balance	BALCE	Board Measurement	BM
Bank	BK	Bank Draft	B/D
Bleached	BLCHED	Beginning	BEG
Between	BTWN	Bill of Lading	B/L
C			
Cancellation	CANCELN	Cancelling	CCELG
Cancel(led)	CCEL(D)	Cash on delivery	COD
Can not	CANT	Check	CK
Care of	C/O	Charter Party	C/P
Carton	CTN	Commission	COMM
Charge(d)	CHRG(D)	Corporation	CORP
Colors	COLS	Counter Offer	C/OFFER
Company	CO	Cubic foot	CFT
Confirm(ed)	CFM(D)	Credit	CR
Contract	CONT	Correct	CRT
Confirmation	CFMTN		
D			
Days	DS	Delivered	DELVD
Debit	DR	Department	DEPT
Deliver	DLV	Destination	DEST
Delivery	DLVY	Double	DBL
Delivery order	D/O	Documents against Payment	D/P
Demand draft	D/D		
Difference	DIFFRNE	Documents against Acceptance	D/A
Document	DOC		
Dozen	DOZ	Dollars	DLS
December	DEC		
E			
Each	EA	Estimated time of arrival	ETA
End October	ENDOCT	Estimated time of departure	ETD
Enquiry	ENQRY		
Exchange	EXCHG	Encluding	ENCLDG
Export	EXP	Export licence	EL
F			
Factory	FCTRY	February	FEB
Fearing	FEARG	Feet	FT
Figure(s)	FIG(S)	Following	FOLG
Flight	FLT	Forwarded	FORWDD

Forward	FORWD, FWD	Friday	FRI
Freight	FRT		

G

Gallon	GAL	Guarantee	GURANTE
General	GENRL	Government	GOVT
Good	GD	Greasy	GRSY

H

Hamburg	HAM	How	HW
Highway	HIWAY	However	HWEVR
Hour	HR	Heavy	HVY

I

Immediate(ly)	IMMD(LY)	Including	INCLDG
Import	IMPT	Information	INFMTN
Include(d)	INCLD(D)	Instructions	INSTRCTNS
Import licence	IL	Instead of	I/O
Inform(ed)	INFM(D)	Invoice	INV
Instead	INSTD	Irrevocable	IRREV
Interest	INTRST		

J

January	JAN	July	JUL
Japan	JAP		

L

Letter	LTR	Los Angeles	LA
Light	LITE	Letter of guarantee	L/G
Liter	LIT	Letter of indemnity	L/I
London	LDN		

M

Manager	MGR	Monday	MON
March	MAR	Month	MO
Message	MSG	Maximum	MAX
Meter	MTR, M	Measurement	MEASMT
Middle	MID	Minimum	MIN

N

Negotiate	NEGO	New	NU
Negotiation	NEGN	New York	NY
New Zealand	NZ	Next	NXT
Night	NITE	Number	NR, NO
November	NOV	Net Weight	NW
Negotiating	NEGOTG	Nothing	NIL

O			
October	OCT	Original	ORGNL
Open	OPN	Otherwise	OZWS
Option	OPT	Our Cable	OC
Ordinary	ORD	Our telex number 100	OX–100
Origin	OGN	Order	OD, ODR

P			
Payment	PYMT	Poor	PUR
Pennsylvania	PENN	Private(ly)	PRVT(LY)
Philippine	PHLN	Price	PRC
Piece(s)	PC(S)	Please	PLS
Possible	POSSBL	Purchase	PURCHS

Q			
Quality	QLTY	Quantity	QTY, QNTY
Quotation	QUOT, QUTN		

R			
Refer our letter	ROL	Refer your telex	RYTX
Refer your letter	RYL	Refer your telegram	RYT
Refer your cable	RYC	Regarding	REGRDG
Refer our cable	ROC	Remarks	RMKS
Register	REG	Repeat	RPT
Request(ed)	REQST(D)	Read	RD
Respectively	REPCTVLY	Receipt	RCPT
Receive(d)	RCV(D)	Reference	REF
Refer our telegram	ROT	Reported	RPTD

S			
Sample	SMPL	San Francisco	SF
Saturday	SAT	Shipment	SHIPT
Second	SEC	Singapore	SPR
September	SEPT	Station	STN
Service	SERV	Sorry	SRY
Shipped	SHIPD	Sterling	STG
Specification	SPEC	Stop	STP
Square yard	SQYD	Sunday	SUN

T			
Telex	TLX	Transfer	TRNSFR
Telegraph	TEL	Thanks	TKS
Text	TXT	Tokyo	TKY
Though	THO	Tomorrow	TMW
Through	THRU	Tuesday	TUES

Thursday	THUR	Today	TDY
U			
Unacceptable	UNACPTBL	Understand	UNDSTND
Unreceived	UNRCVD	Urgent	URGT
Unknown	UKWN		
W			
Wednesday	WED	Word(s)	WD(S)
Weight	WT		
Y			
Yard	YD	Your	UR, YR
You	U	Yards	YDS
Your telex	YX, YTLX	Your letter	YL
Yours	URS, YS	Your cable	YC

第三節 國際電話傳真

一、電話傳真的特性

電話傳真 (Telefacsimile)，簡稱傳真 (Fax) 是利用電話傳真機 (Telefacsimile transceiver) 傳遞訊息的一種最進步的通訊方式，電話傳真機簡稱為傳真機，是一種可經由普通電話線路為兩地間傳送或接收各種文件、圖表的通訊設備。傳真機基本上由傳送器、轉換系統及接收器構成。傳送時，傳送的一方利用曝光掃描器，將甲地的文件或圖表轉換為電訊，再透過電信局的普通電話網路，由乙地的傳真機拷貝 (Copy) 出來。

由於傳真機具備可快速而相當正確地傳遞影像資料的功能，目前已取代 Telex。

二、利用電話傳真通訊的優點

電話傳真除了迅速、確實之外，尚有下列優點：

1. 機器操作簡單，與撥普通電話差不多。
2. 可進行不在現場通訊：傳送與接收雙方均可不必在現場仍可連絡。
3. 無論文件、圖表、各種印刷、打字與手稿，均可傳送。

4. 費用低廉：電話傳真按原稿 (A4) 張數計費，以傳送同樣的文稿，使用電話傳真其費用只需 Telex 的 $\frac{1}{12}$。

5. 可以同時將訊息傳遞給很多人，提高工作效率。

三、使用電話傳真應注意的基本事項

1. 傳真稿應力求簡潔。如果傳真稿過度冗長，不僅浪費傳送人的通訊費用，也造成對方收訊人的紙張成本。

2. 傳真稿宜使用原件，手寫稿則要寫得清晰，以增加傳真清晰度。

3. 傳真電訊會有 1% 左右模糊不清的機率，根本收不到的可能性也高達 1%。因此最好將傳真原稿的性質分成三類，採取不同的措施：

   第一類為報價單、訂單及契約等重要文件，除了傳真之外，還需將原件再以航空信立即寄出，這才是正式文件。傳真文件不能視為正式文件。

   第二類是例行來往的文件，比如交貨期、排程表和催貨單等。這類文件可在傳真之後，每週一次整包以航空寄出。

   第三類是比較不重要，與金錢和時間無關的文件。可視情形，傳真之後，不再郵寄。

4. 雖然每家公司的信紙上都印有公司名稱、地址、傳真機號碼等，但字體卻不大，傳真之後，多幾成模糊不清。因此，最好在信尾簽名下面加一行傳真機號碼，例如「FAX: 011–886–2–712–0670」。從美國直接撥號到臺灣的一定是那頭六個數字，很多美國人不知道，這一點小小的方便，不但讓收訊人看得清楚傳送人的傳真機號碼，而且可以節省收訊人查號的時間。

5. 不宜利用傳真機傳送高度機密的文件資料。如非要利用傳真機傳送不可，應在傳送之前，先撥電話給對方收訊人，請他在傳真機旁接收。

6. 看到傳真機傳進來的資料，應馬上交給收件人。

7. 傳送完畢後，不要將傳真原稿放在傳真機內忘了拿走。

8. 通知員工傳真機所使用的電話不能作普通電話之用。

9. 不可忘記隨時補充收訊用紙。

四、傳真封頁 (Telefax transmission sheet, Telefax cover page) 的格式

TOP GUN CO. LTD.

8TH FL. 3 NANHAI RD. TAIPEI

TAIWAN, REPUBLIC OF CHINA

TEL: 886–2528–8844 FAX: 886–2747–0622

E-mail: service@cieca.com.tw

TELFAX TRANSMISSION SHEET

TO:　　　　　　　　　　　　　　DATE:

FROM:　　　　　　　　　　　　Total pages including this page ☐

If you do not receive the above mentioned number of pages. Please telephone or fax immediately.

第四節　電子郵件

一、電傳科技的驚人發展

　　近幾十年來，科技進步一日千里，電腦的發達，啟動了資訊傳播的革命。西元 1840 年代美國科學家 Samuel F. B. Morse 發明了第一部電報機後，開始了電傳 (Teletransmission) 的歷史。跟著是 Cablegram（國際電報）、Telex（電報交換）及 Fax（傳真）等型式的電傳，可說不斷地推陳出新。近十幾年來，E-mail（電子郵件）又開始普及，頗有與傳真並駕齊驅的趨勢。只要有一臺電腦，經過連線後，即可與世界上任何地方

有電腦設備連線的人交談，同時並可透過 Internet（網際網路）上網讀取資訊，或下載 (Down load) 列印出所需要的資料，這可說是資訊傳播界的巨大革新。

二、電子郵件的意義及其特點

電子郵件是一種透過電子設備 (Electronic equipment) 的溝通方法。它是目前最迅速、最經濟、最方便的溝通方式之一。只要有一臺電腦，經過連線後，雙方即可透過電子郵件互相溝通。

電子郵件寫作的特點是：

1.表達在形式上直截了當，自然、平易、非正式 (Informal)，少用陳腔濫調的客套。

2.結構 (Structure) 較為單純、簡單。

3.使用句型 (Patterns) 較為簡潔、生動、不囉嗦。

第三十二章
通函 (Circular Letters)

公司行號有時候為某種目的，需將內容千篇一律的函件寄發給許多客戶或可能的客戶 (Prospective customers)，在這種情形，大都以通函 (Circular letters) 或定型函 (Form letters) 的方式完成任務。

通函與傳單 (Circulars) 略有不同。傳單是一種印就的不具信函格式的廣告品，通常多跟印刷品一樣，以露封寄發。通函則是一種正式的信函，有的用印刷，有的用複寫版，有的用複印機 (Duplicator) 印就。通函有 Letterhead, Addressee, Salutation 等，而且和一般的信件一樣封發。通函的特點就是：為了某一種特殊目的，同時向許多人寄發的內容相同的信件。

現代的商家，往往利用通函達成下述目的：

1. 宣布開業 (Announcement of establishment of a new concern)。
2. 宣布成立新部門或設立分支機構 (Announcement of establishment of new department and/or new branch)。
3. 通知遷移新址、變更電話號碼、FAX 號碼、郵政信箱號碼、電子郵件信箱等 (Notice of change in its address, together with changes in the numbers of telephone, fax, post office box, and its e-mail address)。
4. 通知重要人事的異動 (Change in the responsible personnel of the firm)。
5. 通知漲價。
6. 通知推出新產品 (Announcement of sales of new product)。

有些人以為通函的措詞最容易撰寫，其實一不小心就流於刻板、陳腐、平淡無味。因此撰寫這種信件時，應特別注意，使其具有個性，如果能跳出千篇一律的通函的窠臼，而寫出清新動人的通函，其效果一定很大，即使事實的本身很單純，無法以別出心裁的格調寫出，也應避免使用陳腐的成套，而應以明確、簡潔、直截了當的方式寫出來。

第一節　宣布開業

關於通知開業的通函，應注意事項：因為一般人不輕信他們所不知的事情，所以撰寫開業通函時，必須多用腦筋，以求其具有吸引力與說服力，一封開業通函應包括：

1.商號創立日期、地址、經營項目。

2.有充分的資金。

3.幹部有豐富的業務經驗，可提供更佳的服務。

4.能提供客戶所需要的各種貨品，而且價格具有競爭性。

5.希望惠顧。

有時為便於對方的徵信，也可將自己的往來銀行名稱寫明。

No. 347　宣布開業

Gentlemen:

Establishment of TaiwanTrading Co., Ltd.

We have the pleasure of informing you that on April 1, 20–, we established ourselves as an export house of SUNDRY GOODS under the style of TAIWAN TRADING CO., LTD. at the following address:

48 Wuchang Street, Sec. 1, Taipei, Taiwan

Our staff members were trained in such prominent firms as Central Trading Co., Ltd. Mitsui Shoji K.K. and China Trade and Development Corp., and because of our close connections with leading manufacturers of various kind of general merchandise in Taiwan, we are very well-placed to supply you with high-grade goods at most competitive prices.

Upon hearing from you about your requirements, we shall be happy to send you illustrated catalogs, which will give you a good idea of the kinds of merchandise we handle.

It will be deeply appreciated if you will give us a chance to serve you.

Yours faithfully,

【註】

1. we have the pleasure of informing you that：可以 "we are pleased to inform you that" 代替。

2. we have established ourselves as：我們業已創設（以……為業的……商號）。

3. an export house：出口商號，其他如 "general importers and exporters; manufacturers; agents; representative; dealers; buying agents; selling agents; commission agents, commission merchant; sole agents, wholesalers; detailers" 等均說明商號業務特性。

4. under the style of... = under the name of...：以……的名稱（義）。

5. prominent：著名的。也可以 "outstanding"（卓越的）形容。

6. well-placed：很有利的地位；方便的；順手的。

第二節　宣布成立分支機構或部門

成立分支機構本身就是一件很好的宣傳。在撰寫設立分支機構或新部門的通函時，應：

1. 說明（強調）因業務的蒸蒸日上 (Steady growth of the business)，不得不成立新分支機構或新部門。

2. 強調由於新分支機構或新部門的成立，能給客戶提供更佳、更有效率的服務。

3. 寫明新分支機構（或新部門）的地址，開業日期。

4. 介紹新分支機構或新部門的負責人。

5. 告知嗣後有關通信可逕致該分支機構或部門。

No. 348　宣布成立分公司之一

Dear Sirs,

　　In view of the rapid development of our business in Southeast Asia, we have now decided to open a branch at the following address:

Taiwan Trading Co., Ltd., Singapore Branch

900 Upper Cross Street, Singapore, 1

　　Mr. Yang, manager of the new branch, has been holding a responsible position of a sub-manager of the export department at our Head Office in Taipei for the past three years. His ample experience and unlimited resources will be of more service to you.

　　Since the branch office will open on the 1st of August, we shall be pleased if you will send your future inquiries and orders direct to our Singapore office.

Yours faithfully,

【註】

1. in view of = considering。

2. responsible position：負有責任的職位。

3. ample experience：豐富的經驗。

4. resources：多謀才略。

5. send...direct to：逕寄。

No. 349　宣布成立分公司之二

Dear Sirs,

　　We have pleasure in announcing that, owing to the large increase in the volume of our trade with England, we have decided, for the convenience of our customers, to open a new branch in London at:

15 St., James' Place, E.C. 4

and have appointed Mr. Schroder as manager.

　　Mr. Schroder, who was in London for four years as our traveler and has since been with us holding a responsible for four years, has been granted power of procuration. We feel obliged if you will send your enquiries and orders direct to him in future, instead of sending them to us in Taipei. He is, of course, quite conversant with all details of our class of goods and will be able to answer all enquiries.

　　We thank you for your valued support in the past and hope for a continuance of patronage.

Yours faithfully,

【註】

1. traveler：旅行推銷員。常作 "commercial traveler"。

2. granted power of procuration：賦予代理權。

3. quite conversant with：十分嫺熟於……。

4. our class of goods：本公司貨物種類。

5. a continuance of patronage：繼續惠顧。

No. 350　通知成立新部門

Dear Sirs,

　　In order to render better and speedier services to all our customers, we have recently set up an independent Shipping Division. This new Division will take care of all our shipments to customers and will see to it that deliveries are made always on schedules.

Mr. William Chang has been appointed Chief of the Division. As an old customer of ours, you would have known Mr. Chang quite well. He has been with our firm for the past ten years in charge of matters concerning shipping and delivery. With his long experience and enough knowledge, we are confident that we can render services that would be more satisfactory to all our customers in future.

We take this opportunity to thank you for your confidence in us in the past and look forward to your continual patronage in the years to come.

Faithfully yours,

【註】

1. in order to render... = with a view to rendering...。

2. set up = establish。

3. see to it = take care of it：注意；照拂。

4. on schedule：準時。

5. we are confident = we are pretty sure。

6. in future = in the years to come。

7. 假如此項通知是寫給本地客戶時，可將電話號碼順便告知，例如 "For your information, a new telephone has been installed in the new Division and the number is 321–2965. Mr. Chang can be reached by the new telephone."。

第三節　通知遷移新址

通知遷移新址的通函，其內容只要將何時遷移至何地的事實加以說明即可，這種通知的型式有兩種，一為通告方式，一為私信性質 (Personal touch) 方式，採取後一方式時，最好以下列文句結束：

1. Your continual confidence in us is hereby solicited.

2. Your continual patronage will be appreciated.

3. We wish you to continue patronizing us in the years to come.

No. 351　通知遷移地址

NOTICE OF REMOVAL

Taiwan Trading Co., Ltd. would like to announce that as from July 1, 20– its office will be moved to the following address:

33 Hung Yang Road, Taipei, Taiwan
Telephone: 886–02–2321–1234 (10 lines)
FAX: 886–2528–8844
E-mail: btc@alba.com.tw remain unchanged.

【註】

Notice of Removal = Removal Notice。

No. 352　通知若干部門遷移新址

TAIWAN TRADING CO., LTD.
wish to announce that on and after
November 25, 20–

Their offices for Shipment and Receiving
will be located on the Third Floor of
The World Trade Building
Nanking East Road, Sec. 1
Taipei, Taiwan, R.O.C.

Telephone: 02–2321–1989
FAX: 02–2321–1893

No. 353　通知遷移新地址

Dear Sirs,

NOTICE OF REMOVAL

We are pleased to inform you that owing to steady growth of our firm in the past years and in view of facilitating business expansion, we have decided to move our office to the following address:

TAIWAN TRADING CO., LTD.
Huai Ning Building

Room 201

29–31 Hsiang Yang Road

Taipei, Taiwan

Our E-mail, FAX and telephone numbers remain the same, as shown above.

Your continual patronage will be appreciated.

Yours faithfully,

【註】

本例係採 personal touch 方式，一般而言，效果較佳。

第四節　通知人事異動

No. 354　通知重要幹部人事異動

Dear Sirs,

We are pleased to inform you that our Board of Directors has announced the changes of responsible officers of our company. The purpose of such change is to streamline the organization as well as the operations of our company.

With a view to facilitating your contacting persons in charge of the divisions, the following information may be helpful to you.

Mr. Charles Chang has been transferred from Domestic Sales Division to the Foreign Division, and Mr. William Ling has been promoted to take Mr. Chang's place. Mr. George Wu, who has just returned from Germany, has been appointed as Production Manager of our factory. Mr. Wu is an expert in the manufacture of products of our company, and his wide experience and knowledge will surely enable him to improve the quality of our products and reduce the production costs.

With the reshuffle of our personnel, we can assure you that you will, from now on, receive our products of much better quality and at much reduced prices. In order to avail yourselves of our better services, we request you to write us for more specific information.

You are cordially invited to send us your enquiries.

Faithfully yours,

【註】

1. board of directors：董事會。

2. streamline：使現代化；改進。

3. reshuffle：改組。cf. a reshuffle of cabinet：內閣改組。

4. avail yourselves of：利用。

cf. You should avail yourself of the books in the library.（你應利用圖書館的書。）

I will avail myself of your kind invitation and come this evening.（承您好意相約，今晚定往奉陪。）

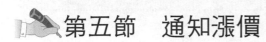第五節　通知漲價

No. 355　通知漲價之一

Gentlemen:

Owing to the rapid and unprecedented rise in cost of raw materials, as a result of the oil crisis, from which our products are manufactured, we are compelled to announce that from and after October 20 our previous price list will be cancelled and replaced by the new one which we enclose for your reference.

All orders received up to October 20 will still be accepted by us at old prices, after which time the new quotations will come into force.

The ever-increasing demand for our products and the continued strong-tone in the market for the raw materials make us bound to look forward to still higher rates within a relative short time. We, therefore, advise you to lose no time in covering your season's requirements.

Yours sincerely,

【註】

1. rapid and unprecedented Rise：（原料價格的）上漲迅速及空前。

2. oil Crisis：石油危機。cf. energy crisis：能源危機。

3. from and after = on and after = as from。

4. new quotation：新定價。

5. come into force：生效。

6. ever-increasing demand：日益增加的需要。

7. continued strong-tone in the market for the raw materials：原料在市場上漲風繼續熾烈。Market-tone：市況。

8. relative short time：相當短的時期內。

9. lose no time：勿失機會。

10. covering：置辦；購儲。

No. 356　通知漲價之二

Dear Sirs,

　　Steady rise in price over the past few months has been a matter of common experience and it will come to you as no surprise that our own production costs have continued to rise with the general trend. Recent devaluation of the currency has been one of the important factors on raising the prices of imported raw materials, of which we are large users. A recent national wages award has added to our labor costs, increased still further by constantly rising overheads.

　　Up to the present, we have been able to absorb rising costs by economies in other phases, but now we find that we can no longer do so. Therefore increases in our prices are inevitable. The new prices will come into force on August 1st and revised price-lists are now being prepared. As soon as they are ready, we shall send copies to all our customers.

　　We can only say how sorry we are that these increases should have been compelled us, but we can assure you that they will not amount to an average of more than about ten percent. As general prices have risen by nearly twenty percent since the last previous issue of our price-list ten months ago, we hope you will not feel that our own increases are unreasonable.

Faithfully yours,

【註】

1. a matter of common experience：一個共同的感受。

2. general trend：一般趨勢。

3. devaluation of currency：通貨貶值。

4. national wage award：指全國性工資調整的裁定。

5. overheads：經常費用。

6. absorb：負擔。

7. by economies in other phases：在其他方面節省開支。

8. inevitable = unavoidable。

9. are now being prepared：現在正準備中。

10. unreasonable：不合理；無理。

第六節　有關通函的有用例句

一、宣布開業開頭句

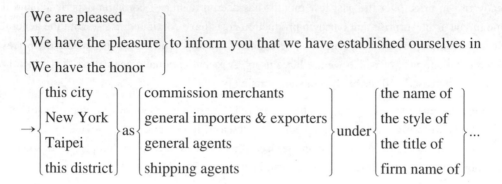

二、宣布開業文中

1. It is our intention to confine ourselves to the wholesale business of electronics.

2. The new firm will devote its attention principally to commission business, in which the shipping of textiles will form an important feature.

3. Feeling confident of our ability to conduct any transactions, and to execute any orders committed to my charge in a speedy, economical, and satisfactory manner, we solicit the favor of your commands.

4. The thorough knowledge and trade experience which I have gained in this branch of business during a ten years' engagement as an employee and manager of prominent firms in this branch both at home and abroad, will enable me to cope with all reasonable requirements.

5. It will be our chief aim to gain the confidence of our business friends by conscientious fulfilment of our obligation and by careful attention to the orders of our friends.

6. The new firm will devote its attention principally to commission house.

7. The excellent reputation which your firm enjoys here, renders us extremely desirous of entering, if possible, into business relations with you, and we therefore offer you

our services for any purchases you may have to make in this area.

8. Having for many years been in almost daily intercourse with the principal manufacturers of our country, and being always well informed about the newest patterns, I flatter myself that you could put my services to profit.

9. We can promise you immediate attention to all orders no matter how small, and it is our firm intention to do our utmost to meet our clients' wishes with regard to delivery as on all other points.

三、宣布開業結尾

We hope to receive your inquiries when in the market.

四、關於設立分支機構

1. As the volume of our trade with Singapore constantly increasing, we have this day opened a new branch in Singapore.

2. For the convenience of our customers, we have decided to open a new branch in New York, and have appointed Mr. Kao as manager.

3. The rapid development of our business in Taichung has compelled us to open a branch at the following address:

4. Our business having become so extended, we have established a branch under the style of...

5. $\begin{cases} \text{We have entrusted the management of our new branch to} \\ \text{We have placed the management in the hands of} \end{cases}$ Mr....., who has, for years past, held a responsible position in our firm.

6. We have appointed Mr..... $\begin{cases} \text{to our manager} \\ \text{as manager} \end{cases}$ Mr.....has been

→ $\begin{cases} \text{connected with our firm} \\ \text{with us} \end{cases}$ for many years, and, as he is thoroughly conversant with the manufacture of our goods, our clients can rely upon the exact and prompt execution of any orders placed in his hands.

五、關於遷移新址

1. We are pleased to inform you that in consequence of the increase of our business, we have been obliged to occupy larger premises, and that we have moved our office to:

2. In consequence of the rapid expansion of our business, we have found it necessary to take larger and more convenient quarters, which we will occupy early next month. We would therefore ask you to address future correspondence to our new office at:

3. We take pleasure in announcing that on and after October 1, we will be located at, ..., to which address please forward all future communications.

六、關於漲價

1. Even a cursory of the market will show you that the change is universal, and we would point out that we have raised prices no further than the circumstances necessitate.

2. Unfortunately, this growing expense had made it necessary for us to raise our selling price. But in order to facilitate office arrangement we have decided to adjust the matter by withdraw the discount on certain orders and not by changing catalog prices.

3. We wish to assure once more that we intend to adhere to our policy of providing a high grade product at a competitive price.

4. We trust that you will realize that this step was taken under pressure, and that no alternative course was open to us.

第三十三章
求才與求職 (Position Vacant & Position Wanted)

由於對外貿易的發展，經營貿易的公司行號越來越多。外國廠商也紛紛前來設立分支公司行號。這些公司行號在在需要諳熟英文的人才。他們招聘人才，在很多場合都是利用報章雜誌刊登「求才」廣告，藉以選聘適當人才。而且這些公司為配合需要，多以英文刊登啟事，應徵求職的人也需要以英文應徵。因此，本書特闢一章討論如何撰寫求才廣告 (Want ads) 以及如何撰擬應徵信 (Letter of application for job)。

第一節　求才廣告的寫法

求才廣告 (Want ads; Advertisement for position vacant) 是廣告的一種，廣告的英文為 "Advertise"，而 "Advertise" 一字則起源於拉丁字 "Adverto"，"Adverto" 的意思是「引起注意」(Turn attention to)。所以求才廣告的文字必須能引人注意 (Attract attention)，激起適當的人才前來應徵。一般而言，求才廣告的內容，應包括下列各項：

1. 徵求人才的公司行號名稱：通常以求才的公司行號為主詞，並以醒目的字體放在最上方。但如不願將其名稱標明時可以類如："A leading trading company"，"A newly-established German electronic company" 的詞句代替，以免推薦、保舉、關說麻煩，同時可向報館暫租一信箱，在廣告上說明回信請寄某報第幾號信箱，或請應徵人將信逕寄第幾號郵政信箱 (P.O. Box)。

2. 徵求那一種人才：說明徵求那一種人才，有時將此一項以醒目的字體放在最上方，以引起讀者的注意。

3. 說明應徵人 (Applicant candidate) 應具備的資格，例如性別、年齡、學歷、經驗、技術等。

4. 必要時，說明待遇優厚、福利好、工作環境佳、升遷機會多等。

5.說明應徵方法：例如應檢送履歷表、相片、希望待遇。

6.寄信地址以及截止期限等。

No. 357　徵求經理秘書

WANTED—SECRETARY TO MANAGER

A leading American Trading Company in Taipei requires female secretary to the manager. All applicants must be college graduates with a bachelor degree and have complete understanding of spoken and written English, and accounting. Applicants are requested to apply in their own handwriting giving personal data including past experience and salary expected and including one recent photo to P.O. Box 1234, Taipei.

　All applications will be treated confidentially.

【註】

1. wanted：「被需要」之意，也即「求才」之意，這是求才廣告中常用的開頭字。

　求才、求職的標題，常以下面的句子表現：

POSITION VACANT

SITUATION VACANT ｝求才；招聘；遺缺需人。

POSITION WANTED

SITUATION WANTED ｝求職。

2. a leading American Trading Company：一規模宏大的美國貿易公司。

3. female secretary：女秘書。cf. male secretary：男秘書。

　junior secretary, assistant secretary, senior secretary, executive secretary, confidential secretary （機要秘書）。

4. with a bachelor degree：具有學士學位。

5. have complete understanding of spoken and written English, and accounting：具有完全瞭解說、寫的英文能力及諳會計。

6. applicants are requested：應徵者必需……。

7. to apply in their own handwriting：以親筆寫應徵信。

8. personal data：個人資料，類似的有：résumé（簡歷），personal history, curriculum vitae, autobiography 等。

9. salary expected：希望待遇。

10. recent photo：近照。cf. bust photo：半身照。

　"photo" 為 "photograph" 的略語。

11. will be treated confidentially：將予秘密處理。

   cf. will be kept confidential：將予保密。

      will be held strictly confidential：將嚴予保密。

      will be kept in strict confidence：將嚴予保密。

No. 358　徵求會計主任

> ### CAREER OPPORTUNITIES
>
> A foreign invested manufacturing company located in Taoyuan has following openings:
>
> Chief Accountant
>
> 1)Male, age under 35.
> 2)College graduate majored in Accounting or related field.
> 3)At least 3 years experience in General/Cost Accounting.
> 4)Good command of English.
>
> Jr. Accountant
>
> 1)Male, age under 30.
> 2)College graduate majored in Accounting or related field.
> 3)Good command of English.
>
> Please send résumé (Name/Address in Chinese) with recent photo to 中壢第 2–6 號信箱 Att. I.R. manager.
>
> （註明應徵職位及希望待遇）

【註】

  1. career opportunity：就業機會，"career" 為職業、事業之意，類似的標題有 "excellent opportunity"; "opportunity available"; "challenging opportunity"。

  2. a foreign invested manufacturing company：一外人投資的製造公司。

  3. located：位於。

  4. has following openings：有下面的空缺。

  5. major in：主修。

  6. related field：相關部門，指 "Related Course"「相關科目」而言。

  7. general/cost accounting：普通會計及成本會計。

  8. good command of：精通。He has good command of English.（他精通英文。）

  9. Jr. accountant：初級會計員。"Jr." 為 "Junior" 的縮寫，高級會計員為 "Sr. accountant"。

  10. résumé [ˌrɛzuˋme]：簡歷，也作 "resume"。

  11. Att. I.R. manager：勞資關係部經理洽，I.R. 為 Industrial Relations（勞資關係）的簡寫。

No. 359　徵求行銷經理

A billion dollar multi-national German corporation in cosmetic product field is activating its operations in Taiwan and requires a:

MARKETING MANAGER

Responsibilities

　　The selected candidate will be responsible overall for attaining the corporate marketing objectives in Taiwan. He will formulate and supervise the execution of marketing plans and also be responsible for sales and budget allocation.

Qualifications

- Chinese, male, between 30 and 40 years of age
- Degree in Business Administration, Marketing or other related studies preferred
- Minimum 4 years experience in marketing consumer goods preferably with cosmetics sales experience
- A good knowledge of retail trade, finance, and logistics
- Good command of English, both spoken and written
- Dynamic, confident and vigorous personality

The successful candidate will receive overseas training on appointment. Interested candidates are invited to submit detailed resume in English (name and address in Chinese) marked "CONFIDENTIAL-MM" to:

華成企業管理服務有限公司
SGV-SOONG & CO.
P.O. Box 1539, Taipei

【註】

1. a billion dollar multi-national corporation：一個億萬元的多國性公司。

2. cosmetic product：化粧品。

3. activate its operations：拓展業務。

4. marketing manager：行銷經理。

5. selected candidate：入選的候選人（應徵人）。

6. overall：全盤的。

7. attaining the corporate marketing objectives：達成公司的行銷目標。

8. formulate：規劃；設計。

9. execution of marketing plans：執行行銷計畫。

10. budget allocation：預算分配。

11. degree in Business Administration：企業管理學位。

12. preferred：較好；最好……；更佳。

13. retail trade：零售業。

14. logistics：後勤。

15. dynamic, confident and vigorous personality：有幹勁、有自信、活潑的個性。

16. successful candidate：入選的（成功的、錄用的）候選人（應徵人）。

17. receive overseas training：接受國外訓練。

18. interested candidate：感興趣的候選人，意指「對此職位有意的人」。

No. 360　徵求秘書

SECRETARY

Wanted by a large U.S. company invested new branch in Taiwan. Female, college graduate, independent and capable, good command of English and typing. Comfortable working condition, prosperous future, attractive and challenging position, please send English resume, state salary expected with recent photo to 6–4, Palace Building, 490, Tun-Hwa S. Road, Taipei.

【註】

1. independent and capable：獨立自主又能幹的。

2. comfortable working condition：舒適的工作環境。

3. prosperous future：有良好的前途。

4. attractive and challenging position：吸引人、具有挑戰性的職位。

No. 361　徵求電氣工程師及生產工程師

PHILIPS ELECTRONICS INDUSTRIES (TAIWAN) LTD.

For its expansion in colour tube production of Chupei Plant offers excellent opportunities for

1. Electrical engineers 2. Production engineers

Requirements:
- Male, completed military service
- University/college graduate
- With at least 2 years experience in production, maintenance, quality control, factory engineering, industrial engineering or equivalent, familiar with PC.

GOOD COMMAND OF ENGLISH BOTH IN WRITING AND SPEAKING

Please send resume in English, name and address in Chinese, with 2-inch photo and expected salary to Personnel Department, Philips Chupei Plant（新竹竹北飛利浦人事部）. Please mark the job applied for on the envelope. If the candidates are qualified to be accepted, they can report for duty after Chinese Lunar New Year.

【註】

1. electrical engineers：電氣工程師。

2. production engineers：生產工程師。

3. complete military service：服役完畢。

4. maintenance：維護。

5. quality control：品質管理。

6. factory engineering：工廠工程。

7. industrial engineering：工業工程。

8. report for duty：報到；到差；復命。"report" 解做報到時，往往用反身式 (reflexive)，向某人報到時加 "to"。cf. report for duty at the office：到辦公室上班，report to the manager：向經理復命，report himself to head office：前往總公司報到。

9. Chinese Lunar Year：中國陰曆年。

No. 362　徵求進出口助理

IMPORT/EXPORT
ASSISTANT WANTED

Energetic young man preferably with 1 or 2 years working experience in Import/Export field is required. Good command of English and typing ability are necessary. Challenging and excellent opportunity to develop with the company. Please send application in English (with name and address in Chinese & English) with full details and recent photo marked "CONFIDENTIAL" to: General Manager, P.O. Box 19–279, Taipei.

【註】

1. assistant：助理；助手。

2. energetic：精力充沛的；有幹勁的。

3. preferably：最好是。

4. are necessary = are essential。

5. challenging：挑戰性的；考驗的。

6. marked "CONFIDENTIAL"：註明「密」字樣。

No. 363　徵求打字員

SITUATION VACANT
Typist/Clerk

Applications are invited for newly created position of female Typist/Clerk in the Export Department of Taiwan Trading Co., Ltd. Applicants must be proficient in speaking, reading and typing English. Preference will be given to applicants with English-language stenography qualifications. Starting salary in the vicinity of NT$30,000 monthly. With attractive fringe benefits. Send application in English, in own handwriting, with full details of experience and qualifications and enclose recent bust photo to Export Manager, 1 Shing Shen South Rd., Sec. 1, Taipei. Applications close Friday 15th October, 20–.

【註】

1. newly created position：新空缺。

2. proficient in：精通於……。

3. preference：優先選擇；優先考慮。

4. stenography = shorthand：速記，stenographer：速記員。

5. starting salary = commencing salary。

6. in the vicinity of：左右；大約。

7. fringe benefits：福利（如壽險、醫療保險等）。

8. applications close Friday：星期五截止申請（收件）。

No. 364　徵求行政秘書

ADMINISTRATIVE SECRETARY

Familiar with PC.
A sound working knowledge of English, both spoken and written.
Independent correspondence ability.
Minimum four years experience in secretarial work.
Attractive salary and pleasant office.
Only thoroughly competent persons need apply.
Mail curriculum vitae in English with recent photo to:
　　P.O. Box 1234, Taipei.
All replies will be held strictly confidential.

【註】

1. PC：為 personal computer（個人電腦）的縮寫。

2. a sound working knowledge of English: 良好的運用英文能力。

3. independent correspondence ability: 獨立處理文牘能力。

4. secretarial work: 秘書工作。

5. pleasant office: 舒適的辦公室。

6. only thoroughly competent persons need Apply: 限於適任（有資格）的人才應徵。

7. curriculum vitae: 履歷表。

No. 365　徵求會計員

CHALLENGING OPPORTUNITY
FOR ACCOUNTING PERSONNEL

A large U.S. invested Electronic Company has openings for high position accounting personnel. Applicants should be:

Fluent in English, both spoken and written.

University/college graduate major in accounting or equivalent education.

Fully conversant in government reporting and taxation.

With at least six years practical experience in cost/tax accounting.

Please send personal data in English with recent photo to:
P.O. Box 4321, Taipei before end of October, 20–

【註】

1. accounting personnel: 會計人員。

2. has openings: 有空缺。

3. fully conversant in: 精通；熟悉……。

4. practical experience: 實際經驗。

No. 366　徵求採購經理

PURCHASING MANAGER WANTED

A leading trading firm needs Purchasing Manager. Candidates must have experience in purchasing grains, such as wheat, soybeans, maize and barley, and have thorough knowledge of all procedures and documentation for importing, speak and write fluent English. Interested candidates should send an English resume, with recent photo to: Personnel Manager, P.O. Box 1234, Taipei.

【註】

grains: 穀類，如小麥、黃豆、玉米、大麥等。

No. 367　徵求速記秘書

STENO-SECRETARY

A large trading company has immediate opening for Steno-secretary, male or female. Applicants must be college graduates with more than four years working experience, good command of English as well as sound shorthand and typing abilities. Some knowledge of German will be helpful. Experienced in garments/sweater business is preferable but not essential. Salary will be commensurate with qualifications and previous experience of applicant. Please send resume and recent photo to P.O. Box 4321, Taipei. Reply to qualified applicants only.

【註】

1. immediate opening：馬上可就職的空缺。

2. some knowledge of German will be helpful：略通德文將有助益。

3. be commensurate with：與……相應；視……而定。

4. reply to qualified applicants only：只對合格的應徵人回信，猶如中文求才廣告中的「合則回」或「合則約」。

No. 368　徵求雜貨出口人才

WHO WANTS TO JOIN US?

We are a young team in the export department of a leading international trading company.

　Our sundry division is growing considerably and we are in urgent need of a young man, capable in handling this line on his own. Past experience and English, Mandarin and Taiwanese in reading and writting are a necessity.

Please send your application with your resume and a recent photo to P.O. Box 22102, Taipei.

【註】

1. sundry division：雜貨部門。

2. handle...on his own：照他自己的意思處理（由自己負責之意）。

第二節　應徵求職信的寫法

求職信 (Employment application letter) 是私人信函中最重要的一種，一封良好的就業應徵信可以決定一個人的一生事業。因此，對於如何撰寫求職信，對於求職的人而言，甚為重要。

本來，求職信就是一封推銷信──推銷自己。不用說，以英文應徵某一職位時，求才的公司，必以應徵人所寫英文的好壞作為決定取採的重要標準。求職信既然是推銷信的一種，那麼求職信應包括推銷信的四要素，即：

1. Attention：開頭部分，包括如何得悉求才的訊息並表示願意應徵：使用精緻的信紙，繕打清楚，遣詞用字適當，拼字正確，標點無誤。

2. Interest：說明自己的學歷、經驗、訓練、特殊技能，表示自己對求才公司能有貢獻，但不可過分誇張，例如：

I graduated in 2003 from the Department of Business Administration of Fujen Catholic University with A-1 rating. During my senior year at Fujen, I passed examination in Business English sponsored by China Productivity Center. Because of my knowledge of English, business administration and business connections, I am confident that I can do an outstanding job for you.

3. Desire：敘述應徵的理由，使求才公司覺得你有前途、潛力，而引起採用的欲望。

It is my earnest desire to work in a trading company like yours, because I wish to put to practical use my knowledge of Business Management and Business Communication I acquired in the University.

I am enclosing a letter of recommendation from Professor Ling of Fujen Catholic University. If you entertain my application favorably, I will do my utmost to justify the confidence you may repose in me.

4. Action：最後，打動求才公司的心，而採取行動。為此，表示希望能會晤面談，並將電話號碼寫上。

May I have the opportunity of meeting and giving you more details about my background? I would appreciate it very much if you could let me know when it would convenient to interview you. My telephone is 12345678.

　　求職信可分為兩種，一種是對於公開求才的應徵信 (Solicited letter of application)；另一種是得自他人的消息或毛遂自薦的應徵信 (Unsolicited letter of application)。無論屬於那一種，應徵信都應特別強調 "You attitude"，在求職信中雖然需將自己描寫一番，但應從求才公司的立場考慮。換句話說，不僅說明你的能力 (Qualifications) 可適合該職位所要求的條件，而且也應表對求才公司的業務有濃厚的興趣。綜上所述，一封完善的 Solicited letter of application，應包括下列各項：

　　1.首先敘述應徵的由來（看到求才廣告或經人介紹）以及表明寫此信的目的。

　　2.次之，說明自己的學歷、經歷、訓練、特殊才能。

　　3.強調自己的學歷、經驗、才能符合該職位。

　　4.表明對求才公司的業務有濃厚興趣，希望能為求才公司貢獻自己的才能。

　　5.提供履歷表、畢業證書、照片、推薦信。

　　6.必要時提供備詢人 (Reference)。

　　7.表示希望賜予面談的機會。

　　8.其他。

　　以上各項，可視情形而決定增減，學經歷、訓練等應力求具體明確，可在信中說明，也可另附履歷表或自傳，尤宜將自己的特殊才能具體說明。

　　假如求職者是剛自學校畢業，初次應徵工作，又無工作經驗，那麼學歷特別重要。因此，在應徵信中宜含蓄地，但有自信地說明，能將理論靈活地應用到實際工作，為該求才公司貢獻一技之長。

　　對於報紙「無機關或公司名稱的求才廣告」(Blind advertisement) 應徵信稱呼用 "Gentlemen" 或 "Dear Sirs"。

　　以下 No. 369 至 No. 382 為 Solicited letter of application。

No. 369　應徵女秘書職位

Gentlemen:

　　In reply to your advertisement in today's *China Times* regarding a secretary to manager, I wish to apply for the position.

I am confident that I can meet your special requirements indicating that the candidate must have a good knowledge of English and accounting, for I graduated from Business Administration Department of Fujen Catholic University last year.

In addition to my study of Business English and Accounting while in the university, I also have had the secretarial experience for two years in Formosa Trading Co., Ltd. The main reason for changing my job is to gain more experience with a superior trading firm like yours. I believe that my education and experience will prove useful for the work in your office.

Enclosed please find my curriculum vitae, certificate of graduation, letter of recommendation from the Dean of Business Administration of the University and one recent photo. With respect to salary, although it is difficult for me to estimate how much my services would be worth to your company, I should think NT$30,000 a month, which is my present monthly salary, a satisfactory beginning salary. I shall be much obliged if you will give me an opportunity for a personal interview.

Sincerely yours,

【註】

1. in reply to：敬覆。

2. apply for the position：應徵這一職位。

3. meet your special requirements：符合您的特別需要。

4. candidate：候選人，也可以 applicant 代替。

5. change job：換工作。

6. a superior trading firm：優越的貿易公司。

7. certificate of graduation = diploma。

8. letter of recommendation：推介書信。

9. dean：系主任。

10. personal interview：面談；面試。

No. 370　履歷表

CURRICULUM VITAE

Name in Full: Wang Ta-wei

Date of Birth: December 7, 1980

Family Relation: Eldest daughter

Permanent Domicile: 58 Tung San Street, Taipei

Present Address: Same as Permanent Domicile

Educational Background: Entered Provincial Hsin Chu Girl Middle School in September, 1992, finished in July, 1998.

Entered Fujen Catholic University Business Administration Department September 1998, finished in September 2002.

Working experience: As stated in the letter of application.

Rewards: Won a prize for three years'regular attendance at the Provincial Hsin Chu Girl Middle School. Won the second place in the Intercollegiate English Speech Contest sponsored by the International Rotary Club Taipei Branch in 1999.

Personal details: Age: 23

Height: 165 cm

Weight: 54 kgs

Health: first class

Marriage: single

Hobbies: reading, stamp collection and sports.

I hereby declare upon my honor the above to be a true and correct statement.

【註】

1. in full: 全名，中國人的姓名不必將姓氏 (surname) 放在名字後面，但如果是英文名字，則應將姓氏放於名字後，如：Charles Chang, George Liu 等。另中國人之名字，第二個字的拼字必須小寫，如：Chang Chin-yuan。

2. Permanent Domicile: 永久通訊處（Domicile 有戶籍的意義）。

3. Rewards: 獎勵。

4. three years' regular attendance: 3 年全勤上學（attendance 出席上課）。

5. second place: 第 2 名。

6. intercollegiate English Speech Contest sponsored by the International Rotary Club Taipei Branch: 由國際扶輪社臺北分社主辦的大專院校校際英語演講比賽。

7. personal details: 個人的資料。

8. declare upon my honor: 以個人的榮譽保證。

No. 371　個人資料

PERSONAL DATA

Name: Elizabeth M. Chang

Address: 1 Wuchang Street, Sec. 1, Taipei

Tel: 2321–1234

Age: 24 years   Date of Birth: Aug. 8, 1980

Place of Birth: Tainan   Marital Status: Single

Height: 5 feet 6 inches   Weight: 120 lbs.

Health: Excellent   Dependents: two; parents

Education:

  College: National Taiwan University, B.A. in Economics, 1999–2003

  High School: North Gate High School, Tainan, 1996–1999

Technical Skills:

  Stenography: 100 WPM

  Typing: 60 WPM

  Office Machine: Mimeograph   Adding and Calculating

                  Offset Duplicator   Personal Computer

                  Fax   Filing Systems

Working Experience:

  July, 20– to present   Taiwan Trading Co., Ltd. Secretary to President

  January, 20– to June, 20– Ark Company   Computer literature

Special interests and activities:

  Golf: active amateur

  Swimming

  Piano

Reference:

  Dr. Peter Pan, Professor of National Taiwan University

【註】

  1. marital status：婚姻狀況。"single" 為未婚，"married" 為已婚。

  2. dependent：家屬。

  3. WPM：為 "words per minute" 的縮寫。

    stenography：100 WPM 為速記每分鐘 100 字之意，education 也可以 "studies pursued" 或 "school attended" 代替，"entered" 為入學，"(was) graduated from" 為畢業於……，如係「結業」則用 "finish"。

  4. computer literature：電腦操作。

No. 372　應徵會計主任職位

I.R. Manager
P.O. Box 2–6
Chung Li

Dear Sir,

In response to your Advertisement in the *United Daily News* of October 2, I am writing you this letter to apply for the position of Chief Accountant. The following are my education, training, and experience in this particular field for your evaluation.

I was graduated, the third among 50–odd students, from Accounting Department of Chengchi University in 1983 with a B.A. degree. Right after completing my military service, I joined a manufacturing firm in 1986 and have been working in the same firm until now. During the past seven years, I first worked in the General Accounting Division to take care of matters pertaining to accounting and taxation and was later transferred to the Cost Accounting Division in charge of cost analysis and control. In March 1993, I was promoted to the position of Senior Accountant to assume the responsibility to supervise part of the Division. It is a pleasure for me to tell you that my supervisory ability has won high praise from the firm's top management.

Besides Accounting, I am particularly interested in English and have taken great pains to study it in the past ten years.

Consequently, I can now speak and write English very fluently and my ability in this regard would be adequate to meet the requirements of the job I am applying for.

As instructed, I am enclosing one recent bust photo of mine. I shall be much obliged to be granted an interview at any time that is convenient to you.

Your favorable response will be deeply appreciated.

Sincerely yours,

【註】

1. for your evaluation = for your deliberation。

2. I was graduated, the third among 50–odd students. （在 50 多個學生中，我以第 3 名畢業。）

3. B.A. degree：文學士學位。

4. military service：兵役。

5. pertaining to = relating to = concerning to。

6. top management：最高管理單位（的人）。

7. take great pains：費了很大的力；盡了很大的力。

8. bust photo：半身照片。

No. 373　應徵行銷經理職位

SGV-SOONG & CO.
P.O. Box No. 1539
Taipei

Sir,

Your advertisement in the *United Daily News* for the position of one Marketing Manager prompts me to offer you my qualifications for this vacancy.

As required, I am enclosing one copy of my resume from which, I am sure, you will be able to evaluate my qualifications. As to English, I can write and speak fluently and am quite confident I have the capabilities required of the position so far as this language is concerned.

With regard to the required experience, I am pleased to inform you that I have had experience in selling consumer goods for more than six years. I am now in charge of the Sales Section of Taiwan Cosmetics Company which is one of the leading manufacturers of cosmetics in Taiwan.

Should my qualifications meet with your approval and an interview be arranged for me, I can be reached by telephone 2321−1234 between the hours of 9:00 a.m. and 5:30 p.m., Monday through Friday.

Sincerely yours,

No. 373−1　履歷表

RESUME

Name: Charles C. Y. Wang（王昌陽）

Sex: Male

Date of Birth: October 8, 1956

Address: 1 Hangchow S. Road, Sec. 1, Taipei（臺北市杭州南路 1 段 1 號）

Marital Status: Married

Education: Graduated from the Department of Business Administration, Fujen Catholic University in 1980 with B.A. degree.

Experience: Salesman, Sales Section, Taiwan Cosmetics Company, 1991−1997; Chief of Sales Section, Taiwan Cosmetics Company since 1997.

Business skills: Personal computer
　　　　　　　　Market research
　　　　　　　　Credit analysis

【註】

　1. resume 也可寫成 résumé，簡歷表之意。

　2. market research: 市場研究。

　3. credit analysis: 信用分析。

No. 374　應徵秘書職位

6–4 Palace Building
490, Tung-Hwa S. Road,
Taipei

Gentlemen:

　In response to your advertisement as inserted in today's *China Times* for a secretary, I am submitting this application for your deliberation.

　I am a graduate of the Department of Commerce of Tamkang University. Since my graduation in 2000, I have been working with the Taiwan Trading Company as secretary. During the past three years, my main responsibility is to type and take dictation in the Import/Export Division and I have acquired much experience in all sorts of secretarial work. I can write and speak English fluently and am able to type and take dictation at 60/80 WPM respectively. I am also strong in PC skills, familiar with Cobol, and JAVA. From March 2003, the company also has assigned me to draft correspondence and, consequently, I am now able to handle correspondence independently.

　As I am quite certain that my qualifications would fit your requirements, I shall, therefore, appreciate it very much if you will give me an opportunity for an interview. At present, I am receiving NT$30,000 monthly. The salary at which I should desire to commence is NT$35,000 per month. If you give me a chance I will accept whatever salary you think reasonable.

　Enclosed please find my resume and recent photo. Please hold this letter of application as strictly confidential.

Yours very truly,

【註】

　1. deliberation = consideration。

　2. take dictation: 聽寫。

　3. assign me to draft correspondence: 指派我草擬信件。

　4. fit your requirements: 適合你的需要。

No. 375　應徵電氣工程師職位

Philips Electronics Industries
(Taiwan) Ltd.

Att: Personnel Department, Philips Chupei Plant

Dear Sirs,

Your advertisement in January 15th issue of the *Central Daily News* has considerably aroused my interest in the position of Electrical Engineer, for which I am writing this letter to you for your consideration. The following is a resume of my educational background, training and experience.

I was graduated from the Department of Electrical Engineering of National Taiwan University in 1998 and completed my military service in 1999. In June, 1999, I was admitted to the Ta Tung Company through very keen competition. Since then, I have been working in the Electronics Testing Division. In the more than three years' working, I gained much training and experience in electronics testing, both of which, I am sure, would help me render very efficient services to your company if I have the honor to be employed.

For your information, I can read, write, and speak English and Japanese fluently and believe that my ability in these respects would be enough to handle the jobs you may assign to me.

As for salary, I hesitate to state a definite amount, but, as long as you have requested me to, I should consider NT$50,000 a month, which is my present monthly salary, satisfactory.

Enclosed please find my recent photo. Should you find my qualifications meet with your requirements, I shall appreciate your giving me an opportunity for an interview. You can reach me by telephone 2321–1234 between the hours 8:00 a. m. and 6:00 p. m., Monday thru Friday.

Sincerely yours,

【註】

1. issue：發行。today's issue：今天發行（的報紙）。

2. arouse my interest：引起我的興趣。

3. electrical engineering：電氣工程。

4. admit：進（某某公司服務）。

No. 376　應徵生產工程師職位

Philips Electronics Industries

Dear Sirs,

I read with immense interest your advertisement in the *China Times* of January 15 that your company is in need of Production engineers. Since my qualifications would meet the requirements of this job, I am submitting my application for your deliberation.

In 1998, I was graduated, cum laude, from the Department of Mechanical Engineering of National Cheng Kung University in Tainan. Right after completion of my military service in 1999. I was employed by Taiwan Machinery Corp. and have been working in this Corp. until now. During the past four odd years, I served as supervisor to look after all lines of production and am now quite experienced in this particular job. Moreover, I can read English and German technical materials, write technical reports in English and German and speak fluently these foreign languages. I am sure these abilities of mine would enable me to meet any challenge required of this position.

As required, I am enclosing one 2-inch photo and eagerly awaiting giving me an opportunity for an interview at your early convenience. With regard to salary, I think this matter should be left in your hands, as I am certain we can arrive at a satisfactory arrangement.

Sincerely yours,

【註】

1. immense interest：莫大的興趣。

2. cum laude：以優等（畢業）。

3. left in your hands：委由您決定；留在您手中。

4. arrive at：達成。

No. 377　應徵進出口助理職位

General Manager
P.O. Box 19-279
Taipei

Dear Sir,

Your advertisement in the *China Post* for an Import/Export assistant, indicates that your company is in need of a competent man, preferably with 1 or 2 years' working experience in Import/Export field. Please consider me an applicant for the position because I possess all the qualifications you required.

In 2000, I was graduated, the second place from the Department of International Trade, Tung Hai University. Right after completion of my military service in 2001, I joined a firm which has been engaged in international trade for many years. I am now still working in this firm and my main responsibility is to handle and process all the paper work concerning import/export. In addition, I also have good communication skills in written and spoken English, familiar with PC.

I am quite confident that both my education and experience would fit your requirements fairly well. Should my application receive your favorable consideration, please grant me an interview. You may write me at the above address or reach me by phone at 2321–1234 in care of Mr. H. H. Yang. I am sure an interview will convince you I am the right man for the job. Enclosed please find my recent photo.

Sincerely yours,

【註】

1. competent man：適任的人。

2. second place：第 2 名。

3. be engaged in：從事於……。

4. convince you：使你相信。

5. right man：適當的人。

No. 378　應徵電腦鍵入員職位

Manager

Taiwan Trading Co., Ltd.
1 Shing Shen South Rd., Sec. 1
Taipei

Dear Sir,

I read with immense interest your advertisement in the *China Post* of May 4 that there is a newly created position of female English key-in operator in your Export Department. Considering my education, training and experience, I am quite confident that I would be the right person for the job.

Right after my graduation from the Department of Commercial Correspondence of Providence University of Arts and Science in 2001, I was employed through very keen competition among 100–odd applicants by the Formosa Enterprises, Ltd. as key-in operator and have been working in the firm for the past two years. During this period, I have acquainted myself with all the routine clerical work of foreign trade.

So far as my capability is concerned, I can key-in at 70 WPM and am proficient in speaking and reading English. Besides keying, I can operate fax machine.

I shall be much obliged if you would extend my application your favorable consideration and, should my qualifications be satisfactory to you arrange an interview with me at any time that is convenient to you. As required, I am enclosing one recent photo of mine.

Please keep my application in strict confidence.

Sincerely yours,

【註】

1. key-in operator：電腦鍵入員。

2. Providence University of Arts and Science：靜宜大學。

3. acquainted myself with：熟悉；諳熟。

4. routine clerical work：例行的書記工作。

5. capability：與 "ability" 互通，指智力（或體力）的「能力」，後面接 "of" 或 "for"。"capability" 常指天生或潛在的能力，在此意義上，與 "ability" 不同，後者常可指學到的能力。

6. be proficient in：精通；諳熟。

7. arrange an interview with me：為我安排面談。

No. 379　應徵行政秘書職位

P.O. Box 1234
Taipei

Dear Sirs,

In your advertisement for an Administrative Secretary in today's *China Post* you emphasize the fact that only a thoroughly competent person need apply for the position. Here are my reasons for believing I meet this requirement.

I am a female, aged 30, single and was graduated, cum laude, from the Taipei Municipal College of Commerce in 1990. Right after my graduation, I went to the States and completed all the work of secretarial course at the Columbia University and stayed there for 2–odd years.

I returned Taiwan in 1993 and joined the Esso Eastern Inc., first as assistant secretary and then promoted to the position of Senior Secretary in 1995 and have been holding this job ever since. So far as language is concerned, I am well versed in both Chinese and English and able to translate the former into the latter and vice versa. Besides, I can speak Japanese fluently. Since I have had long years of experience in stenography, I can take dictation and type at 90 and 70 WPM respectively. I can also operate computer quite well.

If you are satisfied with my qualifications, I shall be glad to have the priviledge of an interview at your convenience. You can reach me by phone 2321–1234 during the hours of 8:00 a.m. and 5:30 p.m., Monday through Friday. As instructed, I am enclosing curriculum vitae and one recent photo of mine. In the meantime, please hold my application strictly confidential.

Sincerely yours,

【註】

　　1. emphasize：強調。

　　2. Taipei Municipal College of Commerce：臺北市商業專科學校。

　　3. secretarial course：秘書課程。

　　4. am well versed in：精通於；對……很有造詣。

　　5. vice versa：反過來也是一樣。

No. 380　應徵會計員職位

P.O. Box 4321
Taipei

Gentlemen,

　　Your Want Ads in the classified advertisement column of the *United Daily News* of April 10 prompts me to write this letter of application for the position of Accounting Personnel.

　　As called for in the advertisement, I am submitting and enclosing one copy of my personal data for your evaluation. From the data, you probably will find my qualifications which, I believe, would meet most, if not all, of the requirements. Enclosed please also find my recent bust photo.

　　So far as English is concerned, I can both write and speak this language fluently. I am not only fully conversant in government reporting and taxation but also have five–odd years' experience in writing accounting/financial reports in English. Therefore, I am quite sure that my long years' experience in this particular field would help me do my job better and more effectively.

　　Should my education and experience meet with your approval, it will be deeply appreciated if you will arrange for an interview in order that you will be able to evaluate further my qualifications.

　　I am expecting to meet you soon.

Sincerely yours,

【註】

　　1. classified advertisement column：分類廣告欄。

　　2. as called for：依（廣告）所要求。

　　3. if not all：即使不是全部。

　　4. government reporting：政府申報。

No. 381　應徵採購經理職位

Personnel Manager

P.O. Box 1234
Taipei

Dear Sir,

I response to your Want Ads as inserted in the *Economic Daily News* of October 12, I am writing you this letter to apply for the position of Purchasing Manager. The following are my education, training and experience.

I was graduated from Department of Commerce, National Taiwan University in 1986. Right after completion of my college education and military service, I was employed by Taiwan Trading Co., Ltd. in 1988 and worked in this company for five years. During this period, I first served as salesman and then as Section Chief of the Sales Section. In charge of the importation and distribution of wheat and maize. In 1993, I left this company and joined Mutsui Shoji K. K. and have been working in this firm as a deputy manager of Import Department in charge of importation and distribution of various kinds of grains. My years' experience has made me thoroughly versed in the import procedures of grains and know, and is known to, most of the retailers across the island. The well-established connections of mine would, I am sure, help very much the promotion of sales of imported grains.

Should you find my education, training and experience satisfactory, it will be highly appreciated if you will arrange for an interview with me at whatever time that is convenient to you. A recent photo of mine is enclosed as instructed.

Sincerely yours,

【註】

1. distribution：配銷。

2. maize：玉米，又稱為 "corn" 或 "Indian corn"。

3. deputy manager：副理。

4. retailers across the island：全島的零售商。

5. promotion of sales：推銷。

No. 382　應徵秘書職位

P.O. Box 4321
Taipei

Gentlemen,

Your advertisement in the *China Daily News* has aroused my immense interest and I am writing this letter to apply for the position of a secretary. The following are my qualifications.

In June, 1999, I graduated, the second among 60-odd students, from the Department of Commercial Correspondence of Tamkang University. Right after my graduation, I joined the Far East Garment Company as assistant secretary and was promoted to the position of secretary in 2002. During the period when I was working in this firm, I have acquired much experience in the secretarial work of garment/sweater business. Besides, my responsibility also includes preparation of quotation/offer sheets and operation of computer and fax machine. I can speak and write English fluently and have some knowledge of German. I believe my knowledge of these languages is adequate to meet all the requirements of your company.

As required, I am enclosing one recent photo of mine and if you are satisfied with my qualifications, I shall appreciate your arranging an interview with me. You can reach me by the address given below:

Miss Judy Liu
123 Hangchow S. Road, Sec. 1
Taipei

Please keep this application in strict confidence.

Yours sincerely,

【註】

preparation of quotation/offer sheets：繕製報價單。

第三節　自薦求職信的寫法

自學校畢業後，為了求職，常常需寫自薦信。這種自薦信並非應求才的廣告而寫，但如寫得充實而有力，往往比起憑求才廣告的應徵更有希望。因為應求才廣告的應徵信，數目較多，不易引起收信人的注意；而非憑求才廣告的自薦信，在同一時期內，不至於很多，所以容易引起收信人的注意。這種自薦信，除開頭更應著重引人注意之外，其他部分，與根據求才廣告而寫的應徵信大同小異。

No. 383　自薦求職信之一──電腦鍵入員

Dear Sir,

Mrs. Dorothy Wang, secretary to one of your directors, has told me that you have a vacancy for a key-in operator and I should like to offer myself for the post.

I am 24 years old and a graduate of the Department of Commercial Correspondence of Ming Chuan University. I graduated with a good academic record, with "A" levels in English, Accounting and International Trade and then took an intensive one-year secretarial course at the Li-shing Girls Secretary Training Center, passing the test of operating computer.

I am now a key-in operator with Taiwan Trading Co., Ltd. and have spent two very happy years there, but the firm is small and I wish to widen my experience.

Mr. H. K. Chen, my former principal, has written a letter of recommendation which is enclosed and has kindly agreed to give further information which you may request. In case, as I hope, you are interested in my application, you will, of course, be able to get more information about me from my present employers.

I enjoy the kind of work I am doing, but wish to continue it in circumstances that offer better prospects. I shall be glad to call for an interview at any time.

　　　　　　　　　　　　　　　　　　　　　　　　　　　　Yours faithfully,

【註】

1. secretary to one of your directors：貴公司某董事的秘書。

2. a vacancy for：……的空缺。

3. to offer myself for the post：應徵此一職務。

4. a good academic record：良好的學業成績。

5. intensive one-year secretarial course：1 年的密集秘書課程。

6. widen my experience：擴大工作經驗；增加經驗。

7. letter of recommendation：保（推）薦信。

8. in circumstances that offer better prospects：在能提供更佳發展的工作環境中。

9. call for an interview at any time：在任何時間均可前來面談。

No. 384　自薦求職信之二──研究員

Dear Sir,

For the past three years I have been a statistician in the research unit of ABC Co., Ltd., Taipei, and now wish to make a change. My only reason is to widen my experience and at the same time improve my prospects. It has occurred to me that a large and well-known organization such as yours might be able to use my services.

I am twenty-eight years of age and in excellent health. I thoroughly enjoy working on investigations, particularly when statistical work is involved. At the National Taiwan University I specialized in marketing and advertising and was awarded an MBA degree for may thesis on "Statistical Investigation in Research".

Although I have had no experience in consumer research, I am familiar with the methods employed and fully understand their importance in the recording of buying habits and trends. I should like to feel that there is an opportunity to use my services in this type of research and that you will invite me to call on you. I could then give you further information and bring letter of recommendation.

Yours faithfully,

【註】

1. statistician: 統計員。
2. research unit: 研究部門。
3. make a change: 換環境。
4. improve one's prospects: 擴大眼界。
5. it has occurred to me: 我覺得。
6. statistical work is involved: 與統計有關的。
7. specialized in: 主修。
8. marketing: 行銷。
9. MBA degree: 商學碩士學位。
10. thesis: 論文。
11. consumer research: 消費者研究。
12. buying habits and trends: 購買者習性及趨向。
13. I should like to feel...call on you.: 我覺得如果您讓我來拜會您的話，相信您一定會覺得這是一個讓我效力的最好機會。

No. 385　自薦求職信之三——會計員

Dear Mr. Barton:

It is my ambition to become an accountant in a successful manufacturing company such as yours. For the past several years I have been studying and working...equipping myself with essential qualifications for a more responsible position and bright future.

My present position requires considerable accounting knowledge; however the opportunities for advancement are quite limited and I should like to make a change in the near future. Richland Chemical Company has grown steadily since its organization and, undoubtedly, affords its employees many opportunities to prove their ability.

The following qualifications, I believe, fit me to fill satisfactorily a responsible position with your company.

Experience:

3 years ······ commercial banking

5 years ······ accounting ······ with present employer

Education:

High school graduate, 1 year business college (night school), 2 year commercial banking course (night school), 4 years Taiwan University (night school) ······ general and advanced accounting.

Personal:

25 years of age, single, Chinese, enjoy good health.

I will gladly give you references and additional information regarding my character and ability if I may have the privilege of an interview. May I hear from you?

Telephone ······ 2321–1234

Sincerely yours,

【註】

1. ambition：抱負。

2. equip myself with...for：使我自己具備······以便（擔任更重要的工作及有光明的前途）。

   cf. equip oneself for...：充實自己以便······。

3. advancement：昇遷。

4. make a change：變動，意指更換工作。

5. afford：提供。

No. 386　自薦求職信之四──貿易人員

Gentlemen:

On my graduation from college this fall, I am desirous of securing a position that will offer me opportunity in the field of foreign trade. Knowing something of the scope and enterprise of your huge export department, I thought perhaps you would kept me in mind for a possible opening.

I am strong and alert, familiar with PC, and shall be twenty years of age in July next year. At present I am a student in the college of..., but I shall graduate from the college this coming July, finishing the requirements in three years. I have had no business experience, but my college record has been good, a copy of my antecedents is enclosed for your reference.

Dr. A. B. Chien, President of the College of..., will be glad to tell you more about my character and ability. I shall be glad to call at any time for an interview.

Very truly yours,

【註】

1. secure a position：謀取一職，"position" 也可以 "job" 代替。

2. keep me in mind = have me mind = bear me in mind = remember：記住我；把我記在心裡。

3. finish the requirements：修完必修課程。

4. college record：大學成績。

5. antecedents：經歷；履歷表。

6. President：校長。

No. 387　自薦求職信之五——秘書

Dear Mr. Williams:

Someday in the future you may have need for a new secretary.

Here is why I should like to offer myself for the job, and here is why I am so much interested in obtaining it. For one thing, I know that you do an enormous variety of work very fast and very well. This offers a real challenge to whoever works for you. It is the kind of challenge I like to meet because, with all due modesty, I have so trained myself in secretarial work that only exacting problems are interesting to me.

As to my mechanical abilities, I can take dictation at the rate of 180 words a minute, type at the rate of 70 words per minute, and operate computer.

I cannot think of any job in which I would be so useful as that of secretary to you, since, in addition to my business training and experience, I could put to work for you and your organization the know-how of practical, everyday business handling.

Yours very truly,

【註】

1. someday in future：將來有一天。

2. for one thing：首先；一則（在申述理由時用之）。

　　for one thing I haven't money, for another...：一則我沒錢，二則……。

3. an enormous variety of work：各式各樣的工作。

4. a real challenge：真正的考驗。

5. it is the kind of challenge I like to meet：我願意嘗試這考驗。

6. with all due modesty：我敢說（客氣地說）。

7. exacting problems：需要特別注意的問題；費力的問題；難以處理的問題。

8. mechanical abilities：技術上的本領。

9. I cannot think...to you：我想不出有任何的職位可以比做你的秘書，更能顯出我的本事。

10. business training：商業訓練。

11. I could put to work...business handling：我能用我的技巧和知識來替貴公司處理每天的業務。

No. 388　自薦求職信之六──秘書兼電腦鍵入員

Gentlemen:

I figured it out this way:

Every firm uses one or more secretaries and key-in operators. They are, so to speak, "Standard equipment" in every office.

So I learned secretarial science which covers computer literature and communication skill in written and spoken English.

I am twenty-one years old and have a college education.

I am what might be called "experienced beginner", and have no exaggerated ideas as to what I should expect as a starting salary.

If you believe it possible that you could find a place for me in your organization, I will appreciate an opportunity to show you samples of my work, and to give you any information you may wish.

Please do tell me when I may see you.

Yours very truly,

【註】

1. I figured it out this way：我的想法是這樣的。

2. so to speak：可以說。

3. standard equipment：標準裝備。

4. secretarial science：商業文書科。

5. cover：包括。

6. have a college education：受過大學教育，即大學畢業。

7. I am what might be called：我可以說是……。

8. experienced beginner：有經驗的開始者。

9. have no exaggerated ideas as to...：對於……沒有太大的奢望。

10. show you samples of my work：向您出示我的工作實例。

No. 389　自薦求職信之七──秘書

Dear Sir,

Could you use a dependable secretary?

During the past two years I have been with the Smith Trading Company of this city. Because our office was small, I performed many different duties, this gave me an excellent understanding of the routine of an office.

I can take shorthand, operate a computer, type rapidly and accurately, act as a receptionist, and write letters dealing with routine situations. The enclosed personal record will give you complete details about my education and personal qualifications.

May I come in to see you at your convenience?

Sincerely yours,

【註】

　　1. dependable：可靠的。

　　2. the routine of an office：一個辦公室的例行工作。

　　3. act as receptionist：當接待員。

　　失了業的中年人求職比年輕人要難得多，因此，這種求職信應力求別出心裁，才有成功的希望。

No. 390　中年人的求職信

Gentlemen:

Will you give work to a man aged forty-two?

In December, I lost a job as Purchasing Manager. The reason did not concern my ability or character.

A shock came when applications for other positions were turned down with the comment "too old".

A man in perfect health, with twenty years experience and a very special ability for purchasing work was to stay idle.

My service covers every phase of foreign and domestic purchasing. No assignment is too small or too large to be given proper attention.

And as far as age goes, what does it matter as long as gray hair tops gray matter?

I ask your help. May I have a talk with you?

> You can leave a message at phone 2123–3210, or write—–and I will come running.

【註】

1. the reason did not concern my ability or character：理由與我的能力或品行無關。

2. shock：震驚。

3. turn down：退回。

4. a man in perfect health：體格健全的人。

5. stay idle：閒散；呆著無事做；失業。

6. cover：包括。

7. every phase of：……的各方面。

8. no assignment is too small or too large to be given proper attention.：職務無分大小，都盡力以赴。

9. As Far As Age Goes：至於年齡。

10. As Long As Gray Hair Tops Gray Matter：祇要白頭髮的人能夠克服困難的工作。

11. You can leave a message at phone 2123–3210, or write...（你可以打電話到 2123–3210 或函……。）

12. I will come running：我將奔跑前來聆教。

No. 391　在報紙上刊登「求職」廣告

> POSITION WANTED
>
> Very capable secretary-stenographer, many years' experience, can hold executive position, open for engagement. Best references.
>
> Replies to Box 123, China Post.

【註】

1. position wanted：待聘；求職。

2. can hold executive position：能夠擔任行政事務。

3. open for engagement = is open for engagement（徵求職務）

    　　　　　　　 = seeks position

    　　　　　　　 = seeks employment

4. best references：有最佳備詢人之意。

　　要別人來聘，就該將自己的長處寫出來，例如上面的廣告待聘的人是一位極能幹的秘書兼速記員，而且有多年的經驗，並且能夠擔任行政事務。

第四節　推薦信的寫法

推薦信 (Letter of recommendation) 是推薦人向求職人的未來雇主 (Prospective employers) 所寫有關求職人能力、品德及為人的信，也就是介紹信的一種。

撰寫推薦信時，應兼顧到被推薦人及收信人（雇主）雙方。換言之，推薦人固應寬大為懷，為被推薦人美言，但也應有是非感，不可因太熱心，導致收信人誤會 (Misleading)。須知因推薦人的陳述不實或誇大，往往會引起嚴重的後果。因為雇主可能因誤信推薦人的陳述，而將主要的職位給與求職人，以致雇主受到損失。

由於推薦人對收信人所負責任相當大，所以寫推薦信時應就被推薦人的能力、品德及為人等作公正的敘述，不可有偏頗，言過其實。須知如因推薦有誤，致收信人因而遭受損失時，推薦人即使不負法律上責任，也須負道義上的賠償責任。

推薦信可分為不指名的一般推薦信 (General recommendation) 與指名的特別推薦信 (Special recommendation) 兩種。前者是寫給未指定收信人的 (To who it may concern)，這種推薦信比較正式，持信人（即被推薦人）可視情形向任何人提出，指名的推薦信則是寫給某特定人的信。

不指名的一般推薦信 (General letter of recommendation)，不指名誰是收信人，通常以 "To whom it may concern:" 或 "To whom it may concern:" 做為 Salutation，且其後面通常用 Colon (:)，信的結尾沒有 Complimentary close。這種推薦信類似證明書，效果較差，它的第一句往往是 "This is to certify that...", 或 "This is to testify that...", 整個信中不應有 "You" 的字眼。

一封完善的推薦信的內容應包含下列各項：

1.被推薦者的全名 (Full name)：僅寫 Mr. Hang, Miss Wu 是不可以的。

2.說明認識多久了？(How long have you known him?)

3.認識程度。(How well do you know him?)

4.說明與被推薦人的關係 (Relationship)：同事、主管、師生關係。

5.被推薦人的表現 (Performance)：工作表現、學習表現。

6.結論：無保留地或有保留的推薦、普通推薦或極力推薦。

No. 392　不指名一般推薦信——推薦翻譯人員

To whom it may concern:

This is to certify that Mr. George M. Cheng worked under me as a translator for three years. He left me two years ago on account of his family's removal to Taipei. I was very sorry to lose him.

Mr. Cheng did fine work in both English-Chinese and Chinese-English translation, and I find that he has improved a great deal since he left me. He possesses a keen intellect and aims at perfection in doing any piece of work. He is industrious, painstaking, and above all, punctual.

I recommend him without reserve for translation work.

William Taylor

【註】

1. to whom it may concern：逕啟者。

2. English-Chinese and Chinese-English Translation：由英譯中和由中譯英。

3. any piece of work：任何一件工作。

4. painstaking：刻苦耐勞的。

5. above all：尤其（是 adverbial phrase）。

6. punctual：守時。

7. without reserve：沒有保留地；完全地。

下面是一封不用 "This is to certify that..." 開頭的一般推薦信。

No. 393　不指名一般推薦信——推薦速記打字員

To whom it may concern:

The bearer, Miss Wang I-min was in our employ as steno-typist from May 20, 1989 to June 30, 1992, during which period she rendered satisfactory services. On account of her efficiency her salary was raised in March 1990, to NT$25,000 a month. She left our services at her own request.

She is painstaking and conscientious worker. Her character and habits are entirely appreciable.

We can confidently recommend her to any one who desires her services.

Taiwan Trading Co., Ltd.

K. Kang, President

【註】

1. the bearer：持信人。

2. in our employ：受雇於我們。

3. render satisfactory services: 服務成績令人滿意。

4. at her own request = of her own accord = by her own wish: 自願地；出於自願；出於本意。

5. conscientious worker: 盡職的工作者。

6. Her character and habits are entirely appreciable. (她的品性與習慣也甚可貴。)

至於指名的特別推薦信，其型式與普通信一樣，Salutation 用 "Dear...", Complimentary close 則用 "Very truly yours," 或 "Yours very truly," 等詞句，而且大都封口逕寄，間亦有交本人面投者。書寫特別推薦信的動機，或為應本人的請求，或為應公司的詢問（這種情形即為後述的 Letters of reference），求才公司對於特別推薦信較為重視，以其所述較為切實。因此寫特別推薦信時，切忌籠統、空泛，特別推薦信的末尾，有時有 "If you care to write me further, I shall be glad to reply." 等句，這有兩種作用：一為加強對於被推薦人的保證，是積極的，一為保留不能直說之語，而須待對方秘密詢問者，是消極的。

No. 394　指名的特別推薦信

Dear Mr. Ward:

　　Mr. Richard Lin has requested that I write you regarding his work during the three years he has been in our employ.

　　During this time he has performed his work in a highly satisfactory manner. He has been punctual for all appointments and has executed his assignment efficiently. I can truthfully say he is conscientious and ambitious. In closing I am glad to recommend him unqualifiedly for a position with your company.

Very truly yours,

【註】

1. truthfully: 誠實地。

2. unqualifiedly: 無條件地。

形容被推薦人的用語有：

honest	誠實	responsible	負責
sincere	誠懇	aggressive	有衝勁
faithful	忠實	patient	有耐心
enthusiastic	熱情	conscientious	盡責
cooperative	合作	initiative	主動

cheerful	樂觀	amicable	和藹
reliable	可靠	diligent	勤勉
dynamic	有幹勁		

No. 395　指名的特別推薦信

Dear Mr. Chang:

　　I learn that Mr. H. K. Lee is being considered for a place in your employ, and am pleased to recommend him.

　　Mr. Lee has until recently been employed by the Ta Ming Textile Factory here as a treasurer. Family conditions make it necessary for him to obtain an employment in Taipei.

　　I have known him for several years. He is thorough and reliable and particularly capable as an accountant. His service here has proved satisfactory. He merits your confidence. I shall be grateful to you if you will give him a favorable consideration.

　　Should further particulars concerning Mr. Lee be required, I shall be very much pleased to communicate the same to you.

Yours sincerely,

【註】

1. family conditions: 家庭關係。

2. employment: 職業。

3. thorough: 縝密的；周到的。

4. capable: 能幹的；有能力的。

5. merits: 應得。

6. particulars: 詳情（通常用複數）。

7. communicate: 報告；通知。

第五節　諮詢信的寫法

　　有些雇主於雇用職員時，往往請求職的人提供三人或更多的備詢人 (Reference)，以便雇主向他們打聽有關求職人的人品 (Character)、能力 (Ability) 或資格 (Qualifications) 等。在此情形下，求職人有提供適當的熟人作為備詢人，並且應先取得其同意。因為備詢人實際上就是身分保證人。

撰寫請求別人作備詢人的信 (Request to use name as reference) 時，應注意下列各點：

1.請求別人作備詢人時，應注意禮貌。

2.告訴他你所應徵的是那一機構的那一種職位。

3.如他對你不太熟悉時，應將你自己的情形向他說清楚。

4.必要時附上貼妥郵票寫明回郵地址的信封。

No. 396　請求作備詢人

Dear Mr. Liang:

　　I am applying for a position as clerk with Taiwan Trading Co., Ltd. and shall greatly appreciate it if you will allow me to use your name as a reference.

　　From September, 19– to June, 20–, I worked in the school bank, I also was a student of yours in bookkeeping.

　　The training I received under your supervision would valuable in this position, as many of its duties...typing, filing, handling money, and writing receipts...are similar to those I performed in the school bank.

　　I shall consider it a priviledge to use your name as a reference and am enclosing a stamped, self-addressed envelope for your reply.

Yours very truly,

【註】

1. allow me to use your name as a reference.：允許我請您作備詢人。

2. school bank：指學校的實習銀行。

3. stamped, self-addressed envelope：貼妥郵票，寫好回郵地址的信封。

　　雇主向備詢人打聽有關求職人的資格 (Qualifications)、品行 (Character)、能力 (Ability) 等時所寫的諮詢信 (Letter of reference) 應力求扼要 (Brief)、具體 (Specific) 及謙恭 (Courteous)，避免囉嗦 (Wordy)。

No. 397　諮詢信之一

Mr. George Chiu, Manager
Taiwan Trading Co., Ltd.

Dear Mr. Chiu:

Mr. Thomas Wang has applied for a position in the Export Department of our Company and has given your name as a personal reference.

We understand that Mr. Wang has been in your employ for the past three years and, therefore, shall appreciate your supplying us with any information concerning Mr. Wang's character and ability while he was employed by your company.

Please be assured that any information that you may give us will be kept in strict confidence.

Very truly yours,

【註】

supply one with = furnish one with = provide one with：供給；提供。

No. 398　諮詢信之二

Dear Mr. Teng:

Miss Judy Ling has made application for the position of assistant secretary to the manager of our Export Department. She has suggested that we write you regarding her character and ability to do similar work with your Firm for the past two years.

It will, therefore, be appreciated if you will provide us with information concerning therewith at your early convenience. You may rest assured that any comments you may give us will be held strictly confidential.

Very truly yours,

【註】

1. therewith = with that = with it：與此。

2. you may rest assured that：你儘可放心。

No. 399　備詢人答覆諮詢人之一

Dear Mr. Liu:

I have received your letter of March 25, 2003, asking us to furnish you with information regarding Mr. Wang's character and ability and am pleased to give you the following for your reference.

Mr. Wang joined our company in December 1999, right after his completion of military service. He was placed in our Export Section and handled our export business for two-odd years. During this period, we found his work quite satisfactory.

In May, 2002, as a reward for the efficient manner in which he performed his duties, we promoted him to the position of Chief of Export Section. This position has imposed upon him added burdens and responsibilities which he has undertaken in a most praiseworthy and satisfactory way.

Much to our regret, Mr. Wang, owing to personal reasons, had to resign from this company. Nevertheless, I do always have full confidence in Mr. Wang's integrity and ability.

I hope that the above information will be useful to you.

Yours sincerely,

【註】

1. I have received...：用 "I" 比較 personal，用 "We" 則比較 business（事務性）。

2. was placed in：被安置在……；被派在……。

3. perform = fulfil = carry out：完成；履行。

4. impose upon：加諸於……。

5. added burdens：更多的責任。

6. praiseworthy = commendable：值得稱讚的。

7. owing to personal reasons：由於個人的理由。

8. have full confidence in：對……有充分的信賴。

9. integrity = honesty = sincerity：誠實；廉正；誠懇。

No. 400　備詢人答覆諮詢人之二

Dear Mr. Chen:

I have received your letter of April 5, 2003, requesting our comments on Miss Ling's character ability as an assistant secretary.

As requested, we are pleased to inform you that Miss Ling has served as junior secretary to our Deputy General Manager during the past three years. During this period, she has distinguished herself as a conscientious worker and has always done her job impeccably well. Miss Ling resigned from our company on account of personal reasons and we are feeling very regret for her leaving us.

As Miss Ling is a responsible girl and has pleasant character, I am convinced that she will do her best if she would be employed by your firm.

Yours very truly,

【註】

1. deputy general manager：副總經理。

2. distinguish herself = make herself famous。

3. conscientious worker：盡責的工作者。

4. impeccably = faultlessly：無瑕疵地。

No. 401　備詢人答覆諮詢人之三

Dear Sirs,

Miss H. K. Chang, applicant for the position of accountant of your firm, has been a student of mine at various times during the period from 2000 to 2002.

During this three-year period, she has shown herself to be quiet, reliable, persistent, not very brilliant, but industrious and diligent.

In appearance, manners, and speech, she is likewise quietly behaved; she is not aggressive enough, yet attractive. My acquaintance with her limits me to these general terms, though I believe that you will be able to get adequate information on her other qualifications from other references.

Yours very truly,

【註】

1. persistent：不屈不撓的。

2. brilliant：有才氣的；卓越的；優秀的。

3. aggressive = vigorous：有衝勁的。

No. 402　備詢人答覆諮詢人之四

Dear Mr. Ford:

I am glad to give you a commentary on the qualifications and character of Mr. Richard Chao whom you inquired in your letter of May 5, 2002.

Mr. Chao entered our employ in June 1995, following his graduation from college. He was placed in our accounting department and served us as one of our assistant accountants for three years. During this time we found his work highly satisfactory.

In 2000, as a reward for the efficient manner in which he fulfilled his duties, we promoted him to the position of accountant. This position has imposed upon him added burdens and responsibilities which he has undertaken in a most commendable manner.

Mr. Chao is a man of unusual ability and energy. He is a certified public accountant and possesses a thorough knowledge of accounting problems such as confront the average business house to-day. His command of English enable him to write with unusual clearness and force. He has pleasing manner and is well liked by his associates. For the past years we have hope we might advance Mr. Chao again, but condition have been such that we find ourselves unable to do this. However, since I believe Mr. Chao is deserving of a promotion, I do not hesitate to recommend him for the position of chief accountant in your company. Our only regret is that we cannot offer him a comparable position within our own firm at this time.

Very truly yours,

【註】

1. certified public accountant：會計師。
2. pleasing manner：令人喜愛的舉止。
3. well liked by his associates：受同事喜愛。
4. deserve a promotion：值得擢升。
5. comparable position：相應的職位。

第六節　通知面談、感謝錄用等信的寫法

No. 403　通知應徵人前來面談

Dear Sir,

　　With reference to your application in answer to the advertising in the *China Post*, I should be pleased to see you here at 10:30, July 10, 2003, if it is convenient to you.

　　Please bring your diploma and transcript of academic record with yourself.

Yours faithfully,

【註】

Transcript of Academic Record：成績單（學業）。

No. 404　通知應徵人空缺已由別人遞補

Dear Sir,

　　In reply to your application of May 5, I regret that I have already filled the vacancy and that therefore no object would be served by your calling upon me.

Yours faithfully,

【註】

    have filled the vacancy：空缺已遞補（本信的中譯：應徵函奉悉，敝處所懸位置，已予遞補，不勝歉疚，恐勞玉趾，用特奉覆）。

No. 405　錄用後的謝函

Dear Mr. Chen:

    I am very happy to accept your offer of a position in your company. As a matter of fact, I consider it quite an honor to become affiliated with your company.

    I shall be able to report for work on June 1. I am looking forward with much anticipation to this date.

Respectively yours,

【註】

    1. as a matter of fact = in fact。

    2. become affiliated with：加入，也可以 "be affiliated with" 代替。

    3. I shall be able to report for work：我能（在……月……日）到職。

    4. I am looking forward with much anticipation to this date.：我渴望著這一天的早些來臨。

No. 406　通知錄用請在 6 月 1 日到職上班

Dear Mr. Wang:

    I am writing to confirm the offer we made to you when you called on us yesterday of the post of chemical engineer as from June 1 next.

    Your duties will be as outlined at the interview, but more particulary you will work under the direction of the production manager.

    The appointment carries a commencing salary of NT$80,000 a month, rising by annual increments of NT$6,000 to NT$7,000 and entitles you to two weeks' leave of absence each year. The appointment may be terminated at any time by either side giving two months' notice in writing.

    Please be good enough to confirm that you accept the appointment on the terms stated and that you will be able to commence your duties on June 1, 2004.

Yours sincerely,

【註】

1. work under the direction of: 在……指導下工作，意指「由……督導」。
2. commence your duties: 開始上班。

No. 407　覆謝錄用並願在 6 月 1 日上班

Dear Mr. Spencer:

First, may I thank you for the kindness and courtesy shown me when I called for interview on May 10. I am pleased to accept the post of chemical engineer on the terms stated in your letter of May 15, and confirm that I can take up my duties on June 1, 2004.

I feel very happy to have obtained this appointment and assure you that I shall do everything I can to make a success of my work.

Yours sincerely,

【註】

take up my duties: 接下我的職務，即「到職」或「上班」之意。

第七節　有關求職、推薦的有用例句

一、開頭句

1. $\begin{Bmatrix} \left.\begin{matrix} \text{In} \\ \text{With} \end{matrix}\right\} \text{reference} \\ \text{In} \begin{Bmatrix} \text{reply} \\ \text{response} \\ \text{answer} \end{Bmatrix} \\ \text{Replying} \\ \text{Referring} \end{Bmatrix}$ to your advertisement in $\begin{Bmatrix} \text{today's} \\ \text{yesterday's} \\ \text{April 10th's} \end{Bmatrix}$

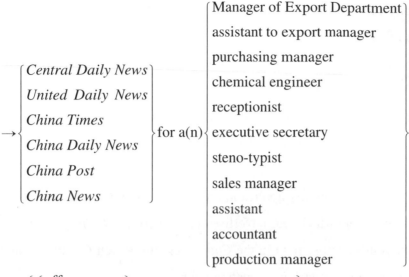

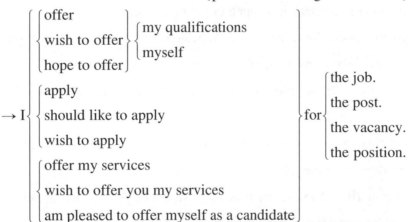

註：a. job：工作；b. post：職務；c. vacancy：空缺；d. position：職位。

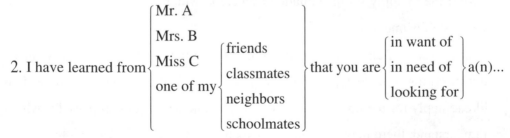

3. Will you consider the contents of this letter in connection with your advertisement for a(n)...?

4. I shall be grateful if you would consider my qualifications for...advertised in today's *China Post*.

5. I am writing this letter to you because of your reputation in business circles. I am hoping that there might be a suitable place for me on your staff.

6. Please consider me an applicant for the position of...which you advertised in yesterday's *United Daily News*.

7. I am applying for the position of..., advertised in today's...

8. I am applying for the position as assistant accountant which you have advertised in the *China Post*.

9. Please consider this letter my application for the position of bookkeeper in your Accounting Department which was advertised in April 10th's *China Times*.

10. I have seen your advertisement in the *China News* of March 6, and I should like to apply for the job of secretary now open in your company.

11. In your advertisement for a(n)..., you indicated that you require the services of a(n)...Please consider me as an applicant for the position.

12. Your advertisement in this morning's *China Times* for a(n)...prompts me to offer you my qualifications for this position.

13. I understand that there may soon be an opening for a(n)...in your company and I should like to apply for the job.

14. This is for your files. I know there is no position in your office at present, but I should like to be considered when there is one.

15. Permit me to apply for the situation as a clerk, which I have heard is now vacant in your establishment.

16. I am 23 years of age and an accounting major at the...college. I have also had training in typing, bookkeeping and shorthand. As I shall graduate next June, I should like to apply for the position of assistant accountant in your company for which job, I understand, there may be an opening soon.

17. I am writing to inquire whether you will need the services of a young woman with educational training in college (and some part-time experience).

18. Perhaps there is a position in your organization for an attractive, and conscientious secretary-stenographer.

19. Mr. A tells me that there is a vacancy for...Here are my qualifications.

20. Mr. B has suggested that I apply to you for the position of...to be available next month.

21. Mr. C of your company has informed me that you are looking for a(n)...with at least three years' experience.

二、關於年齡或學歷的說明

1. I am...years $\begin{cases} \text{of age} \\ \text{old} \end{cases}$, and $\begin{cases} \text{graduated from} \\ \text{was graduated from} \\ \text{a graduate of} \\ \text{shall graduate from} \end{cases}$ ... $\begin{cases} \text{college.} \\ \text{university.} \end{cases}$

2. I shall graduate $\begin{cases} \text{in June} \\ \text{this coming summer} \\ \text{next month} \end{cases}$ from $\begin{cases} \text{college.} \\ \text{university.} \end{cases}$

3. I $\begin{cases} \text{graduated} \\ \text{was graduated} \end{cases}$ $\begin{cases} \text{in 2004.} \\ \text{last year.} \\ \text{last summer.} \\ \text{this summer.} \\ \text{a year ago.} \\ \text{two months ago.} \end{cases}$

4. $\begin{cases} \text{I majored in} \\ \text{My major was} \\ \text{I took up} \end{cases}$ $\begin{cases} \text{Accounting.} \\ \text{Economics.} \\ \text{Foreign Language and Literature.} \\ \text{Marketing.} \\ \text{International Trade.} \\ \text{Business Administration.} \\ \text{Chemical Engineering.} \\ \text{Law.} \\ \text{Secretarial Science.} \end{cases}$

5. I have a degree of
- Bachelor of Commerce. (B.C.)
- Bachelor of Business Administration. (B.B.A.)
- Bachelor of Accountancy. (B. Acc.)
- Bachelor of Arts. (B.A.)
- Bachelor of Science. (B. S.)
- Master of Business Administration. (M. B. A.)
- Master of Science. (M.S.)
- Ph. D.

6. I attended...Primary School, ...Junior Middle School, ...Senior Middle School, and... College.

7. I entered...School in 1998, changed to...School in 1999, and finished in 2000.

三、關於經歷及專長的說明

1. I

have a
- good（充分）
- fair（相當）
- slight（約略）
- sound（徹底）
- thorough（完全）

knowledge of（瞭解）
- accounting.
- marketing.
- shorthand.
- operating PC.
- export/import procedures.
- English.
- taxation.

am well acquainted with（熟諳）

konw（懂）

am strong in PC skills and familiar with the use of SAP.

am familiar with PC and have good communication skill in written and spoken English.

2. I
- can
  - fulfil the ordinary duties of
    - office work.
    - accounting.
  - take dictation in
    - English.
    - Spanish.
- am able to
  - take dictation
  - write shorthand
  - type

  at 80 WPM.

3. For the past...
- months
- years

,

→ I
- have been
  - in the office of
  - in the employment of
  - employed by
  - with

  ...Co. as
  - key-in operator.
  - secretary.
  - accountant.
  - typist.
- have
  - served as an
    - administrative secretary
    - assistant secretary

    to Mr....
  - worked as...in...company.

4. I have had...years' experience
- in
  - my present post.
  - a trading company.
- with a factory as an accountant.

5. Since my graduation from...

→ I
- have been
  - in the employ of
  - employed in

  Co. as a(n)...
  - engaged in（職務）under the employment of...Co.
- have been working for a
  - government
  - stated-owned

  enterprise as a(n)...

6. I have just completed my military service.

四、學歷、經歷不在信中說明而以附件處理時

1. Enclosed are a $\left\{\begin{array}{l}\text{personal data sheet}\\\text{resume}\\\text{curriculum vitae}\\\text{personal record sheet}\end{array}\right\}$ and two copies of letters of recommendation.

2. The details of my education and experience are given on the

→ enclosed $\left\{\begin{array}{l}\text{personal record.}\\\text{resume sheet.}\\\text{personal data sheet.}\end{array}\right.$

五、關於能適任的表示

1. I $\left\{\begin{array}{l}\text{believe}\\\text{think}\\\text{feel}\\\text{am sure}\end{array}\right\}$ that $\left\{\begin{array}{l}\text{I am competent to meet the requirements}\\\text{I shall be able to do the job}\\\text{I can fill satisfactorily the position}\\\text{my experience has been of the kind}\\\text{my ability can meet the requirements}\end{array}\right\}$

→ $\left\{\begin{array}{l}\text{as advertised.}\\\text{which you have specified.}\\\text{which you advertised.}\\\text{called for by the advertisement.}\end{array}\right.$

2. I am confident that my experience and references will show you that I can fulfil the particular requirements of your secretary position.

3. I am sure I can readily adapt myself to the routine of your office.

4. I feel quite certain that as a result of the course in filing which I completed at...College, I can install and operate efficiently a filing system for your organization.

5. Your advertisement interested me because it offered advancement in a field for which I am particularly qualified by experience and attitude.

六、關於待遇

1. I should consider NT$25,000 a month
   - a fair
     - starting salary.
     - initial salary.
   - satisfactory
     - compensation.
     - remuneration.
   - satisfactory.

2. The salary at which I should
   - desire to commence is NT$20,000
   - require would be NT$30,000

   - a month.
   - per month.
   - monthly.

3. I should require
   - a starting
   - a commencing

   salary of NT$20,000 per month.

   a salary of NT$25,000 a month to begin with.

4.
   - In regard to
   - With regard to
   - Regarding

   salary, I think it is better to leave it to you to fix after experience of my capacity.

   （至於薪水，在考驗我的能力之後，聽由你裁奪。）

5. Although it is difficult for me to say what compensation I should deserve, I should consider NT$...a month a fair starting salary.

6. As much as I should like to join your company, it would not be advisable for me to change my position for less than NT$..., which is my present
   - monthly
   - yearly

   salary.

7. At present, I am receiving NT$...monthly. The salary at which I should desire to commence is NT$...per month. If you give me a chance, I will accept whatever salary you think reasonable.

8. I feel it is too bold for me to state what my salary should be. My first consideration is to satisfy you completely. However, while I am in probation, I should consider NT$...a month satisfactory compensation.

註：a. too bold：太大膽；b. in probation = serving apprenticeship：試用期間。

9. I hesitate to state a definite salary, but, as long as you have requested me to, I should consider NT$...a month satisfactory.

　　註：a. hesitate to state：難於提出；b. as long as：既然。

七、關於面談

1. May I have $\begin{cases} \text{an interview?} \\ \text{the opportunity to discuss this matter further with you?} \end{cases}$

2. I shall be pleased if you will be good enough to favor me with an opportunity for a personal interview.

3. If you $\begin{cases} \text{wish} \\ \text{ask} \\ \text{desire} \end{cases}$, I shall be $\begin{cases} \text{pleased} \\ \text{glad} \\ \text{happy} \end{cases}$ to $\begin{cases} \text{call for an interview} \\ \text{call in person} \\ \text{have a chance} \\ \text{have an opportunity} \end{cases}$

$\rightarrow$ to $\begin{cases} \text{tell you more about myself and my qualifications.} \\ \text{demonstrate my ability.} \\ \text{prove my ability.} \\ \text{talk with you so that you may judge my qualifications further.} \end{cases}$

　　註：a. call for an Interview：接受會見；b. call in Person：親往拜訪；c. demonstrate my ability：證明本人的能力。

4. $\begin{cases} \text{I assure you that if} \begin{cases} \text{appointed} \\ \text{successful} \\ \text{my application be successful} \end{cases} \\ \text{Should you} \begin{cases} \text{consider my application favorably} \\ \text{entertain my application favorably} \\ \text{think favorably my application} \\ \text{kindly entertain my application} \\ \text{give me a trial} \\ \text{select me to fill the vacancy} \end{cases} \end{cases}$,

$$\rightarrow \text{I would} \begin{cases} \text{endeavor} \\ \text{do my best} \\ \text{do my utmost} \end{cases} \text{to} \begin{cases} \text{satisfy your requirements.} \\ \text{give you satisfaction.} \\ \text{afford you every satisfaction.} \\ \text{gain your confidence.} \\ \text{deserve your confidence.} \\ \text{please you.} \end{cases}$$

註：a. if appointed; if successful; if my application be successful 均可譯為「如蒙錄用」；

b. should you consider my application favorably; entertain my application favorably; think favorably my application; kindly entertain my application 均可譯為「如蒙考慮」；

c. should you give me a trial：如蒙試用。

$$5. \text{You can} \begin{cases} \text{reach me by phone at 2321--1234} \\ \text{reach me by dialing phone 2321--1234} \\ \text{call phone number 2321--1234 and ask for me} \end{cases} \text{between}$$

$$\rightarrow \begin{cases} \text{the hour of 7--9 a.m. and 6--10 p.m.} \\ \text{eight and six during the day.} \end{cases}$$

6. I can be reached by telephone 2123–3214 between the hours of 9:00 a.m. and 5:00 p.m., Monday through Friday.

八、推薦信用語

㈠開頭句

$$1. \text{It is} \begin{cases} \text{my} \\ \text{a real} \end{cases} \text{pleasure to recommend Mr. George Liu...}$$

$$2. \text{I am} \begin{cases} \text{pleased} \\ \text{happy} \end{cases} \text{to recommend Miss...}$$

3. I am pleased to recommend without reservation Miss...

4. It is a great pleasure to recommend to you Mr. H. K. Ling as a worthy candidate for...

5. This is to recommend Miss Nancy Liu for your favorable consideration of her application for...

6. Mr. H. H. Wang, my former student at National Taiwan University, has asked me to

write in the interest of his application for...

7. Miss Judy C. Hang, presently an accountant at this company, has asked me to write in support of her application for...

8. I have the pleasure of recommending my former student Mr. Robert S. Wang for...

9. $\begin{Bmatrix} \text{In my capacity as} \\ \text{In the capacity of} \end{Bmatrix}$ $\begin{Bmatrix} \text{President of Taiwan University} \\ \text{Dean of Department of Commerce} \end{Bmatrix}$, I would like to

$\rightarrow$ recommend $\begin{Bmatrix} \text{Mr. William H. Lin for...} \\ \text{most enthusiastically Mr. A who has applied for...} \end{Bmatrix}$

(二)學業、能力、品格的描述

1. Mr. Lin graduated with honors and received a B.A. degree from this University.

2. During the academic year of 1990–1991, Miss Wang took my course in Business Administration at Fujen Catholic University and achieved outstanding scores at the semester finals.

   註：at the semester finals：在期終考試（注意：finals 為多數）。

3. He is qualified to conduct correspondence, and is expert and accurate at calculations.

4. He has a good command of the English language.

5. As to his English proficiency, he is able to speak, read and write the language efficiently.

6. He is honest, sincere and faithful, always adhering to high ethical standards.

   註：adhering to high ethical standards：遵守高度的倫理標準。

7. In his performance of duties, he has displayed a high sense of responsibility and always helped whenever he could.

   註：always helped whenever he could：只要他能力所及經常幫助他人。

8. As to his character and personality, he is honest, reliable, responsible, and cooperative.

9. After his undergraduate study and military service, he worked for two years with an industrial firm, handling the quality control of electronics production.

10. In the performance of her duties, she displayed profound knowledge in market surveying, which proved greatly helpful to our sales programs.

㈢結尾句

1. I recommend him without reservation and will appreciate your assistance to his application.

2. It would be greatly appreciated if you would favorably consider his application.

3. Your favorable consideration of his application will be most appreciated. If further information regarding this ambitious young man is required, please feel free to let me know.

4. In consideration of the above facts, I recommend him with reservation.

第三十四章
貿易社交信 (Social Letters in Foreign Trade)

第一節　貿易社交信的特色

英文貿易社交信雖不屬於本書範圍，但是做貿易之餘，人與人之間難免也有社交。因此，特闢一章介紹英文貿易社交信 (Social letters in foreign trade; Social correspondence in foreign trade) 的寫法，諒對讀者不無益處。

社交信與商業書信，在性質上不同，因為：

1. 商業書信主要在於傳遞有關商業上的信息；而社交書信則主要在於連絡感情，表示關心以及體貼。

2. 社交書信通信的雙方都是個人；而商業書信通信的雙方則為公司行號，即使以個人名義發出或寄給個人，仍是代表公司行號傳遞信息。

3. 社交書信的發信人與收信人必定彼此認識，雖然熟識程度或彼此關係可能有深淺的不同。而商業書信的發信人與收信人則可能素不相識，從未晤面。

因此，社交書信的寫法，在遣詞用字或體裁方面，與商業書信略有不同，社交書信的內容及語氣方面，多較富感情、親切而不必太拘禮。

然而，社交書信與商業書信一樣，最起碼包括 Heading, Inside address, Salutation, Body of the letter 及 Complimentary close 五大主要部分。這五大主要部分，原則上雖與商業書信的寫法沒有什麼不同，但仍須注意下列各點：

1. Heading：社交書信既是個人與個人間的通信，通常不用公司或機關的信紙。不過如係業務上認識的朋友，而其通信雖然是私人的社交性質，實質上則與公務多少有關係的，仍可使用公司或機關的信紙，就 Heading 而言，即使是寫給親密的朋友，也應將 Heading 完整地寫出。Heading 可寫在信紙右上方。

2. Inside address：寫給親友時，通常不必寫 Inside address，如要寫上 Inside address,

則可仿商業書信的寫法。

3. Salutation：社交書信的 Salutation 比起商業書信，較非正式 (Informal)。通常在 "Dear" 後面加上見面時所用的稱呼，例如見面時稱他為 Mr. Chang，則社交書信的 Salutation 可以用 "Dear Mr. Chang" 或 "My dear Mr. Chang"；如見面時稱他為 "Charles"，則 Salutation 可用 "Dear Charles"。另外，在社交書信的 Salutation，除了第一字及名詞須大寫外，其餘的不需大寫。例：

Dear Dr. Chang

My dear Dr. Chang（dear 不必大寫）

再者，在標點方面，社交書信的 Salutation 後面的標點，通常係用逗點（,）或驚嘆符號（!）而不用冒號（:），這與商業書信略有不同。

4. Body of the letter：語氣及措詞宜親切 (Affectionate)、友善 (Friendly)、親密 (Intimate)。

5. Complimentary close：商業書信中的 Complimentary close 多用 "Faithfully yours," 或 "Yours faithfully," 等詞句，但在社交書信則視寫信人與收信人關係的深淺，用不同的詞句。例如 "Sincerely yours,", "Yours sincerely,", "Yours very sincerely,", "Sincerely," 等。但 "Sincerely yours,", "Cordially yours,", "Cordially," 等在社交信中，不管關係深淺都可通用。要好的朋友之間或親戚之間，則可用下面任何一種："Affectionately yours,", "Lovingly yours,", "With love," 等字樣。

大致說來，商業社交信有下列幾種：

1. Letters of introduction。

2. Letters of invitation。

3. Letters of thanks。

4. Letters of congratulations。

5. Letters of sympathy。

6. Letters of New Year's greetings。

第二節　介紹信的寫法

進出口廠商為了拓展業務往往有機會出國旅行訪問，在出國之前，如能由相關的人士或公司行號，尤其由往來銀行寫信介紹，請求對方予以協助，那麼辦起事來方便得多，介紹信應 Brief 而 Sincere，因介紹信並無保證 (Indorsement) 之意在內。介紹信言過其分，即變成保薦信矣。介紹信的內容應包括：

1. 被介紹人的名字。

2. 介紹人與被介紹人的關係。

3. 介紹的理由（目的）。

4. 有關被介紹人的背景等。

5. 結束時表示如能惠予協助當感激不盡。

No. 408 銀行為其客戶寫介紹信

Mr. Charles C. Y. Chang   May 6, 20–
Senior Vice President
Central Trust of China
49 Wu-Chang, St., Section 1
Taipei, Taiwan

Dear Mr. Chang,

　　We have the pleasure of introducing Mr. Archie Gann and Mr. Frank Ricciardi of Columbus, Ohio. Mr. Gann is President and Mr. Ricciardi is Marketing Manager for Dorcy Cycle Corp., a valued client of our bank.

　　They will be visiting Taiwan from March 17 to March 20 on business. Although this letter of introduction is not to be construed as an authorization for extension of credit, any courtesies or assistance you may provide will be appreciated. They may desire to discuss export financing with someone at your bank. Mr. Gann and Mr. Ricciardi will identify themselves with appropriate travel documents and a copy of this letter.

　　Thank you very much for your cooperation in this matter and we shall look forward to the opportunity to assist you in the future.

                                                              Sincerely,

                                                              A. Z. Sofia
                                                              Senior Vice President

AZS/sah

【註】

　　1. a valued client of our bank：敝行的重要客戶。

2. on business：為了業務；因商務。

3. is to be construed as an authorization for extension of credit：被解釋為貸款的授權書。

4. any courtesies or assistance you may provide will be appreciated：如蒙照顧及協助,將不勝感激。

any courtesy you may show to him will be considered $\left\{\begin{array}{l}\text{a favor}\\\text{as shown}\end{array}\right\}$ to myself：如蒙照顧,將感同身受。

5. export financing：出口融資。

6. identify oneself with：以……做為識別之用。"identify oneself with" 另有「認為同一,認同」之意,例如：He identified himself with the Labor party.（他認同工黨。）

No. 409　為客戶寫介紹信

Dear Mr. Chang,

　Please permit me to introduce Mr. A, the bearer of this letter, Manager of Taiwan Trading Co., Ltd., with whom we have had business relations for many years and is believed to be one of the most trustworthy firms in our city.

　He is going to visit your city for the purpose of getting acquainted with the general business customs and conditions in order to organise a new branch there.

　As this is his first trip to your city, I sincerely ask you to render him all the assistance possible and facilitate the purpose of his trip.

Yours faithfully,

【註】

1. the bearer of this letter：持信人。

2. get acquainted with：熟悉。

3. render him all the assistance：給他一切的協助。

注意：introduce 該把被介紹的人做 object,不該把收信人 (you) 做 object,例如 introduce him to you 不可改作 introduce you to him,又如 introduce to you my cousin 不可改作 introduce you to my cousin。

　將介紹信正本交給被介紹人後,另把介紹信副本寄給介紹信的收信人,是一種既合乎禮貌,又使大家方便的做法。

No. 410　寄介紹信副本

Dear Mr. Yano,

We enclose a copy of a letter of introduction to you we are handing to Mr. C. Y. Chang, who will be leaving early next week for a business trip to Far East. He expects to be in Tokyo for a short time only, but while there will be interested in contacting woolen, cotton and silk piece goods manufacturers and exporters.

Mr. Chang and the company bearing his name have been well and favorably known to us for a number of years. He is a man of substantial means. We bespeak for him the courtesies of your office. Mr. Chang will be amply supplied with funds and this letter is not to be construed as an authority to extend credit.

Yours faithfully,

【註】

1. business trip = business tour：商務旅行。

2. a man of substantial means：很富有的人。

3. bespeak for him：為他請求。

No. 411　對於介紹人的覆信

Dear Mr. Wang,

We have received your letter of September 1, enclosing a copy of a letter of introduction which you have given to Mr. Andrew F. Parker of the American Merchandise Company.

It will be a great pleasure for us to meet Mr. Parker upon his arrival and to assist him in any way we can in order to make his trip enjoyable and successful.

Yours sincerely,

【註】

1. to meet：接見。

2. enjoyable and successful 也可以 "pleasant and fruitful" 代替。

No. 412　通知訪問

Dear Mr. Liu,

I am pleased to inform you that I am scheduled to leave on January 20 and arrive in Taipei on January 25 by PAA flight No. 2.

I look forward with pleasure to seeing you and all your colleagues at your company and will contact you upon arrival in your city.

　　With kindest personal regards,

Yours sincerely,

【註】

　　PAA: Pan American Airline：泛美航空公司。

No. 413　歡迎來訪

Dear Mr. Mu,

　　Thank you for your letter of January 2, 20–. It's certainly good to know that you will be visit us in the latter part of this month.

　　We are very happy to have you with us. Please let us know your travel plan, so that arrangements can be made to make your stay here both enjoyable and productive.

　　Our greatest sage, Confucius said, "It's always a pleasure to greet a friend from afar." As a faithful supporter of our company, you are certainly our friend and a very good friend and will be accorded our very warm hospitality.

　　You must have heard of the beautiful scenery and delicious food here. Come and enjoy them yourself. We will take good care of you.

Yours sincerely,

【註】

　　1. happy to have you with us：能跟您在一起很高興；承蒙光臨不勝愉快。

　　2. so that arrangements can be made：以便能安排一切。

　　3. productive：有收穫；不虛此行。

　　4. It's always...from afar.（有朋自遠方來，不亦樂乎。）

　　5. Come and enjoy them yourself!（請您自己親自來享受吧!）

　　6. We will take good care of you.（我們會好好地照顧您。）

第三節　感謝信的寫法

　　出國考察業務如承國外廠商接待，那麼回國後，應由本人或公司出面，向接待廠商寫信致謝，這種感謝信的內容應包括：

　　1.平安地回到本國。

　　2.表示訪問對方時承蒙接待。

　　3.表示收穫良多。

　　4.希望貴我雙方關係能更加密切。

No. 414　由本人出面致謝

Dear Mr. Brown,

　　Our visit to your country was most enjoyable, thanks to you and your colleague. You were all so generous with your time, and so patient with us and our endless questions!

　　Bill and I want you to know how much we appreciate all you did to make our stay interesting and enjoyable.

　　We hope that when you visit Taiwan, you will plan on visiting us. We'd welcome the opportunity to return your hospitality.

　　Thanks to all of you, for a truly unforgettable interlude in your country.

Yours sincerely,

【註】

　　1. you were all so generous with your time：那麼慷慨花那麼多時間陪我們。

　　2. so patient with us and our endless questions：不厭其煩地替我們解答問題。

　　3. how much...and enjoyable.：多麼地感激你，使我們停留的期間，感到愉快而有趣。

　　4. We'd welcome...hospitality.：希望有機會來報答你的款待。

　　5. Thanks to...in your country.：謝謝你們大家，給我們在貴國種種無法忘懷的一切。

No. 415　由本人出面致謝

Dear Mr. Wang,

　　I am now back in Taiwan and am taking this opportunity of expressing to you my sincere appreciation for your kindness shown to me on the occasion of my visit to your city, especially for a very nice lunch at which I was able to enjoy a good conversation with you.

　　My visit to your country, although very short, was most fruitful, informative, and enjoyable to me. I shall always indebted to you for your courtesy which will be long remembered with the greatest pleasure.

With my kind personal regards to yourself and Mr. _____ .

Yours faithfully,

【註】

1. fruitful: 獲益良多。

2. informative: 有益的。

3. I shall always...long remembered: 對於您的好意將永不忘懷。

No. 416　由本人出面致謝

Dear Mr. Chang,

I truly enjoyed meeting with you during my recent visit to Taipei with Jayson Mugar. I returned only yesterday from a tour of slightly over one month to the Southeast Asian area. My visit to Taipei was certainly one of the highlights of my trip.

I look forward to working with you and your fine organization in the future and hope that you will contact me directly if there is anything which we can do for you.

Again, it was a great pleasure meeting with you and I thank you very much for your time.

With best regards,

Sincerely,

【註】

1. highlights: 最主要的部分；最有趣的事；最精彩的場面。

2. with best regards: 謹致問候；耑此敬頌近安（用於信尾），也可以 "with kind regards" 代之。

No. 417　由本人出面致謝

Dear Mr. Chang,

Now that I have returned to Manila, I wish to take this opportunity to thank you and your colleague for the courteous reception which you accorded me during my recent visit to Taipei.

It was indeed a pleasure meeting with you, and I very much enjoyed over conversation regarding matters of mutual interest.

I look forward to returning to Taipei in the near future and to renewing our acquaintance.

Meanwhile, I send you my best wishes and kindest regards.

Sincerely yours,

【註】

1. accord：給與。cf. We accorded him a hearty welcome. （我們給他熱誠的歡迎。）

2. renew our acquaintance：重溫舊交。

3. I send you my best wishes and kindest regards：耑此敬頌福安。

No. 418　由公司出面致謝

Dear Mr. Chen,

　　Mr. M. H. Lee has just returned from his visit to...and has told us of the friendship and generous welcome which you extended to him when he called at your firm. We place a high value on these personal discussions.

　　On behalf of this company, I would like to express our appreciation of your kind reception of our representative.

Yours faithfully,

【註】

1. generous welcome：慷慨的歡迎，也可以 "warm welcome" 代替。

2. place a high value：珍視；重視。

3. express our appreciation of...：對於（您的親切接待）謹致謝意。

No. 419　由公司出面致謝

Dear Sir,

　　Mr. K. Kan, Managing Director for our Foreign Department, has now returned from his trip to your country, arriving here on 24th November last.

　　We take this opportunity of expressing to you our very sincere thanks for the kind reception and assistance you have given him, without which his visit to your country for the first time under a rather demanding itinerary would not have been as successful and enjoyable.

　　There is no doubt that the firsthand knowledge he gained of your market conditions during this trip will serve the useful purpose of intensifying our efforts to promote business with your market and also of enhancing the spirit of co-operation between our two companies.

　　Thank you again for all that you did for Mr. Kan and with our kindest regards in which he joins most heartily.

Yours faithfully,

【註】

1. demanding itinerary: 緊湊的日程。

2. firsthand: 第一手的；直接的。

3. enhance: 提高。

No. 420　由公司出面致謝

Dear Sirs,

This is to express our sincere appreciation for the courtesies and assistance you have kindly extended to our Mr. Ma when he recently stopped over at your port on his way from Indonesia.

We learn with a great pleasure that during his brief stay in your port he was able to discuss with you various aspects of our business relationship with you and at the same time obtain valuable information on the current business conditions in your market. We have no doubt that the personal contact he made with you will contribute not a little to our mutual understanding and promotion of business.

For our part we place a high value on these personal discussions.

On behalf of this Company, the writer would like to express our appreciation of your kind reception of our representative.

Yours faithfully,

【註】

1. stop over: 中途停留。

2. mutual understanding: 雙方的瞭解。

第四節　邀請信的寫法

邀請信 (Letter of invitation) 分為二種，一為正式的 (Formal)，一為非正式的 (Informal)。正式的邀請信，與其說是信，不如說是請帖。邀請信的發出數量較多時，大多用正式的；規模小的則用非正式的。

現在先介紹正式的邀請信：

正式的邀請信如上所述實際上就是請帖。它沒有 Heading, Inside address, Salutation, Complimentary close 和 Signature。正像中文請帖中沒有什麼「鑒」和什麼「安」，型式和文字都很呆板，不過也有若干點應注意，一不當心，就弄出笑話。下面是邀請參加酒會的正式邀請信。

No. 421　邀請參加酒會

> Mr. Stewart H. Cole
> Vice President and General Manager
> of Chemical Bank, Taipei Branch
> And Mrs. Cole
> request the honor of your presence
> at a reception
> on Friday, October Eighth, 20–
> from six to seven o'clock
> at International Reception Hall, Grand Hotel, Taipei
> to meet
> Mr. Donald C. Platten, Chairman of the Board
> Chemical Bank, New York
> and Mrs. Platten
>
> R.S.V.P. (Regret only)
> Tel: 2321–2211, Miss Helen Wang

值得注意的有：

1. 全文是一句 Simple sentence，將 "Mr. Stewart H. Cole and Mrs. Cole" 做主詞，也即邀請人（發信人）。如 Mr. Stewart H. Cole 的職銜不加時，應用 "Mr. & Mrs. Stewart H. Cole"。"request" 為動詞。主詞用第三人稱 (Third person)，不用第一人稱。動詞用現在式 (Present time)，如邀請人為單獨一人時，不用 "I request"，而應寫成 "Mr....requests"，如果邀請人由公司出面時，用 "...Company requests"。

2. 雖然是一句 Simple sentence，注意分成幾行：

(1)邀請人：可能的話，邀請人排成一行，如：

Mr. & Mrs. Chang，但如需加上職銜，而又很冗長的話，自不必堅持排成一行，如上面的實例即排成四行。

(2)邀請詞及被邀請人：接著邀請人的下面就是邀請詞及被邀請人，其詞句有下列幾種表現法：

① request the pleasure of your presence（恭請光臨）

② request the pleasure of

　　Mr....

presence

③ request the pleasure of the company of

Mr....

④ request the pleasure of your company

（pleasure 可用 honor 代替）

3.邀請目的：放在被邀請人的下一行，例如：

at a reception（歡迎會，酒會）

如請吃午餐 (Luncheon) 或晚宴 (Dinner)，雞尾酒會 (Cocktail party) 用 at luncheon 或 at dinner，"at" 的後面無 "a" 字。

4.日期寫成一行，放在邀請目的的下面。須先寫明星期幾，而後寫上某年某月某日，星期幾前面有 "on" 年月日，在很正式的場合都拼出 (Spell out)。

5.時間寫成一行，放在日期下面，例如：

在酒會時："from 6:00 p.m. to 7:00 p.m." 這是說「客人可以從下午 6 點到 7 點隨時來」（酒會並不是在客人到齊以後才有的，是隨到隨吃的），在很正式的場合，from 6:00 p.m. to 7:00 p.m. 應寫成 "from six to seven o'clock"。

邀請午餐或晚宴時，時間前面用 "at"，如："at six o'clock"。

6.邀請地點：寫在時間下面一行。必要時，應寫上詳細地址。

7.如邀請的目的，除單純的午餐、晚宴以及酒會之外，尚另有理由時，可將其理由寫在：

(1) "at a reception"，"at dinner" 或 "at luncheon" 下面。

(2)邀請信最上面。

(3)邀請信最下面。

邀請理由有：

(1)介紹

to meet

Mr. Donald C. Platten

(2)介紹

to introduce

Mr. Stewart H. Cole

(3)歡迎

in honor of

Mr. Gordon T. Wallis

(4)歡迎

on the occasion of the visit of

Mr. Davis Rockefeller

(5)慶祝開業典禮

in celebration of the opening of their (his) new office

(6)慶祝開業典禮

at the opening ceremony of their (his)...branch office

8.在左下角常有 "R.S.V.P." 或 "Please reply" 或 "Please respond" 字樣,意思為「請答覆是否出席」, "R.S.V.P." 為法文 "Répondez S'il Vous Plaît" 的縮寫,即「請答覆」之意。括弧中 "Regret only" 是指「只有在不能出席時答覆」之意。

9.為便於答覆,有時還寫上電話號碼及接洽人名稱。

No. 422　邀請晚宴

Mr. & Mrs. Richard T. Wang
request the pleasure of the company of
Mr. & Mrs. S. Y. Liu
at dinner
to introduce
Mr. Joseph A. Nice
On Friday, October 1, 20–
at Six O'clock
at International Reception Hall, Grand Hotel

R.S.V.P.

No. 423　邀請午宴

Mr. Charles K. Chen
requests the honor of your Company
at Luncheon

On Friday, the twenty-eighth of October, 20–
At 12:30 p.m.
at the Ambassador Hotel, Four Springs Garden

R.S.V.P.

　　非正式的邀請信，猶如普通的友誼信 (Friendly letter)，內容通常很簡單，但語氣要親切，例如：

No. 424　邀請晚宴

Dear Mr. Lin,

　　Our president, Mr. W. L. Stones, is giving a dinner party in honor of Mr. George L. Smith at the Grand Hotel on Wednesday evening June 18, six o'clock.

　　We would be pleased to have you join us at that time and look forward to the pleasure of your company.

Yours sincerely,

　　答覆正式的邀請信（即請帖），該用正式的；答覆非正式的邀請信，該用非正式的。不論答覆那一種，該說明出席或不出席，不可說「也許可以來」，「希望能來」，「或許不能來」，「恐怕不能來」等不確定的話，不能出席時，宜說明不能出席的理由。

　　下面是正式答覆的幾個例示。

No. 425　覆接受邀請

Mr. C. Y. Chang
accepts with much pleasure
the kind invitation of
Mr. and Mrs. Stewart H. Cole's
reception
on Friday, October eighth, 20–
from six to seven o'clock
at International Reception Hall, Grand Hotel
to meet
Mr. Donald C. Platten, Chairman of the Board
Chemical Bank, New York
and Mrs. Platten

No. 426　謝絕邀請

> Mr. C. Y. Chang
> regrets that a previous engagement
> prevents him from accepting
> Mr. and Mrs. Stewart H. Cole's
> kind invitation to a reception
> on Friday, October 8, 20–
> from 6:00 p.m. to 7:00 p.m.
> at International Reception Hall, Grand Hotel
> to meet
> Mr. Donald C. Platten, Chairman of the Board
> Chemical Bank, New York
> and Mrs. Platten

注意下列各點：

1. 接受邀請的答覆用 "accepts with (much) pleasure"「（極）樂意接受（邀請）」或 "is pleased to accept"；謝絕邀請的答覆用 "regrets that..."（當然如主詞是多數，則該用 accept, are 和 regret）。

2. 接受邀請的答覆依照請帖寫明星期幾，年月日和時刻和地點；但謝絕時可以將時刻省略。

3. 接受邀請的答覆裡的 "is pleased to accept" 不可改作 "will be pleased to accept"。

4. 謝絕的答覆裡的 "prevents" 不可改作 "will prevent"。

5. 謝絕的答覆裡的 "prevents him from accepting" 也可改寫成 "prevents his accepting"。

6. 謝絕的答覆裡的 "regrets that...from accepting" 也可改寫成 "regrets that on account of a previous engagement he is unable to (cannot) accept"。

7. 答覆非正式的邀請信，最簡單的便是把正式的覆信句子改成 First Person 的語氣。

No. 427　接受邀請

> Dear Mr. and Mrs....,
>
> 　　Thank you for your kind invitation to a reception on Friday, October 8, 20– from 6:00 p.m. to 7:00 p.m., at International Reception Hall, Grand Hotel.

I shall be honored to attend.

Sincerely yours,

【註】

I shall be honored to attend.：本人將一定出席。

No. 428　謝絕邀請

Dear Mr. and Mrs....,

I deeply regret that I am unable to accept your kind invitation to a reception on Friday, October 8, 20– as I have a previous engagement for that evening.

I am very sorry to miss the pleasure of meeting Mr. and Mrs. Cole of whom I have heard so much.

Thank you all the same for your invitation.

Sincerely yours,

【註】

1. miss the pleasure of meeting：無緣拜識。

2. Thank you all the same for your invitation.（雖無法參加，但承邀請仍深表感謝。）

　all the same = just the same：同樣地。

No. 429　邀請派對

Dear Mr. Liu,

We are having a small party on December 24 to see in the New Year and we should be delighted if you could join us here at Seven O'clock. Supper will be served at eight.

Yours Sincerely,

No. 430　覆參加派對

Dear Mr. Ho,

I shall be pleased very much to join your party on December 24 at Seven O'clock.

It is always enjoyable to see out the Old Year and greet New Year with good friends.

Thanks a lot for inviting me to your celebration.

Yours sincerely,

【註】

1. to see out the Old Year and greet the New Year：送舊年迎新年。

2. Thanks a lot for...to your celebration.（謝謝邀我參加你們的慶祝盛會。）

No. 431　謝絕參加派對

Dear Mr. Ho,

Thank you so much for your note, asking me to your New Year party. Unfortunately, I have already promised to join the Smiths that evening to celebrate, and so I must regretfully decline your kind invitation.

May we take this opportunity of wishing you "A Happy and Prosperous New Year."

【註】

promised to join the Smiths that evening to celebrate：已答應 Smith 夫婦在那晚參加他們的慶祝派對。

第五節　祝賀信的寫法

No. 432　祝賀就任新職

Dear Mr. Yu,

It gives us a great pleasure to learn that you have been appointed Manager of the Export Department of your highly esteemed firm.

On this occasion we should like to send you our sincerest congratulations on this honourable appointment wishing you every success in the future.

We hope that with your friendly cooperation the amicable relations we have had the pleasure to maintain with your firm will be further intensified in the future.

Yours sincerely,

【註】

1. highly esteemed firm：寶號。

2. on this occasion = at this time。

3. send you our sincerest congratulations：謹祝賀。"send" 可以 "extend" 代替。

4. honourable appointment：光榮的任命；榮任。

5. wishing you every success：祝您事事順遂。"wishing you every success" 也可以 "and to extend every good wish for your success in your new responsibilities" 代替。

6. intensified：加強。

No. 433 祝賀昇遷

Dear Mr. Lin,

Heartiest congratulations on your promotion to the import manager of the Taiwan Textile Company. I know how very happy you and your family must be. You should consider this promotion a well-deserved reward for your many years of conscientious and untiring efforts in behalf of your company.

I rejoice with you in your good fortune and I look forward to a more prosperous relations between our two firms as you begin with your new post.

Sincerely,

【註】

1. heartiest：最熱忱的。

2. a well-deserved reward：應得的報償。

3. rejoice with：為（你的好運）而高興。

cf. to rejoice at (or in)：為……而快樂。例：I rejoice at your success.

4. good fortune：好運。

5. look forward to 可以 "anticipate" 代替。

6. prosperous：昌隆；興隆。

No. 434 祝賀膺任新職

Dear Mr. Randall,

I congratulate you most sincerely on your recent advancement to the presidency of your company. And I congratulate your firm on having a new chief who knows thoroughly every department of the business, and who is eminently fitted to cope with the problems of today and the future.

In looking back, the twenty-five years you have been associated with Holmes & Sampson seen to me only a few short summers. Our work for your company all this time has been so interesting and pleasant that a quarter of a century has gone by very fast indeed.

By the very nature of our association with your company we know how well you have filled every position you have occupied on your way to the top. Your steady advancement from shipping room to the president's chair has been won by hard work and by the exercise of keen foresight and business intelligence.

These words of felicitation are not from me alone. Mr. Reynolds and Mr. Stevens are figuratively looking over my shoulder. They ask to be remembered to you.

Sincerely yours,

【註】

1. advancement to：晉升……。

2. presidency：總裁職位。

3. cope with：應付；克服。

4. associate with：進（公司服務）。

5. association with：聯繫；與……交往。

6. They ask to be remembered to you.（他們要我代為向你致意。），也可以 "They wish to be re-membered to you." 代替。

No. 435　慶賀晉升

Dear Mr. Platten,

Congratulations! We hope your new position as President of Taiwan Trading Company will bring you much happiness and satisfaction.

You are the most important part of our business. Without customers such as you, not a wheel would turn in our plant...we just could not exist.

Many times I have wished for the opportunity to shake your hand and say "many thanks" for your orders. Since this cannot be done frequently enough, I present the spirit of this letter as a token of our big debt to you for just being a good customer. As you grow and prosper, I should like to feel that the products of ours contributed a small but important part to your prosperity.

In the future, I hope we can serve you better and more often. To this end, we continue to expand our plants, to enlarge our engineering and research facilities...all to produce more and better products to enable industry to broaden our margin of profits.

Again I say "thanks", the opportunity to serve you is one of our life's great pleasures.

Sincerely,

【註】

1. You are the most important part of our business.（您是我們業務中最重要的一部分。）

2. not a wheel would turn in our plant：我們的工廠將無法運轉。

3. shake your hand：握您的手。

4. I present the spirit of this letter as a token of our big debt to you：我用這封信向您表示我們的感激。

5. I should like to feel that...：我認為……。

6. to this end：為此。

7. the opportunity...pleasures：有機會為您服務是我們一生中最榮幸的事。

No. 436　覆謝祝賀

Dear Mr. Chambers,

I take great pleasure in receiving your congratulatory letter dated May 10, 20– on the occasion of my new appointment.

I would like to take this opportunity to express my personal appreciation for the support and co-operation which you extended to me during the term of my past office. I sincerely hope that you will continue the same to us.

With my personal best regards,

Sincerely yours,

【註】

1. congratulatory letter：祝賀信。

2. 第一段句子可以下列各種句子代替：

(1) Thank you very much for the greeting which you extended to me on May 10, on the occasion of my new appointment.

(2) Thank you for your letter of May 10 conveying congratulations on the occasion of my new appointment.

(3) I am very much obliged to you for your cordial congratulations.

(4) It was most kind of you to write me such a cordial letter...

(5) Thank you so much for your kind letter of congratulations...

(6) A thousand thanks for your congratulations on my new appointment...

3. I sincerely hope that...to us. 可以下面句子代替：

I hope that you will continue to give me the same support as you have always done to my predecessor.

No. 437　祝賀設立新公司

Dear Sirs,

We are very pleased to receive the announcement of your opening of new office in Tokyo.

Please accept our warmest wishes on this memorable occasion and convey our words of congratulations to your new office for further expansion of your Far East business.

With kindest regards,

Yours sincerely,

【註】

memorable occasion：值得紀念的日子。

No. 438　祝賀創業 25 週年紀念

Dear Sirs,

It gives me great pleasure to extend to you and your associates our warmest congratulations on the occasion of the twenty-fifth anniversary of your institution, which you will celebrate on May 1st.

The history of your institution has been a distinguished one and we are proud to serve as one of your suppliers in U.S.A. We wish you every success in the years to come which we confidently expect will see the continued growth of your fine Institution and the expansion of our cordial and valued relationship.

With best regards,

Sincerely yours,

No. 439　覆謝祝賀

Dear Sirs,

On behalf of the Taiwan Trading Company I wish to thank you for your congratulations and good wishes upon the twenty-fifth anniversary of the founding of this company.

Your letter symbolizes fine spirit of friendship between our two organizations, and it is most sincerely appreciated.

With best regards,

Yours sincerely,

【註】

symbolize: 象徵。

No. 440　誌往來（交易）週年的信

Dear Mr. Wang,

This day marks another milestone in our business relations. Just one year ago today you opened an account with us.

We are ever mindful of this day, and I personally wish you to know that we are today even more interested in serving you than we were the first day we did business with you. And it is because you are such a good customer that we hope to continue to justify your confidence in us.

Is there anything that we can do to improve our service? To make it more acceptable? And to your greater satisfaction?

Sincerely,

【註】

1. This day marks...business relations.（今天是我們業務關係上的另一個里程碑。）

2. Just one year...account with us.（一年前的今天您成了我們的往來戶。）

3. we are ever mindful of this day：我們永遠記得這一天，"ever" 為永遠之意。

4. wish you to know：希望您能瞭解。

5. we hope to continue to justify...in us：我希望繼續對得起您對我們的信賴（心）。

本信簡單明瞭、動聽、熱情，頗能引起收信人的好感。

第六節　慰問信 (Letters of Sympathy) 的寫法

No. 441　慰問病人 (Sympathy upon illness)

Dear Mr. Johnson,

We were very sorry to learn from Mr. Brown of your recent illness. I do hope you will be well on the road to recovery by the time of this letter reaches you.

My colleagues join me in sending you our regards and best wishes.

Sincerely yours,

【註】

寫信慰問病人時，應力求簡單，不可太囉嗦 (wordy)。

"I do hope you will be well on the road to recovery...reaches you." 也可以 "We extend our sincere best wishes for your speedy recovery." (祈禱早日康復) 代替。well on the road to recovery 為「正在復元中」之意。

No. 442　答謝慰問

Dear Mr. Chang,

This is to thank you for your very kind letter of inquiry during my recent illness. I should have answered it earlier, but the doctor did not allow me to write anything till this morning.

It is at time like this that one really appreciates the kindness of friends. I am feeling very much better now, and hope to be coming home soon. Thank you once more for your kind letter.

Sincerely,

【註】

1. I should have answered it earlier: 早該答覆。

2. It is at time...your kind letter.: 這一段的意思是：「一個人唯有在這種時期才能真正瞭解友誼的可貴，我現在已經好多了。可望於短期內回家，對於你親切的信，再申謝忱。」

No. 443　死亡通知

Taipei, July 5, 20–

We announce with deep sorrow
the death of our Chairman of
the Board of Directors,
Mr....
on July 1, 20–
Taiwan Trading Co., Ltd.

寫弔慰信要注意：謹慎其事，含意悲悼，既不能說得過分痛苦悲哀，也不能謹致表面上的同情。要之，不外乎措詞要能流露真情摯意為上。

No. 444 弔慰信 (Letters of condolence)

Dear Sirs,

We were deeply grieved to hear from you of the death of distinguished Chairman, Mr....

We fully realize how much the loss of Mr.... will be felt by all of you and would like to extend our deepest sympathy on this sorrowful occasion.

Yours sincerely,

【註】

1. 第一段可以 "We were shocked to learn that you had lost your Chairman, Mr...." 代替，均可譯成「奉聞貴公司董事長××先生逝世靈耗，不勝震驚（悲傷）」。

2. we fully realize...felt by all you：我們十分曉解您們是如何地悲傷。

3. and would like to extend...occasion 可以 "And please accept my sincere sympathy and that of my entire organization."（本人及公司全體人員敬表誠摯的同情致慰問之忱）代替。

No. 445 故人遺族謝函

The members of the family of the late John Smith gratefully acknowledge your expression of sympathy on their recent bereavement.

It is deeply appreciated.

【註】

這是由故人的遺族向弔慰者發出的謝函。本信以第三人稱來寫，所以可作為印刷卡片或新聞啟事的文章。

No. 446 向弔慰者致謝函

The Board of Directors and
the General Manager of
Taiwan Trading Co., Ltd.
acknowledge with thanks
the many expressions of sympathy addressed
to them upon the decease of
their esteemed Chairman, Mr....

July 14, 20

第七節　賀年信的寫法

No. 447　賀新年贈桌曆

Dear Mr. Smith,

Hearty congratulations to you on the advent of a bright and prosperous New Year.

I am assure you of my deep appreciation of your patronage during the past year and solicit a continuance of your favors.

As a small token of my best wishes, I am sending you a desk calendar by separate airmail, which you will please accept.

Sincerely,

【註】

1. as a small token of my best wishes：聊表祝福之意。

2. which you will please accept：敬請笑納。

No. 448　覆謝贈桌曆

Dear Mr. Nixon,

Thank you very much for desk calendar which you sent me. It's simply beautiful.

I very much appreciate your thinking of me. You can be sure that your sentiments will be reciprocated.

As I turn over each page of the calendar, I, too, will think of you everyday.

As a token of my appreciation, I'm sending you under separate cover some preserved Chinese beef. It's nothing very much, and I hope you will like it.

My very good wishes to you,

Sincerely yours,

【註】

1. it's simply beautiful：真精美。

2. your thinking of me：想到我；想念我。

3. You can be sure...reciprocated. （你可以確信你的情意將會得到回報的。）

4. preserved beef：經加工便於保存的牛肉（如牛肉乾、牛肉罐頭等）。

5. it's nothing very much：這是小小的一點東西。

6. My very good wishes to you.（特向您祝福；祝您萬事如意。）

No. 449　覆謝贈約會簿

Dear C. Y.,

　You were most kind to send me the attractive appointments book for 2004. The diary is most useful and I appreciate your thoughtfulness.

　I send you and all friends at Taiwan Trading Company my very best wishes for a happy and prosperous year in 2004.

Sincerely,

【註】

1. attractive appointments book：精美的約會簿。

2. 第二段可譯成：「謹祝臺灣貿易公司全體同仁新年快樂萬事如意」。

第八節　有用的介紹信例句

一、介紹信開頭句

a	We have the pleasure of / We have much pleasure in / We take great pleasure in / It gives us pleasure to / have this opportunity of	introducing to you Mr....who
b	This is to / This will / This letter will / This letter will serve to / This letter may serve to / It is with real pleasure that we	introduce to you Mr....who

c The bearer of this letter, Mr....,

The bearer of these lines, Mr....,

This letter will be delivered (handed) to you by Mr....who

will be in your city shortly.

is visiting your city on business.

is making a business trip to your city.

is spending a few weeks in your district.

is coming to London in the interests of his company.

will be in your city for a few days on his way to Europe.

will be leaving early next week for a business trip to the U.S.

二、被介紹人姓名、身分

Mr. K. Wang is an officer of The Bank of Taiwan,

Mr. G. Kan is chief of Cotton Section of The Dow Spinning Co., Ltd., Taipei, one of the top-ranking mills in this country,

Mr. T. Ong, export manager of Tai Spinning Co., Ltd., Taipei,

Mr. M. Yang, president of Taiwan Trading Co., Ltd.,

→ with which we are closely connected

with whom we enjoy a very close connection

whose products we are exporting as agents

who are our principal bankers for many years,

三、介紹的目的

→ is proceeding to

is visiting

is coming to

$\rightarrow$ { your port to open up a branch office.

your city in order to form fresh connections and open new markets for his products.

Paris to attend International Wool Conference to be held there in August.

Australia for the purpose of studying the market there for textiles.

London to explore the possibility of having his business represented in United Kingdom.

your country for the purpose of extending the commercial relations of his firm with leading firms there.

四、希望予以協助

1. { We shall appreciate

We shall much appreciate

We would appreciate }

$\rightarrow$ { anything you may be able to do to assist him.

any courtesies (and assistance) which you may extend to him.

any advice or assistance you may render him.

any assistance you may be able to accord Mr....

any assistance you may render (extend) to Mr....

any assistance and courtesy to Mr....

any courtesies or facilities that may be extended to him in accomplishing his mission.

any information you can afford him, or introduction to houses in his line of business which you can give him.

2. Anything you may be able to do to assist him

$\rightarrow$ { will be very much appreciated.

will be greatly appreciated by us.

will be appreciated not only by his company but also by us.

五、預誌謝意

1. You will favour us by giving him any advice or assistance which it may be in your

power to render during his stay.

2. We should feel very much obliged if you would kindly furnish him with any advice of which he may stand in need, or do him any other service which you may have the opportunity to render.

3. We feel sure Mr. Yoshida will appreciate any opportunities that you may give him to become acquainted with the businessmen and business conditions of your city.

4. We would appreciate your rendering him every assistance, and giving him every information which he may require, or which may seem appropriate to ensure the success of his journey.

5. We need scarcely assure you that we shall appreciate to the full whatever you may do for him.

6. We are sure you will assist our representative to the utmost and thank you beforehand for your aid; it will be but an addition to the obligations your kindness has so frequently laid upon us.

7. We shall consider any attention shown to Mr. Goto as a personal favour, which we shall be pleased to reciprocate whenever you allow us the opportunity.

第九節 國外訪問歸國後感謝信例句

一、報告平安返回本國的開頭句

1. Having now returned to Japan, I would like to express my thanks...

2. Now that I am once again back in Japan, I would like to...

3. I have safely returned to Osaka on 5th September.

4. I arrived back safely yesterday evening, my first thought is to again thank you for...

5. The first thing I wish to do, now that I have returned to Tokyo, is to express to you my most grateful thanks...

6. I arrived in Osaka on the 12th instant and am feeling quite well, apart perhaps from being a little tired after our numerous activities.

7. As you will see, I am now back in Japan and am taking this opportunity of thanking you for...

8. This is the first opportunity I have of thanking you for...

9. After having been on my way home through a lot of interesting countries for a considerably long time I am now back in my own country and wish to express my thanks for...

10. After quite an uneventful journey back from your country, the desk at our office here has now claimed me, but, before settling down to normal duties, I wish to express to you...

11. This is merely a hurried note to let you know that we arrived home safely last Friday after a pleasant but rather protracted journey.

12. I have been so extremely busy since my return from your country that...

13. I am only now able to sit down and tell you how much I appreciate the kindness...

etc.

二、感謝承蒙接待、協助

Example: I wish to express my sincere appreciation for your courtesies and assistance extended to me during my recent visit to your city.

1.

We wish to express to you our sincere/great appreciation

We wish to express our thanks/appreciation

We wish to express our very sincere thanks

We wish to express our heartfelt appreciation

We wish to take this opportunity of expressing to you our appreciation

We are taking this opportunity of thanking you

We would like to take this opportunity of thanking you

We would like to avail ourselves of this opportunity to let you know how grateful we are for your kindness

We thank you for...

We do thank you for...

We wish to convey to you our thanks...

We deeply appreciate

We would like to write and thank you most sincerely/cordially

We should like to thank you very much indeed

We are writing to thank you

We are writing to say how we enjoyed

The first thing I wish to do is to express our most grateful thanks

My first thought is to again thank you and your colleagues

I am only now able to sit down and tell you how much I appreciate

This letter is written for the express purpose to thank you

Words fail me in trying to thank you

2. (*Preceded by "for" where necessary*)

the kindness/generosity

the kindness and hospitality

many kindness and courtesies

the wonderful kindness

your very great personal kindness

the exceptional kindness

all the kindness

the courtesies

your courtesies and assistance

your courtesies and thoughtfulness

the very many courtesies

the hospitality and friendship

the hospitality, courtesy, and assistance

the most lavish hospitality you bestowed on us

the most wonderful hospitality and cooperation

the extremely generous help and kindness

the most effective cooperation

everything you have done for me

the wonderful welcome and assistance

extended to me

shown to me

accorded us

which you showed to me

you extended to me

which we received at the hands of all whom we met at your company

that were so freely extended to me

which has been bestowed on me

三、表示此次旅行獲益良多

1. This trip was both enjoyable and successful.

   a pleasant and successful trip

2. Altogether our visit to your country, although far too short, was a most fruitful one, and to me most informative and enjoyable.

3. My journey to Japan will remain an unforgettable memory.

4. My visit will be one which I shall long remember with the greatest pleasure.

5. Unfortunately, our stay in your country was very short, and....

6. I was not able to see nearly as much as I would have liked.

7. Nevertheless, we were able to enjoy some of its highlights.

8. I enjoyed very much indeed my visit to your country, but I am quite certain that we would not have seen nearly as much or enjoyed it nearly so well had it not been for the attention that your staff gave us and the arrangements they made for our entertainment.

9. Your personal kindness and help to us were instrumental in making our visit exceptionally pleasant for us both.

10. Your wonderful hospitality and many courtesies extended to me had made my trip most enjoyable and a memorable one.

11. We feel sure that his trip to your country will be instrumental in cementing our friendship and increasing our understanding on mutual problems.

12. My visit to you in Singapore was of tremendous value to me and I trust that it will have results to our mutual satisfaction.

13. We are quite firmly of the opinion that the opportunities of exchanging personal views on such visit do a great deal to assist our mutual interests to advantage.

14. We feel sure that personal visits of this nature assist in further cementing our long standing and highly valued relationship with your Company.

15. The personal contacts made from the exchange of visits are mutually beneficial to the pleasant and long relationship with your Company.

16. I am quite sure that this personal contact with the directors and heads of Departments will prove most valuable in our further dealings.

17. We sincerely hope both of our firms will benefit in the future from his visit to your country.

四、結尾句

1. Once again, I would like to thank you for the wonderful welcome extended to me.

2. With best personal regards to you and your colleagues.

3. I trust I shall have the pleasure of meeting you once again in the near future, if not in New Zealand, then I hope in Taiwan.

4. With very many thanks and kind personal regards to yourself, and would you please also convey my regards to Mr....

5. Many thanks again for all that you and your Company did for me, and with warmest regards.

6. Permit me to assure you again of my sincere gratitude for all you did to make this visit one of my most memorable trips.

7. Once more, my sincere and heartiest thanks for all that you did for me during my visit in your country.

8. With kindest regards and looking forward to a long future of constantly increasing business and very close relations between our two firms, I am.

9. With very kindest personal regards and wishing you world-wide success in business.

10. With kind personal regards, I remain.

11. Many thanks again, Sir, for all your kindness.

12. Thank you again, Sir, and with warmest personal regards.

13. Our kindest regards to yourself and to all with whom we were associated in Australia.

14. I trust that at some date in the future I will have an opportunity of again meeting you.

15. Should you ever come to Taiwan, I certainly hope that you will call me and arrange

to visit our office.

16. I trust you are well and hope that in the not too distant future you will re-visit Taiwan when we can repay a little of your hospitality. My wife joins me in sending our kindest regards to you and we would ask that you convey our best wishes to Mr....

17. We both join in sending you our regards and best wishes with the hope that it may not be too long before we all meet again.

第三十五章
標點符號的用法

標點符號的使用，其目的在於使文字清楚、易讀、易解。以下將標點符號用法加以介述。

一、Period (·) —— 句點

1.用在每一敘述句 (Declarative sentence) 或祈使句 (Imperative sentence) 之後。

例： We mailed you our check on Thursday. (Declarative)

Write us as soon as you reach New York. (Imperative)

但句尾的字如是縮寫字，則不要再加句點。例如：

The meeting will be held on May 15, at 10 a.m.

注意： 如果是驚嘆句或疑問句，則句尾字即使是縮寫字，應加驚嘆號或疑問號，

例： May I come to see you at 2:00 p.m.?

Get up, it is already 8:00 a.m.!

2.用在縮寫字起首字母 (Initial) 後面。

例： Mr. 為 Mister 的縮寫，October 的縮寫為 Oct.。

但有些縮寫字往往不加句點，例如 CIF 是 Cost, Insurance, Freight 的縮寫，但在 "C" "I" "F" 的後面往往不加句點。類似情形尚有 FOB, CFR, FAS, BOFT（國貿局）, FBI, USA 等等。

有些縮寫字已不再視為縮寫字，所以在後面不加句點。

例： Exam ——→ Examination

Ad ——→ Advertisement

3.引用句中，中間部分省略時，用三點句點，末尾省略時，用四個句點。

例： The letter read: "We can not...until after you have filled out the questionnaire."

The letter read: "We will visit your country...."

4.用於元與角分之間。

例： $4.99, $.08

二、Comma（,）──逗點

1. 用於複合句 (Compound sentence) 中接續詞之前。

   例：I plan to visit your country early next month, and I hope at that time to have a talk with you at your office.

   但複合句很短而意義明確時，可省略逗點。

   例：Miss Wang is a secretary but Mr. Wang is a manager.

2. 複合句的主詞為同一字，而以 "and" 接另一動詞時不加逗點，但以 "but" 連接時，應加逗點。

   例：We received your cable of May 10 and have shipped the goods today.

   We received your cable of May 10, but were unable to effect the shipment in time.

3. 用於分隔一系列（三個以上）同等名詞 (Coordinate nouns)、形容詞、動詞或副詞、片語或子句。

   例：I spoke to the receptionist, the secretary, the clerk, and the manager.

   We plan, fabricate, erect, and sell our own portable building.

   We want a careful, intelligent, conscientious clerk.

   但如名詞前有兩個以上的非同等形容詞時，不用逗點。

   例：The secretary wore a large green eyeshade.

   （green 修飾 eyeshade，而 large 則修飾 green eyeshade）

4. Jr., etc. 出現在句子中間時，其前後應加逗點。

   例：Mr. James Johnson, Jr., is our salesman.

5. 用於敘述性子句（即非限制性的 Nonrestrictive）前或後。

   例：Mr. Chang, who had been a faithful employee, retired last week.（子句在句子中間）

   Always faithful in carrying out his duties, Mr. Chang retired last week.（子句在句首）

   Mr. Chang retired last week, even though he would have liked to serve the company for more years.（子句在句尾）

6.敘述性子句用逗點，限制性 (Restrictive) 子句不用逗點。

> 例：We showed your sample to our maker, who submitted us this offer.（我們把你的樣品向我們的製造商出示，他們提出了本報價。）—— 敘述性用法。
>
> We showed your sample to our maker who submitted us this offer.（我們把你的樣品出示給向我們提出報價的製造商。）—— 限制性用法。
>
> Employees who are careless should be discharged.

7.同位語 (Appositive)，如取消也無妨，則前後要加逗點；如取消則意義不明顯時，則前後不用逗點。

> 例：Mr. Chang, a faithful fellow, has been with the company since 1930.
>
> The term letter of credit used frequently in business correspondence
>
> 職位或學位等，其前後均加逗點。
>
> Mr. Chang, Vice President of our company, will visit your city next month.

8.獨立子句、片語，或放在文首的副詞子句，以逗點分隔。

> 例：My fiancé, much to my regret, will go abroad.
>
> Frankly speaking, to write a letter is not so difficult as you think.
>
> If our proposals are acceptable, please write us before the end of this month.
>
> If this offer is accepted, we shall give you another order next month.

9.一系列名詞（三個以上），最後兩個名詞雖由 "and" 連接，但不相關聯時，"and" 前面，用逗點隔開。

> 例：Please note the delivery dates for tobacco, cotton, corn, and wheat.
>
> （如寫成 corn and wheat，則變成三種交貨期）

10.下列引導性措詞 (Introductory expressions)，出現在句子開頭時，用逗點隔開。

Accordingly,	In short,
Actually,	In general,
After all,	Meanwhile,
Again,	Moreover,
Also,	Next,
As a matter of fact,	No,
As a rule,	Nevertheless,

Besides,	Of course,
By the way,	On the contrary,
Consequently,	On the other hand,
First,	Otherwise,
For example,	Perhaps,
For instance,	Secondly,
Fortunately,	So,
Further,	Still,
Furthermore,	Strictly speaking,
Hence,	Frankly speaking,
However,	Then,
In any case,	To be sure,
In brief,	To say the truth,
Indeed,	Uufortunately,
In fact,	Well,
In other words,	Without doubt,
In the first place,	Yes,
In the meantime,	Yet,

11.用於分隔表示對比的字詞。

　例：The factory workers, Not the office staff, Are on strike.

12.下面過渡性 (Transitional) 的單字或片語，出現在句子中間時，其前後均加逗點。

Also	Too
Therefore	Again
As a rule	In the first place
As we see it	However
As it were	If any
As you know	In addition
By the way	In fact
For example	In brief

Indeed	In turn
Finally	Namely
i.e.	Say
e.g.	Of course
viz.	On the other hand
Still	That is

例：Our business is very good, that is, in the light of present times.

We have established a sizable market, i.e., in U.S.A., Japan, and South Africa.

13.在引用他人談話時，引用句的前或後，應以逗點分隔。

例：He said, "I must go home now."

"Today", he cried, "I am a free man."

14.用於分隔年與月份之後的日期。

例：July 8, 1978.

15.用於分隔地址（省、市、鎮、街、巷等）。

例：100 Wuchang Street, Section 1, Taipei, Taiwan.

16.用於直接稱呼時。

例：Mr. Chang, please come to my office.

17.用於分隔句子中的釋義短句。

例：Your last letter, although mailed on the 18th, did not reach us until the 28th.

18.用於輕微的感嘆詞後面。

例：Well, I didn't think we'd get such a large order.

19.用於句子中有省略字時。

例：One of the officers has been with us for fifteen years; the other, for ten. ("has been" is omitted before "for ten")

This girl is a secretary; the other, a bookkeeper. ("is" is omitted before "a bookkeeper")

20.用於分隔數目字，以便閱讀。

例：NT $123,456　　　　　123,456,789 dozen

但年代、電話號碼、門牌號碼、頁數、L/C 號碼等不用逗點隔開。

例： 1982   1234   Park Avenue

Page 1234   L/C No. 1256

三、Semicolon（；）──分號

1.同等子句 (Coordinate clause) 不用連接詞時，以分號分隔。

例： The last day of the month is always a busy day for bank employees; it often forces them to work overtime.

2.以下面的連接性副詞連接複合句時，在這些連接性副詞前面應用分號。

However	Otherwise
Accordingly	For
Nevertheless	Hence
Besides	Consequently
Then	Likewise
Thus	Notwithstanding
Therefore	

例： I saw no reason for moving; therefore I stayed still.

Your letter requesting cancellation of your order 123 was received only this morning; otherwise this order would have been shipped today.

However 當做 Conjunctive adverb 時，其後面須加逗點。

3.用於分隔一連串的名詞。

例： The new members of the Board and their home town follow: Mr. C. H. Chen, Taipei, Taiwan; Mr. K. H. Wang, Tainan, Taiwan; Mr. M. N. Chia, Seattle, U.S.A.

4.下面字詞前面，可以分號分隔。

Namely	For instance
As	i.e.
That is	e.g.
For example	viz.
That is to say	

例： He should be given the job; that is, he has the necessary qualifications.

5.引號後面，如還有文字時，在引號後面用分號。

　　例：The cable read: "Cannot ship before October 10"; consequently we had to disappoint our customer.

四、Colon（:）──冒號

1.用於表示依序詳細說明或列舉時。

　　例：Four articles are on the desk: a stamp pad, a bottle of ink, an abacus and fountain pen.

　　　　We have sent you today the following shipping documents:

　　　　The terms and conditions are as follows:

　　　　Commercial invoice in two copies

　　　　Bill of lading in two copies

　　　　Packing list in three copies

2.用於直接引述 (Direct quotation)（也可用逗點）。

　　例：Our manager said: "These data must be kept confidential."

　　　　Our motto is: "First come, first served."

3.用在 Appositive phrase 或 Clause 之前。

　　例：Our chief requirement is this: we need good quality and prompt shipment.

4.用於分隔時和分。

　　例：Time: 10:18 p.m.

5.用在 Salutation 之後。

　　例：Dear Sirs:

　　　　但也可用逗點。

　　例：Dear Sirs,

五、Apostrophe（'）──縮寫記號，所有格記號

　1.用於表示所有格。

　⑴單數名詞：其所有格加 "'s"。

　　　例：my father's hat

⑵多數名詞語尾有 "s" 時；其所有格，只在 "s" 後面加 " ' "。

　例：The shareholders' appeals

⑶多數名詞語尾無 "s" 時，其所有格應加 "'s"。

　例：Men's suits

　　　Children's coats

⑷單數固有名詞以 -s, -x, -ch 或 -sh 為語尾時，其所有格加 "'s" 或只加 " ' "。

　例：Mr. Fields's territory　　Mr. Fields' territory

　　　Miss Cox's records　　Miss Cox' record

　但加 "'s" 的較通行。多數固有名詞時，其所有格加 Apostrophe。

　例：The Clarks' letters

⑸縮寫字的所有格，Apostrophe 應加在 Period 後面。

　例：A. I. B.'s regulations

　　　CTC's policy

⑹表示聯合的所有格時，在最後一名詞加 "'s"。

　例：Smith and Jones's opinions

⑺複合名詞的所有格，在最後一字加 "'s"。

　例：My father-in-law's overcoat

⑻非生物的所有格以用 "of" 表示為佳。

　　　Poor: The shipment's date

　　　Better: The date of shipment

　　　Poor: The contract's signing

　　　Better: The signing of the contract

　但與時間，度量衡或人格化有關的習慣用語，不在此限。

　　　A month's stay　　　For pity's sake

　　　A day's work　　　A dime's worth

　　　A mile's length

⑼在公司或社團名稱裡，常將所有格符號省略。

　例：The Farmers Bank of China

2.用於表示省略了字母或數字。

例：Let's (= Let us) have a drink.

　　can't (= can not)

　　don't (= do not)

　　I was born in U.S. in the spring of '42 (= 1942).

　　It's (= It is) just eight o'clock (= of the clock).

3.用於表示字母單字或數字的多數。

例：There are four o's and three m's in this sentence.

　　Do not use too many we's in a business letter.

　　Mind your P's and Q's.（當心說話或舉動。）

六、Quotation mark (" ") ──引號

1.雙引號用於直接引用語。

例：Our manager said, "Everybody must come to office at 8:00 a.m."

2.雙引號用於表示強調或意義特殊的字。

例：The term "Super" is used freely today.

3.引用語中有引用語時，直接引用語用雙引號，引用語中的引用語用單引號。

例：Our manager always quote the words of our chairman, "Never forget that con-

　　fucius said, 'Do not do to others what you do not want others to do to you.'"

4.雙引號用以引用書中章節、雜誌中的論文名稱。

例："Writing effective collection letters " is a very informative chapter.

5.用於船名。

例：S.S. "Hapeh"

七、Hyphen (-) ──連字號

1.用於複合單字。

例：father-in-law

　　well-known firm

2.數字由 21 至 99 的數目，用文字寫出而當做形容詞用時，用連字號。

例：twenty-one orders

eighty-eight invoices

3.用於接頭語。

例： pro-America

re-elect

self-educated

4.當分數用文字寫出時。

例： five-sixths

5.當一個字分做兩行寫的時候，在第一行的末尾加連字號。

例： ......quad-

plicate.

6.當兩複合形容詞形容同一名詞時，第一形容詞後面只加連字號。

例： This molasses is sold in one-and two-quart bottles.

7.在多數場合，ex, vice 的接頭語都加連字號。

例： ex-president

vice-chairman

八、Question mark（?）——問號

1.用於任何直接問句 (Direct question) 之後。

例： What is your opinion?

形式上是疑問句但實際上係禮貌的請求句，不用問號。

例： Will you please send us a catalog.

2.對某一敘述無把握或有所疑惑時，在其後面以（?）表示。

例： The check was for US$1,928.31(?).

九、Exclamation mark（!）——驚嘆號

1.用在感嘆字句之後。

例： Wait! Don't release that shipment now!

2.用於呼告格之後。

例： O Liberty! O Liberty! What crimes are committed in thy name.

（啊，自由啊！啊，自由啊！在你的名義下，不知犯了多少罪呀。）

3.用於形式上為問句而內容為感嘆句的句子。

例：How can you believe such a wild assertion!

十、Dash (─) ──破折號

1.用於表示思路或結構 (Structure) 的突然破裂 (Sudden break)。

例：ABC company ordered ten of our mahogany coffee tables─I forget the style numbers─for one of their hotel accounts.

2.用在短同位格片語之前，以加強語氣。

例：We called on the Fargo Company─the principal dealers of electrical appliances in Taiwan.

3.用於分隔含有逗點的同位格措詞。

例：Three of our trucks─the '69 chevrolet, the '69 Ford, and the '70 G. M. C.─are in excellent conditions.

4.用於分隔重複字或重複片語，以加強語氣。

例：We must have your check within ten days─ten days, not a day later!

5.當 as, namely, that is 等詞用於介紹一系列項目時，在這些詞後面加破折號。

例：There are three steps in our sales training program, namely─

　　1. A six-month period in the factory

　　2. A six-month period in the offices

　　3. A six-month period on the road working with an experienced salesman.

6.用在限制或彙述前面一系列詞語的同位格子句前面。

例：Courtesy, reliability, and knowledge─these qualities we demand of all our employees.

7.某一時間至另一時間或某一日至另一日之間，用破折號分隔以示其起迄。

例：Our personnel offices are open from 8:00 a.m. ─ 4:30 p.m.

8.用於表示說話時的遲疑。

例：I hope so─but─that is─well, I wil give it a trial.

　　　　（我希望是那樣……但是……那就是……啊，我要來試一下。）

十一、Parenthesis (( )) ——圓括弧

1. 以插入句，說明或補充解釋時，用圓括弧隔開。

例：This report (I think Mr. Smith submitted it last week) covers the period from March 1 to June 15.

2. 用於括住參考或指示。

例：The revised price pages have been prepared for all our accounts (see pages 8 to 16 for wholesaler discounts) effective May 1.

3. 用於括住同位格數字 (Figures)。

例：Our wheelbarrows list for twenty dollars ($20) in lots of one (1) dozen.

We received fifty (50) orders for our TV sets this month.

4. 用於括住一系列項目之前的數字或字母 (Letters)。

例：After the orders have been registered, they are classified as follows:

(1) F. I. (Fill Immediately); (2) HFC (Hold for Confirmation) and (3) CAR (Credit Approval Required).

十二、Bracket (〔 〕) ——方括弧

用於作者，或註釋者所加的說明或補充。

例：In that summer 〔2004〕 business dropped 20 per cent.

國際貿易實務詳論　張錦源／著

　　買賣的原理、原則為貿易實務的重心，貿易條件的解釋、交易條件的內涵、契約成立的過程、契約條款的訂定要領等，均為學習貿易實務者所不可或缺的知識。本書對此均予詳細介紹，期使讀者實際從事貿易時能駕輕就熟。國際間每一宗交易，從初步接洽開始，經報價、接受、訂約，以迄交貨、付款為止，其間有相當錯綜複雜的過程。本書按交易過程先後作有條理的說明，期使讀者對全部交易過程能獲得一完整的概念。除了進出口貿易外，對於託收、三角貿易、轉口貿易、相對貿易、整廠輸出、OEM 貿易、經銷、代理、寄售等特殊貿易，本書亦有深入淺出的介紹，為坊間同類書籍所欠缺。

國際貿易實務　張盛涵／著

　　國際貿易實務是一門以貿易操作實務為基礎，經由業界長期間的摸索歸納，形成的一套協助貿易進行的相關商業慣例與法律。本書於各章節之貿易實務主題中，展示其運作原理，協助讀者了解貿易實務規範的背後原理，培養讀者面對龐雜貿易事務時，具有洞悉關鍵，執簡馭繁的能力。此乃本書主要特色之一。本書第二個主要特色，則係來自於重要性與日俱增並已成為我國主要貿易對手的中國大陸。書中對於貿易術語的說明，若兩岸的用語有所差異，均予標示並陳，以利讀者明悉對岸用語，彌縫雙方差異。

國際貿易原理與政策　王騰坤／著

　　本書採取國際多元化教學目標來撰寫，除了一般國際貿易理論的架構外，輔以貿易政策執行與國際經貿組織的探討，並系統性地說明國際貿易發生的問題，從古典二分法的觀點到現代國貿理論，來進一步的認識與了解。本書特色：⑴理論完備，面面俱到：特別納入了國際政治經濟學之相關概念，同時提供國際貿易新理論觀點，完整無遺漏。⑵創新架構，宏觀思考：本書將相關理論融會綜合，比較各個思維觀點下理論模型的變化情況，以更宏觀的角度帶領讀者了解理論基礎。⑶資訊匯總，知識導引：將目前國際貿易組織及貿易體系架構資訊最完整的呈現，同時讓讀者理解其中過往由來以及目前最新發展情勢、各國各式貿易爭端之處理方式，省去讀者必須花的資料收集成本，以系統性及具比較性的方式整理歸納。

保險學概要　袁宗蔚／著、鍾玉輝／修訂

　　本書初版於民國五十三年，頗受社會喜愛，經多次再版印刷，至今已逾四十載仍為大家所重視，堪稱價值非凡。其固因作者享有保險泰斗之盛名外，內容詳實，扼要掌握保險精華，行文流暢用字簡潔，使讀者易於融入保險領域，更是本書特色。此次修訂目的在使本書更能符合時代，繼續擁有活用之價值，俾造福後進學子。修訂者跨足產業及學術兩域，教學與實務經驗豐富，見譽中臺灣。修正內容主要為分論之火災保險、海上保險、責任保險、汽車保險、社會保險等隨時代變遷，因實務需要應更改之各章，其他部分有內容不符時宜者亦經修改殆遍。

《貨幣與金融體系》　　賈昭南／著

　　本書特色：⑴鑑往知來：總覽貨幣與金融體系的特徵並引述其發展歷史，使讀者能夠全方位掌握其現況與未來發展趨勢。⑵數據解析：引用資訊經濟學理論介紹金融機構的特徵，使讀者更深入瞭解貨幣與金融體系的重要性與其在經濟體系中的地位。特別以我國金融體系的相關統計數據為例，說明國內現況。⑶國際視野：介紹歐美日等先進國家的貨幣與金融體系發展現況，供讀者相互比較並加深印象。⑷本書適用對象：本書適用於貨幣銀行相關課程的教學與自習使用，為進入貨幣與金融體系領域提供入門基礎。

《貨幣銀行學》　　許光華、許可達、嚴宗銘／著

　　本書特色：⑴理論完備：將貨幣銀行學中的基本理論，加以整理闡述，並說明凱因斯以後，新古典學派的重貨幣學派及後凱因斯學派與杜賓等的貨幣需求理論，且論及當代貨幣理論的發展；同時亦說明財政政策的所得與利率效果，闡明政府施行擴張性財政政策的預算來源。⑵實務導引：本書各章篇末，均依據該章的學習目的，蒐集與其相關的實務性議題，整理剖析，作為學習的參考與驗證。⑶彈性的課程設計：授課老師可依課程規劃的需要加以取捨，本書各章節之安排，如以較為簡略之講授方式，適於一學期之課程；若採較深入之講授，亦適用於一學年之課程需要。

《經濟學》　　賴錦璋／著

　　本書特色：⑴化抽象為具體，看得見摸得著：本書利用大量生活狀況實例，帶出經濟學的觀念，將經濟融入生活，讓學生從生活體悟經濟。⑵用筆幽默風趣，內容好讀好記：作者用輕鬆幽默的筆調，平易近人的語言講解經濟學，讓經濟不再是經常忘記。⑶精闢講解重點，理論涵蓋全面：內容涵蓋個體及總體經濟學的重要議題，並將較困難章節標示，學生可視自身需求選擇閱讀內容。⑷經濟現況說明，統計數據佐證：介紹臺灣各發展階段的經濟狀況，更透過歷年實際的統計數據，幫助學生學習如何運用統計資料分析經濟現況，不只給學生魚，更教學生以後如何自行釣魚。

《總體經濟學》　　楊雅惠／編著

　　總體經濟學是用來分析總體經濟的知識與工具，而如何利用其基本架構，來剖析經濟脈動、研判經濟本質，乃是一大課題。一般總體經濟學書籍，皆會將各理論清楚介紹，但是缺乏實際分析或是案例，本書即著眼於此，除了使用完整的邏輯架構鋪陳之外，另外特別在每章內文中巧妙導入臺灣之經濟實務資訊，如民生痛苦指數、國民所得統計等相關實際數據。在閱讀理論部分後，讀者可以馬上利用實際數據與實務接軌，這部分將成為讀者在日後進行經濟分析之學習基石。

管理學　張世佳／著

　　本書係依據技職體系之科技大學、技術學院及專校學生培育特色所編撰的管理用書，強調管理學術理論與實務應用並重。除了各種基本的管理理論外，亦引進目前廣為企業引用的管理新議題如「知識管理」、「平衡計分卡」及「從 A 到 A$^+$」等。透過淺顯易懂的用語及圖列式的條理表達方式，來闡述管理理論要義。此外，本書配合不同章節內容引用國內知名企業的本土管理個案，使學生在所熟識的企業情境下，研討各種卓越的管理經驗，強化學生實務應用能力。

行銷管理——觀念活用與實務應用　李宗儒／編著

　　本書以宏觀的角度探討行銷管理涵蓋之範圍，從行銷的基本概念出發，用深入淺出的方式呈現行銷管理之核心概念。行銷學科的發展與個案探討，密不可分，因此本書有系統的網羅並整理國內外行銷相關書籍，如期刊論文、專書、雜誌等及網際網路上之相關個案與知識，其目的在於讓讀者有一系統化的概念，以助其建立行銷架構與應用。同時亦將目前許多新興的議題融入書中，每一章節以簡單的實務案例作為引言，使讀者可以更清楚章節內介紹的理論觀念；並提出學習目標，以及在章節最後列出思考與討論的題目，使讀者可以前後呼應，更加融會貫通。

策略管理學　榮泰生／著

　　本書的撰寫整體架構是由外（外部環境）而內（組織內部環境），由小（功能層次）而大（公司層次）。使讀者能夠循序漸進掌握策略管理的整體觀念，見樹又見林；並參考了最新版美國暢銷「策略管理」教科書的精華、當代有關研究論文，以及相關個案，向讀者完整的提供最新思維、觀念及實務。目前資訊科技及通訊科技在策略管理上所扮演的角色日益重要，因此在有關課題上均介紹最新科技的應用。為了增加讀者的學習效果及實際應用能力，本書在每章中均有複習題可幫助讀者徹底了解該章的重要觀念，以及練習題讓讀者就所選定的公司作實際運用，使得讀者能夠「實學實用」，並訓練讀者的判斷、思考及整合能力。

《財務管理》　戴欽泉／著

　　全球化經營的趨勢下，企業必須對國際財務狀況有所瞭解，方能在瞬息萬變的艱鉅環境中生存。本書最大特色在於對臺灣及美國的財務制度及經營環境作清晰的介紹與比較，並在闡述理論後，均有例題說明其應用，以協助大專院校學生及企業界人士瞭解財務管理理論、財務分析、財務決策，以及財務有關的企業經營之內容。財務管理是動態的，本書融合了財務管理、會計學、投資學、統計學、企業管理觀點，以更宏觀的角度分析全局，幫助財務經理以全盤化的思考分析，選擇最適當的財務決策，以達成財務（企業）管理的目標——股東財富極大化。

《期貨與選擇權》　　陳能靜、吳阿秋／著

　　本書以深入淺出的方式介紹期貨及選擇權之市場、價格及其交易策略，並對國內期貨市場之商品、交易、結算制度及其發展作詳盡之探討。除了作為大專相關科系用書，亦適合作為準備研究所入學考試，與相關從業人員進一步配合實務研修之參考用書。

《生產與作業管理》　　張百棧／著

　　生產與作業管理是企業運作裡不可或缺的一環，只要是與企業之運作流程、效率及品質有關之事項均涵蓋其中。在少量多樣化生產趨勢下，生產與作業管理也越顯得繁瑣而複雜，舉凡從訂單之承接、原物料的採購、生產加工、製程管理、到產品產出，都需要生管人員投入相當心力，以確保公司能在適當時間依照顧客要求，交給一定數量的貨品或服務。全書利用淺顯易懂的範例來說明各生產管理系統的概念，並藉由實務上之案例探討讓讀者加深印象，在每個章節後面均收錄了近年來各大公、私立研究所考試之考題，期望本書不但能夠幫助學生進一步瞭解生產與作業管理在實務上的應用，更能幫助為了升學而做準備之研究所考生。

《行銷研究──觀念與應用》　　黃俊堯、黃士瑜／著

　　行銷研究旨在協助行銷者瞭解顧客與潛在顧客的態度與行為，掌握競爭市場動態，進而提升行銷資源配置的效率，協助評估行銷策略與作為的實效。不同於一般的行銷研究教科書，本書以實務操作導向出發，強調行銷研究的主要概念與分析方法，並以大量業界的實例或其改寫配合理論說明，希望能讓所有有興趣於行銷研究的讀者可以比較完整地瞭解行銷研究的原理與實作。全書內容可供大專院校行銷研究課程做為教材，亦可提供行銷研究業界或企業組織行銷部門在新進人員訓練或者內外部溝通時加以參考。